LabVIEW Graphical Programming

LabVIEW Graphical Programming

Richard Jennings
Fabiola De la Cueva

Fifth Edition

New York   Chicago   San Francisco
Athens   London   Madrid   Mexico City
Milan   New Delhi   Singapore   Sydney   Toronto

Library of Congress Control Number: 2019938180

McGraw-Hill Education books are available at special quantity discounts to use as premiums and sales promotions, or for use in corporate training programs. To contact a representative please visit the Contact Us page at www.mhprofessional.com.

LabVIEW Graphical Programming, Fifth Edition

Copyright ©2020 by McGraw-Hill Education. All rights reserved. Printed in the United States of America. Except as permitted under the United States Copyright Act of 1976, no part of this publication may be reproduced or distributed in any form or by any means, or stored in a data base or retrieval system, without the prior written permission of the publisher.

1 2 3 4 5 6   QVS   23 22 21 20 19

ISBN   978-1-260-13526-8
MHID   1-260-13526-8

The pages within this book were printed on acid-free paper.

Sponsoring Editor
Robert Argentieri

Editorial Supervisor
Donna M. Martone

Acquisitions Coordinator
Elizabeth Houde

Project Manager
Revathi Viswanathan
Cenveo® Publisher Services

Copy Editor
Manish Kumar

Proofreader
Lisa McCoy

Indexer
Arc Indexing

Production Supervisor
Lynn M. Messina

Composition
Cenveo Publisher Services

Art Director, Cover
Jeff Weeks

Information contained in this work has been obtained by McGraw-Hill Education from sources believed to be reliable. However, neither McGraw-Hill Education nor its authors guarantee the accuracy or completeness of any information published herein, and neither McGraw-Hill Education nor its authors shall be responsible for any errors, omissions, or damages arising out of use of this information. This work is published with the understanding that McGraw-Hill Education and its authors are supplying information but are not attempting to render engineering or other professional services. If such services are required, the assistance of an appropriate professional should be sought.

Contents at a Glance

1	Roots	1
2	LabVIEW Fundamentals	61
3	Data Acquisition	147
4	LabVIEW Object-Oriented Programming	207
5	Why Would You Want to Use a Framework?	273
6	Unit Testing	401
7	Developing in LabVIEW for Teams	487
8	Enterprise and IoT Messaging	551
	Abbreviation	589
	Index	591

Contents

Foreword	xv
Preface	xix
Acknowledgments	xxi

1 Roots ... 1
- LabVIEW and Automation ... 1
  - Virtual Instruments: LabVIEW's Foundation ... 3
  - Why Use LabVIEW? ... 6
- The Origin of LabVIEW ... 7
  - Introduction ... 8
  - A Vision Emerges ... 8
  - All the World's an Instrument ... 10
  - A Hard-Core UNIX Guy Won over by the Macintosh ... 11
  - Putting It All Together with Pictures ... 12
  - Favoring the Underdog Platform for System Design ... 14
  - Ramping up Development ... 15
  - Stretching the Limits of Tools and Machine ... 16
  - Facing Reality on Estimated Development Times ... 17
  - Shipping the First Version ... 19
  - Apple Catches up with the Potential Offered by LabVIEW ... 19
  - LabVIEW 2: A First-Rate Instrument Control Product Becomes a World-Class Programming System ... 23
  - Info-LabVIEW List: The Beginnings of an Ecosystem ... 24
  - The Port to Windows and Sun ... 25
  - LabVIEW 3 ... 26
  - LabVIEW 4 ... 27
- LabVIEW Continues to Improve ... 28
  - LabVIEW 5 ... 28
  - The LabVIEW RT Branch ... 30

LabVIEW FPGA 30
LabVIEW 6i 31
NI Forums: The Beginning of the Ecosystem 33
LAVA Forums: The LabVIEW Ecosystem Continues to Grow ... 33
LabVIEW 7 33
LabVIEW Champions 34
LabVIEW 8 34
LabVIEW 8.2 36
LabVIEW Tools Network 37
LabVIEW 8.6 37
LabVIEW Switches to Yearly Releases 38
LabVIEW 2009 38
CLA Summits, LabVIEW Community-Driven Events 39
Idea Exchange 40
LabVIEW 2010 40
LabVIEW 2011 41
LabVIEW 2012 42
LabVIEW 2013 43
LabVIEW 2014 45
LabVIEW 2015 45
The Knights of NI 46
LabVIEW 2016 48
LabVIEW 2017 49
LabVIEW 2018 51
GDevCon—The Ecosystem Independent from NI 53
Future Versions of LabVIEW 54
LabVIEW NXG 1.0–3.0 54
Dr. T and Jeff Kodosky Inducted into the National Inventors
    Hall of Fame 56
LabVIEW Release Timeline 56
LabVIEW Handles Big Jobs 58
CERN 58

2 LabVIEW Fundamentals 61
Dataflow 61
LabVIEW under the Hood 62
The Parts of a VI 62
How VIs Are Compiled 63
The LabVIEW Environment 67
Front Panel 67
Controls 68
Property Nodes 69

Block Diagram 70
Looping 71
    While Loops 71
    For Loops 72
Shift Registers 73
    Uninitialized Shift Registers 75
Variables 77
    Local Variables 77
    Global Variables 78
    State Machines 81
SubVIs 83
Data Types 84
    Numeric Types 85
    Strings 87
    Parsing Strings 88
    Spreadsheets, Strings, and Arrays 92
    Arrays 95
    Clusters 103
    Waveforms 106
    Data-Type Conversions 109
    Conversion and Coercion 110
    Intricate Conversions and Type Casting 111
    Flatten To String (... Do *what?*) 114
    Enumerated Types (Enums) 115
    Get Carried Away Department 116
Timing 116
    Where Do Little Timers Come From? 117
    Using the Built-In Timing Functions 118
    Intervals 118
    Timed Structures 120
    Execution and Priority 123
    Timing Guidelines 123
    Absolute Timing Functions 124
    High-Resolution and High-Accuracy Timing 126
Synchronization 127
    Polling 127
    Events 128
    Occurrences 131
    Notifiers 134

	Queues	136
	Me and You, Rendezvous	141
	Benchmarking Code	142
3	**Data Acquisition**	**147**
	Inputs and Outputs	147
	Origins of Signals	147
	Transducers and Sensors	148
	Actuators	150
	Categories of Signals	151
	Connections	156
	Sampling Signals	173
	Sampling Theorem	173
	Filtering and Averaging	175
	About ADCs, DACs, and Multiplexers	177
	Digital-to-Analog Converters	182
	Triggering and Timing	183
	Throughput	187
	Writing a Data Acquisition Program	188
	Bibliography	204
4	**LabVIEW Object-Oriented Programming**	**207**
	What, Where, When, and Why	207
	Background	207
	What?	208
	Where and When?	219
	Why?	224
	How?	226
	HAL: Hardware Abstraction Layers	243
	MAL: Measurement Abstraction Layers	244
	Actor Framework: The Most Recognizable LVOOP Architecture	244
	SOLID Principles of Object-Oriented Design	245
	SRP: Single Responsibility Principle	247
	OCP: The Open-Closed Principle	250
	LSP: The Liskov Substitution Principle	252
	ISP: The Interface Segregation Principle	255
	DIP: The Dependency Inversion Principle	263
	Caveats	268
	Accessors Get/Set Instead of Read/Write	269
	Classes in LabVIEW NXG	269
	References	269

5 Why Would You Want to Use a Framework? 273
What? 274
    What Is a Process? 274
    What Is an Abstraction Layer? 274
    What Is a Framework? 275
When Do You Make the Decision to Break Your Application into
   Parallel Tasks? 275
What Are the Design Decisions That Need to Be Nailed Down
   at the Beginning to Make a Parallel Design Successful? 276
What Project-Level Enforcement Is Available to Help Make
   Better Decisions? 278
    Project Libraries 279
Frameworks for LabVIEW 281
    Advantages of Using a Framework 282
    Disadvantages of Using a Framework 283
    Framework versus Design Pattern 283
    The Contract between the Framework and the Programmer 283
Why Not Make Your Own Framework? 284
    Criteria to Evaluate Frameworks 285
Key Components 286
    Interprocess Communication 287
    Module Initialization 287
    Stop Processes Gracefully 288
    Error Handling Strategy 288
Sharing Modules 289
    Configuring Source Code Control Repository Dependencies . . . 290
    Packaging Modules Using VIPM 290
    Packaging Modules in PPLs 292
DQMH 298
    Use Cases 298
    What Is DQMH? 298
    How to Use DQMH 312
    DQMH: Behind the Scenes 343
    Other DQMH Tools 351
    Sharing Reusable DQMH Modules 352
Actor Framework 354
    Use Cases 354
    What Is Actor Framework? 354
    How to Use Actor Framework 366
    Actor Framework: Behind the Scenes 387

Other Actor Framework Tools 391
Advanced Actor Framework 392
Sharing Reusable Actors 398
References 399

6 Unit Testing 401
What Is Unit Testing? 402
Unit Testing as Code Documentation 403
What Makes a Good Unit Test? 403
What Are Assertions? 404
Writing VIs with Testing in Mind 405
Test Harness versus Automated Test Frameworks 406
What Are the Automated Test Frameworks Available for LabVIEW? ... 406
What Is the Difference Between Black Box and White Box Testing? ... 407
What Is TDD? 407
What Is Regression Testing? 408
Getting Started with Unit Testing 408
Test Coverage 411
Test Coverage Example 411
Test Vectors 419
What about Testing Classes? Public versus Private VIs in Libraries 424
Example of Unit Testing for a LabVIEW Class 425
When Are Setup and Teardown Required? 447
Should You Add Test Cases That Are Designed to Fail? ... 449
What about Testing When the Expected Output Is an Array
 of NaN (Not a Number)? 449
When Would You Create Unit Tests for a DQMH Public API? 449
Unit Testing for a DQMH Module 450
What about RT? 464
Example of Unit Testing for LabVIEW RT 464
What about FPGA VIs? 466
Example of Unit Testing for LabVIEW FPGA 470
Unit Tests for the GUI? 475
Unit Test Reporting 475
Example of Assertions 475
Assertions with Caraya 477
Assertions with AssertAPI 479
Unit Testing and LabVIEW NXG 482
Tool Comparison 483
Unit Testing Tool Alternatives for LabVIEW 484
References 485

7 Developing in LabVIEW for Teams ... 487
- Where Is Your Team At? ... 488
- What Is the Problem You Are Trying to Solve? ... 490
- What Is Technical Wealth? ... 491
- From Model to Code ... 494
  - What Is a Model? ... 494
  - Wasn't LabVIEW Supposed to Remove the Need for Modeling? ... 494
  - Why and When to Use Models ... 497
  - Types of Models ... 497
  - Modelling Tools ... 501
- Source Code Control—The Developer's Time Machine ... 501
  - What Is Source Code Control? ... 502
  - Centralized Source Code Control ... 503
  - Distributed Source Code Control ... 506
  - Branch Merge versus Code Merge ... 509
  - LabVIEW Compare ... 509
  - LabVIEW Merge ... 513
  - Good Practices for Source Code Control ... 514
  - Establish a Source Code Workflow ... 518
  - How to Select the Source Code Control Tool and the Workflow for Your Team ... 522
- Workstations ... 523
  - Virtual Machines ... 525
  - Workstations Setup ... 526
  - Build Server ... 526
  - Test Computer ... 527
- LabVIEW Style Guidelines ... 527
- Code Review Process ... 528
  - Code Review Frequency ... 530
  - VI Analyzer Tests Configuration ... 531
  - VI Analyzer Report ... 532
  - Prepare Code and Documentation for a Code Review ... 533
  - LabVIEW Compare for Code Reviews ... 533
  - Code Review Checklist ... 535
  - Conducting a Code Review ... 535
  - Postreview Actions ... 536

  CASE Tools 536
    Why Do You Want to Use LabVIEW VI Scripting? 537
    Common Areas That Use LabVIEW Scripting 537
    Five Steps to Become a VI Scripting Ninja 538
  Deployment, Continuous Integration, and Continuous Delivery 542
    Build Engine 542
    Build Environment 546
  Other Tips 548
  References 549

8 Enterprise and IoT Messaging **551**
  MQTT Messaging Protocol 555
  Install a Message Broker 558
  MQTTDQMH Client 561
    Getting Started 561
  MQTTDQMH Application 567
  Messaging in the Cloud 571
  Toolkits 572
  AWS IoT 573
    Enabling the IoT Message Broker and CloudWatch 574
    Create Certificates and Keys 574
    IoT Policies 577
    DynamoDB 578
    IoT Rules 580

 Abbreviation **589**

 Index **591**

Foreword

The first version of LabVIEW shipped in October 1986. We were a small, scrappy team with a startup mindset. We personally duplicated and packaged the first 50 sets of floppy disks and manuals and hand-carried them to shipping! When I reflect on everything that has happened since then with LabVIEW, I am pleased by how successful LabVIEW has been, and I continue to be amazed by how integral it has been to my life and how much time I have invested in developing LabVIEW. But I am also humbled when I realize that we are still working on a number of the original ideas and still dreaming of some of the farther out, fuzzier ideas. It has taken decades to reach the point where we can begin exploring these ideas seriously, and it inevitably makes me wonder how much time is left—because, after all, I am a grandfather now. I still come to work each day excited by the opportunity of putting my inventor's hat on. It is exhilarating.

The original goal for LabVIEW was to make it easier for scientists and engineers to automate their test and measurement systems without requiring a team of programmers, similar to the way spreadsheet programs helped financial analysts. LabVIEW successfully accomplished that goal. The chapter on LabVIEW fundamentals presented in this book will guide you through the graphic syntax of LabVIEW. You won't have to worry about your code failing to compile due to a missing semicolon.

There have been countless examples where domain experts, engineers, scientists, technicians, and even medical researchers have successfully automated their measurement systems. They sped up their research and discovery, reduced test times and cost, and reduced time to market for new products.

LabVIEW has a long history of continuous improvement and innovation. Each version introduced new capabilities while preserving backward compatibility. Among the most notable advances were the introduction of Real Time and FPGA support. Without having to be a VHDL expert, a LabVIEW programmer can address higher performance requirements by building diagrams that run on the FPGA.

The complexity of the systems being built today continues to increase. There is a need for more parallelism, more physical I/O, tighter timing and synchronization, and more distributed components. Long gone are the "lone-wolf programmer" days, because more complex systems require teams of LabVIEW programmers. How do we deal with this complexity? One way is good software engineering—to institute good policies and processes and maintain thorough test suites and documentation. This book has chapters about unit testing and how to best organize your team to be successful working with LabVIEW. Large test systems are successfully being built today using these software engineering practices. Frameworks can also help by organizing and constraining designs that follow approved patterns. This book dedicates a chapter to LabVIEW frameworks, focusing on two of the most popular frameworks: Actor Framework and DQMH.

Another powerful way to deal with complexity is to raise the level of abstraction we use to design systems. Many systems being built today are inherently complex because they are solving complex problems. But we do not want to add artificial and unnecessary complexity on top of that. By raising the level of abstraction in our software and focusing on higher-level concepts, we can reduce artificial complexity. When driving a car, if you want to accelerate and merge into highway traffic, using the higher level of abstraction of an automatic transmission is simpler than the standard transmission where you need an extra arm and a leg for the gear shift and clutch. You will find more details regarding levels of abstraction and how to model your systems in both the chapters on LabVIEW object-oriented programming and developing in LabVIEW for teams.

We continue to innovate with better high-level abstractions. A couple of years ago we introduced the channel wire in LabVIEW, raising the level of abstraction for communicating parallel processes. It is easier, more visible, and more understandable than using the lower-level language elements. And we have more ideas where that came from; there is more we want to do to further raise the level of abstraction in LabVIEW. We have always said that LabVIEW starts with making the impossible possible, and only later making the possible easy. During the NIWeek 2018 Keynote, I described the research we are doing on displaying different loops running on different devices (FPGA, Real Time, and network targets) on the same block diagram. This is just one example of the research we are doing to raise our ability to abstract complex ideas.

In the chapter on developing in LabVIEW for teams, Richard and Fabiola will guide you through the process of modeling your application before drawing a single wire. One area of research for us is bringing that whiteboard exercise to LabVIEW. We often start our design by drawing our device under test, with all the instruments we will be using. This high-level sketch tends to get lost or never upgraded. Imagine if we could drop an abstract node in LabVIEW to represent all the different elements of our sketch. Preserving this as part of the project could also be a launching point for the rest of our application. This would make the abstraction even clearer. This would be in essence a semantic zoom, where we could navigate through all the different levels of abstraction in our code, from the high-level sketch to the lowest area where the code configures the DAQ board and acquires the signal. Changes to the lowest levels would be reflected at the highest levels as needed. Such a rich design environment would enable end users to create systems that are easier to understand and maintain over time.

Working at multiple levels of abstraction is a powerful way to tame complexity. It enables the progressive disclosure and abstraction of semantic detail so you can better design your measurement system and confidently evolve it as requirements change.

I am very pleased with the efforts to grow the G community. Events like GDevCon, organized by groups independent from NI, and new initiatives like Gcentral, expand the options for LabVIEW programmers to collaborate and share solutions to our common challenges. NI announced the LabVIEW Community Edition, a free, fully capable version for noncommercial use. This free version, in combination with the community efforts, will surely increase the number of engineers, scientists, and technicians who will benefit from the higher levels of abstraction that graphical programming brings.

Our world is getting more complex, requiring more complex test and measurement systems. Because of this, we need to continue creating and using more sophisticated tools that reduce artificial complexity and support higher levels of abstraction. We work every day to refine, research, and implement new ideas to do this. As we further develop this vision, we continue to deliver advances in LabVIEW. By empowering you with innovative tools to build the systems you need, we know the future is in good hands.

Jeff Kodosky
Austin, Texas
September 1, 2019

Preface

It's been almost 30 years since I, Richard Jennings, fell in love with LabVIEW. As a laser repairman at Lawrence Livermore National Laboratory, my job sometimes required writing down a reading on a power meter every 15 minutes, all night long. If I could just get the computer to take those readings, I could go do something else. Fortunately, Gary W. Johnson, the original author of this book, pushed the idea of "data acquisition crash carts" at LLNL. Crash carts were a portable cart with a Mac IIx loaded with NI plug-in cards, and there was one nearby. Someone showed me how to use it, and I was amazed! I used assembly and Basic in school, but now I could SEE how to program. Wiring icons together on the block diagram and displaying data on the front panel was incredibly intuitive. I quickly got lost in the manuals and the examples and started automating everything I could. Eventually, I got a job as a full-time LabVIEW programmer, a career path I never envisioned coming out of school. In 2000 Gary asked me to coauthor the third edition of *LabVIEW Graphical Programming*, and in 2005 I took an entrepreneurial leave from Sandia National Laboratories to move back home to San Antonio, Texas, to work on the fourth edition and start a LabVIEW consulting business.

Over the years I got to know Fabiola De la Cueva as one of the best LabVIEW programmers around. Fabiola's company, Delacor, helps companies apply professional software development methods to large LabVIEW projects and teams. I asked Fab to bring her knowledge to you in this fifth edition of *LabVIEW Graphical Programming*. We asked Gary if he wanted to come on the adventure with us, but recently retired from LLNL, he said, "Now, every day is Saturday," and encouraged us to "kick the whale down the beach."

It's been almost 20 years since I, Fab, fell in love with LabVIEW in 2000 when I started work at National Instruments. My first reaction when I saw LabVIEW was "who has been looking into my brain?" This is how I always visualized programming! Back in college, I would draw boxes with an arrow to model a While Loop around the brackets on my printed C code. Finally, I didn't have to translate the diagrams in my head to text; I could go directly to code. Back then I thought I was a hardware engineer. Even with how much I enjoyed programming in LabVIEW, I took the hardware career route at NI. I left NI in 2006 to pursue a master's degree in electrical engineering, specializing in medical devices at the University of Texas. When I got there, word got around that I knew LabVIEW, and next thing I knew, I was developing LabVIEW programs for pay. I finished my master's but did not start a hardware company, as I had planned. Instead, I founded Delacor and I have dedicated my time ever since to learning LabVIEW, teaching it to others, and helping LabVIEW teams apply good software engineering practices to their work. In 2014 the Delacor team and I began working on a toolkit to help people who don't eat, breathe, and sleep LabVIEW as we do create large applications. It is called the Delacor Queued Message Handler (DQMH). What started as a way to show others what type of code we could make became a framework that a lot of LabVIEW programmers are using. I continue to be humbled by the trust that other programmers have put in our work. I was surprised when Richard, one of my LabVIEW heroes, asked me to write this book with him. I had always said I would never write a book, but I couldn't say no to one of my heroes, so here we are. How is that for getting out of my comfort zone?

The target audience for this book is intermediate to advanced LabVIEW developers, like you, who want to take your skills and applications to the next level. We cover the core concepts of LabVIEW in Chapter 2 "LabVIEW Fundamentals," including the LabVIEW execution engine, the parts of a VI, and benchmarking performance. Chapter 3, "Data Acquisition," covers the basics of interface hardware, signal conditioning, and analog/digital conversion. More than half the LabVIEW questions that coworkers ask us to turn out to be hardware- and signal-related. Information in this chapter is vital and will be useful no matter what software you may use for measurement and control.

Most LabVIEW applications start out as a tool written by a single developer, but when that tool becomes a runaway success and the team starts increasing in size, then development can get bogged down. Delacor brings professional software development practices to companies that need to deliver a professional LabVIEW product to market. Chapter 4, "LabVIEW Object-Oriented Programming," starts with who should use LV OOP and why. OOP is not the answer for every team, but it does have advantages when you want to enforce a programming methodology on a development team. Chapter 5, "LabVIEW Frameworks," covers the two most popular application frameworks in use today: the Delacor Queued Message Handler and the Actor Framework. Fabiola has lots of tips on how you should approach modeling your application along with the advantages and disadvantages of using a framework. In Chapter 6, "Unit Testing," we look at unit testing in LabVIEW as a practice to create modular, testable code. Unit testing is a design practice to help you write robust, testable code that works as designed. Chapter 7, "Developing In LabVIEW for Teams," covers the processes and tools Delacor uses every day to bring professional software development to companies around the world. Finally, we close in Chapter 8, "Enterprise and IoT Messaging," with some tips and examples for building applications that communicate with enterprise message brokers and with Amazon Web Services' Internet of Things (IoT) message broker.

All of the source code used in this book is freely available on GitHub at https://github.com/LGP5. We hope you find it useful.

Richard Jennings and Fabiola De la Cueva

Acknowledgments

First, we want to thank the people who taught us LabVIEW. Beginning with Gary W. Johnson, the original author of this book, thanks for trusting the future of your book to us. Nancy Henson opened the door to advanced architectures and LabVIEW object-oriented programming for Fabiola when she taught her the "Advanced Architectures in LabVIEW" course. Jeffrey Travis and Jim Kring introduced new approaches with their book and gave Fabiola her first consulting opportunities. Justin Goeres taught Fabiola all she needed to know about the worst-named feature in LabVIEW, the user events. His public/private events messaging architecture was a big influence in what became DQMH. Steve Watts taught Fabiola software engineering approaches to LabVIEW, first via his book, and later with long virtual and in-person chats. Darren Nattinger taught Fab all she knows about scripting and continues to advise her in the art of moving away from brainless programming. Brian Powell continues to teach Fabiola about humble programming.

Thanks to all the presenters at GDevCon, CLA Summits, and NIWeek. They continue to show us how much more there is to learn. Thanks to those who invest their free time sharing their knowledge via the NI forums, the LAVAG online community, blog posts, podcasts, and the LabVIEW Wiki; their time and commitment to the G community does not go unnoticed. Thanks to all the LabVIEW programmers who create free tools and share them with the community; we are looking forward to GCentral making the process of sharing, finding, and collaborating in LabVIEW projects a straightforward task.

We reached out to several experts in the LabVIEW community to review the new content for the fifth edition of this book. Their feedback varied from harsh comments to words of encouragement. In the end, we take responsibility for what ended up in this book, but we couldn't have done it without them. Some reviewed large sections of the different chapters, while others provided guidance on the general ideas. Others discussed with us different approaches to teaching new concepts to LabVIEW programmers. In no particular order, we would like to thank our LabVIEW friends: Brian Powell, Stephen Loftus-Mercer, James McNally, Adriaan Rijllart, Peter Horn,

Peter Bokor, Darren Nattinger, Mikael Holmstrom, Andrei Zagorodni, Samuel Taggart, Oliver Wachno, Tanja Wachno, Ernesto García Ruiz, Dmitry Sagatelyan, Jeffrey Habets, Ching-Wa Yu, Jarobit Piña Saez, Huaxin Gong, Peter Adelhardt, Eric Jensen, Luc Des Ruelles, Olivier Jourdan, Matthias Baudot, Christopher Farmer, Joerg Hampel, Justin Goeres, Chris Relf, Michael Avialiotis, Dave Snyder, Matt Pollock, Val Brown, Paul Morris, and Benjamin Celis.

Thanks to Eric Reffet from NI for verifying that we were not violating any NDAs when talking about future developments in LabVIEW. Thanks to Shelley Gretlein and Allie Verlander from NI for providing a more recent picture of the LabVIEW team. Thanks to Hugo Andrade for sharing with us some of the stories behind the birth of LabVIEW RT and LabVIEW FPGA. Thanks to Michael Phillips from NI for taking the time to talk about the history of LabVIEW RT. Thanks to Tony Vento for reviewing the chapter on the history of LabVIEW. He was there almost from the beginning, and we were sad to see him leave NI.

Special thanks go to James McNally, Peter Bokor, Sam Taggart, and Oli Wachno, who reviewed more than one chapter and provided detailed feedback in their areas of expertise. We know some of them put aside their own projects to review the content and provide timely feedback. We are forever grateful.

Fabiola wishes to mention Steve Watts particularly because he reviewed all of the new content in this book and provided words of wisdom and laughter when she was ready to quit. If you follow Steve's blog, you will recognize some of our discussions that became posts in his blog and informed sections in this book. Fabiola also wants to thank Richard for trusting her with this new edition and all those Saturdays spent writing, editing, and discarding content.

Thanks to the Delacor team for keeping the business going while Fabiola was busy writing and researching content for this book. Also, without them, DQMH would not be the popular framework it is today. Thanks to the DQMH Trusted Advisors for their feedback and support. Credit also goes to our spouses, Patty Jennings and Luis F. Orozco, whose patience during this project cannot be overstated. Particularly, thanks to Luis F. Orozco, for providing support as a husband, a friend, and a colleague. We couldn't have done it without him.

Finally, thanks to Jeff Kodosky and Dr. T for inventing LabVIEW and giving us the perfect tool to create a professional career around it. The old saying is true: when you do what you love, you don't work a day in your life! Thanks to Jeff for taking the time off from his continuous research to write the foreword for this book and for his dedication to the G community. We are honored to have him as a member of that community, and we look forward to seeing more of his ideas come to life in LabVIEW.

Thanks to you for trusting us and reading this book.

Happy wiring!

CHAPTER 1

Roots

LabVIEW and Automation

We live in a time where automation is part of everyday life. You or someone you know has already automated their whole house and controls it via a smart speaker. We carry in our pockets more computing power than was needed to send a human to the moon. Oh, and it can make phone calls, too! Computers are supposed to make things easier, faster, or more automatic, that is, less work for the human user. LabVIEW is a unique graphical programming system that makes computer automation a breeze for the technician, scientist, or engineer working in many areas of laboratory research, industrial control, and data analysis. You have a job to do—someone is probably paying you to make things happen—and LabVIEW can be a real help in getting that job done, provided that you apply it properly. Does this approach always result in an improvement in productivity or quality? Sometimes we actually spend more time doing the same old thing. We also become slaves to our computers, always fussing over the setup, installing new (necessary?) software upgrades, and generally wasting time.

You must avoid this trap. The key is to analyze your problems and see where LabVIEW and specialized computer hardware can be used to their greatest advantage, then make efficient use of existing LabVIEW solutions. As you will see, many automation problems have already been solved for you, and the programs and equipment are readily available. There are no great mysteries here, just some straightforward engineering decisions you have to make regarding the advantages and disadvantages of computer automation.

Let's take a pragmatic view of the situation. There are many operations that beg for automation. Among them are the following:

- Long-term, low-speed operations such as environmental monitoring and control
- High-speed operations such as pulsed power diagnostics in which a great deal of data is collected in a short time
- Repetitive operations, such as automated testing and calibration, and experiments that are run many times
- Remote or hazardous operations where it is impractical, impossible, or dangerous to have a human operator present
- High-precision operations that are beyond human capability
- Complex operations with many inputs and outputs
- In all these cases, please observe that a computer-automated system makes practical an operation or experiment that you might not otherwise attempt. Automation may also offer some additional advantages such as these:
    - It eliminates operator-induced variations in the process or data collection methods. Repeatability is drastically improved because the computer never gets tired and always does things the same way.
    - It increases data throughput because you can operate a system at computer speed rather than human speed.
- There are some disadvantages hiding in this process of computer automation, however:
    - Automation may introduce new sources of error through improper use of sensors, signal conditioning, data conversion, and occasionally through computational (for example, round-off) errors.
    - Misapplication of any hardware or software system is a ticket for trouble. For instance, attempting to collect data at excessively high rates results in data recording errors.
    - Reliability is always a question with computer systems. System failures (crashes) and software bugs plague every high-tech installation known, and they will plague yours as well.

Always consider the cost-effectiveness of a potential automation solution. It seems as if everything these days is driven by money. If you can do it cheaper-better-faster, it is likely to be accepted by the owner, the shareholders, or whoever pays the bills. But is a computer guaranteed to save you money or time? If I have a one-time experiment in which I can adequately record the data on an oscilloscope or with my pencil, then taking days to write a special program makes no sense whatsoever.

One way to automate simple, one-time experiments is to build a LabVIEW crash cart, which is much like the doctor's crash cart in an emergency room. When someone has a short-term measurement problem, I can roll in my portable rack of equipment. It contains a computer with LabVIEW, analog interface hardware, and some programmable instruments. I can quickly configure the general-purpose data acquisition program, record data, and analyze them—all within a few hours. You might want to consider this concept if you work in an area that has a need for versatile data acquisition. Use whatever spare equipment you may have, recycle some tried-and-true LabVIEW programs, and pile them on a cart. It doesn't even matter what kind of computer you have since LabVIEW runs on Windows, Macintosh, and Linux. The crash cart concept is simple and marvelously effective.

Automation is expensive: The costs of sensors, computers, software, and the programmer's effort quickly add up. But in the end, a marvelous new capability can arise. You are suddenly freed from the labor of logging and interpreting data. The operator no longer has to orchestrate so many critical adjustments. And data quality and product quality rise. If your situation fits the basic requirements for which automation is appropriate, then by all means consider LabVIEW as a solution.

Sometimes there is pushback to use LabVIEW. Software programmers are no longer rare, but they normally know text languages and might think they can put together an automation program faster than a LabVIEW programmer can. In my experience, that is not the case. With LabVIEW you can get up and running pretty quickly and end up with an executable with a decent user interface faster than with a text programming language. Additionally, LabVIEW does an excellent job optimizing the use of multiple cores, multithreading, and memory use. When a text-based programmer tells me that LabVIEW is not good, my favorite example is to show them multiple loops running in parallel updating the front panel. I ask them how long it would take them to write the same code in their text language of choice, and frequently that is the end of the discussion. I don't even have to show them the connection with hardware.

Virtual Instruments: LabVIEW's Foundation

LabVIEW made the concept of virtual instruments (VI) a practical reality. The objective in virtual instrumentation is to use a general-purpose computer to mimic real instruments with their dedicated controls and displays, but with the added versatility that comes with software. Instead of buying an oscilloscope and a spectrum analyzer, you can buy one high-performance analog-to-digital converter (ADC) and use a computer running LabVIEW to simulate all these instruments and more. (See Fig. 1.1.) The VI concept is so fundamental to the way that LabVIEW works that the programs you write in LabVIEW are in fact called VIs. You use simple instruments (subVIs) to build more complex instruments, just as you use subprograms to build a more complex main program in a conventional programming language.

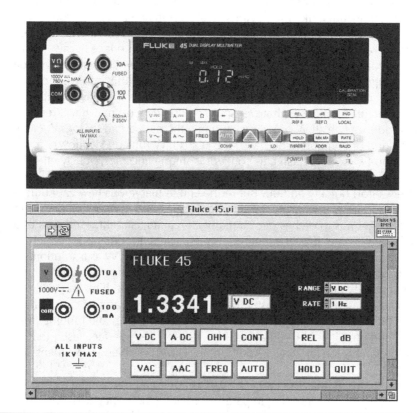

FIG. 1.1 *This virtual instrument (bottom) is a customized version of the real instrument, but with only the features that you need.*

Virtual versus real instrumentation. It used to be that the major drawback in using a computer for implementing virtual instruments was that the computer had only one central microprocessor. An application that used multiple instruments could easily overburden the processor. A standalone instrument, however, may contain any number of processors, each dedicated to specific processing tasks. This is no longer the case; computers now come with multiple cores, and LabVIEW does an excellent job in optimizing their use. One drawback still present is that the operating system on the computer (for example, Windows) still has to take care of multiple tasks, including reacting to the end-user actions.

Various technologies from National Instruments (NI) have addressed these issues. First, a version of LabVIEW—LabVIEW RT[1]—allows you to run independent programs on multiple dedicated, real-time processors that serve to unload tasks from the desktop host. Embedded systems running LabVIEW RT don't have to take care of end-user actions; they are generally headless and can be dedicated 100 percent to running your program. Second, LabVIEW permits you to link all kinds of computing platforms, as well as other software applications, over a variety of networks, thus further distributing the workload. Third, plug-in boards now contain their own processors or FPGAs.[2] FPGAs are a good example of reconfigurable hardware that finds its way onto plug-in boards. Many plug-in data acquisition boards also have sophisticated direct memory access (DMA), timing, and triggering capabilities that can span multiple boards, resulting in improved synchronization and signal coupling between boards. These developments, along with the development of more capable operating systems and computer architectures, have brought parallel processing capabilities to computers, making them more sophisticated platforms for instrumentation and data acquisition applications. However, sophistication comes at the expense of complexity, because you must have greater knowledge of these hardware components and their interconnection than is required to use a standalone instrument with similar capabilities. Virtual instrumentation software is essential for turning these sophisticated hardware combinations into usable instrumentation systems.

Virtual instrumentation offers the greatest benefit over real instruments in the areas of price/performance, flexibility, and customization. For the price of a dedicated high-performance instrument, you can assemble a computer-based system with the fundamental hardware and software components to design virtual instruments targeted for specific applications. The hardware may be plug-in boards, external instruments, or a combination of both. In either case, a software interface can be as complicated or as simple as needed to serve the application. You can simplify the operation of a complex standalone instrument with virtual instruments that focus on controlling only subsets of the instrument's full capabilities. I, for one, get lost in the buttons and menus on the panels of many modern instruments and find a simple VI a welcome relief because I get to put in only the functionality I care about.

Although LabVIEW has existed since 1986, the virtual instrumentation and block diagram concepts embodied in its design are still at the leading edge of instrumentation and computer science technology today. The cost of developing test program software continues to rise with the increasing complexity of the devices being tested and the instruments needed to test them. Software modularity, maintainability, and reusability—key benefits of LabVIEW's

---

[1] LabVIEW Real Time.
[2] Field-programmable gate arrays.

hierarchical and homogeneous structure—are critically important to reducing the burden of individual software developers. Reusing routines that you have written and sharing them with others can save a great deal of time and make programs more reliable.

Why Use LabVIEW?

I use LabVIEW because it has significant advantages over other languages and other control and data acquisition packages.

- My productivity is simply better in LabVIEW than with other programming languages. I've measured a factor of 5 compared with C on a small project. Others have reported improvements of 15 times. Quick prototypes as well as finished systems are now routinely delivered in what used to be record time.

- The graphical user interface is built in, intuitive in operation, and simple to apply.

- LabVIEW's graphical language—G—is a real programming language, not a specialized application. It has a few intrinsic limitations. It is a compiled language too.

- There is only a minimal performance penalty when it is compared with other programming languages, and LabVIEW does some things better. No other graphical programming system can make this claim.

- Programmer frustration is reduced because annoying syntax errors are eliminated. Have you ever gotten yourself into an argument with a C compiler over what is considered "legal"? Or made a seemingly minor error with a pointer and had your machine crash?

- Rapid prototyping is encouraged by the LabVIEW environment, leading to quick and effective demonstrations that can, within reason, be reused or expanded into complete applications.

- Sophisticated built-in libraries and add-on toolkits address specific needs in all areas of science, engineering, and mathematics.

- Programming in LabVIEW is fun. I would never say that about C (challenging, yes, but fun, no).

- The immediacy of seeing a block diagram and being able to understand what it does at a glance is a great benefit. It also means that spaghetti code will be easier to spot, too!

Like any other tool, LabVIEW is useful only if you know how to use it. And the more skilled you are in the use of that tool, the more you will use it. After many years of practice, I'm really comfortable with LabVIEW. Yet I continue to learn and improve every day. Solving easy problems with LabVIEW is easy. Solving more complex problems with LabVIEW is easier than in C, but it is still difficult to solve them. My friend Brian Powell[3] says that programming in C is like learning to play the violin. You would not want anyone listening to you for a long time, and it would take a lifetime to master. Programming in LabVIEW is like learning to play the guitar. You could be entertaining your friends pretty quickly, but it would still take a lifetime to master!

The Origin of LabVIEW

A computer scientist friend of mine relates this pseudo-biblical history of the computer programming world.

In the beginning, there was only machine language, and all was darkness. But soon, assembly language was invented, and there was a glimmer of light in the programming world. Then came Fortran, *and the light went out.*

This verse conveys the feeling that traditional computer languages, even high-level languages, leave much to be desired. You spend a lot of time learning all kinds of syntactical subtleties, metaphors, compiler and linker commands, and so forth, just to say "Hello, world." And heaven forbid that you should want to draw a graph or make something move across the screen or send a message to another computer. We're talking about many days or weeks of work here. It's no wonder that it took so many years to make the computer a useful servant of the common person. For the working technician, scientist, or engineer, these classical battles with programming languages have been most counterproductive. All you wanted to do is make the darned thing display a temperature measurement from your experiment, and what did you get?

```
(beep) SYNTAX ERROR AT LINE 1326
```

Thanks a lot, oh mighty compiler. Well, times changed in a big way because of LabVIEW. At last, we and other working troops have a programming language that eliminates that arcane syntax, hides the compiler, and builds the graphical user interface right in. Please note that LabVIEW is a compiled language, and contrary to what some people assume, LabVIEW is not interpreted. The graphics on the block diagram are the code. There is no fooling around; just wire up some icons and run. And yet the thought of actually programming with pictures is so incredible when contrasted with ordinary computer languages. How did they do it? Who came up with this idea?

Here is a most enlightening story of the origins of LabVIEW. It's a saga of vision; of fear and loathing in the cruel world of computer programming; of hard work and

---

[3] Brian Powell worked at National Instruments for 26 years and was a key member of the LabVIEW software development team since LabVIEW 1.1.

long hours; and of breakthroughs, invention, and ultimate success. The original story, *An Instrument That Isn't Really*, was written by Michael Santori of NI and has been updated for this book.[4]

Introduction

Prior to the introduction of personal computers in the early 1980s, nearly all laboratories using programmable instrumentation controlled their test systems using dedicated instrument controllers. These expensive, single-purpose controllers had integral communication ports for controlling instrumentation using the IEEE-488 bus, also known as the General-Purpose Interface Bus (GPIB). With the arrival of personal computers, however, engineers and scientists began looking for a way to use these cost-effective, general-purpose computers to control benchtop instruments. This development fueled the growth of NI, which by 1983 was the dominant supplier of GPIB hardware interfaces for personal computers (as well as for minicomputers and other machines not dedicated solely to controlling instruments).

So, by 1983, GPIB was firmly established as the practical mechanism for electrically connecting instruments to computers. Except for dealing with some differing interpretations of the IEEE-488 specification by instrument manufacturers, users had few problems physically configuring their systems. The software to control the instruments, however, was not in such a good state. Almost 100 percent of all instrument control programs developed at this time were written in the BASIC programming language because BASIC was the dominant language used on the large installed base of dedicated instrument controllers. Although BASIC had advantages (including a simple and readable command set and interactive capabilities), it had one fundamental problem: like any other text-based programming language, it required engineers, scientists, and technicians who used the instruments to become programmers. These users had to translate their knowledge of applications and instruments into the lines of text required to produce a test program. This process, more often than not, proved to be a cumbersome and tedious chore, especially for those with little or no prior programming experience.

A Vision Emerges

NI, which had its own team of programmers struggling to develop BASIC programs to control instrumentation, was sensitive to the burden that instrumentation programming placed on engineers and scientists. A new tool for developing instrumentation software programs was clearly needed. But what form would it take? Dr. Jim Truchard and Jeff Kodosky, two of the founders of NI, along with Jack MacCrisken, who was then a consultant, began the task of inventing this tool. (See Fig. 1.2.) Truchard was in search of a software tool that would markedly change the

---

[4] © 1990 IEEE. Reprinted, with permission, from *IEEE Spectrum*, vol. 27, no. 8, pp. 36–39, August 1990.

way engineers and scientists approached their test development needs. A model software product that came to mind was the electronic spreadsheet. The spreadsheet addressed the same general problem Truchard, Kodosky, and MacCrisken[5] faced—making the computer accessible to nonprogrammer computer users. Whereas the spreadsheet addressed the needs of financial planners, this entrepreneurial trio wanted to help engineers and scientists. They had their rallying cry—they would invent a software tool that had the same impact for scientists and engineers that the spreadsheet had on the financial community.

FIG. 1.2 *(Left to right): Jack MacCrisken, Jeff Kodosky, and Jim Truchard, LabVIEW inventors.*

In 1984, the company, still relatively small in terms of revenue, decided to embark on a journey that would ultimately take several years. Truchard committed research and development funding to this phantom product and named Kodosky as

---
[5] Jack MacCrisken passed away in January 2018.

the person to make it materialize. MacCrisken proved to be the catalyst—an amplifier for innovation on the part of Kodosky—while Truchard served as the facilitator and primary user.

Dr. T, as he is affectionately known at NI, has a knack for knowing when the product is right.

Kodosky wanted to move to an office away from the rest of the company so he could get away from the day-to-day distractions of the office and create an environment ripe for inspiration and innovation. He also wanted a site close to the University of Texas at Austin so he could access the many resources available at UT, including libraries for research purposes and, later, student programmers to staff his project. There were two offices available in the desired vicinity. One office was on the ground floor with floor-to-ceiling windows overlooking the pool at an apartment complex. The other office was on the second floor of the building and had no windows at all. He chose the latter. It would prove to be a fortuitous decision.

All the World's an Instrument

The first fundamental concept behind LabVIEW was rooted in a large test system that Truchard and Kodosky had worked on at the Applied Research Laboratory in the late 1970s. Shipyard technicians used this system to test Navy sonar transducers. However, engineers and researchers also had access to the system for conducting underwater acoustics experiments. The system was flexible because Kodosky incorporated several levels of user interaction into its design. A technician could operate the system and run specific test procedures with predefined limits on parameters, while an acoustics engineer had access to the lower-level facilities for actually designing the test procedures. The most flexibility was given to the researchers, who had access to all the programmable hardware in the system to configure as they desired (they could also blow up the equipment if they weren't careful). Two major drawbacks to the system were that it was an incredible investment in programming time—over 18 work-years—and that users had to understand the complicated mnemonics in menus in order to change anything.

Over several years, Kodosky refined the concept of this test system to the notion of instrumentation software as a hierarchy of virtual instruments. A virtual instrument (VI) would be composed of lower-level virtual instruments, much like a real instrument was composed of printed circuit boards, and boards were composed of integrated circuits (ICs). The bottom-level VIs represented the most fundamental software building blocks: computational and input/output (I/O) operations. Kodosky gave particular emphasis to the interconnection and nesting of multiple software layers. Specifically, he envisioned VIs as having the same type of construction at all levels. In the hardware domain, the techniques for assembling ICs into boards are dramatically different than assembling boards into a chassis. In the software domain, assembling statements into subroutines differs from assembling subroutines into programs, and these activities differ greatly from assembling concurrent programs into systems. The VI model of homogeneous structure and interface, at all levels, greatly simplifies the construction of software—a necessary achievement for improving design productivity. From a practical point of view, it was essential that VIs have a superset of the properties of the analogous software components they were replacing. Thus, LabVIEW had to have the computational ability of a programming language and the parallelism of concurrent programs.

Another major design characteristic of the virtual instrument model was that each VI had a user interface component. Using traditional programming approaches, even a simple command-line user interface for a typical test program was a complex maze of input and output statements often added after the core of the program was written. With a VI, the user interface was an integral part of the software model. An engineer could interact with any VI at any level in the system simply by opening the VI's user interface. The user interface would make it easy to test software modules incrementally and interactively during system development.

In addition, because the user interface was an integral part of every VI, it was always available for troubleshooting a system when a fault occurred. (The virtual instrument concept was so central to LabVIEW's incarnation that it eventually became embodied in the name of the product. Although Kodosky's initial concerns did not extend to the naming of the product, much thought would ultimately go into the name LabVIEW, which is an acronym for *Laboratory Virtual Instrument Engineering Workbench*.)

A Hard-Core UNIX Guy Won over by the Macintosh

The next fundamental concept of LabVIEW was more of a breakthrough than a slow evolution over time. Kodosky had never been interested in personal computers because they didn't have megabytes of memory and disk storage, and they didn't run UNIX. About the time Kodosky started his research on LabVIEW, however, his brother-in-law introduced him to the new Apple Macintosh personal computer. Kodosky's recollection of the incident was that "after playing with MacPaint for over three hours, I realized it was time to leave and I hadn't even said hello to my sister." He promptly bought his own Macintosh. After playing with the Macintosh, Kodosky came to the conclusion that the most intuitive user interface for a VI would be a facsimile of a real instrument front panel. (The Macintosh was a revelation because DOS and UNIX systems in 1983 did not have the requisite graphical user interface.) Most engineers learn about an instrument by studying its front panel and experimenting with it. With its mouse, menus, scroll bars, and icons, the Macintosh proved that the right interface would also allow someone to learn software by experimentation. VIs with graphical front panels that could be operated using the mouse would be simple to operate. A user could discover how they work, minimizing documentation requirements (although people rarely documented their BASIC programs anyway).

Putting It All Together with Pictures

The final conceptual piece of LabVIEW was the programming technique. A VI with an easy-to-use graphical front panel programmed in BASIC or C would simplify operation but would make the development of a VI more difficult. The code necessary to construct and operate a graphical panel is considerably more complex than that required to communicate with an instrument.

To begin addressing the programming problem, Kodosky went back to his model software product—the spreadsheet. Spreadsheet programs are so successful because they display data and programs as rows and columns of numbers and formulas. The presentation is simple and familiar to businesspeople. What do engineers do when they design a system? They draw a block diagram. Block diagrams help an engineer visualize the problem, but only suggest a design. Translation of a block diagram to a schematic or computer program, however, requires a great deal of skill. What Kodosky wanted was a software-diagramming technique that would be easy to use for conceptualizing a system, yet flexible and powerful enough to actually serve as a programming language for developing instrumentation software.

Two visual tools Kodosky considered were flowcharts and state diagrams. It was obvious that flowcharts could not help. These charts offered visualization of a process, but to really understand them, you have to read the fine print in the boxes on the chart. Thus, the chart occupies too much space relative to the fine print yet adds very little information to a well-formatted program. The other option, a state diagram, is flexible and powerful, but the perspective is very different from that of a block diagram. Representing a system as a collection of state diagrams requires a great amount of skill. Even after completion, the diagrams must be augmented with textual descriptions of the transitions and actions before they can be understood.

Another approach Kodosky considered was dataflow diagrams. Dataflow diagrams, long recommended as a top-level software design tool, have much in common with engineering block diagrams. Their one major weakness is the difficulty involved in making them powerful enough to represent iterative and conditional computations. Special nodes and feedback cycles have to be introduced into the diagram to represent these computations, making it extremely difficult to design or even understand a dataflow diagram for anything but the simplest computations. Kodosky felt strongly, however, that dataflow had some potential for his new software system.

By the end of 1984, Kodosky had experimented with most of the diagramming techniques, but they were all lacking in some way. Dataflow diagrams were the easiest to work with up until the point where loops were needed. Considering a typical test scenario, however, such as "take 10 measurements and average them," it's obvious that loops and iteration are at the heart of most instrumentation applications. In desperation, Kodosky began to make ad hoc sketches to depict loops specifically for these types of operations. Loops are basic building blocks of modern structured programming languages, but it was not clear how or if they could be drawn in a dataflow concept. The answer that emerged was a box—a box in a dataflow diagram could represent a loop. From the outside, the box would behave as any other node in the diagram, but inside it would contain another diagram, a subdiagram, representing the contents of the loop. All the semantics of the loop behavior could be encapsulated in the border of the box. In fact, all the common structures of structured programming languages could be represented by different types of boxes. His structured dataflow diagrams were inherently parallel because they were based on dataflow. In 1990, the first two U.S. patents were issued, covering structured dataflow diagrams and virtual instrument panels. (See Fig. 1.3.)

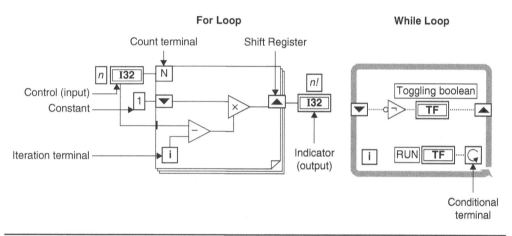

FIG. 1.3 LabVIEW For Loop and While Loop programming structures with Shift Registers to recirculate data from previous iterations.

Kodosky was convinced he had achieved a major breakthrough but was still troubled by a nagging point. There are times when it is important to force operations to take place sequentially—even when there is no dataflow requiring it. For example, a signal generator must provide a stimulus before a voltmeter can measure a response, even though there isn't an explicit data dependency between the instruments. A special box to represent sequential operations, however, would be cumbersome and take up extra space. During one of their design discussions, Truchard suggested that steps in a sequence were like frames in a movie. This comment led to the notion of having several sequential subdiagrams share the same screen space. It also led to the distinctive graphic style of the Sequence Structure, as shown in Fig. 1.4.

14 LabVIEW Graphical Programming

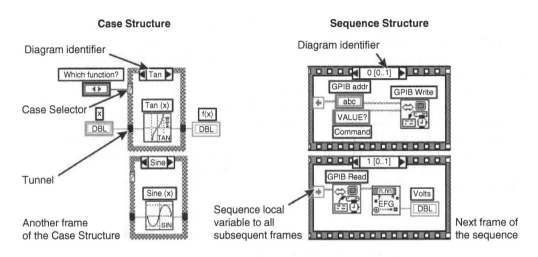

FIG. 1.4 *LabVIEW Case Structures are used for branching or decision operations, and Sequence Structures are used for explicit flow control.*

After inventing the fundamentals of LabVIEW block diagramming, it was a simple matter of using MacPaint to produce pictures of VI panels and diagrams for several common applications. When Kodosky showed them to some engineers at NI, the impact was dramatic. The engineers understood the meaning of the diagrams and correctly guessed how to operate the front panel. Of equal importance, the reviewers expressed great confidence that they would be able to easily construct such diagrams to do other applications. Now, the only remaining task was to write the software, facetiously known as *SMOP: small matter of programming*.

Favoring the Underdog Platform for System Design

Although Kodosky felt that the graphics tools on the Macintosh made it the computer of choice for developing LabVIEW, the clearer marketing choice was the DOS-based IBM PC. The Macintosh could never be the final platform for the product because it wasn't an open machine, and salespeople would never be able to sell it because the Macintosh was considered a toy by many scientists and engineers. Politics and marketing aside, it wasn't at all clear that you could build a system in which a user draws a picture and the system runs it. Even if such a system could be built, would it be fast enough to be useful? Putting marketing concerns aside momentarily, Kodosky decided to build a prototype on the Macintosh prior to building the real system on a PC.

Kodosky's affinity for the Macintosh was not for aesthetic reasons alone. The Macintosh system ROM contains high-performance graphics routines collectively known as QuickDraw functions. The Macintosh's most significant graphics capability is its ability to manipulate arbitrarily shaped regions quickly. This capability makes animated, interactive graphics possible. The graphics region algebra performed by the Macintosh is fast because of the unique coordinate system built into QuickDraw: the pixels are between, not on, the gridlines. In addition, the graphics display of the Macintosh uses square pixels, which simplifies drawing in general and rotations of bitmaps in particular. This latter capability proves useful especially for displaying rotating knobs and indicators on a VI front panel.

The operating system of the Macintosh is well integrated. It contains graphics, event management, input/output, memory management, resource and file management, and more—all tuned to the hardware environment for fast and efficient operation. Also, the Macintosh uses Motorola's 68000 family of microprocessors. These processors are an excellent base for large applications because they have large uniform address space (handling large arrays of data is easy) and a uniform instruction set (compiler-generated code is efficient). Remember that this was 1985: the IBM PC compatibles were still battling to break the 640-kilobyte barrier and had no intrinsic graphics support. It wasn't until Microsoft released Windows 3.0 in 1991 that a version of LabVIEW for the PC became feasible.

Ramping up Development

Kodosky hired several people just out of school (and some part-time people still in school) to staff the development team. Without much experience, none of the team members were daunted by the size and complexity of the software project they were undertaking, and instead they jumped into it with enthusiasm. The team bought 10 Macintoshes equipped with 512 kilobytes of memory and internal hard-disk drives called HyperDrives. They connected all the computers to a large temperamental disk server. The team took up residence in the same office near campus used by Kodosky for his brainstorming session. The choice of location resulted in 11 people crammed into 800 square feet. As it turned out, the working conditions were almost ideal for the project. There were occasional distractions with that many people in one room, but the level of communication was tremendous. When a discussion erupted between two team members, it would invariably have some impact on another aspect of the system they were inventing. The other members working on aspects of the project affected by the proposed change would enter the discussion and quickly resolve the issue. The lack of windows and a clock also helped the team stay focused.

They worked for long hours and couldn't afford to worry about the time. All-nighters were the rule rather than the exception, and lunch break often didn't happen until 3 p.m. There was a refrigerator and a microwave in the room so the team could eat and work at the same time. The main nutritional staples during development were Double Stuf Oreo cookies and popcorn, and an occasional mass exodus to Armin's for Middle Eastern food.

The early development proceeded at an incredible pace. In 4 months' time, Kodosky put together a team and the team learned how to program the Macintosh. MacCrisken contributed his project management skills and devised crucial data structure and software entity relationship diagrams that served as an overall road map for software development. They soon produced a proof-of-concept prototype that could control a GPIB instrument (through a serial port adapter), take multiple measurements, and display the average of the measurements. In proving the concept, however, it also became clear that there was a severe problem with the software speed. It would take two more development iterations and a year before the team would produce a viable product. (See Fig. 1.5.)

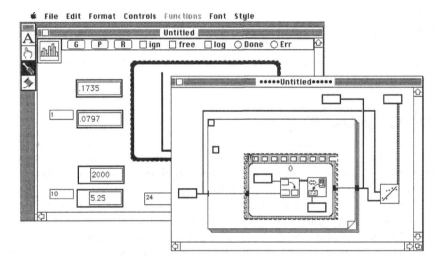

FIG. 1.5 *Screenshots from a very early prototype of LabVIEW. (Thanks to Paul Daley of LLNL who discovered these old LabVIEW versions deep in his diskette archive.)*

Stretching the Limits of Tools and Machine

After finishing the first prototype, Kodosky decided to continue working on the Macintosh because he felt the team still had much to learn before they were ready to begin the real product. It was at this time that the team began encountering the realities of developing such a large software system. The first problems they encountered were in the development tools. The software overflowed some of the internal tables, first in the C compiler and then in the linker. The team worked with the vendor to remedy the problem. Each time it occurred, the vendor expanded the tables. These fixes would last for a couple months until the project grew to overflow them again. The project continued to challenge the capabilities of the development tools for the duration of the project.

The next obstacle encountered was the Macintosh jump table. The project made heavy use of object-oriented techniques, which resulted in lots of functions, causing the jump tables to overflow. The only solution was to compromise on design principles and work within the limits imposed by the platform. As it turned out, such compromises would become more commonplace in the pursuit of acceptable performance.

The last major obstacle was memory. The project was already getting too large for the 512-kilobyte capacity of Macintosh, and the team still hadn't implemented all the required functions, let alone the desirable ones they had been hoping to include. The prospects looked dim for implementing the complete system on a DOS-based PC, even with extensive use of overlaying techniques. This situation almost proved fatal to the project. The team was at a dead end, and morale was at an all-time low. It was at this opportune time that Apple came to the rescue by introducing the Macintosh Plus in January 1986. The Macintosh Plus was essentially identical to the existing Macintosh except that it had a memory capacity of one megabyte. Suddenly, there was enough memory to implement and run the product with most of the features the team wanted.

Once again, the issue of the marketability of the Macintosh arose. A quick perusal of the DOS-based PC market showed that the software and hardware technology had not advanced very much. Kodosky decided (with approval by Dr. Truchard after some persuasion) that, having come this far on the Macintosh, they would go ahead and build the first version of LabVIEW on the Macintosh. By the time the first version of LabVIEW was complete, there would surely, they thought, be a new PC that could run large programs.

Facing Reality on Estimated Development Times

The initial estimates of the remaining development effort were grossly inaccurate. The April 1986 introduction date passed without a formal software release. In May, in anticipation of an imminent shipping date, the team moved from their campus workroom to the main office, where they could be close to the application engineers who did customer support. This event caused much excitement but still no product.

It was at this point that the company became overanxious and tried to force the issue by prematurely starting beta testing. The testing was a fiasco. The software was far from complete. Many bugs were encountered in doing even the simplest and most common operations. Development nearly ground to a halt as the developers spent their time listening to beta testers calling in the same problems. (See Fig. 1.6.)

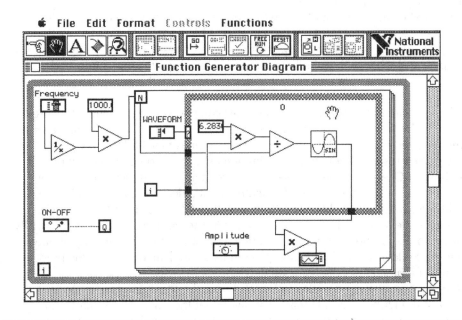

FIG. 1.6 *This diagram screenshot is from LabVIEW beta 0.36 in June 1986. The familiar structures (While and For Loops and Case Structure) had appeared by this time.*

As the overall design neared completion, the team began focusing more on details, especially performance. One of the original design goals was to match the performance of interpreted BASIC. It was not at all clear how much invention or redesign it would require to achieve this performance target, making it impossible to predict when the team would achieve this goal. On most computational benchmarks, the software was competitive with BASIC. There was one particular benchmark, the Sieve of Eratosthenes, that posed, by nature of its algorithm and design, particular problems for dataflow implementations. The performance numbers the team measured for the sieve benchmark were particularly horrendous and discouraging—a fraction of 1 second for a compiled C program, 2 minutes for interpreted BASIC, and over 8 hours for LabVIEW.

Kodosky did his best to predict when the software would be complete, based on the number of bugs, but was not sure how the major bugs would be found, much less fixed. Efficiently testing such a complex interactive program was a vexing and complex problem. The team finally settled on the "bug day" approach. They picked a day when the entire team would stop working on the source code and simply use LabVIEW. They would try all types of editing operations, build as many and varied VIs as they could, and write down all the problems they encountered until the whiteboards on every wall were full. The first bug day lasted only 3 hours. The team sorted the list and for the next 5 weeks fixed all the fatal flaws and as many minor flaws as possible. Then they had another bug day. They repeated this process until they couldn't generate even one fatal flaw during an entire day. The product wasn't perfect, but at least it would not be an embarrassment.

Shipping the First Version

In October 1986, the team figured out how to bypass some of the overhead in calling a subVI, producing some improvement in performance (for all but the sieve benchmark, which was better, but still 20 times slower than BASIC). The decision was made to ship the product. The team personally duplicated and packaged the first 50 disks and hand-carried them to shipping. Version 1.0 was on the streets.

The reaction to LabVIEW was, in a word, startling. The product received worldwide acclaim as the first viable, visual, or graphical language. There were many compliments for a well-designed product, especially from research and development groups who had had their Macintosh-based projects canceled by less-adventurous CEOs and marketing departments. Interestingly enough, the anticipated demand of the targeted BASIC users did not materialize. These people were apparently content to continue programming as they had been doing. Instead, LabVIEW was attracting and eliciting demands from customers who had never programmed at all but who were trying to develop systems considered extremely difficult by experienced programmers in any language. Yet these customers believed they could successfully accomplish their application goals with LabVIEW.

Apple Catches up with the Potential Offered by LabVIEW

Shortly after shipment of LabVIEW began, the company received its first prototype of the Macintosh II. This new version had many design features that promised to legitimize the Macintosh in the scientific and engineering community. The most important of these features was the open architecture of the new machine. Previous Macintosh versions did not have the capability to accept plug-in expansion boards. The only mechanisms available for I/O were RS-422 serial and SCSI (Small Computer Systems Interface) ports. NI sold standalone interface box products that converted these ports to IEEE-488 control ports, but performance suffered greatly.

The open architecture of Macintosh II made it possible to add not only IEEE-488 support but also other much-needed I/O capabilities, such as analog-to-digital conversion and digital I/O. The Macintosh II used the NuBus architecture, an IEEE standard bus that gave the new machine high-performance 32-bit capabilities for instrumentation and data acquisition that were unmatched by any computer short of a minicomputer (the PC's bus was 16 bits). With the flexibility and performance afforded by the new Macintosh, users now had access to the hardware capabilities needed to take full advantage of LabVIEW's virtual instrumentation capabilities. Audrey Harvey (now retired) led the hardware development team that produced the first Macintosh II NuBus interface boards, and Lynda Harrell (now founder and CEO of G Systems, an NI Alliance Partner) wrote the original LabDriver VIs that supported this new high-performance I/O. With such impressive new capabilities and little news from the PC world, NI found itself embarking on another iteration of LabVIEW development, still on the Macintosh.

Effective memory management turned out to be the key to making this graphical language competitive with ordinary interpreted languages. The development team had used a literal interpretation of dataflow programming. Each time data (a wire) leaves a source (a node), the Macintosh memory manager is called to allocate space for the new data, adding tremendous overhead. Other performance factors involved effective interpretation of the VI and diagram hierarchy and the scheduling of execution among nodes on the diagram. It became obvious that memory reuse was vital, but a suitable algorithm was far from obvious. Kodosky and MacCrisken spent about four intensive weekend brainstorming sessions juggling the various factors, eventually arriving at a vague algorithm that appeared to address everything. The whiteboard was covered with all sorts of instantiation diagrams with "little yellow arrows and blue dots" showing the prescribed flow of data. Then, one Monday morning, they called in Jeff Parker (now a LabVIEW consultant) and Steve Rogers (now a chief architect at NI) and introduced them to this magnificent algorithm. The two of them proceeded to implement the algorithm (to the point that Kodosky admittedly doesn't understand it anymore). They kept the famous whiteboard around for about a year, occasionally referring to it to make sure everything was right. LabVIEW 1.1 included these concepts, known collectively as inplaceness.

While the development team was working with the new Macintosh, they were also scrambling to meet some of the demands made by customers. They were simultaneously making incremental improvements in performance, fixing flaws that came to light after shipment began, and trying to plan future developments. As a result of this process, LabVIEW progressed from version 1.0 to 1.2 (and the sieve progressed to 23 seconds). LabVIEW 1.2 was a very reliable and robust product. Gary, for one, wrote a lot of useful programs in version 1.2 and can't recall crashing. (See Fig. 1.7.)

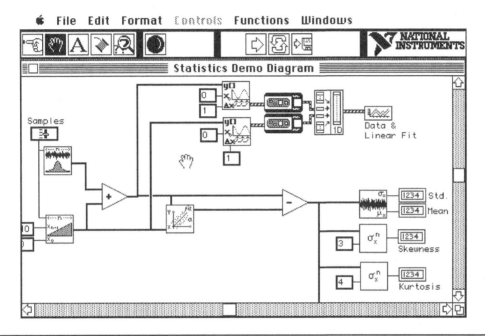

FIG. 1.7 *This is LabVIEW 1.2. It ran only in black and white, and you couldn't move an object once it was wired. Somehow, we early users managed to get a lot accomplished and enjoyed ourselves, at that.*

The Macintosh II gave LabVIEW a much-needed and significant boost in performance. The improvement, however, was short-lived. The internal architecture of LabVIEW 1.2 was showing signs of distress, and the software was apparently abusing the Macintosh resource manager as well as its memory manager. It was becoming clear that the only real way to enhance capabilities was with a complete redesign. At the least, a redesign could incorporate new diagram analysis algorithms and a fast built-in compiler that would eliminate performance problems once and for all. The major objective of the redesign was to achieve execution performance within a factor of two of compiled C.

FIG. 1.8 *This is how we got to know the LabVIEW 1 development team: the About LabVIEW dialog box had these way-cool portraits. Below is the team celebrating the 35th anniversary of LabVIEW at NIWeek 2011.*

LabVIEW 2: A First-Rate Instrument Control Product Becomes a World-Class Programming System

Even as the plans for the next-generation LabVIEW were becoming firm, customers were fast approaching, and exceeding, the limits of LabVIEW 1.2. Some users were building systems of VIs using up to 8 MB of memory (then the limit on a Macintosh II). The mega-applications took up to 30 minutes to load into memory. Users reeled at the nefarious "Too many objects" error message. And members of the development team often shuddered when they thought of the huge number of allocated structures and the complexity of their interconnections needed to make such a system work.

The decision to redesign LabVIEW brought with it a new set of pressures. In responding to some of the customer demands, the company had to admit that version 1.2 was at its design limits. As a result, work began on a new version, already becoming known as LabVIEW 2. Once the word was out, the development team was pressured into predicting a release date. Despite Kodosky's best intentions, the scope of the redesign resulted in several missed shipping deadlines. He realized that the design of a hierarchical dataflow compiler with polymorphic functions and sophisticated memory management was a new science—and it took a while.

LabVIEW 2 was designed with formalized object-oriented programming (OOP) techniques, on the insistence of Jeff Parker. OOP has many advantages over common procedural languages, and in fact, it was an enabling technology in this case. Unfortunately, OOP tools for C language development were in a rather primitive state in the 1988 time frame, so the team wrote their own spreadsheet-based development tools to automatically generate source code files and to keep track of objects and methods.

The development team released an alpha version of the software in late 1988. Cosmetically, this version appeared to be in excellent shape. The compiler was working well and showed great increases in performance. There were also many enhancements in the editing capabilities of the product. All these positive signs resulted in an air of excitement and of imminent release. Unfortunately, the team had a long list of items they knew had to be fixed to produce a technically sound product. Over a year elapsed before the team released the final product. In January 1990, LabVIEW 2 shipped to the first eager customers. I have to say that being a beta tester was a real thrill: the improvement in speed and flexibility was astounding.

The LabVIEW 2 compiler is especially notable not only for its performance but also for its integration into the development system. Developing in a standard programming language normally requires separate compilation and linking steps to produce an executable program. The LabVIEW 2 compiler is an integral and invisible part of the LabVIEW system, compiling diagrams in a fraction of the time required by standard compilers. From a user's point of view, the compiler is so fast and invisible that LabVIEW 2 is every bit as interactive as the previous interpreted versions. (See Fig. 1.9.)

FIG. 1.9 *The team that delivered LabVIEW 2 into the hands of engineers and scientists (clockwise from upper left): Jack Barber, Karen Austin, Henry Velick, Jeff Kodosky, Tom Chamberlain, Deborah Batto, Paul Austin, Wei Tian, Steve Chall, Meg Fletcher, Rob Dye, Steve Rogers, and Brian Powell. Not shown: Jeff Parker, Jack MacCrisken, and Monnie Anderson.*

Info-LabVIEW List: The Beginnings of an Ecosystem

Info-LabVIEW[6] started as an independent mailing list to discuss LabVIEW. Tom Coradeschi made the first post on February 14, 1991. He created the list when LabVIEW could only run on Macintosh computers. The list has been running since then and is considered a LabVIEW institution. You can still subscribe to it and get the e-mails. The majority of LabVIEW programmers use LabVIEW for Windows or Linux, but questions for the Mac still appear regularly. Info-LabVIEW is not the only independent LabVIEW forum, but they were the first. It was the beginning of an ecosystem that extends beyond NI. This is also the place where LabVIEW rusty nails could be found. These were tips, tricks, and hacks for LabVIEW. If you like history and want to find out more, read Greg McKaskle's Rusty Nails post.[7]

---

[6] info-labview.org/Info-LabVIEW
[7] McKaskle, Greg. "Re: Complete list of Labview Options" info-labview.org/ILVMessages/1999/05/27/Info-LabVIEW_Digest_1999-05-27_025.html

The Port to Windows and Sun

The next major quest in the development of LabVIEW was the portable or platform-independent version. Dr. Truchard (and thousands of users) had always wanted LabVIEW to run on the PC, but until Windows 3.0 came along, there was little hope of doing so because of the lack of 32-bit addressing support that is vital to the operation of such a large, sophisticated application. UNIX workstations, on the other hand, are well suited to such development, but the workstation market alone was not big enough to warrant the effort required. These reasons made Kodosky somewhat resistant to the whole idea of programming on the PC, but MacCrisken finally convinced him that portability itself—the isolation of the machine-dependent layer—is the real challenge. So, the port was on.

Microsoft Windows 3 turned out to be a major kludge with regard to 32-bit applications (Windows 3 itself and DOS are 16-bit applications). Only in Appendix E of the Windows programming guide was there any mention whatsoever of 32-bit techniques. And most of the information contained in that appendix referred to storage of data and not applications. Finally, only one C compiler—Watcom C—was suited to LabVIEW development. But before Watcom C became available, Steve Rogers created a set of glue routines that translate 32-bit information back and forth to the 16-bit system function calls (in accordance with Appendix E). He managed to successfully debug these low-level routines without as much as a symbolic debugger, living instead with hexadecimal dumps. This gave the development team a 6-month head start. Rogers summed up the entire situation: "It's ugly." Development on the Sun SPARCstation, in contrast, was a relative breeze. Like all good workstations, the Sun supports a full range of professional development tools with few compromises—a programmer's dream. However, the X Windows environment that was selected for the LabVIEW graphical interface was totally different from the Macintosh or Windows toolbox environments. A great deal of effort was expended on the low-level graphics routines, but the long-term payoff is in the portability of X Windows–based programs. Development on the Sun was so convenient, in fact, that when a bug was encountered in the Windows version, the programmer would often do his or her debugging on the Sun rather than suffering alone on the PC. Kodosky reports, "The Sun port made the PC port much easier and faster." LabVIEW 2.5, which was released in August 1992, required rewriting about 80 percent of LabVIEW 2 to break out the machine-dependent, or manager, layer. Creating this manager layer required some compromises with regard to the look and feel of the particular platforms. For instance, creating floating windows (such as the LabVIEW Help window) is trivial on the Macintosh, difficult on the PC, and impossible under X Windows. The result is some degree of least-common-denominator programming, but the situation has continued to improve in later versions through the use of some additional machine-dependent programming. (See Fig. 1.10.)

FIG. 1.10 *"Some new features are so brilliant that eye protection is recommended when viewing."* The team that delivered LabVIEW 2.5 and 3.0 includes (left to right, rear) Steve Rogers, Thad Engeling, Duncan Hudson, Kevin Woram, Greg Richardson, Greg McKaskle; (middle) Dean Luick, Meg Kay, Deborah Batto-Bryant, Paul Austin, Darshan Shah; (seated) Brian Powell, Bruce Mihura. Not pictured: Gregg Fowler, Apostolos Karmirantzos, Ron Stuart, Rob Dye, Jeff Kodosky, Jack MacCrisken, Stepan Riha.

LabVIEW 3

The LabVIEW 2.5 development effort established a new and flexible architecture that made the unification of all three versions in LabVIEW 3 relatively easy. LabVIEW 3, which shipped in July 1993, included a number of new features beyond those introduced in version 2.5. Many of these important features were the suggestions of users accumulated over several years. Kodosky and his team, after the long and painful port, finally had the time to do some really creative programming. For instance, there had long been requests for a method by which the characteristics of controls and indicators could be changed programmatically. The Attribute Node addressed this need. Similarly, Local Variables made it possible to both read from and write to controls and indicators. This is an extension of strict dataflow programming, but it is a convenient way to solve many tricky problems. Many subtle compiler improvements were also made that enhance performance, robustness, and extensibility. Additional U.S. patents were issued in 1994 covering these extensions to structured dataflow diagrams such as globals and locals, occurrences, attribute nodes, execution highlighting, and so forth.

LabVIEW 3 was a grand success, both for NI and for the user community worldwide. Very large applications and systems were assembled containing thousands of VIs. The LabVIEW Application Builder permitted the compilation of true standalone applications for distribution. Important add-on toolkits were offered by NI and third parties, covering such areas as process control, imaging, and database access. Hundreds of consultants and corporations have made LabVIEW their core competency; speaking as one of them, the phones just keep on ringing.

LabVIEW 4

Like many sophisticated software packages, LabVIEW has both benefited and suffered from feature creep: the designers respond to every user request, the package bulks up, and pretty soon the beginner is overwhelmed. April 1996 brought LabVIEW 4 to the masses and, with it, some solutions to perceived ease-of-use issues. Controls, functions, and tools were moved into customizable floating palettes, menus were reorganized, and elaborate online help was added. Tip strips appear whenever you point to icons, buttons, or other objects. Debugging became much more powerful. And even the manuals received an infusion of valuable new information. I heard significant, positive feedback from new users on many of these features.

LabVIEW 4 enabled the assembly of very large applications consisting of thousands of VIs, including standalone executables, courtesy of the LabVIEW Application Builder. Most important, NI realized that LabVIEW was a viable development platform with regard to creating new features and even extending LabVIEW itself. The first such extension was the DAQ Channel Wizard, a fairly sophisticated application that provided a comprehensive user interface for managing channel configurations on plug-in boards. Development time was drastically reduced compared to conventional C-based coding. Audrey Harvey began the Channel Wizard development in LabVIEW, showing key features in a prototypical fashion. When it became apparent that her prototype met most of the perceived needs of the user community, it was handed off to Deborah Batto-Bryant, who perfected it for release with LabVIEW 4.0. This strategy has since been followed many times. These LabVIEW-based features, when well designed, are indistinguishable from their C-based counterparts. The underlying plumbing in LabVIEW 4 was not significantly changed from LabVIEW 3, though some advanced R&D activities were certainly underway. The team called the next generation LabVIEW 10, referring to some distant, future release, and split off a special group dedicated to new features such as undo, multithreading, and real-time capability. While some of these features (including a simple form of undo) almost made it into LabVIEW 4.1, they had to wait until the next version.

LabVIEW Continues to Improve

LabVIEW 5

For LabVIEW 5, Jeff Kodosky's team adopted completely new development practices based on a formalized, milestone-driven scheme that explicitly measured software quality. (See Fig. 1.11.) Bugs were tracked quantitatively during all phases of the development cycle, from code design to user beta testing. The result, in February 1998, was the most reliable LabVIEW release ever, in spite of its increased complexity.

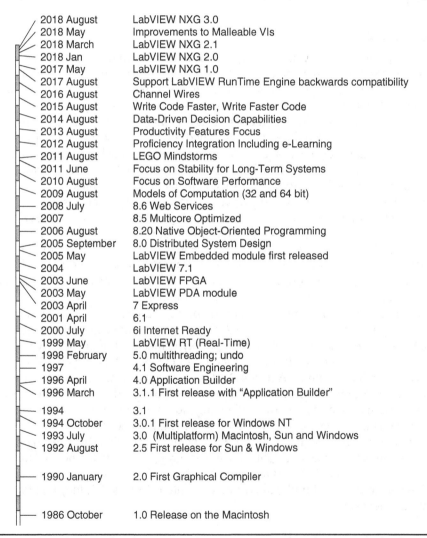

FIG. 1.11 *The LabVIEW development milestones up to LabVIEW 2018 and LabVIEW NXG 3.0.*

The quality improvement process began with a thorough scrubbing of LabVIEW 4—a laborious but necessary task—prior to adding new features. Then the fruits of the prior LabVIEW 10 efforts and other advanced concepts were incorporated. Ultimately, a good deal of the application was rewritten.

Undo was positively the most-requested feature among users, all the way back to the dawn of the product. But it was also considered one of the most daunting. Behind the scenes, LabVIEW is an extremely complex collection of objects—orders of magnitude more complicated than any word processor or drawing program. The simple-minded approach to implementing undo is simply to duplicate the entire VI status at each editing step. However, this quickly leads to excessive memory usage, making only a single-level undo feasible. Instead, the team devised an incremental undo strategy that is fast, memory efficient, and supports multiple levels. It's so unique that it's patented. Of course, it took a while to develop, and it involved a complete rewrite of the LabVIEW editor. That was step 1.

Step 2 required rewriting the LabVIEW execution system to accommodate another advanced feature: multithreaded execution. Applications that support multiple threads use the scheduling capabilities of the computer's operating system to divide CPU time between various tasks or threads. For instance, the user interface thread can run independently of the data acquisition thread and the data analysis thread, and all of those threads can have different priorities. While the potential performance enhancements are great, multithreading has traditionally been a fairly difficult programming challenge tackled only by skilled programmers. The LabVIEW 5 team brought multithreaded development to the masses by providing a simple, low-risk interface.

As any programmer can tell you, the simpler a feature appears, the more complex is its underlying code—and that certainly describes the conversion of LabVIEW to multithreading. The major task was to evaluate every one of the thousands of functions in LabVIEW to determine whether it was thread-safe, reentrant, or needed some form of protection. This evaluation process was carried out by first writing scripts in PERL (a string manipulation language) that analyzed every line of code in the execution system, and then having the programmers perform a second (human) evaluation. As it turned out, about 80 percent of the code required no changes, but every bit had to be checked. Kodosky said the process was ". . . like doing a heart and intestine transplant."

The third and final step in the Great Rewrite involved the compiler. As you know by now, LabVIEW is unique in the graphical programming world, in that it directly generates executable object code for the target machine. Indeed, that compiler technology is one of the crown jewels at NI. At the time, one of the latent shortcomings in the compiler was the difficulty involved in supporting multiple platforms. Every time a new one was added, a great deal of duplicated effort was needed to write another platform-specific compiler process. The solution was to create a new platform-independent layer that consolidated and unified the code prior to object code generation.

Now adding a new platform involves only the lowest level of object code generation, saving time and reducing the chances for new bugs. It also reduced the code size of the compiler by several thousand lines. It's another of those underlying technology improvements that remains invisible to the user except for increased product reliability.

The LabVIEW RT Branch

As long as any of us can remember, we've been asking for enhanced real-time performance from LabVIEW. It's also been a hot topic of discussion for the LabVIEW team as well, and they finally came up with a solution that was released in the form of LabVIEW RT in May 1999. After years of discussion and conceptual design that hadn't completely gelled into a solution, Jeff Kodosky offered a personal challenge to Darshan Shah: "I bet you can't!" Needless to say, Darshan was up to the challenge.

LabVIEW RT required a number of innovations. First, a specialized real-time operating system (RTOS) was needed to provide faster and more predictable scheduling response than a desktop OS. RTOSs often lack many of the services that standard LabVIEW requires, and they must be arranged for a specific hardware configuration. Darshan chose PharLap TNT Embedded because it supported many Windows Win32 API (application programming interface) calls, which made it much easier to port LabVIEW as well as NI-DAQ. To simplify hardware compatibility problems, the first release of LabVIEW RT ran only on a special DAQ board married to an 80486 processor, which was chosen for its code compatibility and adequate performance at a low cost. It was a reasonable engineering decision for an initial product release.

A second challenge was creating a simple and intuitive way of interacting with the embedded LabVIEW RT processor board. Darshan's solution was to provide a simple configuration dialog that tells LabVIEW to compile and run on either the host machine or the embedded board. This completely shields the user from the underlying platform change and all the real-time complexity that one would normally have to face. Another benefit of this architecture was that its development was synergistic with the other new LabVIEW 5 concepts: the division between the user interface and execution systems and multithreading.

LabVIEW FPGA

Another team at NI that got to enjoy the challenge of bringing new technology to life was the LabVIEW FPGA group. This was a long project, with the initial public demonstration in 1997, the "Pioneer" release in 2002,[8] and the first public release

---

[8] Press release, May 20, 2003, announcing LabVIEW FPGA software and PXI-7831R reconfigurable I/O hardware. http://ua.automation.com/content/national-instruments-labview-and-reconfigurable-io-deliver-fpga-benefits-to-measurement-and-control-engineers

in 2003. This team was small and took advantage of hiring as many interns as they could. Imagine being an intern and getting to work on patentable technology! One of the motivating use cases was to be able to analyze serial data using different types of protocols. The PCI-6810/PXI-6810 Serial Data Analyzer board had about 10 personalities, and the end user would change the personality depending on the protocol they were using. The problem with this approach was that NI became the bottleneck when end users wanted to customize the personalities even more, and they would have to do it using VHDL. NI had used FPGAs as glue logic, but they were slowly making their way into system elements. Could they target FPGAs in LabVIEW? Could they make an instrument like the serial analyzer but programmable? They did that and more. From monitoring serial streams, they moved to empower control engineers to go from having loops on the high kilohertz to loops that could now run at tens of megahertz.

Single-Cycle Timed Loop

Enabling high-speed applications in LabVIEW FPGA introduced the need for a low-level access structure. The LabVIEW FPGA team wanted to take care of this at a high level, but customers needed this feature a lot earlier than the team could implement it as part of the high-level synthesis. A single-cycle timed loop (SCTL)[9] is a special flavor of the LabVIEW timed loop structure. When used with an FPGA target, this loop executes all the code inside within one tick of the FPGA clock. The default FPGA clock is the 40-MHz clock. The developer can use the SCTL with derived clocks to achieve rates other than 40 MHz, but once the application is compiled and running, the clock timing and other properties of the loop cannot be changed.

LabVIEW 6i

June 2000 brought us LabVIEW 6i, with a host of new user-interface features (rendered in 3D, no less; unfortunately, these controls were called "modern" controls, which is funny now that they are still around after almost 20 years and they are still called "modern"!), extensive performance and memory optimization at all levels, and perhaps most important, a very powerful VI server. Introduced in LabVIEW 5, the VI server gives the user external hooks into VIs. For instance, you can write programs that load and run VIs, access VI documentation, and change VI setup information, all without directly connecting anything to the target VI. Furthermore, most of these actions can be performed by applications other than LabVIEW, including applications running elsewhere on a network. This new paradigm effectively publishes LabVIEW, making it a true member of an application system rather than exclusively running the show as it had in the past. You can even export shared libraries (DLLs) in LabVIEW. From the perspective of professional

---

[9] Single-Cycle Timed Loop FAQ for the LabVIEW FPGA Module
http://digital.ni.com/public.nsf/allkb/722A9451AE4E23A586257212007DC5FD

LabVIEW programmers, we believe that LabVIEW 6 has truly fulfilled the promise of graphical programming.

At the nuts-and-bolts level, LabVIEW 6 is now completely developed in C++, in contrast to all the earlier versions, which used C with object extensions created by the LabVIEW team. They stuck with C for so long because C++ compiler effectiveness was, in their judgment, not as good as C. As a point of reference, LabVIEW 6 consists of about 800,000 lines of code. While that's a pretty big number, there are plenty of applications running on your very own computer that are much heftier (including NI-DAQ, but that's another story). Also, the growth rate in LabVIEW's code size has actually diminished over the years due to improved programming techniques. It's always impressive to see continuous improvement in an application's performance and efficiency, rather than pure bloat as new features are added. (See Fig. 1.12.)

FIG. 1.12 *My, how we have grown! This is most of the LabVIEW 6 team.*

NI Forums: The Beginning of the Ecosystem

The NI forums started around 2001. Originally, they were staffed by NI applications engineers, and slowly the community started to take over. I remember that a lot of people at NI were skeptical about the community answering questions, and some people were nervous that this was just going to be more work for the applications engineers. Within a year it was obvious that there were a lot of people out there willing to share their knowledge.

LAVA Forums: The LabVIEW Ecosystem Continues to Grow

LAVA stands for LabVIEW Advanced Virtual Architects. Michael Aivaliotis created the LAVA forums[10] in 2002, and it has become the largest independent online LabVIEW community.[11] It now hosts the LabVIEW Wiki and the LAVA code repositories as well. I remember when I started working as a consultant, LAVA was a daily visit for me, for the longest time as a lurker, but eventually I was brave enough to start asking questions, and every now and then I could even contribute too. The topics in LAVA used to be more advanced than the topics in the NI-hosted LabVIEW forums. If you want to find more rusty nails, LAVA is the place to go. Not only do members find hidden features (like discovering VI Scripting before it was public or special LabVIEW INI tokens), but they also help others push the limits of LabVIEW. With time, the NI-hosted LabVIEW forums and communities started to host advanced discussions as well.

LabVIEW 7

LabVIEW 7 did not have a service pack. NI was ready to release LabVIEW 7, but they had to wait for DAQmx to be ready. By the time DAQmx was ready to ship, LabVIEW 7 had gone through so much testing that they were effectively releasing the service pack. LabVIEW 7 Express released in April 2003 with new features and wizards designed to make programming and getting started easier for people new to LabVIEW. Express blocks make configuring instruments, testing analysis routines, and most common tasks simple. The DAQ Assistant made data acquisition easier for everyone. Now with a few clicks of a mouse anyone can create a fully functioning data acquisition program. You're not limited to simple tasks either; the DAQ Assistant makes complex trigger applications a snap. The LabVIEW team went out of their way to help the novice with new, simplified ways of doing things without fatally compromising the power of LabVIEW. Express VIs have earned a bad reputation over the years because not all of them are created with the same attention to detail, and some of them end up adding a lot of bloat to your code. I only use the File Dialog Express VI and the DAQ Assistant. Before you cringe,

---

[10] lavag.org
[11] vishots.com/about

I only use the DAQ Assistant as a quick way to configure and test my setup, and then I right-click on the assistant and convert to code. Unlike the rest of the Express VIs, DAQ Assistant scripts only the code I need and does not add any bloat.

LabVIEW 7 also brings LabVIEW to new platforms far from the desktop. Targets include a variety of compact computers and operating systems, including Windows CE, Palm OS, and the ARM processor family. New in LabVIEW is a compiler for field-programmable gate arrays (FPGAs). LabVIEW for FPGA continues NI's push of LabVIEW Everywhere. This is more than just marketing hype. With LabVIEW FPGA, engineers are now able to convert LabVIEW VIs into hardware implementations of G code.

LabVIEW Champions

The LabVIEW Champions[12] program started in 2005. I joke that a LabVIEW Champion is someone who loves LabVIEW so much that they are willing to do marketing for free. The official definition is that "a LabVIEW Champion is a top tier, NI platform enthusiast who NI has recognized for their exceptional application development, technical depth and breadth, and leadership and contributions to the LabVIEW community. A LabVIEW Champion is the first to stand up and defend the NI Platform but still has a healthy sense of critical evaluation of the platform, always pushing it to be easier to use and more powerful."[13] We are a very opinionated bunch and are lucky that NI still listens to us. "LabVIEW Champions are also credible technology experts from around the world who inspire others to learn and grow with LabVIEW through active technical community participation and face-to-face interactions."[14] There is a misconception that you have to participate a lot in the forums to be recognized as a LabVIEW Champion, which is not the case. NI uses the Knight of NI recognition for those who post a lot on the forums. LabVIEW Champions do not work for NI and do not speak on NI's behalf. Every year, in December, NI evaluates the list of applicants and decides who will be inducted into the LabVIEW Champions directory the following year.

LabVIEW 8

Released in October 2005, LabVIEW 8 introduced new tools to make LabVIEW developers more productive and application development and integration across a wide range of platforms easier. The LabVIEW project provides a cohesive environment for application developers.

Finally, at long last, we can simultaneously develop applications for multiple targets. Developers using LabVIEW FPGA and RT can develop applications on the

---

[12] https://forums.ni.com/t5/LabVIEW-Champions/ct-p/7029
[13] Recognition and Ranks forums.ni.com/t5/Using-the-NI-Community/Recognition-and-Ranks/ta-p/3698600
[14] Recognition and Ranks forums.ni.com/t5/Using-the-NI-Community/Recognition-and-Ranks/ta-p/3698600

Chapter 1: Roots **35**

host and the target without having to close one environment and switch to another. The project interface provides a relatively painless way to manage development and deployment of large applications by many developers. LabVIEW is the best programming language in test and measurement, and it's time we had some big-time tools to manage our projects. Look for more OOP as LabVIEW includes built-in OO tools. NI continues to take LabVIEW deeper into education and simulation with the educational version of LabVIEW for DSPs and the new LabVIEW MathScript. MathScript is an integral part of LabVIEW that combines dataflow with text-based mathematical programming. And as shown in Fig. 1.13, the LabVIEW team continued to grow.

FIG. 1.13 *LabVIEW keeps growing and so does the development team. This is the team circa 2003. Note the wall behind the team with all the patents NI had until that point. If you visit NI now, you will see that is full and several of the walls around the lobby are full of NI patent plaques. (Photo courtesy of National Instruments Corporation.)*

LabVIEW 8.2

It was released in August 2006. LabVIEW 8.2 brought native OOP to LabVIEW. Although there are still some people who refuse to see LVOOP as part of LabVIEW, classes are now part of the G syntax.

History of Object-Oriented Programming for LabVIEW Programmers

Object-oriented software development started in the 1960s. Objects are now pretty common in text-based programming languages, but it took until the early 1990s with C++ and Java for the software community to fully embrace them. LabVIEW did not gain OO features until 2006. Why the delay? I believe that LabVIEW has always been at the cutting edge of programming, but it was unclear how OO and dataflow fit together, especially with the LabVIEW focus on nonprogrammers. Since the LabVIEW community tends to wait to use tools and paradigms that text-based programmers have been using for a while, NI took its time to figure out that integration.

Why did NI go through the trouble of including this language feature? After all, only a handful of LabVIEW programmers were asking for classes support by that name. However, LabVIEW programmers had asked for more code reuse, lower code maintenance, and better type safety. They were not asking for OO by name, but they were asking for things that could be addressed by implementing OO in LabVIEW.[15]

There were LabVIEW programmers who focused on bringing object-oriented design to LabVIEW. For example, Steve Watts and Jon Conway's book called *Software Engineering Approach to LabVIEW*[16] goes into more detail in describing this LabVIEW component-oriented design. With this approach, LabVIEW programmers could create modular applications where each object was represented with a VI with an uninitialized Shift Register and an enumerator to decide which action to execute. The reference design used was the same as that used for "Functional Global Variables," with the great difference being that the data is private to the component and stored in the uninitialized Shift Register. Another term used for this type of design is the "action engine."[17] These objects were singletons, meaning that you could not have more than one instance of the same object in your application.

In the *LabVIEW for Everyone* book,[18] Jeffrey Travis and James Kring presented a way to use the project libraries introduced in LabVIEW 8.0 as a way to

---

[15] LabVIEW Classes State of the Art, presentation at NI Week 2012 by Stephen Loftus-Mercer.
[16] *A Software Engineering Approach to LabVIEW*, First Edition by Jon Conway and Steve Watts. Prentice Hall 2013.
[17] Community Nugget 4/08/2007 Action Engines by Ben Rayner forums.ni.com/t5/LabVIEW/Community-Nugget-4-08-2007-Action-Engines/td-p/503801
[18] Appendix D dedicated to LabVIEW Object-Oriented Programming in the book *LabVIEW for Everyone. Graphical Programming Made Easy and Fun*, Third Edition by Jeffrey Travis and Jim Kring.

encapsulate the action engines or components. They would make the action engine, and its operation enumerator, private and create public VIs to access each operation.

Before NI was ready to include OO as a native G feature, Endevo (later known as Symbio) created their own version of OO and called it GOOP (G object-oriented programming).

Finally, NI introduced native LabVIEW classes as a new feature in LabVIEW 8.20. Then NI acquired the GDS toolkit from Endevo in 2014 and offered it for free via the LabVIEW Tools Network. Moving forward, for NXG (2017), NI has been investigating ways to make the transition from non-OO to LVOOP more straightforward. As part of this effort, the difference between a cluster and an object is seamless.

LabVIEW Tools Network

The LabVIEW ecosystem had grown so much that third-party companies were extending the capabilities of LabVIEW. LabVIEW programmers had been creating valuable LabVIEW tools for years. NI customers were demanding ways to assess the quality of code that others were selling or sharing for free. On August 6, 2007, NI announced the LabVIEW compatibility program for third-party products and the LabVIEW Tools Network.[19] There were also some people who refused to try anything that was not produced by NI. The LabVIEW Tools Network[20] is the online store for LabVIEW Tools that NI has certified as compatible with LabVIEW. There are other places where LabVIEW programmers can download tools like the LAVA code repository and the NI forums, but those are at the LabVIEW programmer's own risk.

LabVIEW 8.6[21]

LabVIEW 8.6 brought the Breakpoint Manager Window. You can now see where all your breakpoints are at and disable them or delete them in one single location. There were also other new features to improve the developer experience. One little one that I use every time I program is the option to link the input and output terminal of a Case Structure or Event Structure through all the cases. This is one of those little things that can make your day better. You no longer have to go through several case structures, wiring through a wire that connects to Shift Registers that you only use in one or two cases.

Another developer experience enhancement that I use every single day is Quick Drop. Quick Drop is a dialog box to search for a block diagram or front panel object by name and place it on the block diagram or front panel without having to use

---

[19] investor.ni.com/news-releases/news-release-details/ni-announces-labview-compatibility-program-third-party-products
[20] ni.com/labviewtools
[21] Go to ni.com/info and enter code upnote86 to the LabVIEW 8.6 upgrade notes.

the palettes. As the palettes continue to grow with each new version of LabVIEW, Quick Drop is very useful. All you have to do is to press <Ctrl-Space> and the Quick Drop window shows up.

There were some environment and performance enhancements, such as reducing the VI size. And behind the scenes, there were changes to improve the stability of the IDE. I am missing several other features for this version, but these are the ones that stood out for me.

LabVIEW Switches to Yearly Releases

LabVIEW 2009[22,23]

In 2009, NI began naming releases after the year in which they are released. A bug fix is termed a service pack; for example, the 2009 service pack 1 (2009 sp1 for short) was released in February 2010. Service packs are released every 6 months. This meant a predictable release cycle and more uniform scope.

LabVIEW 2009 focused on the advanced LabVIEW programmers. It brought features that increase the performance of your code, reduce the memory requirements for your application, and bring tools for improving software engineering practices.

Data value reference (DVR) and the In-Place Element structure were released in this version of LabVIEW. For advanced LabVIEW programmers, these were great new features. I remember being excited about reducing the amount of data copies. I was also frightened because I could see text-based programmers abusing DVRs, since they are basically their beloved pointers. As you know, in LabVIEW, the data is on the wire. Whenever you branch a wire, there is a chance for a data copy. This is not a concern for the majority of applications written in LabVIEW, but the process of allocating new memory and copying data can consume additional processor time, especially if you are manipulating large amounts of information. DVRs allow you to reference any data type to reduce the amount of memory your application uses. Instead of wires containing the actual value, a reference can point to the location where that is stored.

As mentioned earlier, LabVIEW helps you take advantage of multicore processors by automatically detecting sections of code that can be assigned to different threads. LabVIEW 2009 extended this capability to For Loop structures that could have separate iterations execute in parallel. Of course, there are times that the For Loop cannot be parallelized, for example, when you have a Shift Register and each iteration depends on the results from the previous iterations. You can find out if

---

[22] Five Advanced Features in LabVIEW 2009 http://www.ni.com/newsletter/50840/en/

[23] Go to ni.com/info and enter code upnote9 to the LabVIEW 2009 upgrade notes.

your loops are parallelizable via a tool added to the Tools menu that allows you to find parallelizable loops in your VI.

One more neat feature added in LabVIEW 2009 was native recursion. Starting with this version, we can have a VI call itself as long as all VIs in the VI hierarchy are marked as reentrant.

Another productivity-enhancing tool was opening Quick Drop for you to create your own keyboard shortcut actions. A lot of members of the community jumped at this opportunity, and you can find very useful tools available at the Quick Drop enthusiasts community.[24]

The applications with LabVIEW were getting even larger, and the teams too. More LabVIEW programmers started using source-code control tools and needed a way to call LabVIEW Compare from those tools. NI started distributing LabVIEW Compare as a command-line executable, so it can be called when a source-code control tool wants to compare two LabVIEW files.

Another big change[25] was that in LabVIEW 8.6, the Application Builder saved VIs and library files in a flat list within the application and saved VIs with conflicting file names outside the application in separate folders. In LabVIEW 2009, the Application Builder started storing source files within the application, using a layout similar to the directory structure of the source files on disk. This internal file layout preserves source file hierarchy inside the application, and it means that if you call VIs dynamically, you can use relative paths to ensure the application loads the VIs correctly.

This version of LabVIEW was the first one to be released in both 32-bit and 64-bit versions. The LabVIEW 64-bit version is useful for applications that manipulate large amounts of data because it enables software to store more data in physical memory (RAM) simultaneously.

CLA Summits, LabVIEW Community-Driven Events

The first Certified LabVIEW Architect (CLA) summit took place at NI headquarters on March 8–9, 2010. Nancy Henson was the first chair, and Fabiola De la Cueva was the first co-chair. The theme of the summit back then was advanced error handling. Even though this event was hosted by NI, the agenda was driven by the community, and we only let NI take over a couple of sessions.

The first CLA summit was small, so small that the first social event after the summit was at Brian Powell's home. It was a success, and it continues to happen every year. Nestor Ceron was visiting from the UK when the second CLA Summit took place in Austin and asked me what was going on. Nestor decided Europe should have their own CLA Summit and asked me to help. I suggested to ask Steve Watts to be the first European CLA Summit chair. The first CLA summit in Europe

---

[24] http://ni.com/quickdrop
[25] You can still see the repercussions when you build an executable and need to decide whether to use the file structure from 8.6 and earlier.

was on March 28–29, 2012, at the NI office in Newbury, United Kingdom. These events are free, and the only requirement is for attendants to be CLAs. The summits are often listed as the number-one reason to get certified. As LabVIEW grows, NI no longer holds all of the knowledge. There are more advanced LabVIEW programmers outside of NI than there are inside. I have attended every CLA summit in Austin and in Europe because it is where I get to expand my knowledge of LabVIEW—not only at the presentations but during the discussions as well. I remember one of the dinners at the first CLA summit where I discovered the best LabVIEW debugging feature: Suspend when called! Every year I go back home with homework to continue my LabVIEW education. The location for the CLA summit in Europe changes every year, and it included the best location ever, CERN, where we even got a tour of the different experiments.

Idea Exchange

The LabVIEW R&D team at NI had always listened to customer feedback, but there was no formal way for LabVIEW programmers to give their input. I remember times when some LabVIEW R&D team members would reach out directly to lavag.org and the forums to get our input on certain features. I also knew whom to talk to if I wanted to describe one of my feature requests, but not every LabVIEW programmer had those resources. The idea exchange[26] started in June 2009, and we started to see the benefit starting with LabVIEW 2010, where already some of the idea exchange feature requests became a reality. NI started to include in their release notes how many community ideas were implemented, and the forum members would get a shout-out in the LabVIEW Upgrade notes, noting their forum name and the idea that they had that the LabVIEW team turned into a feature.

LabVIEW 2010[27]

This version included an improved back-end compiler that generated optimized machine code, improving your application's runtime execution up to 20 percent compared to code written in LabVIEW 2009.

This version brought a major feature for those LabVIEW teams: using source-code control tools. You can now separate your compiled code from your source code. I call the * that shows up next to your VI or project name when it changes a "dirty dot." With this version, only the VIs that you actually change get a dirty dot and those that only recompiled don't. You do have to enable this feature via the LabVIEW options menu.

---

[26] You can find all the idea exchanges by visiting ni.com/info and entering code ex3gus.
[27] Go to ni.com/info and enter code upnote10 to the LabVIEW 2010 upgrade notes.

More LabVIEW programmers were creating plug-in type applications and distributing VIs that would be loaded at runtime with an application was a hassle. This version introduced packed project libraries. They package source code into a single file. I did not start using PPLs until LabVIEW 2017, because before then they were LabVIEW version specific, and I didn't want to deal with the maintenance nightmare of having to rebuild the PPLs every time we were upgrading to a new version of LabVIEW.

There were a lot of LabVIEW programmers who were still wary of creating subVIs because they didn't want to take the performance hit associated with a VI call. LabVIEW 2010 brought the option of inlining subVIs to eliminate subVI overhead. You can inline subVIs into their calling VIs to eliminate subVI overhead and increase code optimization in calling VIs. When you inline a subVI, LabVIEW inserts the compiled code of the subVI into the compiled code of the calling VI.

Some of the first ideas from the idea exchange saw the light of day. Wires can now have associated labels, and even if you hide the label, when you hover over the wire, the tip strip shows its name. You can now right-click a cluster constant and select to view it as an icon. The appearance of the boolean constant only showed the actual value (either true or false). The global and local variables also got a makeover to indicate if they were set to read or write by adding an arrow to their icon. String got the option to display their style. The "switcheroo" tool works now for the connector pane; this is when you press and hold the <Ctrl> key and use the positioning tool to select the two terminals you want to switch in the connector pane. Quick Drop got new <Ctrl> key shortcuts, like <Ctrl-P> that replaces the front panel or block diagram object(s) you had selected before opening Quick Drop with the object you select in the dialog box. Also, you can now configure which key corresponds to which keyboard shortcut in Quick Drop.

LabVIEW 2011[28]

The focus of LabVIEW 2011 was on being a stabilization and performance release, and very few new features were allowed. This version was so stable that it was the first version where a lot of people, including NI systems engineers, upgraded without waiting for SP1 to release. Applications solved with LabVIEW were getting larger and larger too.

More ideas from the idea exchange made their way into LabVIEW. Some of these might seem like small improvements, but it is the addition of several little things that make for a nice user experience. NI likes to call us users because we use LabVIEW. I think these features make for a better developer experience because I see us as LabVIEW programmers. Some of these ideas included having a glyph over the terminal to identify when a cluster is a typedef and being able to create a typedef directly from a block diagram constant as opposed as having to convert it to a control first.

---

[28] Go to ni.com/info and enter code upnote11 to the LabVIEW 2011 upgrade notes.

Another feature that you might take for granted is that you no longer need to unbundle the error cluster to get the status component and then wire it into a logic function; you can wire the error wire directly.

One of those times when Darren Nattinger and Stephen Loftus-Mercer from LabVIEW R&D reached out to lavag.org was the feature to improve the experience of creating a subVI.[29] It was one of those features that they knew they had ready to release, but they were getting pushback from their bosses because the idea in the idea exchange was not getting that many votes. It had about 79 kudos, and they needed to get it to more than 160 kudos. Well, you can imagine what happened next; they got more than enough votes for this feature to release. In previous versions of LabVIEW, when you created a subVI from a block diagram selection, you had to clean up the connector pane and front panel of the new subVI manually, and the connector pane was made to fit exactly the number of inputs and outputs that you had wired. With this new improved way, LabVIEW automatically builds the connector pane and front panel of the subVIs following good programming practices, such as using the standard 4×2×2×4 connector pane (unless the subVI requires more terminals), error in and out in the lower corners of the connector pane, references or class terminals in the upper corners of the connector pane, controls aligned on the left side of the front panel, and indicators aligned on the right side of the front panel.

LabVIEW 2012[30]

More features initiated from the idea exchange continued to make their way into LabVIEW. LabVIEW 2012 finally brought the Event structure to the LabVIEW Base Package; this meant that now there was no excuse for the shipping examples to still show polling code when the event driven is much better.

One feature introduced in this version that I use in every application is the option to configure the output tunnel in a loop to concatenate arrays across loop iterations. Before this version, you had to use a Shift Register and the Build Array function. Starting with LabVIEW 2012, all you have to do is right-click on the output tunnel and configure its tunnel mode to concatenating. Along the same lines, you can also configure to omit values that meet a condition you specify by configuring the output tunnel mode to conditional.

Other small features that add up to productivity and readability enhancements are the option to remove broken wires for a selected section of code instead of the entire block diagram. You can add subdiagram labels to document your code within structures.

The Actor Framework passed from being a community project that you could only download from the forums to shipping with LabVIEW. This also meant it got a facelift and more detailed documentation. Stephen Loftus-Mercer now had access to

---

[29] https://lavag.org/topic/13060-do-you-wish-create-subvi-from-selection-worked-better/

[30] Go to ni.com/info and enter code upnote12 to the LabVIEW 2012 upgrade notes.

technical writers and graphic designers to polish the toolkit and make it LabVIEW Release quality.

LabVIEW 2012 was the first version to provide project templates to assist you in creating applications. Before this, we had to rely on very basic VI templates that were great for instructional purposes and for understanding the different design patterns in LabVIEW but were very far from what a complete, professional application should look like. The project templates were not enough. The LabVIEW team also added the sample project templates, which are full applications complete with a build specification built using the project templates as a springboard. I love project templates. There is no longer an excuse of getting the blank VI syndrome, where you are tasked with solving a problem using LabVIEW and you have no idea where to get started. In the old times, the temptation to start throwing spaghetti code until the application worked was too great. Starting from a project template makes it easier to get working code faster, and there is less likelihood of ending with spaghetti code. With this feature they also gave us the power of creating our own project templates. No more excuses for the developers in your team for not adding the things that you know each application has to have.

LabVIEW 2013[31]

I tried using Web services in previous versions of LabVIEW, and although I succeeded, it was a lot more work than I was willing to go through, and it was a pain to debug. LabVIEW 2013 brought major improvements to creating Web services in LabVIEW. Instead of having to create a build specification, you can create a Web service project item and add the files you need right there. Publishing a Web service also got easier; just right-click on the Web service project item and select Publish. Also, if you need to publish a Web service with your LabVIEW application, the build specification takes care of everything for you, and the application automatically publishes the Web service when it runs to a Web server that is specific to the application. Web services no longer run in a separate context from the main LabVIEW instance, so you can implement communication via the APIs that LabVIEW provides. Debugging got a lot easier too by just starting the debugging session by right-clicking on the Web service project item and selecting to start.

At the CLA summit in 2011, Justin Goeres gave a presentation about his public and private events. He showed how you can register for a cluster of events and only handle the events you care about. Then Steen Schmidt pointed out that if the code registered for an event without adding an event case to handle that event, the event structure timer would still reset when any of the events in the cluster of events registered would fire. This led half of the room to say this was a bug and the other half to say it was a feature. Norm Kirchner stood up on the table protesting, and this led to a whole discussion about all the things the CLAs wanted to see improved with

---

[31] Go to ni.com/info and enter code upnote13 to the LabVIEW 2013 upgrade notes.

respect to events.[32] Unfortunately, it was too late to rename the user events, which to me are the worst-named LabVIEW feature. They should be called dynamic events or custom events, but again NI sees us as users, not as LabVIEW programmers. Some of the things we requested were to have a way to enqueue an event at the other end of the event queue. We wanted an easier way to debug events (the only way we had back then was to use Desktop Execution Trace Toolkit), and there were some requests to flush the event queue.

I can tell you that the event timer resetting was a bug, because starting with LabVIEW 2013, the timer only resets when an event fires, if and only if an event structure registered for AND has a case to handle that event. We also got a new feature to view the enqueued events at runtime by right-clicking on an event structure and selecting the Event Inspector Window. We also got priority events; the Generate Event primitive has an input to set the priority of the event. Normal priority enqueues the event at the end of the queue. High priority enqueues it in front of any previously generated event. We got a feature to flush the event queue, and they added an option in the Edit Events dialog box to limit the maximum instances of a particular event in the event queue. If that was not enough, the LabVIEW team added scripting functions for the Event Structure. I was especially excited about those because they made part of DQMH possible. If you were wondering if getting your CLA certification was worth it, now you know it; you might get to see a future feature of LabVIEW discussed right in front your eyes.

The idea exchange–driven features continued to show up in this version too. You can attach comments to block diagram objects, and the cleanup diagram tool keeps them close to each other. You can mark your comments with bookmarks using hashtags (#) and then view them all in the Bookmark Manager. One of my customers jumped up and down in excitement when that feature came up because he was already using hashtags to make it easier for him to find text using the search feature, but now he can see them organized. It gets even better. The Bookmark Manager is written in LabVIEW, and the code is open, so you can create your own version of the Bookmark Manager. The application builder now automatically selects NI software for installers, and there is no longer the need to install sections of the LabVIEW runtime engine that your application does not use.

Finally, the shipping examples got some long-needed love and care. A lot of old-timers would no longer go to the examples because they were outdated or no longer worked as expected. Some of them were not very good at showing good programming practices; others didn't even meet basic style guidelines. Darren Nattinger led a team that set out to clean up the examples, and you can use some of the style guidelines and VI Analyzer tests they used to clean up the examples too.[33]

---

[32] Unfortunately, there is no video or pictures of Norm Kirchner standing up on the table protesting, but there is some audio available via 003 VISP Justin Goeres—CLA summit http://vishots.com/interview-justin-goeres-cla-summit/
[33] LabVIEW Example Programs Style Guidelines https://forums.ni.com/t5/Using-the-NI-Community/Example-Programs-Style-Guidelines/ta-p/3698614

LabVIEW 2014[34]

By this version, it was obvious that NI focused more on getting LabVIEW to be more stable and not so much on new features. There were only a handful of new features, and rumors started to go around that something else was going on. Especially when the most-talked-about feature was the idea exchange–driven feature that brought a new icon to distinguish the 32-bit from the 64-bit version of LabVIEW. (See Fig. 1.14.)

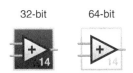

FIG. 1.14 *New system tray icons that indicate the version and bitness of LabVIEW.*

Another idea exchange feature that came out in 2014 is that you can right-click on a tunnel in the case structure and select to replace with Case Selector.

Quick Drop added another keyboard shortcut. This time, if you press <Ctrl-Space> to get to Quick Drop and then <Ctrl-W>, LabVIEW will automatically wire the highlighted objects together.

One controversial feature that came out in LabVIEW 2014 was the option to review and update type definition instances. I welcomed this new feature with open arms because I had been bitten more than once by LabVIEW being too helpful and making assumptions when I modified typedef clusters. Before, when you applied changes from a type definition to all instances of that definition, LabVIEW could lose or incorrectly preserve the default values for those instances. Starting with LabVIEW 2014, LabVIEW will try to automatically preserve the default values for each instance, but if it cannot, it will place the instance in an unresolved state until you manually update it using Review and Update from the Typedef dialog box.

LabVIEW 2015[35]

NIWeek 2015 almost did not have any segments for new LabVIEW features. Finally, Darren Nattinger from LabVIEW R&D was invited on stage to present the new features. He said on stage that this was the best LabVIEW version ever, and he even came up with a brilliant tagline "Write faster code. Write code faster." Darren is the undefeated, now retired from competition, fastest LabVIEW programmer in the world. He also is behind the creation of tools like VI Analyzer and Quick Drop. He knows a thing or two about writing LabVIEW code fast.

---

[34] Go to ni.com/info and enter code upnote14 to the LabVIEW 2014 upgrade notes.
[35] Go to ni.com/info and enter code upnote15 to the LabVIEW 2015 upgrade notes.

This version opened up more possibilities for you to extend the IDE and create your own features. You no longer need to wait for the LabVIEW team to implement some of the features you have been requesting for years. You can add custom items to the shortcut menu of the front panel and block diagram objects by creating your own shortcut menu plug-ins. These plug-ins work at edit time. There is an online community dedicated to sharing some of those plug-ins.[36] If you visit the community, make sure to check out Darren's and AristosQueue's favorite plug-ins lists. Some examples of plug-ins that ship with LabVIEW and that originated in the idea exchange are the option to change an array to an element, the option to size array constants to contents, and the option to transpose 2D arrays.

That was all exciting, but the feature that almost got a standing ovation was the ability to reduce space from the front panel and block diagram. For years we had the option of pressing the <Ctrl> key and dragging the mouse in any direction to make space. LabVIEW 2015 added the option of pressing <Ctrl-Alt> while dragging to reduce space. This is definitely one of my favorite new features in LabVIEW.

One more feature that I liked from this version is that the free labels now detect URLs and convert them into hyperlinks underlined in blue text. You can click and open a hyperlink in the default web browser. Delacor takes advantage of that feature to include links to YouTube videos that explain how certain parts of DQMH work.

Actor Framework got some improvements too. You can now create actor and message classes directly via right-click menus on the Project Explorer. If you get to talk to Allen Smith, he is very proud of this feature, and he is one of the few persons who knows how to use the project provider. Unfortunately, the LabVIEW Project Provider[37] was not designed or built with people outside NI using it. With time, the original developers were gone, and then it was up to David Ladolcetta in the LabVIEW Tools Network team to figure it out and create the documentation from there. Adding right-click menu code to front panel and block diagrams is easy; adding right-click menu code to the Project Explorer is definitely not easy.

Finally, our old and faithful Write to Spreadsheet File.vi and Read from Spreadsheet File.vi got replaced with Write Delimited Spreadsheet.vi and Read Delimited Spreadsheet.vi. The new versions do have an error input, error output, and updated connector pane.

The Knights of NI

The highest rank at the NI forums is Knight of NI. The official definition is that "these contributors have consistently given back to the NI forums for years. They have a

---

[36] ni.com/lvmenus
[37] LabVIEW Project Provider community https://forums.ni.com/t5/LabVIEW-Project-Providers/gp-p/bymqyodmkc

wide breadth of knowledge and many engage with the Community daily! These users are recognizable in the discussion forums by their ranking title and icon."[38]

Dennis Knutson was the first Knight of NI. He was the first NI community member to reach 10,000 posts, and Jeff Kodosky knighted him on March 14, 2007.[39] He was also one of the first LabVIEW Champions. When he reached 10,000 posts, Tony Vento and others at NI discussed how they could recognize him. Tony had been the sales field engineer who introduced Dennis to LabVIEW and gave him a call to see what recognition he would like. Dennis was a very humble man and didn't want anything. Dennis was also a Monty Python fan, and Tony knew this. For you young readers, there is a movie called *Monty Python and the Holy Grail*. In that comedy there is a group of knights called the "Knights Who Say Ni." Well the joke tells itself, so they named Dennis the first "Knight Who Says Ni." NI did not want to officially go with that and they changed the name to The Knights of NI. Dennis passed away, within hours of posting his last reply to the forums, on February 21, 2016.[40] The LabVIEW community is large and full of people who are passionate about LabVIEW; there are also lots of people wanting to help others. I fell in love with LabVIEW, but I stayed because of the community around it. (See Fig. 1.15.)

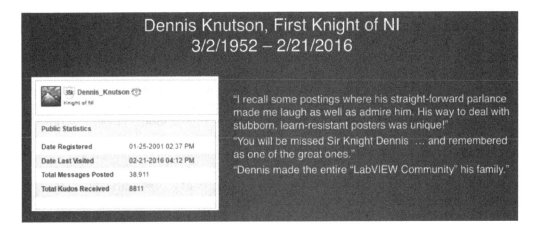

FIG. 1.15 *Tony Vento's tribute to Dennis Knutson delivered at Alliance Day during NIWeek 2016.*

---

[38] https://forums.ni.com/t5/Using-the-NI-Community/Recognition-and-Ranks/ta-p/3698600
[39] https://forums.ni.com/t5/BreakPoint/Congratulations-Dennis/m-p/491863
[40] Rayner, Ben. The start of a new era dawns with the passing of the first Knight forums.ni.com/t5/BreakPoint/The-start-of-a-new-era-dawns-with-The-passing-of-the-first/td-p/3298485

LabVIEW 2016[41]

LabVIEW 2016 improved how you select, move, and resize objects on the front panel or block diagram. When you select objects, the area covered by the selection rectangle displays in gray and a marquee highlights the selected objects. Selected structures appear with a darker background to indicate your selection includes them. By default, when you create a selection rectangle around objects, you must enclose the entire structure or the midpoint of a wire segment to select it. If you press the spacebar when you create the selection rectangle, you select any objects that the selection rectangle touches. To restore the default selection behavior, press the spacebar again. When you move selected objects, the entire selection moves in real time. Certain objects, such as structures, rearrange or resize to accommodate the movement of the selected objects. It is a really neat effect, and you might not have even noticed it. When you resize a structure by dragging the resizing handles, the structure grows or shrinks in real time instead of displaying dashed borders.

Channel Wires (also known as Jeff's wires)

Jeff's wires had been discussed for years! He wanted a better way to represent the communication between loops; finally, in LabVIEW 2016, Jeff, with the help of Stephen Loftus-Mercer, brought this feature to reality and named it channel wires. You can use channel wires to communicate data between parallel sections of code. Channel wires are asynchronous wires that connect two parallel sections of code without forcing an execution order, thus creating no data dependency between sections of code. If you have not used or seen channel wires and you have been programming with LabVIEW for a while, I warn you, they take time to get used to. The first thing to realize is that they are not depicted as wires; they look more like pipes that traverse over the boundaries of two loops. I like that you can see the data flowing from one loop to the other loop. I don't like that, as far as I know, you have to determine at edit time both ends of the communication. I am looking for an opportunity to use them in a real-life application. The LabVIEW geek in me is excited that Jeff continues to look for ways of representing different models of computation and that he feels he is not done with LabVIEW. Last time I talked to him, he said he felt like he was running out of time and there were still lots of things he wanted to accomplish in LabVIEW. (See Fig. 1.16.)

---

[41] Go to ni.com/info and enter code upnote16 to the LabVIEW 2016 upgrade notes.

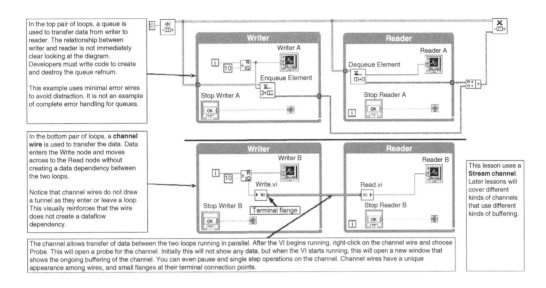

FIG. 1.16 *Channel Basics Lesson 1.vi shows two loops on the top using a queue to communicate data. The two loops on the bottom use the Stream Channel wire instead. The resulting execution is exactly the same, just the graphic syntax is different.*

LabVIEW 2017[42]

NI upgraded to a more aggressive compiler for building both the LabVIEW development environment and the LabVIEW Run-Time Engine. This upgrade reduced aggregate VI load time and VI compile time.

Some of the features of LabVIEW NXG started to make their way into the current LabVIEW. LabVIEW 2017 automatically maintains wire connectivity when you move objects in and out of structures on the block diagram. When an object moving in or out of a structure is connected to an object in the structure, LabVIEW creates or removes tunnels to maintain wire connectivity. This is a cool feature, but you don't always want LabVIEW to do that, so you can turn off this behavior by pressing the <W> key.

One feature advanced LabVIEW programmers welcomed in this version was the option to have read-only access for DVRs. At least one of my customers upgraded from LabVIEW 2015 to LabVIEW 2017 just because of this feature. Right-click on the border node on the right of the In Place Element structure and select to allow

---

[42] Go to ni.com/info and enter code upnote17 to the LabVIEW 2017 upgrade notes.

parallel read-only access. LabVIEW then allows multiple concurrent read-only operations and does not modify the DVR.

LabVIEW 2017 includes the Event Messenger channel template.

Starting with LabVIEW 2017, LabVIEW supports backward compatibility for the LabVIEW Run-Time Engine. For example, versions of LabVIEW later than 2017 can load binaries and VIs built with LabVIEW 2017 without recompiling! This improvement applies to standalone applications (EXEs), shared libraries (DLLs), and packed project libraries (PPLs). This was the reason that finally pushed me to try out PPLs; before that, just imagining having to recompile all the PPLs for each new version of LabVIEW made me look away.

Malleable VIs

Another LabVIEW syntax project that Jeff Kodosky and Stephen Loftus-Mercer have been working on are malleable VIs (.vim). With malleable VIs, you create a VI to perform the same operation on any acceptable data type instead of saving a separate copy of the VI for each data type. A malleable VI is similar to a polymorphic VI, but it is more flexible when determining which data types are acceptable. A polymorphic VI uses a predefined list of acceptable data types. A malleable VI computes whether a data type is acceptable by implementation. You might be already using malleable VIs and not know it. LabVIEW 2017 included several malleable VIs and used an orange background to distinguish them.

Built-in Malleable VIs include the following:

- Array palette
    - Decrement Array Element
    - Increment Array Element
    - Shuffle 1D Array
    - Shuffle 2D Array
    - Sort 2D Array
- Comparison palette
    - Is Value Changed
    - Number to Enum
    - Stall Data Flow

LabVIEW 2018[43]

The LabVIEW R&D team continues to grow. We asked for a picture, and NI no longer had a picture with the whole team. Here is a picture with part of the team. (See Fig. 1.17.)

FIG. 1.17 *Part of the LabVIEW team as of March 2019. Can you recognize some of the faces from the previous pictures?*

More improvements for malleable VIs. The Comparison palette includes the new Assert Type subpalette. Use the Assert Type VIs and function to force a malleable VI (.vim) to accept only data types that meet certain requirements. Use the Type Specialization structure to customize sections of code in malleable VIs for specific data types.

---

[43] Go to ni.com/info and enter code upnote18 to the LabVIEW 2018 upgrade notes.

LabVIEW 2018 allows you to run operations by executing commands using the command-line interface (CLI). For example, you can use the CLI in combination with Continuous Integration tools to automate the build process of LabVIEW applications. Before NI came out with this version that ships with LabVIEW, James McNally from Wiresmith Technology in the United Kingdom had created his own version.[44] His version is available via GitHub,[45] and the big difference is the fact that James's version can provide continuous output while the application is running. Also you can use it in EXEs. Historically, James's version couldn't work when LabVIEW was already open, but the new 2.0 version can. By the way, members of the community were very disappointed when NI did not give James his deserved recognition when they released this feature, but that is a story for another book.

LabVIEW 2018 brought a new Python subpalette to the Connectivity palette. You can call Python code from LabVIEW code. If this node doesn't support what you are looking for, there is also a Python Integration Toolkit for LabVIEW by Enthought at the LabVIEW Tools Network.[46]

Some of the features driven by the idea exchange were more options to create typedefs. You can do it via the File>>New>>Other Files or by right-clicking on My Computer and selecting New>>Type Definition. Before 2018, you had to enter a special ini token to be able to use <Ctrl-B> within a label to make text bold. LabVIEW 2018 enables the option of using <Ctrl-B> to make text bold, <Ctrl-I> to italicize text, and <Ctrl-U> to underline text. We finally got the Coerce To type function added to the Conversion palette. Before that, I had to do some Google search magic to find the snippet that contained the function and that way add it to my block diagram.[47] Unlike the Type Cast, this function does not interpret the input data. Use this function to eliminate a coercion dot, to convert data without a type definition to a compatible type definition or vice versa, and to rename data on the wire. I particularly use it to rename user event refnum wires. By the way, this idea was originally posted on the idea exchange back in 2010, so don't despair. If you post something and it gets lots of kudos, it might eventually get rolled into the product.

Finally, this version added a little syntactic sugar that you might find useful. LabVIEW 2018 introduced error registers to simplify error handling on a For Loop with parallel iterations enabled. Error registers take the place of Shift Registers for error clusters on a parallel For Loop. (See Fig. 1.18.)

---

[44] Bringing the Command Line Interface to LabVIEW by James McNally. https://devs.wiresmithtech.com/blog/bringing-command-line-interface-labview/
[45] LabVIEW-CLI by James McNally. https://github.com/JamesMc86/LabVIEW-CLI
[46] ni.com/labviewtools search for Python. http://sine.ni.com/nips/cds/view/p/lang/en/nid/213990
[47] If you are curious, my search terms were "officially support coerce to type labview."

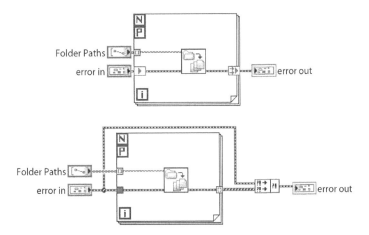

FIG. 1.18 *Error registers automatically merge errors from parallel iterations. LabVIEW preserves the best practice of flowing errors through a Shift Register by automatically converting Shift Registers to error registers when you configure parallel iteration on a For Loop. The loop on the bottom is doing the exact same operation, but it doesn't use the error registers.*

GDevCon—The Ecosystem Independent from NI

In 2017, a group of LabVIEW programmers, led by Steve Watts, set out to create a series of events that is all about world-class graphical programming. The community organizes these events. These events are 100 percent independent of NI. The first GDevCon[48] took place in Cambridge on September 4 and 5, 2018. Tickets sold out within 1 day. They had to change the venue and offer more tickets. The first event was a success, and they are on their way to planning future events.

The idea behind GDevCon is to provide opportunities for LabVIEW programmers of different levels to know each other, learn from some of the leading LabVIEW programmers, and learn more about team-based development and how to use LabVIEW in large applications together with the best software engineering practices. The event caters to teams who want to improve their development and collaboration practices. GDevCon is like a user group of user groups.

[48] gdevcon.com

Future Versions of LabVIEW

Jeff Kodosky and his team continue to look for ways to bring more of that modeling to be done directly in LabVIEW. At one of the NIWeek 2018 keynotes,[49] Jeff showed us some of the projects he is working on. Among them, there were some proofs of concepts where the whiteboard drawing would become actual code. He did say that was far in the future, but he wanted to show us what the goals are he still has for upcoming versions of LabVIEW. In that presentation, Jeff talked about diagrams that display the entire application, from desktop, to CompactRIO (cRIO), to LabVIEW FPGA. Although we are excited about this tool, we couldn't help but cringe at the thought of intertarget spaghetti code.

The other preview was of a more distant tool. This would be a modeling tool that is directly connected to LabVIEW and synchronized with the code. LabVIEW would preserve the storyboard of our high-level sketch as part of the project. Imagine taking a picture of the whiteboard sketch you currently have of your system and being able to click on each part of the diagram and seeing the actual code behind it. Better yet, imagine adding wires to the high-level sketch and seeing them appear in the lower code implementing it. It would be a way to travel through the different levels of abstraction in your application, a semantic zoom.

LabVIEW NXG 1.0–3.0

In the early 2000s, the LabVIEW R&D team started working on what would later become known as LabVIEW NXG. Their goal was to make a radical improvement to LabVIEW, so they chose to start from scratch. This was such a large undertaking that several underlying technologies changed as time passed, and the team had to make major course corrections along the way.

Even if the R&D team had written LabVIEW from scratch a few times before (LabVIEW 1.0, LabVIEW 2.0, LabVIEW 3.0, and LabVIEW 5.0), the G language and LabVIEW tools had advanced so much that this was still a daunting task.

NI started several customer advisory boards in 2014 to solicit feedback from the LabVIEW community and some of their large customers. We got to see previews of what the dream of this LabVIEW revolution was. NI reassured us that they were going to focus on an open platform so others could extend the IDE. As a LabVIEW toolkit developer, this was very appealing to me. I regret not clarifying at the moment if this openness was going to be done for G programmers. I assumed NI knew they were talking to a room full of LabVIEW programmers. NI did not think of clarifying that the extension would be via C# libraries. That said, NI has approached CLAs, the LabVIEW Champions, and large companies with large teams of LabVIEW programmers such as SpaceX. They are listening and see that advanced LabVIEW programmers not only need an improved IDE, we also need all the software

---

[49] Kodosky, Jeff. The Big Picture. youtu.be/D7-ej-cqVql

engineering tools that have been added with time and the advanced programming language features, as well as the possibility of extending the IDE through G.

By 2016 the technology preview had been opened to a larger audience, and some long-standing idea exchange entries, such as the ability to zoom into the block diagram, started to be marked as "in-progress." By NIWeek 2017, the secret was out, and the world of LabVIEW started to learn what many of us had learned under NDA since around 2014. NI officially worked on a new version of LabVIEW called LabVIEW NXG. Version 1.0 was released in parallel with LabVIEW 2017, but it was clear that LabVIEW NXG was not ready for prime time due to its lack of features. In fact, right as NI announced NXG 1.0, they were also announcing the NXG 2.0 beta release, in essence encouraging people to skip NXG 1.0 and go directly to NXG 2.0. (See Fig. 1.19.)

Every Purchase of LabVIEW Includes LabVIEW 2018 and LabVIEW NXG

LabVIEW 2018

LabVIEW 2018 simplifies the design of distributed test, measurement, and control systems decreasing your time to market. Combine LabVIEW 2018 with proven, off-the-shelf customizable hardware from NI which has been used by engineers for over 30 years to develop and deploy custom large-scale industrial and production systems.

We recommend LabVIEW 2018 for the following application areas:
- Design Smart Machines or Industrial Equipment
- Teach Engineering Students

LabVIEW NXG

LabVIEW NXG is the next generation of LabVIEW. Test smarter with LabVIEW NXG by quickly automating your hardware, customizing tests to your specifications, and easily viewing measurement results from anywhere.

We recomment LabVIEW NXG for the following application areas:
- Measure Physical Systems with Sensors or Actuators
- Validate or Verify Electronic Designs
- Develop Production Test Systems

FIG. 1.19 *LabVIEW landing page at ni.com/labview comparing LabVIEW 2018 and LabVIEW NXG.*

The LabVIEW R&D team assured us that NI is not abandoning the current LabVIEW. Indeed, both LabVIEW 2017 and LabVIEW 2018 include new features that make it evident NI has not stopped investing in this product. Jeff and his team continue to work on language features for the current LabVIEW, such as the channel wires and malleable VIs. It is refreshing to see him present at NIWeek on these topics. He is excited to explore new ways to represent different models of computation. The current LabVIEW is ready to test these new concepts that one day will be included in NXG.

LabVIEW serves two audiences. On the one hand, it is a tool that allows technicians, scientists, and engineers to measure and control the world around them. On the other hand, it is a powerful development environment for graphical programmers. The first releases of NXG focused on the first group, making it easier for nonprogrammers to measure the world around them without needing to program. I remember cringing at the marketing message for the first release: "Programming Is Optional." The biggest challenge is that the LabVIEW team had to make the same program that made programming optional be an advanced programming environment for LabVIEW programmers. To their credit, they have started by making changes that will make it easier to connect to software engineering tools like source-code control. For one, the source code of a VI is no longer binary. They are also focusing on making the transition from the nonprogrammer to G programmer more straightforward with decisions like making a cluster and a class be less distinct things and more a common type with different attributes. The cluster type has wide-open data and no methods. Then you can choose to add methods, or encapsulate the data, or enable inheritance. If you add all three of those attributes, you have a class type.

From what I have seen to date, my customers and I will continue to use the current LabVIEW for a while. I see promise in what LabVIEW NXG is doing, and I am looking forward to creating Web Services in the current LabVIEW and then using a LabVIEW NXG Web VI to register them, but we are still a long way from feature compatibility.

Dr. T and Jeff Kodosky Inducted into the National Inventors Hall of Fame

As we were getting ready to send this book to the editor, we got the news that Dr. T[50] and Jeff Kodosky[51] were inducted into the National Inventors Hall of Fame in 2019 for inventing Virtual Instrumentation—LabVIEW™ US Patent No. 4,901,221.

LabVIEW Release Timeline

Starting with LabVIEW 8.0, major releases are timed to happen around NIWeek, the annual NI conference in Austin, Texas. Then there is a bug-fix release about 6 months later. NIWeek used to be every August, but starting in 2017, NI moved it to May. Also in 2017, NI announced LabVIEW NXG 1.0 built using Windows Presentation Foundation (WPF) and representing a new rewrite of the platform.

In 2009, NI began naming the LabVIEW releases after the year they are released.

Name-Version[52]	Date
LabVIEW project begins	April 1983
LabVIEW 1.0 (for Macintosh)	October 1986
LabVIEW 2.0	January 1990

---

[50] https://www.invent.org/inductees/james-truchard
[51] https://www.invent.org/inductees/jeff-kodosky
[52] Table source: https://en.wikipedia.org/wiki/LabVIEW and www.ni.com/en-us/shop/labview/upgrade.html

Name-Version[52]	Date
LabVIEW 2.5 (first release for Sun and Windows)	August 1992
LabVIEW 3.0 (Multiplatform)	July 1993
LabVIEW 3.0.1 (first release for Windows NT)	1994
LabVIEW 3.1	1994
LabVIEW 3.1.1 (first release with "application builder" ability)	1995
LabVIEW 4.0	April 1996
LabVIEW 4.1	1997
LabVIEW 5.0	February 1998
LabVIEW RT (Real Time)	May 1999
LabVIEW 6.0 (6i)	July 2000
LabVIEW 6.1	April 2001
LabVIEW 7.0 (Express)	April 2003
LabVIEW PDA module first released	May 2003
LabVIEW FPGA module first released	June 2003
LabVIEW 7.1	2004
LabVIEW Embedded module first released	May 2005
LabVIEW 8.0	September 2005
LabVIEW 8.20 (native object-oriented programming)	August 2006
LabVIEW 8.2.1	February 2007
LabVIEW 8.5	2007
LabVIEW 8.6	July 2008
LabVIEW 8.6.1	December 2008
LabVIEW 2009 (32 and 64-bit)	August 2009
LabVIEW 2009 SP1	January 2010
LabVIEW 2010	August 2010
LabVIEW 2010 f2	September 2010
LabVIEW 2010 SP1	May 2011
LabVIEW for LEGO MINDSTORMS (2010 SP1 with some modules)	August 2011

(Continued)

Name-Version[52]	Date
LabVIEW 2011	June 2011
LabVIEW 2011 SP1	March 2012
LabVIEW 2012	August 2012
LabVIEW 2012 SP1	December 2012
LabVIEW 2013	August 2013
LabVIEW 2013 SP1	March 2014
LabVIEW 2014	August 2014
LabVIEW 2014 SP1	March 2015
LabVIEW 2015	August 2015
LabVIEW 2015 SP1	March 2016
LabVIEW 2016	August 2016
LabVIEW 2017	May 2017
LabVIEW NXG 1.0	May 2017
LabVIEW 2017 SP1	Jan 2018
LabVIEW NXG 2.0	Jan 2018
LabVIEW NXG 2.1	March 2018
LabVIEW 2018	May 2018
LabVIEW NXG 3.0	Aug 2018

LabVIEW Handles Big Jobs

Some really impressive applications have been written in all versions of LabVIEW, but in the past few years we've seen an absolute explosion in development that leaves us breathless. Even the LabVIEW team members have been known to shake their heads in amazement when shown the size and scope of the latest high-end project from a sophisticated LabVIEW programmer. Here are a few benchmark projects that we've run across.

CERN

CERN has more than 600 LabVIEW users. They use PXI, cRIO, and USB data acquisition hardware, and there are some applications where LabVIEW does not control any hardware at all. Some of these applications are listed here.

PXI	cRIO	PC	Data Analysis and Display
Large Hadron Collider collimator control	Distributed power and transient recording system for the CERN electrical network	LINAC 4 emittance meter	Large Hadron Collider Post-Mortem Analysis framework
Ion and anti-proton injection and extraction kicker magnet control	Seismic recorder for monitoring earthquakes and excavation work in experiment caverns and on the surface	Residual gas analyzer	Large Hadron Collider Dashboard
Full or partial experiment control of experiments (AEGIS, ALPHA, ASACUSA, AWAKE, CAST, CLIC, ISOLDE, MEDICIS, nTOF, COMPASS, TOTEM, …)	Experiment magnet quench protection	RF cavity optical inspection	
Beam spectrum analyzers with RT data treatment	Intermediate frequency beam spectrum analyzer	Hydrostatic level measurement	
LINAC 4 Laser emittance meter	Laser room personal safety protection using the cRIO SIL modules (under development)		
Superconducting magnet and RF cavity test benches			

One of CERN's most interesting applications is their synchronization of PXI systems. Their PXI systems that need timing synchronization with accelerator events as tight as a few nanoseconds, such as kicker magnets, are connected to the CERN timing system.[53] This timing system takes care that all triggers and timing information arrives within a few nanoseconds in all locations around the Large

---

[53] http://accelconf.web.cern.ch/AccelConf/ica03/PAPERS/MP533.PDF

Hadron Collider tunnel and at other accelerator facilities at CERN. The triggers are generally connected to an FPGA board in PXI, where firmware decides which actions to take.

A CERN team[54] is working on closer integration of PXI and cRIO[55] into the CERN timing system to take advantage of all its features and to prepare for the future timing system that is being developed at CERN, called White Rabbit.[56]

---

[54] https://readthedocs.web.cern.ch/display/MTA
[55] http://icalepcs.synchrotron.org.au/papers/wepgf129.pdf
[56] https://en.wikipedia.org/wiki/The_White_Rabbit_Projectint

CHAPTER 2

LabVIEW Fundamentals

Dataflow

The fundamental theme of LabVIEW programming is dataflow and executing nodes based on data dependency. Regular textual programming languages (and most computers) are based on a concept called control flow, which is geared toward making things happen one step at a time with explicit ordering. The LabVIEW language, G, on the other hand, is known formally as a dataflow language. All it cares about is that each node (an object that takes inputs and processes them into outputs) has all its inputs available before executing. This is a new way of programming—one that you have to get used to before declaring yourself a qualified LabVIEW programmer. Although programming with dataflow is easy, it can be constraining because you have to operate within the confines of the dataflow model. This constraint also forces you to write programs that function well. A well-written LabVIEW program flows in a left-to-right and top-to-bottom order, but the execution order is always governed by dataflow. Without dataflow it is impossible to determine execution order. Whenever you break from the dataflow model, you risk writing a program that misbehaves. The dataflow concept should be used to its best advantage at all times.

LabVIEW permits you to have any number of different nodes on a diagram—all executing in parallel. In fact, if you place independent items on the block diagram without any data dependency between them, they will multitask in parallel. Multitasking and parallel processing are powerful concepts made simple in LabVIEW. These capabilities give you great freedom to run various tasks asynchronously with one another without doing any special programming yourself. On the other hand, you sometimes need to force the sequence of execution to guarantee that operations occur in the proper order. Random execution in a test and measurement system is not a good idea.

Dataflow can be forced through a common thread flowing through all the block diagram items. A good common thread is an error code, since just about every operation that you devise probably has some kind of error checking built into it, particularly those that do I/O operations. Each VI should test the incoming error and not execute its function if there is an existing error. It should then pass that error (or its own error) to the output. The default error format is a cluster containing a numeric error code, a string containing the name of the function that generated the error, and an error boolean for quick testing.

LabVIEW under the Hood

Before we go too far, let's pop open the hood and take a look at how LabVIEW operates. This is an advanced topic and one not essential for writing LabVIEW programs, but it helps you write better programs if you have an idea of what is going on. We're going to cover just the highlights. If you want greater in-depth coverage, there are several excellent presentations and applications notes on National Instruments' (NI's) website. Just search for "LabVIEW Compiler."

The Parts of a VI

We see VIs as front panels and block diagrams, but there is a lot that you don't see. Each VI is a self-contained piece of LabVIEW software with four key components (Fig. 2.1):

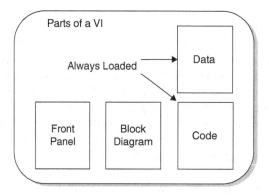

FIG. 2.1 *VI code and data are always loaded into memory when the VI is loaded.*

- **Front panel.** The front panel code contains all the resources for everything on the front panel. This includes text objects, decorations, controls, and indicators. When a VI is used as a subVI, the front panel code is **not loaded into memory unless the front panel is open or the VI contains a Property node or a local variable**.

- **Block diagram.** The block diagram is the dataflow diagram for the VI. The block diagram is not loaded into memory unless you open the block diagram or the VI needs to be recompiled.

- **Data.** The VI data space includes the default front panel control and indicator values, block diagram constants, and required memory buffers. The data space for a VI is always loaded into memory. Be careful when you make current values default on front panel controls or use large arrays as block diagram constants; all that data is stored on disk in your VI.

- **Code.** This is the compiled machine code for your subVI. This is always loaded into memory. If you change platforms or versions of LabVIEW, the block diagram will be reloaded and the code will be recompiled.

Each VI is stored as a single file with the four components noted earlier plus the linker information about the subVIs, controls, .dll, and external code resources that the VI needs. When a VI is loaded into memory, it also loads all its subVIs, and each subVI loads its subVIs, and so on, until the entire hierarchy is loaded.

How VIs Are Compiled

There's an elegant term that the LabVIEW developers use for the code generated from your block diagram—it's called a *clump*. When you push the Run button on a new VI, LabVIEW translates the block diagram into clumps of machine code for your platform. The G compiler makes several passes through the block diagram to order the nodes into clumps of nodes. Execution order within each node is fixed at compile time by dataflow. As the compiler makes its passes, it determines which data buffers can be reused. Then it compiles the clumps into machine code. Here's an example based on a presentation from LabVIEW developer Steve Rogers: The block diagram in Fig. 2.2 gets converted to three clumps of code. The first clump reads the controls and schedules the other two clumps for execution; then it goes to sleep. The other two clumps, one for each For Loop, run to completion, update their indicators, and reschedule the first clump for execution. The first clump then finishes the diagram, displays the data, and exits the VI. The traffic cop responsible for running the clumps of code is the LabVIEW execution engine.

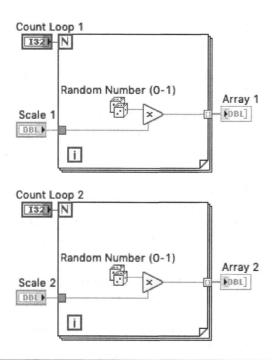

FIG. 2.2 *LabVIEW converts dataflow diagrams to "clumps" of machine code.*

Multitasking, Multithreaded LabVIEW

The LabVIEW execution engine runs each clump of code by using cooperative **multitasking**. The execution engine time slices each clump to make sure all the clumps get a chance to run. Each clump (in Fig. 2.2, For Loop 1 or For Loop 2) either runs to completion or is switched out by the execution engine, and another clump runs. The execution engine proceeds on multitasking each clump of code until either your program ends or someone pushes the Abort button, which is always a nasty way to stop a program! Cooperative multitasking has been successfully used by LabVIEW since its birth, and it is still the process being used today. LabVIEW is also **multithreaded**. A **thread** is an independent program within a program that is managed by the operating system. Each LabVIEW thread contains a copy of the LabVIEW execution engine and the clumps of code that the engine is responsible for running. Within each thread the execution engine still multitasks clumps of code, but now the operating system schedules where and when each thread runs. On a multiprocessor system, the operating system runs threads on different processors in parallel. Figure 2.3 illustrates the multiple threads within LabVIEW's process. The threads communicate via a carefully controlled messaging system. Note that one thread is dedicated to the user interface. Putting the user interface into a separate thread has several advantages:

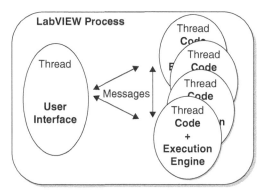

FIG. 2.3 *Multithreading splits the LabVIEW process into multiple threads with one thread dedicated to the user interface. Each thread runs clumps of code, with its own copy of the LabVIEW execution engine. Messages and other protected sharing mechanisms are used to communicate between threads.*

- Decoupling the user interface from the rest of your program allows faster loop rates within your code.
- The user interface doesn't need to run any faster than the refresh rate on most monitors—less than 100 Hz.
- Decoupling the user interface from the rest of the program makes it possible to run the user interface on a separate computer. This is what happens in LabVIEW RT.

Normally, a subVI uses the execution system of its calling VI, but in the VI Properties dialog (Fig. 2.4), you can select a different execution system and therefore ensure a different thread from the calling VI. Running a subVI in a separate execution system increases the chances that it will run in parallel with code in other execution systems on a multiprocessor system. Normally, the disadvantage to placing a subVI in a different execution system is that each call to the subVI will cause a context switch as the operating system changes from one execution system (thread) to another. However, if the VI is autonomous and not called repeatedly by a calling VI, there will only be one context switch when it is originally called, and then the autonomous VI will be merrily on its way in its own execution system.

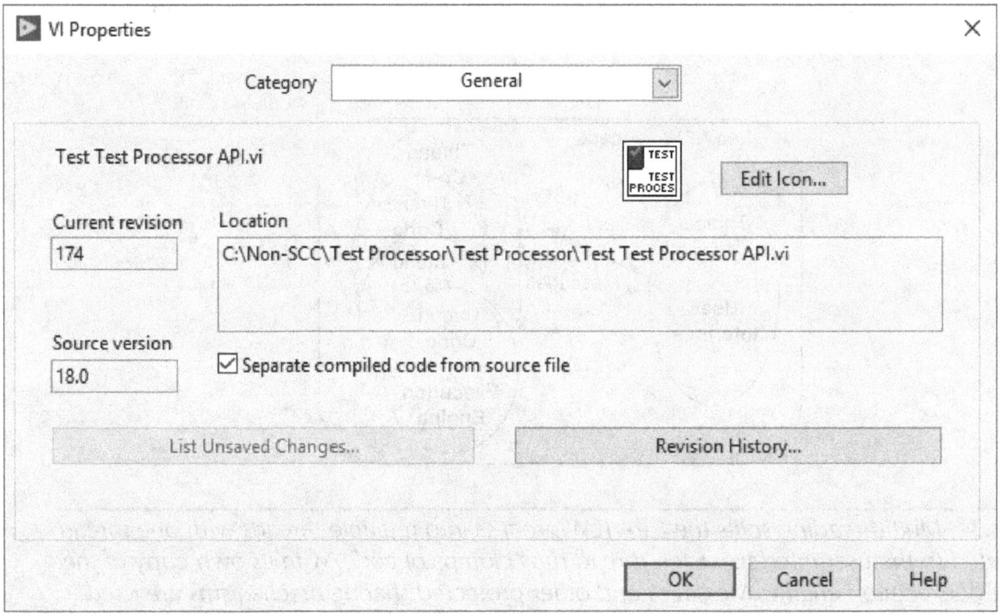

FIG. 2.4 *The default execution system of a VI can be set from the VI Properties dialog.*

What does all this mean? It means the two For Loops in Fig. 2.3 can run in parallel, and you don't have to do anything other than draw them that way. The LabVIEW compiler generates code that runs in parallel, manages all the resources required by your program, spreads the execution around smoothly, and does it all transparently. Imagine if you had to build all this into a test system yourself. National Instruments has spent more than 30 years making sure LabVIEW is the best programming environment in terms of test and measurement.

The LabVIEW R&D team is continuously improving LabVIEW to make it faster and easier to port to new targets. LabVIEW 2009 introduced the Dataflow Intermediate Representation (DFIR), a behind-the-scenes precompiler used to optimize the block diagram before handing the code off to the compiler. Beginning with LabVIEW 2010 the compiler toolchain uses the open-source Low-Level Virtual Machine (LLVM) to enable further optimizations and portability.

You don't need to know the details of the execution engine or multithreading to program in LabVIEW, but we think understanding the concepts will make you a better programmer. Streamline your block diagram to take advantage of dataflow and parallelism, and avoid littering your block diagram with user-interface code, since it forces a context switch to the user-interface thread if the front panel is loaded into memory. However, if the front panel is not loaded into memory, there is no performance hit for front panel controls and indicators.

The LabVIEW Environment

We have already covered the difference between the front panel and the block diagram; but just to reiterate, the block diagram is where you put the nuts and bolts of your application, and the front panel contains your user interface. This is not going to be an exhaustive introduction to LabVIEW—read the manuals and go through the tutorials for that. Throughout this book, we're going to focus on good programming practices, architectures, and aspects that impact how LabVIEW runs your code.

Front Panel

The front panel is the user's window into your application. Take the time to do it right. Making a front panel easy to understand while satisfying complex requirements is one of the keys to successful virtual instrument development. LabVIEW, like other modern graphical presentation programs, has nearly unlimited flexibility in terms of graphical user-interface design. But to quote the manager of a large graphic arts department, "We don't have a shortage of technology, we have a shortage of talent!" Some industries, such as the nuclear power industry, have formal user-interface guidelines that are extremely rigid. If you are involved with a project in such an area, be sure to consult the required documents. Otherwise, you are very much on your own.

Take the time to look at high-quality computer applications and see how they manage objects on the screen. Decide which features you would prefer to emulate or avoid. Try working with someone else's LabVIEW application without getting any instructions. Is it easy to understand and use, or are you bewildered by dozens of illogical, unlabeled controls, 173 different colors, and lots of blinking lights? Observe your customer as he or she tries to use your VI. Don't butt in; just watch. You'll learn a lot from the experience, and fast! If the user gets stuck, you must fix the user interface and/or the programming problems. Form a picture in your own mind of good and bad user-interface design.

One surefire way to create a bad graphical user interface (GUI) is to get carried away with colors, fonts, and pictures. When overused, they quickly become distracting to the operator. (Please don't emulate the glitzy screens you see in advertising; they are exactly what you don't want.) Instead, stick to some common themes. Pick a few text styles and assign them to certain purposes. Similarly, use a standard background color (such as gray or a really light pastel), a standard highlight color, and a couple of status colors such as bright green and red. Human factors specialists tell us that consistency and simplicity really are the keys to designing quality man-machine interfaces (MMIs).

Operators will be less likely to make mistakes if the layouts among various panels are similar. Avoid overlapping controls; multiple overlapping controls containing transparent pieces can really slow down the user interface.

Group logically related controls in the same area of the screen, perhaps with a surrounding box with a subtly different background color. You can see this technique on the panels of high-quality instruments from companies such as Tektronix and Hewlett-Packard. Speaking of which, those real instruments are excellent models for your virtual instruments.

Controls

LabVIEW's controls have a lot of built-in functionality that you should use for best effect in your application. And of course this built-in functionality runs in the user-interface thread. Quite a bit of display work, computational overhead, and user-interface functionality have already been done for you. All LabVIEW's graphs support multiple scales on both the X and Y axes, and all charts support multiple Y scales. Adjust the chart update modes if you want to see the data go from left to right in sweep mode, update all at once in scope mode, or use the default strip chart mode going from right to left. Control properties are all easily updateable from the properties page. Figure 2.5 shows the property page for an OK button. Changing the button behavior (also called mechanical action) can simplify how you handle the button in your application. The default behavior for the OK button is "Latch when released," allowing users to press the buttons, change their minds, and move the mouse off the button before releasing. Your block diagram is never notified about the button press. On the other hand, "Latch when pressed" notifies the block diagram immediately. Experiment with the mechanical behavior; there's a one-in-six chance the default behavior is right for your application.

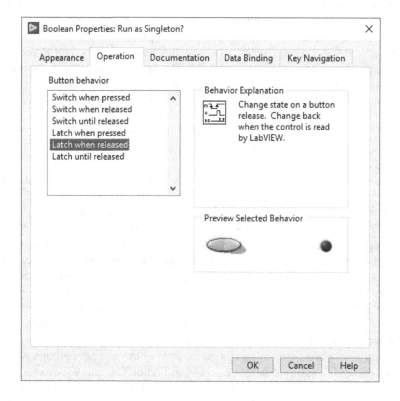

FIG. 2.5 *Control properties are modified through the Property pages. Boolean buttons have six different mechanical actions that affect your block diagram.*

Property Nodes

Property nodes are an extremely flexible way to manipulate the appearance and behavior of the user interface. You create a Property node by popping up on a control or indicator and choosing Create . . . Property Node. This creates a Property node that is statically linked to the control. You do not need to wire a reference to it for it to access the control. The disadvantage is that statically linked Property nodes can only reference front panel objects on their VI. Figure 2.6 shows both a statically linked and a dynamically linked Property node. With a control reference you can access properties of controls that aren't on the local front panel. This is a handy way to clean up a block diagram and generate some reusable user-interface code within a subVI. Every control and every indicator have a long list of properties that you can read or write. There can be multiple Property nodes for any given front panel item. A given Property node can be resized by dragging at a corner to permit access to multiple attributes at one time. Each item in the list can be either a read or a write, and execution order is sequential from top to bottom. We're not going to spend a great deal of time on Property nodes because their usage is highly dependent upon the particular control and its application. But you will see them throughout this book in various examples. Suffice it to say that you can change almost any visible characteristic of a panel item, and it's worth your time to explore the possibilities.

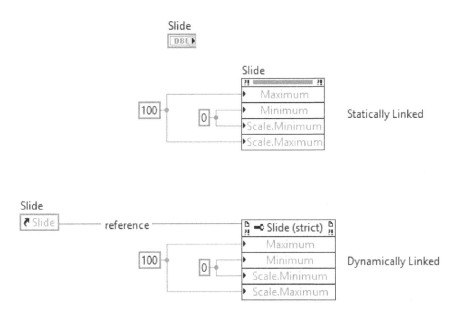

FIG. 2.6 *Statically linked Property nodes can only reference front panel objects within the scope of their VIs. Dynamically linked Property nodes use a control reference to access properties of controls that may be in other VIs.*

There are some performance issues to be aware of with Property nodes. Property nodes execute in the user-interface thread. Remember, in our discussion of dataflow and LabVIEW's execution system, how each node depends on having all its data before it can proceed. Indiscriminately scattering Property nodes throughout your program will seriously degrade performance as portions of your block diagram wait while properties execute in the user interface. A general rule of thumb is to never place Property nodes within the main processing loop(s). Use a separate UI loop to interact with the front panel.

Property nodes can increase the memory requirements of your application. Property nodes within a subVI will cause the subVI's front panel to be loaded into memory. This may or may not be an issue, depending on what you have on your front panel or how many subVIs have Property nodes on them. Try to limit the number of Property nodes and keep them contained with only a few VIs. You could even build a reusable suite of user-interface VIs by using Property nodes and references to front panel items.

One property that you want to avoid is the Value property. This is such a convenient property to use, but it is a performance hog. LabVIEW is optimized for dataflow. Read and write to the actual controls and indicators whenever possible. You can do incredible things with Property nodes; just have the wisdom to show restraint.

Block Diagram

The block diagram is where you draw your program's dataflow diagram. For simple projects you might need only a few Express VIs. These are the "Learn LabVIEW in Three Minutes" VIs that salespeople are so fond of showing. They are great to experiment with, and you can solve many simple projects with just a few Express VIs. It is really easy to take measurements from a DAQ card and log it to disk compared to LabVIEW 2. However, when the time comes to build a complete application to control your experiment, you'll probably find yourself scratching your head and wondering what to do. Read on. In Chapter 5, "LabVIEW Frameworks," we'll talk about design patterns that you can apply to common problems and use to build robust applications.

A large part of computer programming consists of breaking a complex problem into smaller and smaller pieces. This divide-and-conquer technique works in every imaginable situation, from designing a computer program to planning a party. In LabVIEW the pieces are subVIs, and we have an advantage that other programming environments don't—each subVI is capable of being run and tested on its own. You don't have to wait until the application is complete and run against massive test suites. Each subVI can, and should, be debugged as it is developed. LabVIEW makes it easy to iteratively develop, debug, and make all the pieces work together.

Looping

Most of your VIs will contain one or more of the two available loop structures, the **For Loop** and the **While Loop**. Besides the obvious use—doing an operation many times—there are nonobvious ways to use a loop that are helpful to know about.

While Loops

The While Loop is one of the most versatile structures in LabVIEW. With it, you can iterate an unlimited number of times and then suddenly quit when the boolean **conditional terminal** becomes True. You can also pop up on the conditional terminal and select Stop If False. If the condition for stopping is always satisfied (for instance, Stop If False wired to a False boolean constant), the loop executes exactly *one time*. If you put uninitialized shift registers on one of these one-trip loops and then construct your entire subVI inside it, the shift registers become a memory element between calls to the VI. We call this type of VI a **functional global**. You can retain all sorts of status information in this way, such as knowing how long it's been since this subVI was last called (see the section on shift registers that follows).

Graceful Stops

It's considered bad form to write an infinite loop in LabVIEW (or any other language, for that matter). Infinite loops run forever, generally because the programmer told them to stop only when 2 + 2 = 5, or something like that. The hazard in LabVIEW is that the only way to stop an infinite loop is to click the Abort icon in the toolbar, which could leave the I/O hardware in an indeterminate state. At a minimum, you need to put a boolean STOP button on the front panel and wire it to the conditional terminal.

A funny thing happens when you have a whole bunch of stuff going on in a While Loop: It sometimes takes a long time to stop when you press that STOP boolean you so thoughtfully included. The problem (and its solution) is shown in Fig. 2.7. What happens in the left frame is that there is no data dependency between the guts of the loop and the RUN switch, so the switch may be read before the rest of the diagram executes. If that happens, the loop will go around one more time before actually stopping. If one cycle of the loop takes 10 minutes, the user is going to wonder why the STOP switch doesn't seem to do anything. Forcing the STOP boolean to be evaluated as the last thing in the While Loop cures this extra-cycle problem.

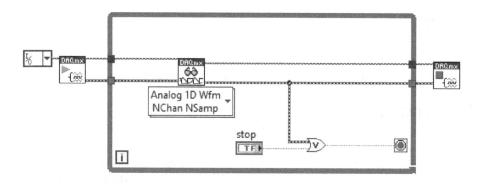

FIG. 2.7 *Evaluate the stop condition at the end of the loop to avoid an extra iteration.*

For Loops

For Loops are great for building and processing arrays and repeating an operation a fixed number of times. Most of the time, you process arrays with a For Loop because LabVIEW already knows how many elements there are, and the **auto-indexing** feature takes care of the iteration count for you automatically: All you have to do is wire the array to the loop, and the number of iterations (count) will be equal to the number of elements in the array. But what happens when you hook up more than one array to the For Loop, each with a different number of elements? What if the count terminal is also wired, but to yet a different number? Figure 2.8 should help clear up some of these questions. *Rule: The smaller count always wins.* If an empty array is hooked up to a For Loop, that loop will *never* execute.

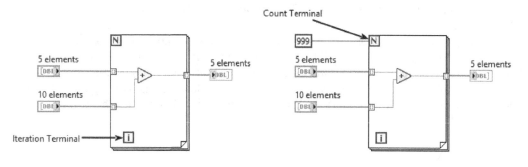

FIG. 2.8 *The smallest array, or count, determines the number of iterations when auto-indexing.*

Shift Registers

Shift registers are special local variables or memory elements available in For Loops and While Loops that transfer values from the completion of one iteration to the beginning of the next. You create them by popping up on the border of a loop. When an iteration of the loop completes, a value is written to the shift register—call this the nth value. On the next iteration, that value is available as a source of data and is now known as the $(n-1)$st value. Also, you can add as many terminals as you like to the left side, thus returning not only the $(n-1)$st value but also $n-2$, $n-3$, and so on. This gives you the ability to do digital filtering, modeling of discrete systems, and other algorithms that require a short history of the values of some variable. Any kind of data can be stored in a shift register—they are **polymorphic**. Figure 2.9 is an implementation of a simple *finite impulse response* (FIR) *filter,* in this case, a moving averager that computes the average of the last four values. You can see from the strip charts how the random numbers have been smoothed over time in the filtered case.

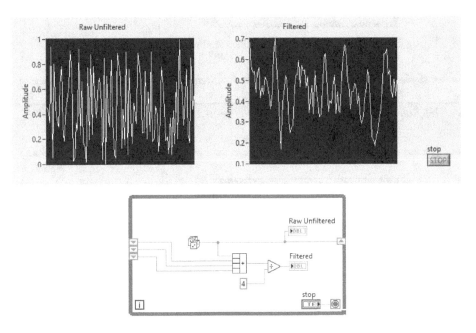

FIG. 2.9 *Using a shift register to perform a simple moving average on random numbers. Note the difference in the graphs of filtered and unfiltered data. This is a very simple case of a finite impulse response filter.*

An alternative method is shown in Fig. 2.10a and 2.10b. A single shift register is used to hold an array. In Fig. 2.10a, the VI is building the array one element at a time until it reaches Size. Once the array reaches Size elements, Rotate 1D Array puts the first element in the last position and Replace Array Subset overwrites the last element with the newest value. If our array had four elements 0, 1, 2, 3, the rotated array is now 1, 2, 3, 0. If the new value is 4, then after Replace Array Subset the array is 1, 2, 3, 4. Note that a TRUE constant is wired to the conditional terminal of the While Loop, so this VI will execute only once. This is a functional global that can be called repeatedly to provide a moving average.

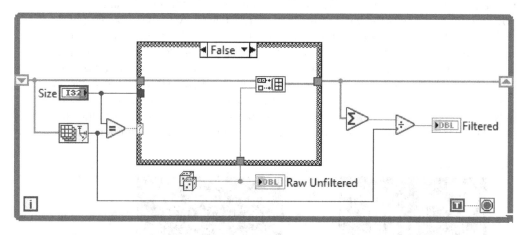

FIG. 2.10A *The Build array concatenates new elements until Size is reached.*

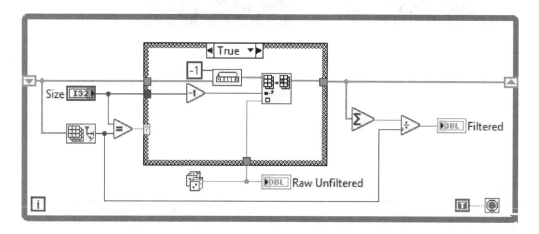

FIG. 2.10B *The Rotate array shifts the array in place by −1 element, then the Replace Array Subset overwrites the oldest element with the newest.*

Figure 2.11 is a very common construct where an array is assembled based on some conditional testing of the incoming data. In this case, it's weeding out empty strings from an incoming array. This program is frequently seen in configuration management programs. In that case, a user is filling out a big cluster array that contains information about the various channels in the system. Leaving a channel name empty implies that the channel is unused and should be deleted from the output array. The Build Array function is located inside a Case Structure that checks to see if the current string array element is empty. To initialize the shift register, create a one-dimensional (1D) string array constant on the diagram by popping up on the shift register terminal and selecting Create Constant.

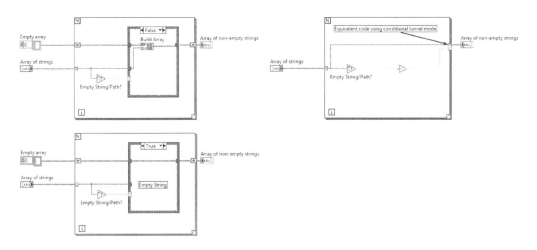

FIG. 2.11 *Weeding out empty strings with a shift register. A similar program might be used to select special numeric values. Notice that we had to use Build Array inside a loop, a relatively slow construct, but also unavoidable.*

Uninitialized Shift Registers

In Fig. 2.11, what would happen if that empty string array constant already had something in it? That data would appear in the output array *ahead* of the desired data. That's why we want an *empty* array for initialization. What happens if the empty string array is left out altogether? The answer is that the first time the program runs after being loaded, the shift register is in fact empty and works as expected. But the shift register then *retains its previous contents until the next execution*. Every time you ran this modified VI, the output string would get bigger and bigger (a good way to make LabVIEW run out of memory, by the way). All shift registers are initialized at compile time as well: Arrays and strings are empty, numerics are zero, and booleans are False. There are some important uses for uninitialized shift registers that you need to know about.

First, you can use uninitialized shift registers to keep track of state information between calls to a subVI. This is an extremely powerful technique and should be studied closely. You will see it dozens of times in this book. This technique, known as **change-of-state detection,** can make a VI act more intelligently to changes in process variables.

A Feedback Node is another kind of shift register that can be used independently of a loop structure. Figure 2.12 shows two methods of doing the same thing. Note that the conditional terminal of the While Loop is wired with a False constant, which means that the contents of the loop execute only once each time the VI is called. Such a construct is of little use as a top-level VI. Instead, its intended use is as a more intelligent subVI—one with **state memory**. That is, each time this subVI is called, the contents of the shift register provide it with logical information regarding conditions at the end of the previous iteration.

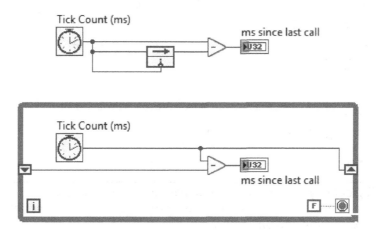

FIG. 2.12 *Two methods for getting elapsed time.*

This allows you to program the subVI to take action based not only on the current inputs but on previous inputs as well. This concept will be extrapolated into powerful **state machines** in subsequent chapters.

The shift register in Fig. 2.12 keeps track of the elapsed time since the subVI was last called. If you compare the elapsed time with a desired value, you could use this to trigger a periodic function such as data logging or watchdog timing. This technique is also used in the PID (proportional integral derivative) Toolkit control blocks whose algorithms are time-dependent. The code using the Feedback Node also keeps track of time between subsequent calls, just in a cleaner way.

Variables

Let's start by reinforcing that G is a dataflow programming language; as such, we don't really need to declare variables. The data is on the wire, and the wire transfers the data throughout the program. We need to always look at ways to get the data from one part of our program to other parts via wires. If this is not possible, then we can look into local variables; if that won't work either, then look at globals; if this is not the way, then look at using references. This is very counterintuitive for text-based programmers; they are used to looking at the reference approach first (as in pointers). We have seen code where a text-based programmer is trying to use LabVIEW and places a sequence structure full of local variables in the first frame; she is declaring her variables. Please, don't do that! Now that we got that out of the way, let's talk about these different approaches.

Local Variables

A local variable allows you to manipulate data within the scope of a LabVIEW diagram. A local variable allows you to read data from or write data to controls and indicators without directly wiring to the usual control or indicator terminal. This means you have unlimited read-write access from multiple locations on the diagram. To create a local variable, pop up on the front panel control or the block diagram terminal and select Create >> Local Variable. Then choose whether you want to read or write data. The local variable behaves exactly the same as the control or indicator's terminal, other than the fact that you are free to read or write. Here are some important facts about local variables:

- Local variables act only on the controls and indicators that reside on the same diagram. You can't use a local variable to access a control that resides in another VI. Use global variables, or better, regular wired connections to subVIs to transfer data outside of the current diagram.

- You can have as many local variables as you want for each control or indicator. Note how confusing this can become: Imagine your controls changing state mysteriously because you accidentally selected the wrong item in one or more local variables. Danger!

- As with global variables, you should use local variables only when there is no other reasonable dataflow alternative. They bypass the explicit flow of data, obscuring the relationships between data sources (controls) and data sinks (indicators).

- Each instance of a local variable requires an additional copy of the associated data. This can be significant for arrays and other data types that contain large amounts of data. This is another reason that it's better to use wires than local variables whenever possible.

- Using a local variable in a subVI causes its front panel to be loaded into memory. It is important to note that local variables operate in the user-interface thread and cause a context switch with each call. Having the front panel in memory and causing a context switch can be a huge hit to your performance, so use local variables sparingly. The one exception to this is on LabVIEW FPGA where a local variable has the same performance as a wire, because underneath they are compiled into the FPGA fabric.

There are three basic uses for local variables: control initialization, control adjustment or interaction, and temporary storage. Managing controls that interact is a cool use for local variables. But a word of caution is in order regarding race conditions. It is very easy to write a program that acts in an unpredictable or undesirable manner because there is more than one source for the data displayed in a control or an indicator. What you must do is to explicitly define the order of execution in such a way that the action of a local variable cannot interfere with other data sources, whether they are user inputs, control terminals, or other local variables on the same diagram. It's difficult to give you a more precise description of the potential problems because there are so many situations. Just think carefully before using local variables, rely on dataflow as much as possible, and always test your program thoroughly.

Global Variables

The difference between built-in globals and the functional globals you make yourself with a subVI is that the built-in ones are not true VIs and as such cannot be programmed to do anything besides simple data storage. However, built-in globals are sometimes faster for most data types. Another advantage of the built-in globals is that you can have all the global data for your entire program present in just one global variable but access them separately with no penalty in performance. With subVI-based globals, you can combine many variables into one global, but you must read and write them all at once, which increases execution time and memory management overhead. Thus, built-in globals are preferred for most applications for performance reasons.

A hazard you need to be aware of when using either type of global variable is the potential for race conditions. A race condition exists when two or more events can occur in any order, but you rely on them to occur in a particular order. While you're developing your VI, or under normal conditions, the order may be as expected and all is well. But under different conditions the order will vary, causing the program to misbehave. Data dependency prevents race conditions from being a general problem in LabVIEW, but global variables provide a way to violate strict dataflow programming. Therefore, it's up to you to understand the pitfalls of race conditions.

An all-around safe approach to avoiding race conditions is to write your overall hierarchy in such a way that a global can only be written from one location.

This condition would be met, for instance, by the client-server architecture where there is one data source (a data acquisition VI) with multiple data readers (display, archive, etc.). A global that is written to one time and read by many is known as a WORM global, Write Once, Read Many. A WORM global is a common use-case for global variables that hold configuration data that is read in at startup and used in multiple locations. Use data flow to make sure that you never read from a global before it is initialized.

This is one of the first rules taught for traditional languages: Initialize the variables, then start running the main program. For example, it would be improper for the data display VI to run before the data acquisition VI because the global variable is initially empty or contains garbage. *Rule: Enforce the order of execution in all situations that use global variables.*

If your application has a global array, there is a risk of excessive data duplication. When an array is passed along through wires on a single diagram, LabVIEW does an admirable job of avoiding array duplication, thus saving memory. This is particularly important when you want to access a single element by indexing, adding, or replacing an element. But if the data comes from a global variable, your program has to read the data, make a local copy, index or modify the array, and then write it back. If you do this process at many different locations, you end up making many copies of the data. This wastes memory and adds execution overhead. A solution is to create a subVI that encapsulates the global, providing whatever access the rest of your program requires. It might have single-element inputs and outputs, addressing the array by element number. In this way, the subVI has the only direct access to the global, guaranteeing that there is only one copy of the data. This implementation is realized automatically when you create functional globals built from shift registers.

Global variables, while handy, can quickly become a programmer's nightmare because they hide the flow of data. For instance, you could write a LabVIEW program where there are several subVIs sitting on a diagram with no wires interconnecting them and no flow control structures. Global variables make this possible: The subVIs all run until a global boolean is set to False, and all the data is passed among the subVIs in global variables. The problem is that nobody can understand what is happening since all data transfers are hidden. Similar things happen in regular programming languages where most of the data is passed in global variables rather than being part of the subroutine calls. This data hiding is not only confusing but also dangerous. All you have to do is access the wrong item in a global at the wrong time, and things will go nuts. How do you troubleshoot your program when you can't even figure out where the data is coming from or going to?

The answer is another rule. *Rule: Use global variables only where there is no other dataflow alternative.* Using global variables to synchronize or exchange information between parallel loops or top-level VIs is perfectly reasonable if done in moderation. Using globals to carry data from one side of a diagram to the other "because the wires would be too long" is asking for trouble. You are in effect making your data accessible to the entire LabVIEW hierarchy. One more helpful tip with any global variable: Use the bookmark feature to place a hashtag (#GlobalConfigAIReadChannelScale) on the block diagram to briefly describe what the global call is doing. When there are many global variables in a program, it's difficult to keep track of which is which, so labeling with a bookmark helps. Opening the bookmark manager from View >> Bookmark Manager… quickly provides a list of all bookmarks in your project. Finally, if you need to locate all instances of a global variable, you can pop up on the global variable and choose Find >> Global References.

Debugging Global Variables

If your program uses many global variables or VIs that use uninitialized shift registers for internal state memory (functional globals), you have an extra layer of complexity and many more opportunities for bugs to crop up. Here's a brief list of common mistakes associated with global variables and state memory:

- Accidentally writing data to the wrong global variable. It's easy to do; just select the wrong item on one of LabVIEW's built-in globals, and you're in trouble.

- Forgetting to initialize shift-register-based memory. Most of the examples in this book that use uninitialized shift registers have a Case Structure inside the loop that writes appropriate data into the shift register at initialization time. If you expect a shift register to be empty each time the program starts, you must initialize it as such. It will automatically be empty when the VI is loaded, but will no longer be empty after the first run.

- Calling a subVI with state memory from multiple locations without making the subVI reentrant when required. You need to use reentrancy whenever you want the independent calls not to share data, for instance, when computing a running average. Reentrant execution is selected from the VI Properties >> Execution menu.

- Conversely, setting up a subVI with state memory as reentrant when you do want multiple calls to share data. Remember that calls to reentrant VIs from different locations don't share data. For instance, a file management subVI that creates a data file in one location and then writes data to the file in another location might keep the file path in an uninitialized shift register. Such a subVI won't work if it is made reentrant.

- Creating race conditions. You must be absolutely certain that every global variable is accessed in the proper order by its various calling VIs. The most common case occurs when you attempt to write and read a global variable on the same diagram without forcing the execution sequence. Race conditions are absolutely the most difficult bugs to find and are the main reason you should avoid using globals haphazardly.

- Try to avoid situations where a global variable is written in more than one location in the hierarchy. It may be hard to figure out which location supplied the latest value.

How do you debug these global variables? The first thing you must do is to locate all the global variable's callers. Open the global variable's panel and select View >> Browse Relationships >> This VI's Callers from the menu, or View >> VI Hierarchy. Alternatively, you can right-click on the global and select Find >> Global References. Note every location where the global is accessed, then audit the list and see if it makes sense.

Keep the panel of the global open while you run the main VI. Observe changes in the displayed data if you can. You may want to single-step one or more calling VIs to more carefully observe the data. One trick is to add a new string control called info to the global variable. At every location where the global is called to read or write data, add another call that writes a message to the info string. It might say, "SubVI abc writing xyz array from inner Case Structure." You can also wire the Call Chain function (in the Application Control palette) to the info string. (Actually, it has to be a string array in this case.) Call Chain returns a string array containing an ordered list of the VI calling chain, from the top-level VI on down. Either of these methods makes it abundantly clear who is accessing what.

Debugging Local Variables

Everything we've said regarding global variables is true to a great degree for local variables. Race conditions, in particular, can be very difficult to diagnose. This is reason enough to avoid using local variables except when absolutely necessary! Execution highlighting and/or single-stepping is probably your best debugging tool because you can usually see in what order the local variables are being executed. However, if your VI has a complicated architecture with parallel loops and timers, the relative ordering of events is time-dependent and cannot be duplicated in the slow motion of execution highlighting. In difficult cases, you must first make sure that you have not inadvertently selected the wrong item in a local variable. Pop up on the problematic control or indicator, and select Find >> Local Variables. This will lead you to each local variable associated with that control.

Be sure that each instance is logically correct: Is it read or write mode, and is it the proper place to access this variable? Your only other recourse is to patiently walk through the logic of your diagram, just as you would in any programming language. See? We told you that local variables are a violation of LabVIEW's otherwise strict dataflow concepts. And what did they get you? The same old problems we have with traditional languages.

State Machines

There is a very powerful and versatile alternative to the Sequence structure, called a state machine, as described in the advanced LabVIEW training course. The general concept of a state machine originates in the world of digital (boolean) logic design where it is a formal method of system design. In LabVIEW, a state machine uses a Case Structure wired to a counter that's maintained in a shift register in a While Loop. This technique allows you to jump around in the sequence by manipulating the counter. For instance, any frame can jump directly to an error-handling frame. This technique is widely used in drivers, sequencers, and complex user interfaces and is also applicable to situations that require extensive error checking. Any time you have a chain of events where one operation depends on the status of a previous operation or where there are many modes of operation, a state machine is a good way to do the job.

Figure 2.13 is an example of the basic structure of a state machine. The state number is maintained as an enumerated value in a shift register, so any frame can jump to any other frame. Shift registers or local variables must also be used for any data that needs to be passed between frames, such as the results of the subVI in this example. One of the configuration tricks you will want to remember is to make your enum a typedef so that changes will propagate to all instances; otherwise, you would have to edit every enum anytime a change is made.

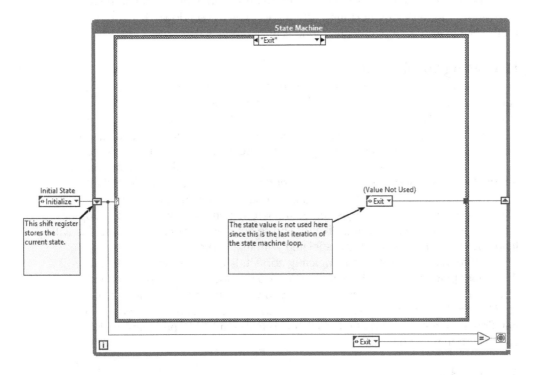

FIG. 2.13 *A basic state machine uses a While Loop with an enumerated typedef to select the next state.*

Each frame that contains an activity may also check for errors (or some other condition) to see where the program should go next. In driver VIs, you commonly have to respond to a variety of error conditions, where you may want to retry communications, abort, or do something else depending on the condition detected. Similarly, in a complex user interface, you may have to handle major changes in display configuration (using Property nodes) or handle errors arising from user data entries. The state machine lets you handle these complex situations.

A state machine can be the diagram behind a flexible user interface, or it can be called as a subVI. In a user interface, the state machine manages various modes of operation. For instance, the user may have a series of mode selections when configuring an experiment. A mode selector switch may directly drive the Case Structure, or you may use Property nodes to selectively hide and show valid mode selection buttons. Depending upon which button is then pressed, the state machine jumps to an appropriate frame to handle that selection.

A favorite use for state machines is intelligent subVIs in such applications as sequencers and profile generators. In those situations, the state machine subVI is called periodically in the main While Loop. Each time it's called, information from the last cycle is available in the shift register, providing the required history: Where have I been? Data passed in through the connector pane provides the current information: Where am I now? And computation within the state machine subVI determines the future: Where should I go next? Those values may drive an I/O device through a subVI within the state machine, or they may be passed back to the calling VI through the connector pane.

SubVIs

A large part of computer programming consists of breaking a complex problem into smaller and smaller pieces. This divide-and-conquer technique works in every imaginable situation from designing a computer program to planning a party. In LabVIEW the pieces are subVIs, and we have an advantage that other programming environments don't—each subVI is capable of being run and tested on its own. You don't have to wait until the application is complete and run against massive test suites. Each subVI can, and should, be debugged as it is developed. LabVIEW makes it easy to iteratively develop, debug, and make all the pieces work together. Refer to the chapter on Unit Testing for formal test capabilities and methods used by the pros.

Approach each problem as a series of smaller problems. As you break it down into modular pieces, think of a one-sentence statement that clearly summarizes the purpose. This one sentence is your subVI. For instance, one might be "This VI loads data from a series of transient recorders and places the data in an output array." Write this down with the labeling tool on the block diagram before you put down a single function; then build to that statement. If you can't write a simple statement like that, you may be creating a catchall subVI. Also write the statement in the VI description to help with documentation. Comments in the VI description will also show up in the context help window. Consider the reusability of the subVIs you create. Can the function be used in several locations in your program? If so, you definitely have a reusable module, saving disk space and memory. If the subVI requirements are almost identical in several locations, it's probably worth writing it in such a way that it becomes a universal solution—perhaps it just needs a mode control.

Each subVI needs an icon. Don't rest content with the default LabVIEW icon; put some time and effort into capturing the essence of what the VI does. You can even designate terminals as being mandatory on the connector pane. They will show up in bold on the context help window. Also create an easily recognizable theme for related VIs, as shown in the instrument driver tree for the Agilent 34401 in Fig. 2.14.

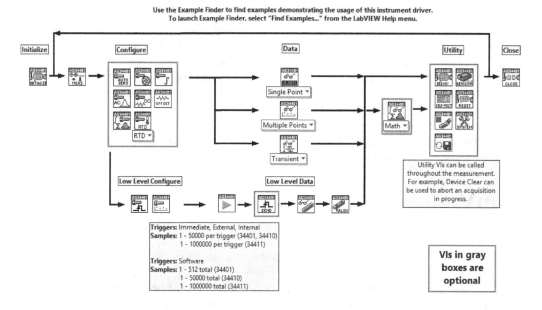

FIG. 2.14 *VI Tree for Agilent 34401.*

Data Types

Our next project is to consider all the major data types that LabVIEW supports. The list consists of **scalar** (single-element) types, such as numerics and booleans, and **structured** types (containing more than one element), such as strings, arrays, and clusters. The LabVIEW *Controls* palette is roughly organized around data types, gathering similar types into subpalettes for easy access. There are, in fact, many ways to *present* a given data type, and that's the main reason that there are apparently so many items in those palettes. For instance, a given numeric type can be displayed as a simple number, a bar graph, a slider, or an analog meter, or in a chart. But underneath is a well-defined representation of the data that you, the programmer, and the machine must mutually understand and agree upon as part of the programming process.

One thing that demonstrates that LabVIEW is a complete programming language is its support for essentially all data types. Numbers can be floating point or integer, with various degrees of precision. Booleans, bytes, strings, and numerics can be combined freely into various structures, giving you total freedom to make the data type suit the problem. **Polymorphism** is an important feature of LabVIEW that simplifies this potentially complicated world of data types into something that even the novice can manage without much study. Polymorphism is the ability to adjust to input data of different types. Most built-in LabVIEW functions are polymorphic. Ordinary virtual instruments (VIs) that you write are not truly polymorphic—they can adapt between numeric types, but not between other data types, such as strings to numerics. Most of the time, you can just wire from source to destination without much worry, since the functions adapt to the kind of data that you supply. How does LabVIEW know what to do? The key is object-oriented programming, where polymorphism is but one of the novel concepts that make this new programming technology so desirable. There are, of course, limits to polymorphism, and we'll see later how to handle those special situations. Please note that this is different than Polymorphic VIs and Malleable VIs (.vim). LabVIEW 2017 introduced a more advanced form of polymorphism, the Malleable VIs. These VIs can adapt each terminal to its corresponding input data type.

Numeric Types

Perhaps the most fundamental data type is the numeric, a scalar value that may generally contain an integer or a real (floating-point) value. LabVIEW explicitly handles all the possible integer and real representations that are available on current 64-bit processors. Figure 2.15 displays the terminals for each representation, along with the meaning of each and the number of bytes of memory occupied. LabVIEW floating-point types follow the IEEE-754 standard, which has thankfully been adopted by all major CPU and compiler designers.

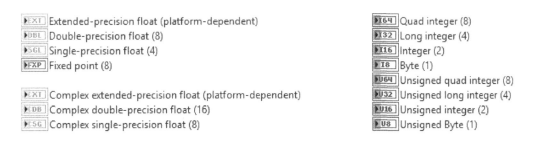

FIG. 2.15 *Scalar numerics in LabVIEW cover the gamut of integer and real types. The number of bytes of memory required for each type is shown.*

The keys to choosing an appropriate numeric representation in most situations are the required **range** and **precision**. In general, the more bytes of memory occupied by a data type, the greater the possible range of values. This factor is most important with integers; among floating-point types, even single precision can handle values up to 10^{38}, and it's not too often that you need such a large number. An **unsigned integer** has an upper range of 2^N-1, where N is the number of bits. Therefore, an unsigned byte ranges up to 255, an unsigned integer (2 bytes) up to 65,535, and an unsigned long integer up to 4,294,967,295. **Signed integers** range up to $2^N - 1$, or about one-half of their unsigned counterparts.

If you attempt to enter a value that is too large into an integer control, it will be coerced to the maximum value. But if an integer mathematical process is going on, **overflow** or **underflow** can produce erroneous results. For instance, if you do an unsigned byte addition of 255 + 1, you get 0, not 256—that's overflow.

Floating-point precision can be a very confusing subject, especially if you're prone to being anal-retentive. While integers by definition have a precision of exactly 1, floating-point types are *inexact* representations. The reason is that while you may be most comfortable entering and displaying a floating-point value in decimal (base-10) notation, the computer stores the value as a binary (base-2) value *with a limited number of bits*. Some rounding must occur, so this base-converted storage is therefore inexact.

For example, enter 3.3 into a single-precision, floating-point control, pop up on the control and set the precision to a large number of digits, and note that the displayed value is 3.2999999523. . . , which is just a hair different from what you entered.

The question is: How important is that rounding error to your application? You should evaluate precision in terms of **significant figures**. For instance, the number 123.450 has six significant figures. According to the IEEE-754 standard, a single-precision float guarantees 7.22 significant figures, a double-precision float gives you 15.95, and an extended-precision float gives you 19.26. When you configure a numeric indicator to display an arbitrary number of digits, all digits beyond the IEEE-specified number of significant figures are insignificant! That is, you must ignore them for the purposes of computation or display. Double-precision values are recommended for most analyses and mathematical operations to reduce cumulative rounding errors. Single-precision values are sufficient for transporting data from most real-world instruments (with simple math), since they give you a resolution of 0.1 part per million, which is better than most real, physical measurements.

IEEE floating-point numbers have a bonus feature: They can tell you when the value is invalid. If you divide by zero, the value is infinity, and a LabVIEW numeric indicator will say **inf**. Another surprisingly useful value is **NaN,** or not-a-number. It makes a great flag for invalid data.

For instance, if you're decoding numbers from a string (see the next section) and a valid number can't be found, you can have the function return NaN instead of zero or some other number that might be mistaken for real data. Also, when you are graphing data, NaN does not plot at all.

That's a handy way to edit out undesired values or to leave a blank space on a graph. You can test for NaN with the comparison function **Not A Number/Path/Refnum**.

Strings

Every programmer spends a lot of time putting together strings of characters and taking them apart. Strings are useful for indicators where you need to say something to the operator, for communications with instruments, and for reading and writing data files that other applications will access. LabVIEW has a nice set of string-wrangling functions built right in. If you know anything about the C language, some of these functions will be familiar. Let's look at some common string problems and their solutions.

Building Strings

Instrument drivers are the classic case study for string building. The problem is to assemble a command for an instrument (usually GPIB) based on several control settings. Figure 2.16 shows how string building in a typical driver works.

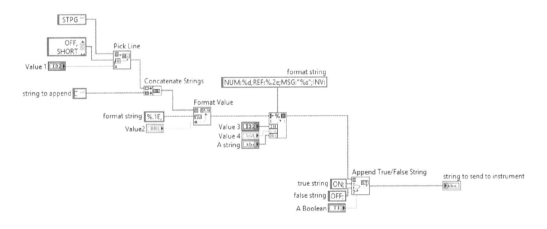

FIG. 2.16 *String building in a driver VI. This one uses most of the major string-building functions in a classic diagonal layout common to many drivers.*

1. The **Pick Line** function is driven by a ring control that supplies a number (0, 1, or 2) that is mapped into the words OFF, SHORT, or LONG. These keywords are appended to the initial command string STPGEN PUL:.

2. The **Concatenate Strings** function tacks on a substring CUR:. If you need to concatenate several strings, you can resize Concatenate Strings (by dragging at any corner) to obtain more inputs.

3. The **Format Value** function is versatile for appending a single numeric value. You need to learn about C-style formatting commands (see the LabVIEW online help for details) to use this function. The percent sign tells it that the next few characters are a formatting instruction. What is nice about this function is that not only does it format the number in a predictable way, but also it allows you to tack on other characters before or after the value. This saves space and gets rid of a lot of Concatenate Strings functions. In this example, the format string %e translates a floating-point value in exponential notation and adds a trailing comma.

4. A more powerful function is available to format multiple values: **Format Into String**. You can pop up on this function and select Edit Format String to obtain an interactive dialog box where you build the otherwise cryptic C-style formatting commands. In this case, it's shown formatting an integer, a float, and a string into one concatenated output with a variety of interspersed characters. Note that you can also use a control or other string to determine the format. In that case, there could be run time errors in the format string, so the function includes error I/O clusters. It is very handy and very compact, similar to its C counterpart sprintf().

5. The **Append True/False String** function uses a boolean control to pick one of two choices, such as ON or OFF, which is then appended to the string. This string-building process may continue as needed to build an elaborate instrument command.

Parsing Strings

The other half of the instrument driver world involves interpreting response messages.

The message may contain all sorts of headers, delimiters, flags, and who knows what, plus a few numbers or important letters that you actually want. Breaking down such a string is known as parsing. It's a classic exercise in computer science and linguistics as well. Remember how challenging it was to study our own language back in fifth grade, parsing sentences into nouns, verbs, and all that? We thought that was tough; then we tackled the reply messages that some instrument manufacturers come up with! Figure 2.17 comes from one of the easier instruments, the Tektronix 370A driver. A typical response message would look like this:

STPGEN NUMBER:18;PULSE:SHORT;OFFSET:-1.37;NVERT:OFF;MULT:ON;

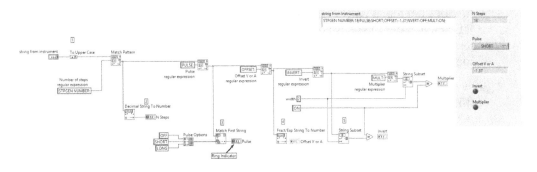

FIG. 2.17 *Using string functions to parse the response message from an instrument. [This is typical of many instruments you may actually encounter.]*

Let's examine the diagram that parses this message.

1. The top row is based on the versatile **Match Pattern** function. Nothing fancy is being done with it here, other than searching for a desired keyword. The manual fully describes the functions of the many special characters you can type into the *regular expression* input that controls Match Pattern. One other thing we did is to pass the incoming string through the **To Upper Case** function. Otherwise, the pattern keys would have to contain both uppercase and lowercase letters. The output from each pattern match is known as *after substring*. It contains the remainder of the original string immediately following the pattern, assuming that the pattern was found. If there is any chance that your pattern might not be found, test the *offset past match* output; if it's less than zero, there was no match, and you can handle things at that point with a Case Structure.

2. Each of the *after substrings* in this example is then passed to another level of parsing. The first one uses the **Decimal String To Number** function, one of several specialized number extractors; others handle octal, hexadecimal, and fraction/scientific notation. These are very robust functions. If the incoming string starts with a valid numeric character, the expected value is returned. The only problem you can run into occurs in cases where several values are run together such as [123.E3-.567.89]. Then you need to use the **Scan From String** function to break it down, provided that the format is fixed. That is, you know exactly where to split it up. You could also use Match Pattern again if there are any other known, embedded flags, even if the flag is only a space character.

3. After locating the keyword PULSE, we expect one of three possible strings: OFF, SHORT, or LONG. **Match First String** searches a string array containing these words and returns the index of the one that matches (0, 1, or 2). The index is wired to a ring indicator named **Pulse** that displays the status.

4. Keyword OFFSET is located, and the substring is passed to the **Fract/Exp String To Number** function to extract the number, which in this case is in fractional form.

5. INVERT is located next, followed by one of two possible strings, ON or OFF. When Gary first wrote this VI, he first used Index and Strip again, with its search array containing ON and OFF. But a bug cropped up! It turned out that another keyword farther along in the string (MULT) also used ON and OFF. Since Match Pattern passes us the *entire remainder* of the string, Index and Strip happily went out and found the *wrong* ON word—the one that belonged to MULT. Things were fine if the INVERT state was ON, since we found that right away. The solution he chose was to use the **String Subset** function, split off the first three characters, and test only those. You could also look for the entire command INVERT:ON or INVERT:OFF.

The moral of the story: Test thoroughly before shipping.

The **Scan From String** function, like the C scanf() function, can save much effort when you are parsing strings (Fig. 2.18). Like its complement **Format Into String**, this function produces one or more polymorphic outputs (numeric or string), depending upon the contents of the Format String input. Input terminals allow you to determine the data format of numeric outputs and to set default values as well.

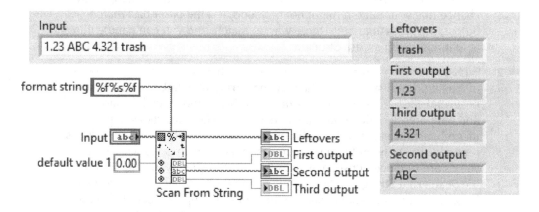

FIG. 2.18 *The Scan From String function is great for parsing simple strings.*

Other difficult parsing problems arise when you attempt to extract information from text files. If the person who designed the file format is kind and thoughtful, all you will have to do is search for a keyword, then read a number. There are other situations that border on the intractable; you need to be a computer science whiz to write a reliable parser in the worst cases. These are very challenging problems, so don't feel ashamed if it takes you a long time to write a successful LabVIEW string-parsing VI.

Dealing with Unprintables

Sometimes you need to create or display a string that contains some of the unprintable ASCII characters, such as control-x or the escape character. The trick is to use the pop-up item **'\' Codes Display**, which works on front panel control and indicators as well as diagram string constants. LabVIEW will interpret one- or two-character codes following the backslash character, as shown in Table 2.1. You can also enter unprintables by simply typing them into the string. Control characters, carriage returns, and so on all work fine, except for the Tab (control-i). LabVIEW uses the Tab key to switch tools, so you have to type \t. Note that the hexadecimal codes require uppercase letters. One other disconcerting feature of these escape codes is this: If you enter a code that can be translated into a printable character (for instance, \41 = A), the printable character will appear as soon as you run the VI. To solve that problem, you can pop up on the string and select **Hex Display**. Then the control simply displays a series of two-digit hexadecimal values without ASCII equivalents.

Escape codes	Interpreted as
\00-\FF	Hexadecimal value of an 8-bit character
\b	Backspace (ASCII BS or equivalent to \08)
\f	Formfeed (ASCII FF or equivalent to \0C)
\n	Newline (ASCII LF or equivalent to \0A)
\r	Return (ASCII CR or equivalent to \0D)
\t	Tab (ASCII HT or equivalent to \09)
\s	Space (equivalent to \20)
\\	Backslash (ASCII\or equivalent to \5C)

TABLE 2.1 *Escape Sequences (\ Codes) for Strings*

There are many conversion functions in the String palette. Figure 2.19 shows just a few of them in action. These string converters are the equivalent of functions such as ASC() and CHR() in Basic, where you convert numbers to and from strings.

The **Scan From String** function is particularly powerful and not limited to hexadecimal conversion. You expand it to suit the number of variables in the source string, then enter an appropriate number of format specifiers in the format string. As with its complement **Format Into String**, you can pop up on this function and select **Edit Format String** to build formatting commands.

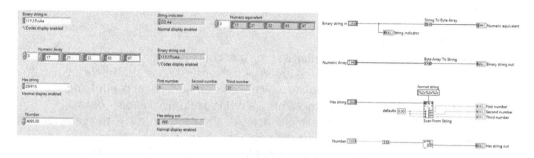

FIG. 2.19 *Here are some of the ways you can convert to and from strings that contain unprintable characters. All these functions are found in the String palette.*

Spreadsheets, Strings, and Arrays

Frequently you will need to write your data to disk for later use by spreadsheet programs and other applications that prefer tab-delimited text. Such a string might look like this:

value<tab> value <tab> value <cr>

Sometimes the delimiter is a comma or other character. No matter, LabVIEW handles all delimiters. A very important portability note is in order here. Most platforms are different with respect to their end-of-line character. A platform-independent constant, End Of Line (🔣) is available from the String function palette that automatically generates the proper character when ported. The proper characters are

PC: carriage return, then line feed (\r\n)
macOS / UNIX: line feed (\n) (ul-end)

A really interesting problem crops up with string controls when the user types a carriage return into the string: LabVIEW always inserts a *line feed*. This might be a portability issue if you plan to carry your VIs from one platform to another.

To convert arrays of numerics to strings, the easiest technique is to use the Array To Spreadsheet String function. Just wire your array into it along with a format specifier, and out pops a tab-delimited string, ready to write to a file. For a one-dimensional (1D) array (Fig. 2.20), tab characters are inserted between each value, and a carriage return is inserted at the end. This looks like one horizontal row of numbers in a spreadsheet. If you don't want the default tab characters between values, you can wire a different string to the delimiter input.

You can read the data back in from a file (simulated here) and convert it back into an array by using the **Spreadsheet String To Array** function. It has an additional requirement that you supply a **type specifier** (such as a 1D array) to give it a hint as to what layout the data might have. The format specifier doesn't have to show the exact field width and decimal precision; plain "%e" or "%d" generally does the job for floating-point or integer values, respectively.

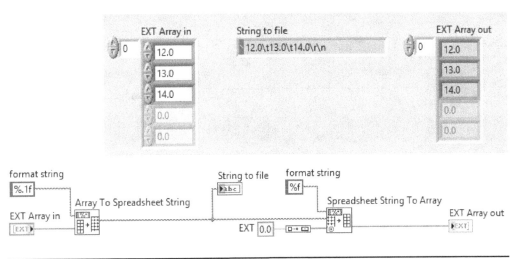

FIG. 2.20 *Converting a 1D array to a spreadsheet-compatible string and back.*

In Fig. 2.21, a two-dimensional (2D) array, also known as a matrix, is easily converted to and from a spreadsheet string. In this case, you get a tab between values in a row and a carriage return at the end of each row. If the rows and columns appear swapped in your spreadsheet program, insert the array function Transpose 2D Array before you convert the array to text. Notice that we hooked up Spreadsheet String To Array to a different type specifier, this time a long integer (I32). Resolution was lost because integers don't have a fractional part; this demonstrates that you need to be careful when mixing data types. We also obtained that type specifier by a different method—an array constant, which you can find in the Array

Function palette. You can create constants of any data type on the diagram. Look through the function palettes, and you'll see constants everywhere. In this case, choose an array constant, and then drag a numeric constant into it. This produces a 1D numeric array. Then pop up on the array and select Add Dimension to make it 2D.

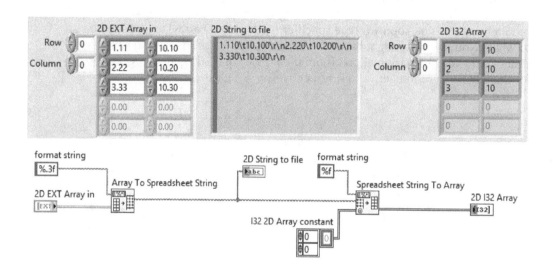

FIG. 2.21 *Converting an array to a table and back, this time using a 2D array (matrix).*

A more general solution to converting arrays to strings is to use a For Loop with a shift register containing one of the string conversion functions and Concatenate Strings. Sometimes, this is needed for more complicated situations where you need to intermingle data from several arrays, you need many columns, or you need other information within rows. Figure 2.22 uses these techniques in a situation where you have a 2D data array (several channels and many samples per channel), another array with timestamps, and a string array with channel names.

The names are used to build a header in the upper For Loop, which is then concatenated to a large string that contains the data. This business can be a real memory burner (and *slow* as well) if your arrays are large. The strings compound the speed and memory efficiency problem.

Chapter 2: LabVIEW Fundamentals

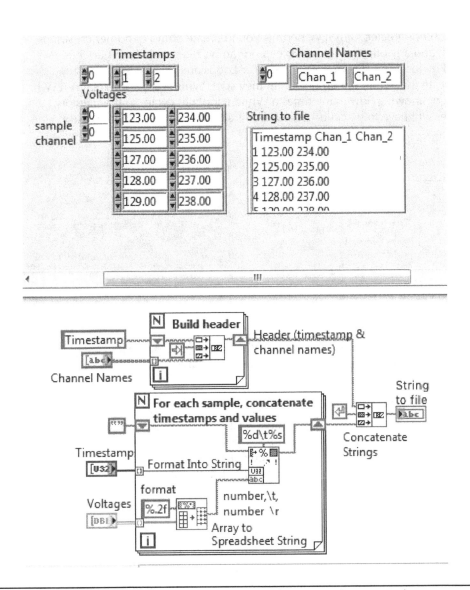

FIG. 2.22 *A realistic example of building a spreadsheet string from a 2D data array, a timestamp array, and a string array containing channel names for the file's header.*

Arrays

Any time you have a series of numbers or any other data type that needs to be handled as a unit, it probably belongs in an array. Most arrays are one-dimensional (1D, a column or vector), a few are 2D (a matrix), and some specialized data sets require 3D or greater. LabVIEW permits you to create arrays of numerics, strings, clusters, and any other data type (except for arrays of arrays). Arrays are often created by loops, as shown in Fig. 2.23. For Loops are the best because they preallocate the required memory when they start. While Loops can't; LabVIEW has no way of knowing how many times a While Loop will cycle, so the memory manager will have to be called occasionally, slowing execution somewhat.

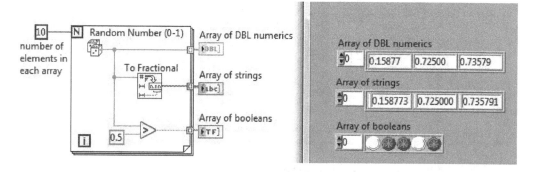

FIG. 2.23 *Creating arrays using a For Loop. This is an efficient way to build arrays with many elements. A While Loop would do the same thing, but without preallocating memory.*

You can also create an array by using the Build Array function (Fig. 2.24). Notice the versatility of Build Array: It lets you concatenate entire arrays to other arrays, or just tack on single elements. It's smart enough to know that if one input is an array and the other is a scalar, then you must be concatenating the scalar onto the array and the output will be similar to the input array. If all inputs have the same dimensional size (for instance, all are 1D arrays), you can pop up on the function and select Concatenate Inputs to concatenate them into a longer array of the same dimension (this is the default behavior). Alternatively, you can turn off concatenation to build an array with $n + 1$ dimensions. This is useful when you need to promote a 1D array to 2D for compatibility with a VI that required 2D data.

To handle more than one input, you can resize the function by dragging at a corner. Note the **coercion dots** where the SGL (single-precision floating point) and I16 (2-byte integer) numeric types are wired to the top Build Array function. This indicates a change of data type because an array cannot contain a mix of data types. LabVIEW must promote all types to the one with the greatest numeric range, which in this case is DBL (double-precision floating point). This also implies that you can't build an array of, say, numerics and strings. For such intermixing, you must turn to **clusters**, which we'll look at a bit later.

Chapter 2: LabVIEW Fundamentals **97**

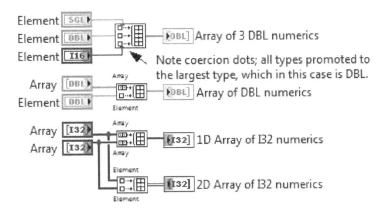

FIG. 2.24 *Using the Build Array function. Use the pop-up menu on its input terminals to determine whether the input is an element or an array. Note the various results.*

Figure 2.25 shows how to find out how many elements are in an array by using the Array Size function. Note that an empty array has zero elements. You can use Index Array to extract a single element. Like most LabVIEW functions, Index Array is polymorphic and will return a scalar of the same type as the array.

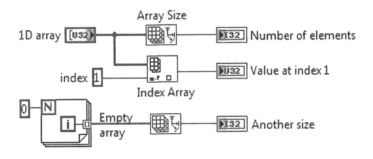

FIG. 2.25 *How to get the size of an array and fetch a single value. Remember that all array indexing is based on 0, not 1.*

If you have a multidimensional array, these same functions still work, but you have more dimensions to keep track of (Fig. 2.26). Array Size returns an array of values, one per dimension. You have to supply an indexing value for each dimension. An exception occurs when you wish to slice the array, extracting a column or row of data, as in the bottom example of Fig. 2.26. In that case, you resize Array Size, but don't wire to one of the index inputs. If you're doing a lot of work with multidimensional arrays, consider creating special subVIs to do the slicing, indexing, and sizing. That way, the inputs to the subVI have names associated with each dimension, so you are not forced to keep track of the purpose for each row or column. Another tip for multidimensional arrays is this: It's a good idea to place labels next to each index on an array control to remind you which is which. Labels such as row and column are useful.

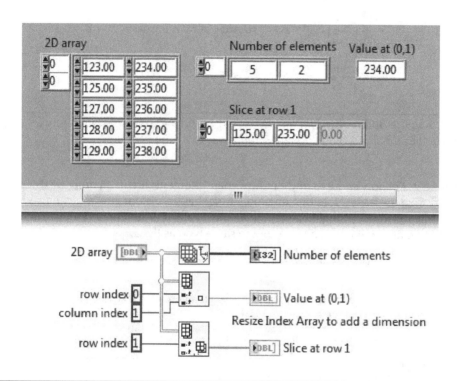

FIG. 2.26 *Sizing, indexing, and slicing 2D arrays are a little different, since you have two indices to manipulate at all steps.*

Besides slicing and dicing, you can easily do some other useful array editing tasks. In Fig. 2.27, the Delete From Array function removes a selected column from a 2D array. The function also works on arrays of other dimensions, such as all array functions. For instance, you can delete a selected range of elements from a 1D array. In that case, you wire a value to the Length input to limit the quantity of elements to delete.

The **Replace Array Subset** function is very powerful for array surgery. In this example, it's shown replacing two particular values in a 3 × 3 matrix. You can replace individual values or arrays of values in a single operation. One other handy function not shown here is **Insert Into Array**. It does just what you'd expect: Given an input array, it inserts an element or new subarray at a particular location.

Chapter 2: LabVIEW Fundamentals

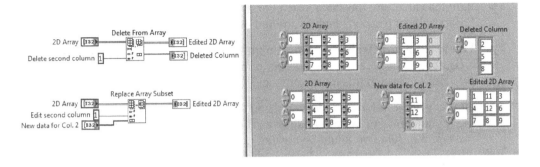

FIG. 2.27 *Two of the more powerful array editing functions are Delete From Array and Replace Array Subset. Like all the array functions, they work on arrays of any dimension.*

Initializing Arrays

Sometimes, you need an array that is initialized when your program starts, say, for a lookup table. There are many ways to do this, as shown in Fig. 2.28.

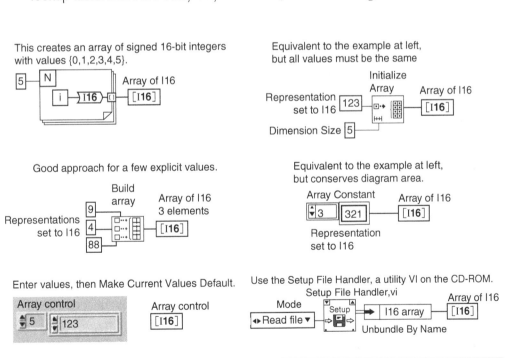

FIG. 2.28 *Several ways of programmatically initializing arrays. For large amounts of data, there are trade-offs between data storage on disk and speed of execution with each one.*

- If all the values are the same, use a For Loop with a constant inside. Disadvantage: It takes a certain amount of time to create the array.

- Use the **Initialize Array** function with the **dimension size** input connected to a constant numeric set to the number of elements. This is equivalent to the previous method, but is more compact.

- Similarly, if the values can be calculated in some straightforward way, put the formula in a For Loop instead of a constant. For instance, a special waveform or function could be created in this way.

- Create a diagram array constant, and manually enter the desired values. Disadvantage: It's tedious and uses memory on disk when the VI is saved.

- Create a front panel array control and manually type in the values. Select **Make Current Value Default** from the control's Data Operations pop-up menu. From now on, that array will always have those values unless you change them. By popping up on the control or its terminal, you can select **Hide Front Panel Control** or position the control off-screen to keep anyone from modifying the data. Disadvantage: The data takes up extra space on the disk when you save the VI.

- If there is lots of data, you could save it in a file and load it at start-up.

A special case of initialization is that of an **empty array**. This is *not* an array with one or more values set to zero, false, empty string, or the like! It contains *zero* elements. In C or Pascal, this corresponds to creating a new pointer to an array. The most frequent use of an empty array is to initialize or reset a shift register that is used to hold an array. Here are some ways to create an empty array:

- Create a front panel array control. Select **Empty Array** from its Data Operations pop-up, and then **Make Current Value Default** from the control's pop-up menu. This is, by definition, an empty array.

- Create a For Loop with the count terminal [N] wired to zero. Place a diagram constant of an appropriate type inside the loop and wire outside the loop. The loop will execute zero times (i.e., not at all), but the array that is created at the loop border tunnel will have the proper type.

- Use the **Initialize Array** function with the **dimension size** input unconnected. This is functionally equivalent to the For Loop with $N = 0$.

- Use a diagram array constant. Select **Empty Array** from its Data Operations pop-up menu.

Note that you can't use the Build Array function. Its output always contains at least one element.

Array Memory Usage and Performance

Perhaps more than any other structure in LabVIEW, arrays are responsible for a great deal of memory usage. It's not unusual to collect thousands or even millions of data points from a physics experiment and then try to analyze or display them all at once. Ultimately, you may see a little bulldozer cursor and/or a cheerful dialog box informing you that LabVIEW has run out of memory. There are some things you can do to prevent this occurrence.

First, read the LabVIEW Technical Note Managing Large Data Sets in LabVIEW; we'll summarize it here, but you can find the full article online at www.ni.com. LabVIEW does its best to conserve memory. When an array is created, LabVIEW has to allocate a contiguous area of memory called a data buffer in which to store the array. (By the way, every data type is subject to the same rules for memory allocation; arrays and strings, in particular, just take up more space.) Figure 2.29 contains a sampling of functions that do and do not reuse memory in a predictable way. If you do a simple operation such as multiplying a scalar by an array, no extra memory management is required. An array that is indexed on the boundary of a For Loop, processed, and then rebuilt also requires no memory management. The Build Array function, on the other hand, always creates a new data buffer. It's better to use Replace Array Element on an existing array, as shown in Fig. 2.30. A quick benchmark of this example showed a performance increase of 30 times. This is one of the few cases where you can explicitly control the allocation of memory in LabVIEW, and it's highly advisable when you handle large arrays or when you require maximum performance.

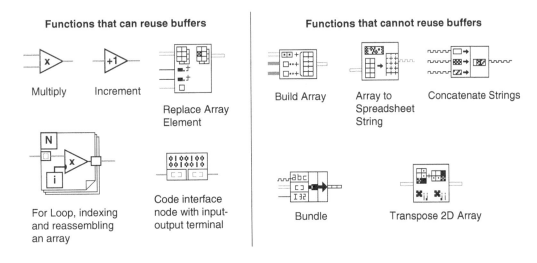

FIG. 2.29 *Here are some of the operations that you can count on for predictable reuse or non-reuse of memory. If you deal with large arrays or strings, think about these differences.*

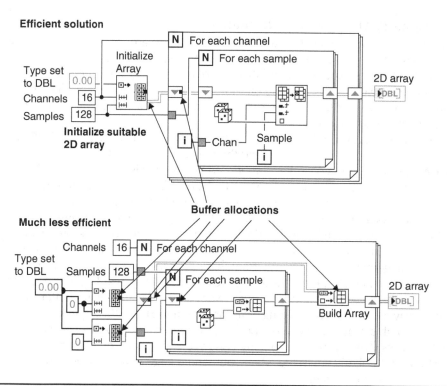

FIG. 2.30 Use Replace Array Element inside a For Loop instead of Build Array to permit reuse of an existing data buffer. This is much faster. The bottom example uses five memory buffers compared to two on the top.

One confusing issue about multidimensional arrays involves keeping track of the indices. Which one is the row and which one is the column? Which one does a nested loop structure act upon first? You can keep track of this by looking at the indicator on the panel, such as the one in Fig. 2.30. The top index, called *channel*, is also the top index on the Replace Array Element function; this is also true of the Index Array function.

So at least you can keep track of that much. By the way, it is considered good form to label your array controls and wires on the diagram as we did in this example. When you access a multidimensional array with nested loops as in the previous example, the outer loop accesses the top index and the inner loop accesses the bottom index. Figure 2.31 summarizes this array index information. The Index Array function even has pop-up tip strips that appear when you hold the wiring tool over one of the index inputs. They call the index inputs column, row, and page, just as we did in this example.

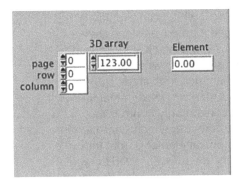 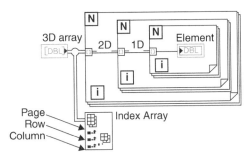

FIG. 2.31 *Organization of array indices.*

All this memory reuse business also adds overhead at execution time because the **memory manager** has to be called. Talk about an overworked manager! The poor guy has to go searching around in RAM, looking for whatever size chunk the program happens to need. If a space can't be found directly, the manager has to shuffle other blocks around until a suitable hole opens up. This can take time, especially when memory is getting tight. This is also the reason your VIs sometimes execute faster the *second* time you run them: Most of the allocation phase of memory management is done on the first iteration or run.

Similarly, when an array is created in a For Loop, LabVIEW can usually predict how much space is needed and can call the memory manager just once. This is not so in a While Loop, since there is no way to know in advance how many times you're going to loop. It's also not so when you are building arrays or concatenating strings inside a loop—two more situations to avoid when performance is paramount. To show buffer allocations on the block diagram select **Tools >> Profile >> Show Buffer Allocations** to bring up the **Show Buffer Allocations** window. Black squares will appear on the block diagram, showing memory allocations. The effect is subtle, like small flakes of black pepper; but if you turn them on and off, the allocations will jump out at you. Figure 2.30 shows what a powerful tool it can be. The best source of information on memory management, *LabVIEW Performance and Memory Management,* can be found online at www.ni.com.

Clusters

You can gather several different data types into a single, more manageable unit, called a **cluster**. It is conceptually the same as a *javascript object* or a *struct* in C. Clusters are normally used to group related data elements that are used in multiple places on a diagram. This reduces wiring clutter—many items are carried along in a single wire. Clusters also reduce the number of terminals required on a subVI. When saved as a **typedef** or **strict typedef** option (right-click on the cluster and select >>

Make Type Def), clusters serve as data type definitions, which can simplify large LabVIEW applications. Saving your cluster as a typedef propagates any changes in the data structure to any code using the typedef cluster. Using clusters is good programming practice, but it does require a little insight as to when and where clusters are best employed. If you're a novice programmer, look at the LabVIEW examples and the figures in this book to see how clusters are used in real life.

An important fact about a cluster is that it can contain only controls or indicators, but not a mixture of both. This precludes the use of a cluster to group a set of controls and indicators on a panel. Use graphical elements from the Decorations palette to group controls and indicators. If you really need to read *and* write values in a cluster, local variables can certainly do the job. We would not recommend using local variables to continuously read and write a cluster because the chance for a race condition is very high. It's much safer to use a local variable to initialize the cluster (just once), or perhaps to correct an errant input or reflect a change of mode. *Rule: For highly interactive panels, don't use a cluster as an input and output element.*

Clusters are assembled on the diagram by using either the Bundle function (Fig. 2.32) or the Bundle By Name function (Fig. 2.33). The data types that you connect to these functions must match the data types in the destination cluster (numeric types are polymorphic; for instance, you can safely connect an integer type to a floating-point type). The Bundle function has one further restriction: The elements must be connected in the proper order. There is a pop-up menu available on the cluster border called Reorder Controls in Cluster...that you use to set the ordering of elements. You must carefully watch cluster ordering. Two otherwise identical clusters with different element orderings can't be connected. An exception to this rule, one which causes bugs that are difficult to trace, occurs when the misordered elements are of similar data types (for instance, all are numeric). You can legally connect the misordered clusters, but element A of one cluster may actually be passed to element B of the other. This blunder is far too common and is one of the reasons for using Bundle By Name.

To disassemble a cluster, you can use the **Unbundle** or the **Unbundle By Name** function. When you create a cluster control, give each element a reasonably short name. Then, when you use Bundle By Name or Unbundle By Name, the name doesn't take up too much space on the diagram. There is a pop-up menu on each of these functions (**Select Item**) with which you select the items to access. Named access has the additional advantage that adding an item to the related cluster control or indicator doesn't break any wires as it does with the unnamed method. *Rule: Always use named cluster access except when there is some compelling reason not to.*

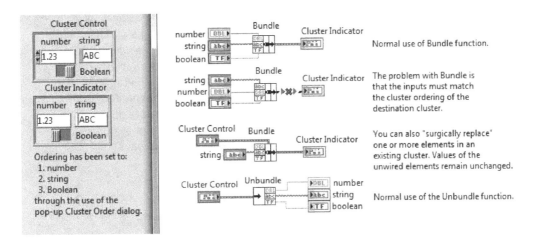

FIG. 2.32 *Using the Bundle and Unbundle functions on clusters.*

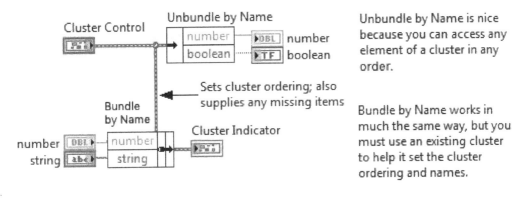

FIG. 2.33 *Use Bundle By Name and Unbundle By Name in preference to their no-name counterparts.*

When you use Bundle By Name, its middle terminal must be wired. The functions of the middle terminal on Bundle By Name are (1) to determine the element ordering and the data types and (2) to set the item names. Even if you have wired all input elements, you must wire the middle terminal because the input elements determine only the data types. The Bundle function does not have this limitation; you need only wire to its middle terminal when you wish to access a limited set of a cluster's elements.

Clusters are often incorporated into arrays (we call them cluster arrays), as shown in Fig. 2.34. It's a convenient way to package large collections of data such as I/O configurations where you have many different pieces of data to describe a channel in a cluster and many channels (clusters) in an array. Cluster arrays are also used to define many of the graph and chart types in LabVIEW. Figure 2.34 also shows the LabVIEW equivalent of an array of arrays, which is implemented as an array of clusters of arrays. Note the difference between this construct and a simple 2D array, or matrix. Use these arrays of cluster arrays when you want to combine arrays with different sizes; multidimensional LabVIEW arrays are always rectangular.

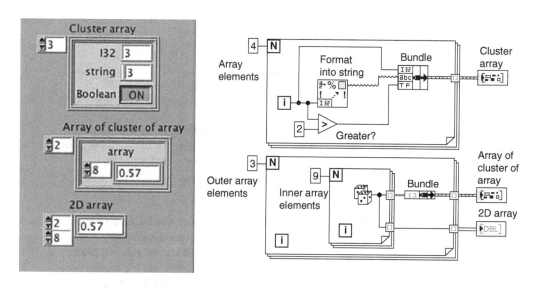

FIG. 2.34 *Building a cluster array (top loop) and building an array of clusters that contain an array (bottom loop). This is distinctly different from a 2D array.*

Waveforms

The **Waveform** is a handy data type. It's a sensible grouping of information that describes the very common situation of a one-dimensional time-varying waveform. The waveform type is similar to a cluster containing a 1D numeric array (the data), timing information (start time and sample interval), and a variant part containing user-definable items, such as the channel's name, units, and error information. Native waveform controls and indicators are available from the I/O control palette, but other indicators, such as graphs, will also adapt directly to waveforms. In the

Programming palette, there's a subpalette, called Waveform, that includes a variety of basic operations, such as building, scaling, mathematical operations, and so forth, plus higher-level functions, such as waveform generation, measurements, and file I/O.

Figure 2.35 shows some of the basic things you might do with a waveform. Waveforms can come from many sources, among them the Waveform Generation VIs, such as the Sine Waveform VI that appears in the figure. The analog data acquisition (DAQ) VIs also handle waveforms directly. Given a waveform, you can either process it as a unit or disassemble it for other purposes. In this example, we see one of the Waveform Measurement utilities, the Basic Averaged DCRMS VI, which extracts the dc and rms ac values from a signal. You can explicitly access waveform components with the Get Waveform Components function, which looks a lot like Unbundle By Name. To assemble a waveform data type from individual components, use the Build Waveform function. Not all inputs need be supplied; quite often, the starting time (t0) can be left unwired, in which case it will be set to zero.

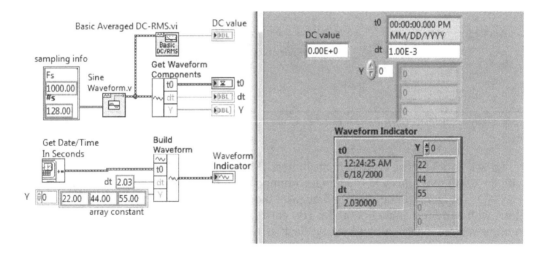

FIG. 2.35 *Here are the basics of the waveform data type. You can access individual components or use the waveform utility VIs to generate, analyze, or manipulate waveforms.*

Thanks to polymorphism, waveforms can be wired to many ordinary functions, but there are some limits and special behavior, as shown in Fig. 2.36. To add a constant offset to each data value in a waveform, you can use the regular Add node.

Another common operation is to sum the data for two waveforms (for instance, to add uniform noise to a synthetic waveform). In this case, the Add node doesn't do what you expect. The result of adding a waveform and a numeric array is counterintuitive: You end up with an array of waveforms with the same number of waveforms as there are elements in the numeric array. Instead, you must use the Build Waveform function to promote your numeric array to the waveform type, and then you use the Add node. Another method is to unbundle the Y array by using Get Waveform Components, add the two arrays with the Add function, and reassemble with Build Waveform. LabVIEW correctly handles the situation where you use the Add node to directly add two waveforms. Both waveforms must have the same dt values. The resulting waveform retains the t0 and attributes of the waveform connected to the upper terminal of the Add node.

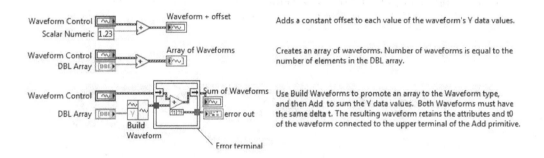

FIG. 2.36 *Polymorphism permits standard operations on waveforms.*

So what about these waveform attributes? That component is primarily for use with the DAQ functions when you have defined channel names and units through the DAQ configuration utilities. In that case, the channel information is passed along with acquired analog data. But you can also insert your own information into this **variant** data field, and read it back as well. Variant data types were originally added to LabVIEW to handle the complicated data that is sometimes required by **ActiveX** objects. Like a cluster, a variant is an arbitrary collection of items, but a variant lets you change the data type (add or delete items) while a VI is running.

In Fig. 2.37, we use Set Waveform Attribute to create a new attribute called My Attribute that is a cluster with a number and a string inside. This information rides along with that waveform until it is either modified or deleted. If you use one of the waveform file I/O utilities, all attributed data will be stored in the file for you. You can also display it in a waveform indicator, as we have in this figure. Pop up on the waveform indicator, and under Visible Items you can choose Attributes. To read the attribute information back into your program, use the Get Waveform Attribute function. If you supply the name of an existing attribute, the function returns exactly what you asked for. If you don't supply a name, it returns all available attributes. In either case, the information comes out in the form of a variant. To deal with that, use the Variant To Data function from the Cluster and Variant palette. This little fellow converts arbitrary variant data to whatever type you wish. In this case, the type is set to a cluster of the right kind, and the information you put in comes back out as desired. What if you give it the wrong data type? Conversion fails, and Variant To Data returns an error cluster that tells you what happened.

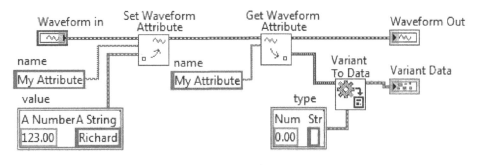

FIG. 2.37 *The stealth component of waveforms, called attributes, can be accessed when you need them. They are variant data types, which require some care during access.*

Data-Type Conversions

Thanks to polymorphism, you can usually wire from source to destination without much worry since the functions adapt to the kind of data that you supply. For instance, in the top part of Fig. 2.38, a constant is added to each value in an array. That's a wonderful time saver compared with having to create a loop to iterate through each element of the array. There are, of course, limits to polymorphism, as the bottom example in Fig. 2.38 shows. The result of adding a boolean to a string is a little hard to define, so the natural polymorphism in LabVIEW doesn't permit these operations. But what if you actually needed to perform such an operation? That's where conversions and type casting come in.

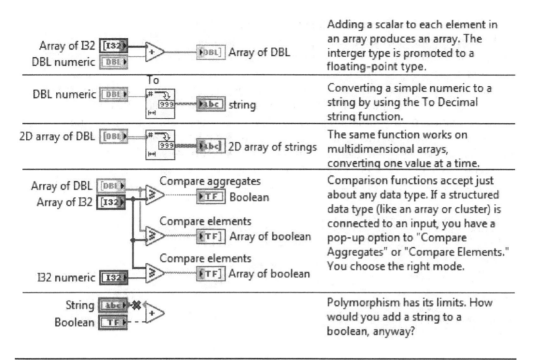

FIG. 2.38 *Polymorphism in action. Think how complicated this would be if the functions didn't adapt automatically. The bottom example shows the limits of polymorphism. How in the world could you add a boolean or a string? If there were a way, it would be included in LabVIEW.*

Conversion and Coercion

Data in LabVIEW has two components, the **data** and **type descriptor**. You can't see the type descriptor; it is used internally to give LabVIEW directions on how to handle the associated data—that's how polymorphic functions know what kind of data is connected. A type descriptor identifies the type of the data (such as a DBL floating-point array) and the number of bytes in the data. When data is **converted** from one type to another, both the data component and the type descriptor are modified in some fashion. For example, [I32]—[DBL]—[DBL], where an I32 (signed integer with 32 bits, or 4 bytes) is converted to a DBL (64-bit, or 8-byte, floating point), the value contained in the I32 is changed into a *mantissa* and an *exponent*. The type descriptor is changed accordingly, and this new data type takes up a bit more memory. For conversion between scalar numeric types, the process is very simple, generally requiring only one CPU instruction. But if aggregate types are involved (strings, clusters, etc.), this conversion process takes some time. First, the value has to be interpreted in some way, requiring that a special conversion program be run. Second, the new data type may require more or less memory, so the system's memory manager may need to be called. By now you should be getting the idea that conversion is something you may want to avoid, if only for performance reasons.

Conversion is explicitly performed by using one of the functions from the Conversion menu. They are polymorphic, so you can feed them scalars (simple numbers or booleans), arrays, clusters, and so on, as long as the input makes some sense. There is another place that conversions occur, sometimes without you being aware. When you make a connection, sometimes a little red dot appears at the destination's terminal: [I32] ⟶ [DBL]. This is called a coercion dot, and it performs exactly the same operation as an explicit conversion function. One other warning about conversion and coercion: *Be wary of lost precision.*

A DBL or an EXT floating point can take on values up to 10^{-237} or thereabouts. If you converted such a big number to an unsigned byte (U8), with a range of only 0–255, then clearly the original value could be lost. It is generally good practice to modify numeric data types to eliminate coercion because it reduces memory usage and increases speed. Use the **Representation** pop-up menu item on controls, indicators, and diagram constants to adjust the representation.

Intricate Conversions and Type Casting

Besides simple numeric type conversions, there are some more advanced ones that you might use in special situations. Figure 2.39 uses an intertype conversion to make a cluster of booleans into an array. Cluster To Array works on any cluster that contains only controls of the same type (you can't arbitrarily mix strings, numerics, etc.). Once you have an array of booleans, the figure shows two ways to find the element that is True. In the upper solution, the Boolean Array To Number function returns a value in the range of 0 to $2^{32} - 1$ based on the bit pattern in the boolean array. Since the bit that is set must correspond to a power of 2, you then take the $\log_2$ of that number, which returns a number between 0 and 32 for each button and −1 for no buttons pressed. Pretty crafty, eh? But the bottom solution turns out to be faster. Starting with the boolean array, use Search 1D Array to find the first element that is true. Search 1D Array returns the element number, which is again a number between 0 and 32, or a large negative number for no buttons.

This number could then be passed to the selection terminal in a Case Structure to take some action according to which switch was pressed.

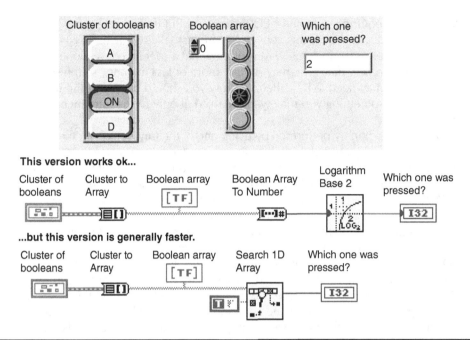

FIG. 2.39 *A cluster of booleans makes a nice user-interface item, but it's hard to interpret. Here are two solutions that find the true bit. The bottom solution turned out to be the faster.*

Figure 2.40 shows a way to use a set of Ring Indicators in a cluster as a status indicator. An array of I32 numerics is converted to a cluster by using the Array To Cluster function. A funny thing happens with this function. How does LabVIEW know how many elements belong in the output cluster (the array can have any number of elements)? For this reason, a pop-up item on Array To Cluster, called Cluster Size, was added. You have to set the number to match the indicator (5, in this case) or you'll get a broken wire.

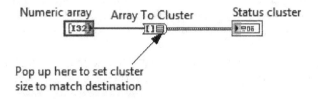

FIG. 2.40 *A numeric array is converted to a cluster of Ring Indicators (which are of type I32) by using Array To Cluster. Remember to use the pop-up item on the conversion function called Cluster Size for the number of cluster elements. In this case, the size is 5.*

One of the most powerful ways to change one data type to another is **type casting**. As opposed to conversions, type casting changes only the type descriptor. *The data component is unchanged.* The data is in no way rescaled or rearranged; it is merely interpreted in a different way. The good news is that this process is very fast, although a new copy of the incoming data has to be made, requiring a call to the memory manager. The bad news is that you have to know what you're doing! Type casting is a specialized operation that you will very rarely need, but if you do, you'll find it on the **Data Manipulation** palette. LabVIEW has enough polymorphism and conversion functions built in that you rarely need the Type Cast function.

The most common use for the Type Cast function is shown in Fig. 2.41, where a binary data string returned from an oscilloscope is type-cast to an array of integers. Notice that some header information and a trailing character had to be removed from the string before casting. Failure to do so would leave extra garbage values in the resultant array. Worse yet, what would happen if the incoming string were off by 1 byte at the beginning? Byte pairs, used to make up I16 integers, would then be incorrectly paired. The results would be very strange. Note that the Type Cast function will accept most data types except for clusters that contain arrays or strings.

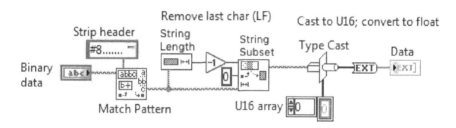

FIG. 2.41 *Lifted from the HP 54510 driver, this code segment strips the header off a data string, removes a trailing linefeed character, then type-casts the data to a U16 integer array, which is finally converted to an EXT array.*

Warning: The Type Cast function expects a certain byte ordering, namely **big-endian**, or most significant byte first, to guarantee portability between platforms running LabVIEW. But there are problems interpreting this data *outside* of LabVIEW. Big-endian is the normal ordering for the macOS and UNIX, but not so on the PC! This is an example where your code, or the data you save in a binary file, may not be machine-independent. Indeed, there is much trouble in type casting land, and you should try to use polymorphism and conversion functions whenever possible.

If you get into serious data conversion and type casting exercises, be careful, be patient, and prepare to explore the other data conversion functions in the Data Manipulation palette. Sometimes, the bytes are out of order, as in the PC versus Mac situation. In that case, try **Swap Bytes** or **Swap Words**, and display the data in a numeric indicator with its format set to hexadecimal, octal, or binary (use the pop-up menu **Format And Precision**). Some instruments go so far as to send the data backward; that is, the first value arrives last. You can use **Reverse Array** or **Reverse String** to cure that nasty situation. **Split Number** and **Join Number** are two other functions that allow you to directly manipulate the ordering of bytes in a machine-independent manner. There is another function called Coerce to Type. It is a combination of coerce and cast. The node coerces the data in the input to the data type on its top input and then it is typecast to that type. It comes handy when wanting to rename user event references or coerce a numeric to an enumerator. This function is supported in LabVIEW 2018 and later, available via Quick Drop, and in the Functions>Numeric>Conversion palette.

Flatten To String (… Do *what?*)

We mentioned before that the Type Cast function can't handle certain complicated data types. That's because LabVIEW stores strings and arrays in *handle blocks,* which are discontiguous segments of memory organized in a tree structure. When such data types are placed in a cluster, the data may be physically stored in many areas of memory. Type Cast expects all the data to be located in a single contiguous area so that it can perform simple and fast transformation.

Occasionally, you need to transmit arbitrarily complex data types over a serial link. The link may be a serial port, a network connection using a protocol such as TCP/IP, or even a binary file on disk, which is, in fact, a serial storage method. Type Cast can't do it, but the Flatten To String function can. What this function does is copy all discontiguous data into one contiguous buffer called the data string. The data string also contains embedded header information for nonscalar items (strings and arrays), which are useful when you are trying to reconstruct flattened data. Figure 2.42 shows Flatten To String in action, along with its counterpart, Unflatten From String. The data string could be transmitted over a network or stored in a file and then read and reconstructed by Unflatten From String. As with all binary formats, you must describe the underlying data format to the reading program. Unflatten From String requires a data type to properly reconstruct the original data.

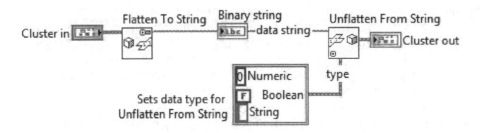

FIG. 2.42 *Use flattened data when you need to transmit complicated data types over a communications link or store them in a binary file. LabVIEW lets you set byte order and optionally prepend the data string's size.*

The most common uses for flattened data are transmission over a network or storage to a binary file. A utility VI could use this technique to store and retrieve clusters on disk as a means of maintaining front panel setup information.

Enumerated Types (Enums)

A special case of an integer numeric is the enum, or enumerated type. It's a notion borrowed from C and Pascal where each consecutive value of an integer, starting with zero, is assigned a name. For instance, red, green, and blue could correspond to 0, 1, and 2. You create an enum by selecting an enum control from the Ring and Enum control palette and then typing in the desired strings. The handiest use for an enum is as a selector in a Case Structure because the name shows up in the header of the Case, thus making your program self-documenting (Fig. 2.43). Enums can be compared with other enums, but the comparisons are actually based on the integer values behind the scenes, not the strings that you see. So you don't need to do any special conversion to compare two dissimilar enums or to compare an enum with an integer.

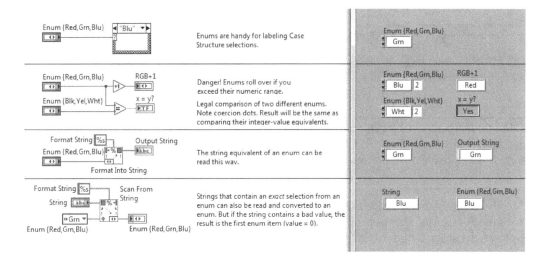

FIG. 2.43 *Enumerated types (enums) are integers with associated text. Here are some useful tips for comparison and conversion.*

Figure 2.43 shows some precautions and tricks for enums. One surprise comes when you attempt to do math with enums. They're just integers, right? Yes, but their range is limited to the number of enumerated items. In this example, there are three values {Red, Grn, Blu}, so when we attempt to add one to the maximum value (Blu, or 2), the answer is Red—the integer has rolled over to 0.

You can also do some nonobvious conversions with enums, as shown in the bottom of Fig. 2.43. Given an enum, you can extract the string value by using **Format Into String** with a format specifier of %s.

Similarly, you can match an incoming string value with the possible string equivalents in an enum by using **Scan From String**. There is a risk of error here, however: The incoming string has to be an exact case-sensitive match for one of the possible enum values. Otherwise, Scan From String will return an error, and your output enum will be set to the value of the input enum. To distinguish this result from a valid conversion, you could add a value to your enum called *Invalid* and handle that value appropriately.

Get Carried Away Department

Here's a grand finale for conversion and type casting. Say that you need to fix the length of a string at three characters and add on a null (zero) terminator. Gary actually had to do this for compatibility with another application that expected a C-style string, which requires the addition of a null terminator. Let's use everything we know about the conversion functions and do it as in Fig. 2.44.

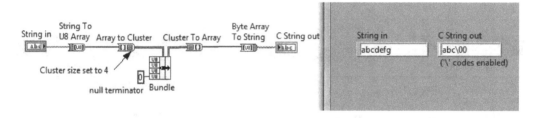

FIG. 2.44 *Making a fixed-length, null-terminated string for compatibility with the C language. The key is the Array To Cluster conversion that produces a fixed number of elements. This is a really obscure example.*

This trick (courtesy of Rob Dye, a member of the LabVIEW development team) uses the Set Cluster Size pop-up setting in an **Array To Cluster** conversion function to force the string length to be four characters, total. The bundler allows parallel access to the whole string, so it's easy to change one character, in this case, the last one. It turns out that this program is very fast because all the conversions performed here are actually type casting operations underneath. The disadvantage of this technique, as compared to using a few string functions, is that it is hard to understand.

Timing

Software-based measurement and control systems have requirements for timing and synchronization that are distinctly different from those of ordinary applications. If you're writing an application that serves only a human operator (such as a word processor), chances are that your only timing requirement is to be fast enough that the user doesn't have to sit and wait for the computer to catch up. Exactly when an event occurs is not so important. But in a control system, there's a physical process that demands regular service to keep those boiling kettles from overflowing. Similarly, a data acquisition system needs to make measurements at highly regulated intervals, lest the analysis algorithms get confused. That's why LabVIEW has built-in timing functions to measure and control time. Depending on your needs, LabVIEW's timing functions can be simple and effective or totally inadequate. And the problems are not all the fault of LabVIEW itself; there are fundamental limitations on all general-purpose computer systems with regard to real-time response, whatever that means (1 second? 0.01 second? 1 nanosecond?). Most applications we've seen work comfortably with the available LabVIEW time measurements that resolve in milliseconds, and many more operate with 1-second resolution. A few applications demand submillisecond resolution and response time, which is problematic owing primarily to operating system issues. Those cases require special attention. So, what's with all this timing stuff, anyway?

Where Do Little Timers Come From?

When LabVIEW was born on the Macintosh, the only timer available was based on the 60-Hz line frequency. Interrupts to the CPU occurred every 1/60 second and were counted as long as the Mac was powered on. The system timer on modern computers is based on a crystal-controlled oscillator on the computer's motherboard. Special registers count each tick of the oscillator. The operating system abstracts the tick count for us as milliseconds and microseconds. It's funny how we instrumentation specialists benefited from the consumer-oriented multimedia timing. Without the need for high resolution time for audio and video, we may well have been stuck at 60 Hz. One might argue that all those ultrafast video accelerators and even the speed of CPUs are driven primarily by the enormous video game market!

Three fine-resolution timers are available within LabVIEW on all platforms: Wait (ms), Wait Until Next ms Multiple, and Tick Count (ms). These functions attempt to resolve milliseconds within the limitations of the operating systems. Two new timing functions, Timed Loops and Timed Sequence Structure, are deterministic programming structures similar to While Loops and Sequence Structures. The determinism and resolution of these two structures are operating system and hardware platform dependent. On a PC running Windows, the Timed Loop has a resolution of 1 millisecond, but under a real-time operating system (RTOS) resolution can be 1 microsecond. Timed Loops also have other exciting features such as the ability to synchronize to a data acquisition task.

Another kind of timer that is available in LabVIEW is the system clock/calendar, which is maintained by a battery-backed crystal oscillator and counter chip. Most computers have a clock/calendar timer of this kind. The timing functions that get the system time are Get Date/Time In Seconds, Get Date/Time String, and Seconds

To Date/Time. These functions are as accurate as your system clock. You can verify this by observing the drift in the clock/calendar displayed on your screen. Unless you synchronize to a network time source, it's probably within a minute per month or so.

Are you curious about the timing resolution of your system? Then write a simple timer test VI like the one shown in Fig. 2.45. A For Loop runs at maximum speed, calling the timing function under test. The initial time is subtracted from the current time, and each value is appended to an array for plotting. The system in Fig. 2.45 has a resolution of 1 millisecond and took approximately 100 nanoseconds per iteration.

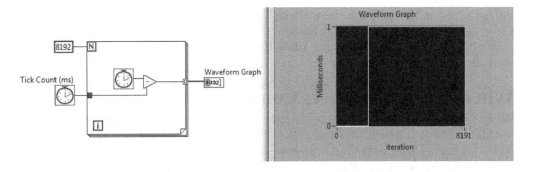

FIG. 2.45 *Write a simple timer test VI to check the resolution of your system's LabVIEW ticker.*

Using the Built-In Timing Functions

There are two things you probably want to do with timers. First, you want to make things happen at regular intervals. Second, you want to record when events occur.

Intervals

Traditionally, if you want a loop to run at a nice, regular interval, the function to use is Wait Until Next ms Multiple. Just place it inside the loop structure, and wire it to a number that's scaled in milliseconds. It waits until the tick count in milliseconds becomes an exact multiple of the value that you supply. If several VIs need to be synchronized, this function will help there as well. For instance, two independent VIs can be forced to run with harmonically related periods, such as 100 millisecond and 200 millisecond, as shown in Fig. 2.46. In this case, every 200 milliseconds, you would find that both VIs are in sync. The effect is exactly like the action of a metronome and a group of musicians—it's their heartbeat. This is not the case if you use the simpler Wait (ms) function; it just guarantees that a certain amount of time has passed, without regard to absolute time. Note that both of these timers work by adding activity to the loop. That is, the loop can't go on to the next cycle until everything in the loop has finished, and that includes the timer. Since LabVIEW executes everything inside the loop in parallel, the timer is presumably started at the same time as everything else. Note that "everything else" must be completed in less time than the desired interval. The timer does not have a magic ability to speed up or abort other functions. Wait Until Next ms Multiple will add on extra delay as necessary to make the tick count come out even, but not if the other tasks overrun the desired interval.

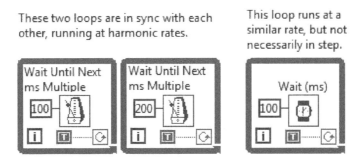

FIG. 2.46 *Wait Until Next ms Multiple (the two left loops) keeps things in sync. The right loop runs every 100 ms, but may be out of phase with the others.*

Both of these loop timers are asynchronous; that is, they do not tie up the entire machine while waiting for time to pass. Any time a call to Wait (ms) is encountered, LabVIEW puts the calling VI into a scheduler queue until the timer expires. If you want your loop to run as fast as possible but not hog all the CPU, place a Wait (ms) with 0 millisecond to wait. The 0-millisecond wait is a special instruction to the LabVIEW scheduler to let other waiting tasks run, but to put yours back in the schedule as soon as possible. This guarantees that other VIs have a chance to run, but still executes your loop as fast as possible. *Rule: Always put a timer in every loop structure. Otherwise, the loop may lock out all other VIs, and even interfere with the responsiveness of the LabVIEW user interface.* The exception to this rule occurs when you know there is some other activity in the loop that will periodically cause LabVIEW to put that part of the diagram in the scheduler queue. For instance, file I/O and many DAQ operations have hidden delays or other scheduling behavior that meets this requirement.

Timed Structures

There are three programming structures in LabVIEW with built-in scheduling behavior: a **Timed Loop**, a **Timed Sequence**, and a combination of the two called a **Timed Loop With Frames**. Timed Loops are most useful when you have multiple parallel loops you need to execute in a deterministic order. LabVIEW's default method of executing code is multiple threads of time-sliced multitasking execution engines with a dash of priorities. In other words, your code will appear to run in parallel, but it is really sharing CPU time with the rest of your application (and everything else on your computer). Deterministically controlling the execution order of one loop over another is an incredibly difficult task—until now.

Timed Loops are as easy to use as a While Loop and ms Timer combination. The Timed Loop in Fig. 2.47 executes every 1 millisecond and calculates the elapsed time. You configure the loop rate by adjusting the period, or dt. You can also specify an offset (t0) for when the loop should start. With these two settings you can configure multiple loops to run side by side with the same period but offset in time from each other. How to configure the period and offset is all that most of us need to know about Timed Loops, but there is much more. You can change everything about a Timed structure's execution, except the name and the timing source, on the fly from within the structure. You can use the execution feedback from the Left Data Node in a process control loop to tell whether the structure started on time or finished late in the previous execution. Time-critical processes can dynamically tune the Timed structure's execution time and priority according to how timely it executed last time. The Left Data node even provides nanosecond-resolution timing information you can use to benchmark performance.

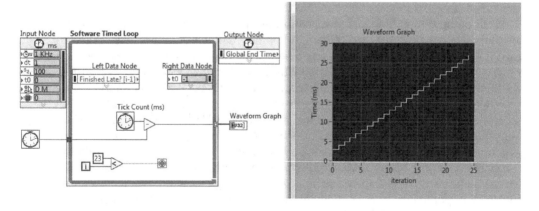

FIG. 2.47 *Timed Loops have a built-in scheduling mechanism. You can configure a Timed Loop's behavior dynamically at run-time through terminals on the Input node, or statically at edit time through a configuration dialog.*

Timing Sources

Timed structures (loops and sequences) run off a configurable timing source. Figure 2.48 shows a typical configuration dialog. The default source is a 1-kHz clock (1-millisecond timer) on a PC. On an RT system, a 1-MHz clock is available, providing a 1-*microsecond* timer. Other configuration options are the Period and Priority. In our dialog, the loop is set to execute every 1000 milliseconds by using the PC's internal 1-kHz clock. Note that you can also use a timing source connected to the structure's terminal. The timing source can be based on your computer's internal timer or linked through DAQmx to a hardware-timed DAQ event with DAQmx Create Timing Source.vi. You can even chain multiple timing sources together by using Build Timing Source Hierarchy.vi. One example in which multiple timing sources could come in handy is if you wanted a loop to execute based on time or every time a trigger occurred.

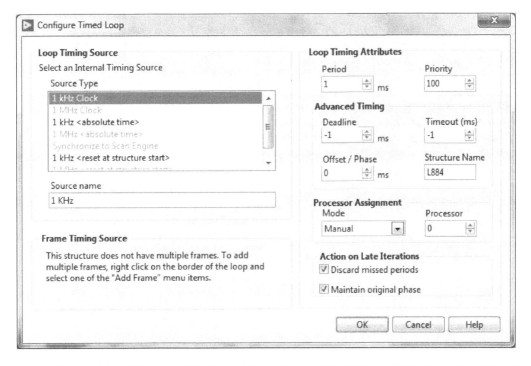

FIG. 2.48 *Configuration dialog for a Timed Loop with a single frame.*

It's important to note that once a timing source is started, it will continue to run even if the timed loop is not running! This can cause unexpected consequences when your Timed Loop starts running as fast as possible to try to catch up to its timing source. To avoid this, use the "<reset at structure start>" feature. This is important if you use a Timed Loop inside another loop. The loop needs to either reset its clock when it starts or be set to discard missed periods. Another method is to clear the timer after exiting the loop, by using Clear Timing Source.vi. This removes the timer but adds unnecessary overhead each time the timer has to be re-created.

Timed Loops With Frames and Timed Sequence structures have the unique capability of running off two timing sources. One source triggers the start of the loop, and a second frame timing source governs each frame. The example in Fig. 2.49 shows a multiframe Timed Loop with dead time inserted between each frame. See Fig. 2.50 for configuration.

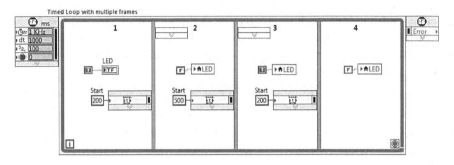

FIG. 2.49 *Frames in a multiframe Timed structure can have start times relative to the completion of the previous frame. The effect is to put dead time between each frame. Frame 2 starts 200 milliseconds after frame 1 completes, frame 3 starts 500 milliseconds after frame 2, and frame 4 starts 200 milliseconds after frame 3.*

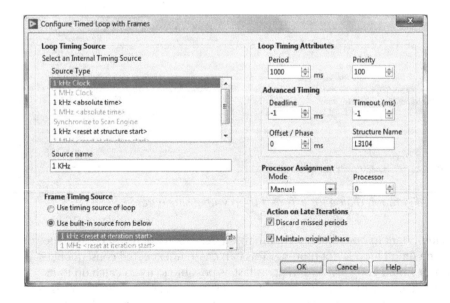

FIG. 2.50 *Multiframe structures can have two timing sources.*

Execution and Priority

Each Timed structure runs in its own thread at a priority level just beneath Time Critical and above High Priority. Timed Loops do not inherit the priority or execution system of the VI they run in. This means that a Timed structure will stop any other code from running until the structure completes. LabVIEW's Timed structure scheduler runs behind the scenes and controls execution based on each Timed structure's schedule and individual priority. Priorities are set before the Timed Loop executes, through either the configuration dialog or a terminal on the Input node. Priority can be adjusted inside the Timed Loop through the Right Data node. The Timed Loop's priority is a positive integer between 1 and 2,147,480,000. But don't get too excited, since the maximum number of Timed structures you can have in memory, running or not, is 128.

Timed structures at the same priority do not multitask between each other. The scheduler is preemptive, but not multitasking. If two structures have the same priority, then dataflow determines which one executes first. Whichever structure starts execution first will finish before the other structure can start. Each Timed Loop runs to completion unless preempted by a higher-priority Timed Loop, a VI running at Time Critical priority, or the operating system. It is this last condition that kills any hope of deterministic operation on a desktop operating system.

On a RTOS, Timed structures are deterministic, meaning the timing can be predicted. This is crucial on an RTOS; but on a non-RTOS (such as Windows), Timed structures are not deterministic. This doesn't mean that you can't use Timed structures on Windows; just be aware that the OS can preempt your structure at any time to let a lower-priority task run. A RTOS runs code based on priority (higher-priority threads preempt lower-priority threads), while a desktop OS executes code based on "fairness"; each thread gets an opportunity to run. This is great when you want to check e-mail in the background on your desktop, but terrible for a process controller. If you need deterministic execution, then use Timed structures on real-time systems such as LabVIEW RT. They are powerful and easy to use.

Timing Guidelines

All LabVIEW's platforms have built-in multithreading and preemptive multitasking that help to mitigate the effects of errant code consuming all the CPU cycles. Even if you forget to put in a timer and start running a tight little loop, every desktop operating system will eventually decide to put LabVIEW to sleep and to let other programs run for a while. But rest assured that the overall responsiveness of your computer will suffer, so please put timers in all your loops.

Here's when to use each of the timers:

- Highly regular loop timing—use Wait Until Next ms Multiple or a Timed Loop
- Many parallel loops with regular timing—use Wait Until Next ms Multiple or Timed Loops

- Arbitrary, asynchronous time delay to give other tasks some time to execute—use Wait (ms)

- Single-shot delays (as opposed to cyclic operations, such as loops)—use Wait (ms) or a Timed Sequence

Absolute Timing Functions

LabVIEW has a set of timing functions that deal with absolute time, which generally includes the date in addition to the time of day. When we are dealing with absolute time, the first thing we need to do is establish our calendar's origin. While every computer platform uses a slightly different standard origin, the LabVIEW designers have attempted to give you a common interface. They set the origin as a time-zone-independent number of seconds that have elapsed since 12:00 a.m., Friday, January 1, 1904, Coordinated Universal Time (UTC). For convenience, we call this epoch time. On your computer, you must configure your time zone (typically through a control panel setting) to shift the hours to your part of the globe. Internally LabVIEW calculates time based on UTC, but displays time according to local time. Using UTC means you can schedule an event on multiple computers to UTC without worrying about time zone offset or daylight saving time.

Epoch time values are available in two formats: as a timestamp and as a cluster known as a date time rec. Timestamp is a high-resolution, 128-bit data type capable of handling dates between 1600 A.D. and 3000 A.D. The upper quad int (64 bits) is the number of seconds since January 1, 1904, and the lower 64 bits is the fractional second. The date/time rec cluster contains nine I32 values (second, minute, hour, day, etc.) and one DBL (fractional second). The seconds value is just what we've discussed: the number of seconds since the beginning of 1904. The date/time rec cluster is useful for extracting individual time and date values, such as finding the current month.

Here is a quick introduction to the absolute time functions and their uses:

- Get Date/Time In Seconds returns a timestamp, which can be converted into the epoch seconds as a DBL float.

- Get Date/Time String returns two separate formatted strings containing time and date. It's nice for screen displays or printing in reports and has several formatting options.

- Format Date/Time String returns a single formatted string containing time and date. You supply a formatting string that allows you to extract any or all possible components of date and time. For instance, the format string %H:%M:%S%3u asks for hours, minutes, and fractional seconds with three decimal digits, with colons between the values. A typical output string would look like this:15:38:17.736.

- Seconds To Date/Time returns a date/time rec cluster. You might use this function when you want to do computations based on a particular aspect of date or time.

- Date/Time To Seconds converts the date/time rec back to seconds. It's useful when you want to combine individual time and date items, such as letting the user sift through data by the day of the week.

Figure 2.51 shows how many of these functions can be connected.

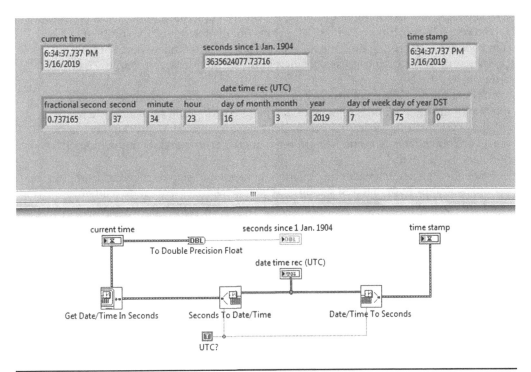

FIG. 2.51 *How to switch between the various formats for system time, particularly getting in and out of the date/time rec cluster.*

Sending Timing Data to Other Applications

Many graphing and spreadsheet programs do a lousy job of importing timestamps of the form 23-Jun-1990 10:03:17. Epoch seconds are OK, but then the other program has to know how to compute the date and time, which requires a big algorithm. Most spreadsheet programs can handle epoch seconds if you give them a little help. In Microsoft Excel, for instance, you must divide the epoch seconds value by 86,400, which is the number of seconds in 1 day. The result is what Excel calls a

Serial Number, where the integer part is the number of days since the zero year and the fractional part is a fraction of 1 day. Then you just format the number as date and time. Again, watch out for the numeric precision problem if you're importing epoch seconds.

High-Resolution and High-Accuracy Timing

If your application requires higher accuracy or resolution than the built-in software timing functions can supply, then you will have to use a hardware solution. The first thing to do is to consider your specifications.

Some experiments simply need a long-term stable time base, over scales of days or even years, to synchronize with other events on other systems. Or you might be concerned with measuring time intervals in the subsecond range where short-term accuracy is most important. These specifications will influence your choice of timing hardware.

For long-term stability, you might get away with setting your computer's clock with one of the network time references; such options are built into all modern operating systems. They guarantee NIST-traceable long-term stability, but there is, of course, unknown latency in the networking that creates uncertainty at the moment your clock is updated.

Excellent time accuracy can be obtained by locking to signals received from the Global Positioning System (GPS) satellites, which provide a 1-Hz clock with an uncertainty of less than 100 nanoseconds and essentially no long-term drift. NI offers the NI-9467 C Series Synchronization Module for CompactDAQ and CRIO.

Perhaps the easiest way to gain timing resolution is to use a National Instruments data acquisition (DAQ) board as a timekeeper. All DAQ boards include a reasonably stable crystal oscillator time base with microsecond (or better) resolution that is used to time A/D and D/A conversions and to drive various countertimers. Through the DAQ VI library, you can use these timers to regulate the cycle time of software loops, or you can use them as high-resolution clocks for timing short-term events. This DAQ solution is particularly good with the onboard counter/timers for applications such as frequency counting and time interval measurement.

But what if the crystal clock on your DAQ board is not good enough? Looking in the NI catalog at the specifications for most of the DAQ boards, we find that the absolute accuracy is ±0.01 percent, or ±100 ppm, with no mention of the temperature coefficient. Is that sufficient for your application? For high-quality calibration work and precision measurements, probably not. It might even be a problem for frequency determination using spectral estimation. In the LabVIEW analysis library, there's a VI called Extract Single Tone Information that locates, with very high precision, the fundamental frequency of a sampled waveform. If the noise level is low, this VI can return a frequency estimate that is accurate to better than 1 ppm—but only if the sample clock is that accurate.

National Instruments does have higher accuracy boards. In the counter/timer line, the PXI-6608 counter/timer module has an oven-controlled crystal oscillator (OCXO) that can be adjusted to match an external reference, and it has a temperature coefficient of $1 \times 10^{-11}/°C$ and aging of 1×10^{-7} per year. With all the NI clock generators, you typically connect them to your DAQ board via real-time system integration (RTSI) bus signals or via programmable function input (PFI) lines. If you're using an external clock generator, the PFI inputs are the only way. There are several DAQ example VIs that show you how to use external clocking for analog and digital I/O operations.

Synchronization

When you build a LabVIEW application, you'll eventually find the need to go beyond simple sequencing and looping to handle all the things that are going on in the world outside your computer. A big step in programming complexity appears when you start handling events. An event is usually defined as something external to your program that says, "Hey, I need service!" and demands that service in a timely fashion. A simple example is a user clicking a button on the panel to start or stop an activity. Other kinds of events come from hardware, such as a trigger, or a message coming in from a network connection. Clearly, these random events don't fall within the simple boundaries of sequential programming.

There are lots of ways to handle events in LabVIEW, and many of them involve creating parallel paths of execution where each path is responsible for a sequential or nonsequential (event-driven) activity.

Since G is intrinsically parallel, you can create an unlimited number of parallel loops or VIs and have confidence that they'll run with some kind of timesharing. But how do you handle those "simultaneous" events in such a way as to avoid collisions and misordering of execution when it matters? And what about passing information between loops or VIs? The complexity and the list of questions just seem to grow and grow.

LabVIEW has some powerful synchronization features, and there are standard techniques that can help you deal with the challenges of parallel programming and event handling. By the way, this is an advanced topic, so if some of the material here seems strange, it's because you might not need many of these features very often (we'll start with the easier ones, so even you beginners can keep reading).

Polling

Let's start off with a straightforward event-handling technique that does not use any esoteric functions or tricks: **polling**. When you are polling, you periodically check the status of something (a flag), looking for a particular value or change of value. Polling is very simple to program (Fig. 2.52), is easy to understand and debug, is very widely used (even by highly experienced LabVIEW dudes!), and is acceptable practice for thousands of applications in all programming languages. The main drawback to polling is that it adds some overhead because of the repeated testing and looping. (If you do it wrong, it can add lots of overhead.)

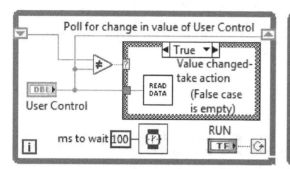 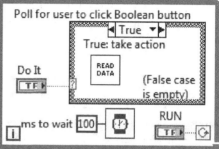

FIG. 2.52 *A simple polling loop handles a user-interface event. The left example watches for a change in a numeric control. The right example awaits a True value from the Boolean button.*

In the left example in Fig. 2.52, we're doing a very simple computation—the comparison of two values—and then deciding whether to take action. There's also a timer in the loop that guarantees that the computation will be done only 10 times per second. Now that's pretty low overhead. But what if we left the timer out? Then LabVIEW would attempt to run the loop as fast as possible—and that could be millions of times per second—and it might hog the CPU. Similarly, you might accidentally insert some complex calculation in the loop where it's not needed, and each cycle of the polling loop would then be more costly.

The other drawback to polling is that you only check the flag's state every so often. If you're trying to handle many events in rapid succession, a polling scheme can be overrun and miss some events. An example might be trying to handle individual bits of data coming in through a high-speed data port. In such cases, there is often a hardware solution (such as the hardware buffers on your serial port interface) or a lower-level driver solution (such as the serial port handler in your operating system). With these helpers in the background doing the fast stuff, your LabVIEW program can often revert to polling for larger packets of data that don't come so often.

Events

Wouldn't it be great if we had a way to handle user-interface events without polling? We do and it's called the **Event** structure. Remember, an event is a notification that something has happened. The Event structure bundles handling of user-interface events (or notifications) into one structure without the need for polling front panel controls, and with a lot fewer wires! The Event structure sleeps without consuming CPU cycles until an event occurs, and when it wakes up, it automatically executes the correct Event case. The Event structure doesn't miss any events and handles all events in the order in which they happen.

The Event structure looks a lot like a Case structure. You can add, edit, or delete events through the Event structure's pop-up menu. Events are broken down into two types: **Notify** events and **Filter** events. A Notify event lets you know that something happened and LabVIEW has handled it. Pressing a key on the keyboard can trigger an event. As a notify event, LabVIEW lets you know the key was pressed so you can use it in your block diagram, but all you can do is react to the event. Figure 2.53 shows a key down Notify event. A Filter event (Fig. 2.54) allows you to change the event's data as it happens or even discard the event. Filter events have the same name as Notify events, but end with a question mark ("?").

The Event structure can save a lot of time and energy, but it can also drive you crazy if you aren't alert. The default configuration of all events is to lock the front panel until the Event case has finished execution. If you put any time-consuming processing inside the Event structure, your front panel will lock up until the event completes. Rule: Handle all but the most basic processing external to the Event structure. Later in this chapter we'll show you a powerful design pattern you can use to pass commands from an event loop to a parallel processing loop with queues.

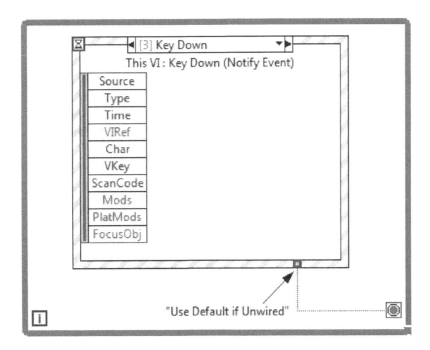

FIG. 2.53 *Notify event. Configured to trigger a notification when any key is pressed.*

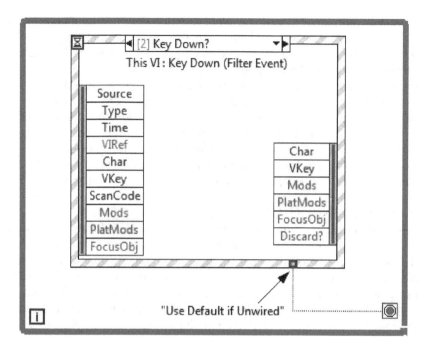

FIG. 2.54 *Filter event. This triggers a notification when any key is pressed, but the key press can be modified or discarded.*

Use events to enhance dataflow, not bypass it. This is especially true with the mechanical action of booleans. A latching boolean does not reset until the block diagram has read its value. If your latching boolean is outside floating on the block diagram, it will never reset. Be sure to place any latching booleans inside their Event case so that their mechanical action will be correct. Changing a control's value through a local variable will not generate an event—use the property "Value (Signaling)" to trigger an event.

The advanced topic of **Dynamic Events** allows you to define and dynamically register your own unique events and pass them between subdiagrams. Dynamic events can unnecessarily complicate your application, so use them only when needed and make sure to document what you are doing and why. Figure 2.55 illustrates configuring a dynamic event to trigger a single Event case for multiple controls. This eliminates manually configuring an Event case for each control—a tedious task if you have tens, or even hundreds, of controls. Property nodes are used to get the control reference to every control on every page of a tab control. The 1D array of references is then used to register all the controls for a value change event. Whenever any control's value is changed, a single Event case will be fired. Inside the Event case we have a reference to the control and its data value as a variant. Also note that all the controls are inside the Event case.

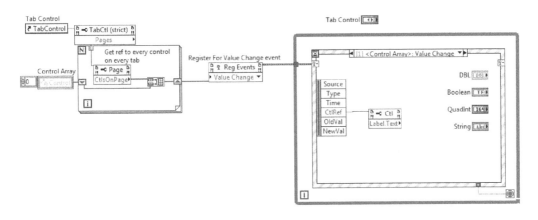

FIG. 2.55 *Use control references for dynamic event registration. A single Event case handles multiple Value Change events.*

Occurrences

An **occurrence** is a synchronization mechanism that allows parallel parts of a LabVIEW program to notify each other when an event has occurred. It's a kind of software trigger. The main reason for using occurrences is to avoid polling and thus reduce overhead. An occurrence does not use any CPU cycles while waiting. For general use, you begin by calling the **Generate Occurrence** function to create an occurrence refnum that must be passed to all other occurrence operations.

Then you can either wait for an occurrence to happen by calling the **Wait On Occurrence** function or use **Set Occurrence** to make an event happen. Any number of Wait On Occurrence nodes can exist within your LabVIEW environment, and all will be triggered simultaneously when the associated Set Occurrence function is called.

Figure 2.56 is a very simple demonstration of occurrences. First, we generate a new occurrence refnum and pass that to the Wait and Set functions that we wish to have related to one another. Next, there's our old friend the polling loop that waits for the user to click the Do It button. When that happens, the occurrence is set and, as if by magic, the event triggers the Wait function, and a friendly dialog box pops up. The magic here is the transmission of information (an event) through thin air from one part of the diagram to another. In fact, the Wait function didn't even have to be on the same piece of diagram; it could be out someplace in another VI. All it has to have is the appropriate occurrence refnum.

132 LabVIEW Graphical Programming

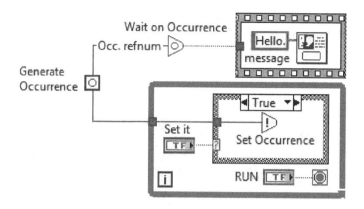

FIG. 2.56 *A trivial occurrence demonstration. The lower loop sets the occurrence when the user clicks the Set It button, causing the Wait On Occurrence function to trigger the dialog box. Terminating the While Loop without setting the occurrence causes the Wait On Occurrence function to wait forever.*

It appears that, as with global variables, we have a new way to violate dataflow programming, hide relationships among elements in our programs, and make things occur for no apparent reason in unrelated parts of our application. Indeed we do, and that's why you must always use occurrences with caution and reservation. As soon as the Set and Wait functions become dislocated, you need to type in some comments explaining what is going on.

Well, things seem pretty reasonable so far, but there is much more to this occurrence game. First, there is a **ms time-out** input on the Wait On Occurrence function that determines how long it will wait before it gets bored and gives up. The default time-out value, −1, will cause it to wait forever until the occurrence is set. If it times out, its output is set to True. This can be a safety mechanism of a sort, when you're not sure if you'll ever get around to setting the occurrence and you want a process to proceed after a while. A very useful application for the time-out is an **abortable wait**, first introduced by Lynda Gruggett (LabVIEW Technical Resource, vol. 3, no. 2). Imagine that you have a While Loop that has to cycle every 100 seconds, as shown in the left side of Fig. 2.57. If you try to stop the loop by setting the conditional terminal to False, you may have to wait a long time. Instead, try the abortable wait (Fig. 2.57, right).

To implement the abortable wait, the loop you want to time and abort is configured with a Wait On Occurrence wired to a time-out value. Whenever the time-out period has expired, the Wait returns True, and the While Loop cycles again, only to be suspended once again on the Wait. When the occurrence is set (by the lower loop, in this example), the Wait is instantly triggered and returns False for the time-out flag, and its loop immediately stops.

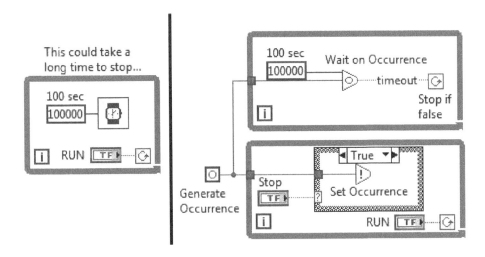

FIG. 2.57 *An abortable wait (right) can be created from a Wait On Occurrence function by using the time-out feature.*

Now for the trickiest feature of the Wait On Occurrence function: the **ignore previous** input. It's basically there to allow you to determine whether to discard occurrence events from a previous execution of the VI that it's in. Actually, it's more complicated than that, so here is a painfully detailed explanation in case you desperately want to know all the gory details.

Each Wait On Occurrence function "remembers" what occurrence it last waited on and at what time it continued (either because the occurrence triggered or because of a time-out). When a VI is loaded, each Wait On Occurrence is initialized with a nonexisting occurrence. When a Wait On Occurrence is called and ignore previous is False, there are four potential outcomes:

1. The occurrence has never been set. In this case, Wait On Occurrence simply waits.

2. The occurrence has been set since this Wait On Occurrence last executed. In this case, Wait On Occurrence does not wait.

3. The occurrence was last set before this Wait last executed, and the last time this Wait was called, it waited on the same occurrence. Wait On Occurrence will then wait.

4. The occurrence was last set before this Wait last executed, but the last time this Wait was called, it waited on a different occurrence. In this case, Wait will not wait!

The first three cases are pretty clear, but the last one may seem a bit strange. It will arise only if you have a Wait On Occurrence inside a loop (or inside a reentrant VI in a loop) and it waits on different occurrences (e.g., out of an array). This can also arise if it is inside a non-reentrant VI and the VI is called with different occurrences. Fortunately, these obscure cases generally don't happen.

Wait On Occurrence behaves in this way due to its implementation. Each occurrence "knows" the last time it was triggered, and each Wait On Occurrence remembers the occurrence it was last called with and what time it triggered (or timed out). When Wait On Occurrence is called and ignore previous is False, it will look at its input. If the input is the same occurrence as last time, it will look at the time of the last firing and wait, depending on whether the time was later than the last execution. If the input is not the same as last time, it will simply look at the time and wait, depending on whether it has ever been fired.

After you've tried a few things with occurrences, you'll get the hang of it and come up with some practical uses. Some experienced LabVIEW users have written very elaborate scheduling and timing programs based on occurrences (far too complex to document here). In essence, you use sets of timers and various logical conditions to determine when occurrences are to be triggered. All your tasks that do the real work reside in loops that wait on those occurrences. This gives you centralized control over the execution of a large number of tasks. It's a great way to solve a complex process control situation. Just hope you don't have to explain it to a beginner.

To sum up our section on occurrences, here is a quote from Stepan Rhia, a former top-notch LabVIEW developer at NI:

> *One may think that occurrences are quirky, a pain to use, and that they should be avoided. One might be right! They are very low level and you often have to add functionality to them in order to use them effectively. In other words, they are not for the faint of heart. Anything implemented with occurrences can also be implemented without them, but maybe not as efficiently.*

Notifiers

As Stepan Rhia indicated, adding functionality around occurrences can make them more useful. And that's exactly what a notifier does: It gives you a way to send information along with the occurrence trigger. Although notifiers still have some of the hazards of occurrences (obscuring the flow of data), they do a good job of hiding the low-level trickery.

To use a notifier, follow these general steps on your diagram. First, use the **Create Notifier VI** to get a notifier refnum that you pass to your other notifier VIs. Second, place a **Wait On Notification VI** in a loop or wherever you want your program to be suspended until the notification event arrives along with its data.

Third, call the **Send Notification VI** to generate the event and send the data. When you're all done, call **Destroy Notifie**r to clean up.

Here's an example that uses notifiers to pass data when an alarm limit is exceeded. We'll follow the numbered zones in Fig. 2.58 to see how it works.

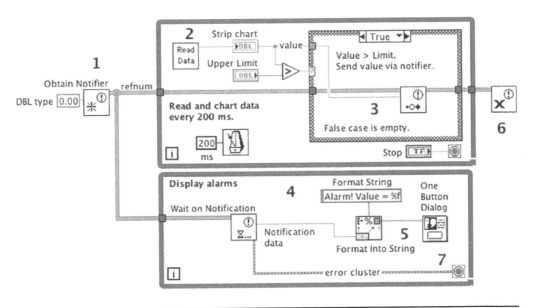

FIG. 2.58 *A notifier sends data from the upper loop to the lower one when an alarm limit is exceeded.*

1. A new notifier is created. Its refnum is passed to two While Loops. The two loops don't have to reside on the same diagram; they just need to have the refnum passed to them, perhaps through a global variable.

2. The upper loop reads a numeric value and displays it on a strip chart every 200 milliseconds. If the value exceeds the Upper Limit (alarm level), we wish to wake up the other loop and send the value along.

3. Inside the True frame of the Case Structure, the value is submitted to the Send Notification VI. Behind the scenes, that VI generates an occurrence and queues up the data.

4. The lower While Loop is waiting for the notification to be set. When it arrives, the Wait On Notification VI returns the data.

5. A message for the user is assembled, and a dialog box pops up.

6. The upper loop stops when the user clicks the Stop button, and the notifier is destroyed.

7. The lower loop stops when the Wait On Notification returns an error because the notifier no longer exists.

If you actually create and run this VI, you'll find one misbehavior. When the notifier is destroyed and the bottom loop terminates, the Wait On Notification VI returns an empty number, which is passed to the dialog box to be displayed as a value of 0; it's a false alarm. To eliminate this bug, all nodes that use the data string should be enclosed in a Case structure that tests the error cluster coming from the Wait On Notification VI. In the remainder of this chapter, you'll see that technique used as a standard practice.

Wait On Notification includes the time-out. Ignore previous features of Wait On Occurrence, and they work the same way. However, most of the time you don't need to worry about them. The only trick you have to remember is to include a way to terminate any loops that are waiting. The error test scheme used here is pretty simple. Alternatively, you could send a notification that contains a message explicitly telling the recipient to shut down.

One important point to keep in mind about notifiers is that the data buffer is only one element deep. **Notifiers are lossy.** Although you can use a single notifier to pass data one way to multiple loops, any loop running slower than the notification will miss data. Use this to your advantage in applications where you want to acquire, analyze, and control at high speed, but display and/or log data at a slower speed.

Queues

Queues function a lot like notifiers but store data in a FIFO (first in, first out) buffer. The important differences between queues and notifiers are as follows:

1. A notifier can pass the same data to multiple loops, but a notifier's data buffer is only one element deep. All notifier's writes are destructive.

2. A queue can have an infinitely deep data buffer (within the limits of your machine), but it's difficult to pass the data to more than one loop. A queue's reads are destructive.

Another important difference between queues and notifiers is the way they behave if no one reads the data. You might think that a queue with a buffer size of 1 would behave the same as a notifier. But once a fixed-size queue is filled, the **producer** loop either can time out and not insert the data or can run at the same rate as the **consumer**.

Figure 2.59 shows the alarm notification application of Fig. 2.58 rewritten using queues. Note that the application looks and acts the same, but with one exception: Because the queue size is infinite, no alarms will be missed if the user fails to press the acknowledgment button in time. Every alarm will be stored in the queue until dequeued. Use a queue when you can't afford to miss any data passed between parallel loops.

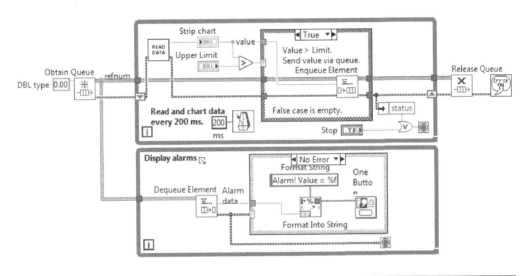

FIG. 2.59 *Alarm notification using queues. No alarms will be dropped because of user inactivity.*

The asynchronous nature of queues makes for a great way to pass messages between two loops running at different rates. Figure 2.60 shows an implementation of this using an Event structure in one loop and a message-driven state machine in the other. The **Queued Message Handler** is one of the most powerful design patterns in LabVIEW for handling user-interface-driven applications.

Semaphores

With parallel execution, sometimes you find a **shared resource** called by several VIs, and a whole new crop of problems pops up. A shared resource might be a global variable, a file, or a hardware driver that could suffer some kind of harm if it's accessed simultaneously or in an undesirable sequence by several VIs. These **critical sections** of your program can be guarded in several ways.

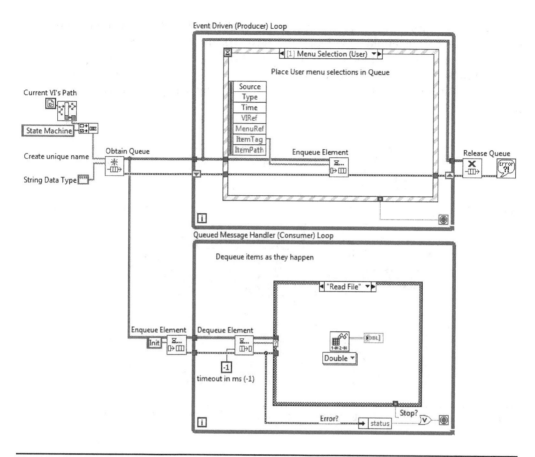

FIG. 2.60 *Queued Message Handler. Each event is placed in a queue and handled in order by the consumer loop.*

One way is to encapsulate the critical section of code in a subVI. The LabVIEW execution system says that a given subVI can only be running under the auspices of one calling VI, unless it is made **reentrant** in the VI Properties dialog. For instance, say that you have a scratch file that several VIs need to read or write. If the open-read/write-close sequence was independently placed in each of those VIs, it's possible that the file could be *simultaneously* written and/or read. What will happen then? A simple solution is to put the file access functions in a subVI. That way, only one caller can have access to the file at any moment.

Global variables are another kind of shared resource that LabVIEW programmers have trouble with. A familiar clash occurs when you're maintaining a kind of global database. Many VIs may need to access it, and the most hazardous moment comes during a read-modify-write cycle, much like the scratch file example. Here, the global variable is read, the data is modified, and then it is written back to the global variable (Fig. 2.61A). What happens if several VIs attempt the same operation? Who will be the last to write the data? This is also known as a **race condition**.

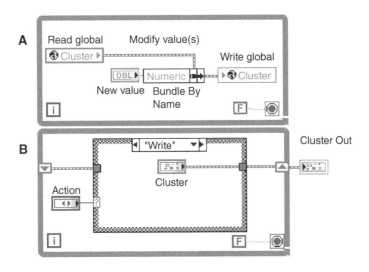

FIG. 2.61 *Global variables are a classic risk for race conditions. You can protect such data by encapsulating it in a subVI or by storing the data in shift registers.*

While built-in LabVIEW globals have no protection, you can solve the problem by encapsulating them inside subVIs, which once again limits access to one caller at a time. Or you can use global variables that are based on shift registers (Fig. 2.61B) and have all the programmed access there in a single subVI.

Computer science has given us a formal protection mechanism. A **semaphore**, also known as a **mutex** (short for mutually exclusive), is an object that protects access to a shared resource. To access a critical section, a task has to wait until the semaphore is available and then immediately set the semaphore to busy, thus locking out access by other tasks. It's also possible for semaphores to permit more than one task (up to a predefined limit) to be in a critical section.

A new semaphore is created with the **Create Semaphore VI**. It has a size input that determines how many different tasks can use the semaphore at the same time. Each time a task starts using the semaphore, the semaphore size is decremented. When the size reaches 0, any task trying to use the semaphore must wait until the semaphore is released by another task. If the size is set to 1 (the default), only one task can access the critical section. If you're trying to avoid a race condition, use 1 as the size.

A task indicates that it wants to use a semaphore by calling the **Acquire Semaphore VI**. When the size of the semaphore is greater than 0, the VI immediately returns and the task proceeds. If the semaphore size is 0, the task waits until the semaphore becomes available. There's a time-out available in case you want to proceed even if the semaphore never becomes available. When a task successfully acquires a semaphore and is finished with its critical section, it releases the semaphore by calling the **Release Semaphore VI**. When a semaphore is no longer needed, call the **Destroy Semaphore VI**. If there are any Acquire Semaphore VIs waiting, they immediately time out and return an error.

Figure 2.62 is an example that uses a semaphore to protect access to a global variable. Only one loop is shown accessing the critical read-modify-write section, but there could be any number of them. Or the critical section could reside in another VI, with the semaphore refnum passed along in a global variable. This example follows the standard sequence of events: Create the semaphore, wait for it to become available, run the critical section, and then release the semaphore. Destroying the semaphore can tell all other users that it's time to quit.

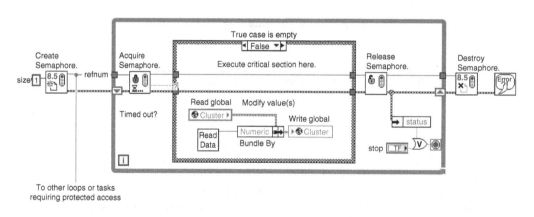

FIG. 2.62 *Semaphores provide a way to protect critical sections of code that can't tolerate simultaneous access from parallel calling VIs.*

Semaphores are widely used in *multithreaded* operating systems where many tasks potentially require simultaneous access to a common resource. Think about your desktop computer's disk system and what would happen if there was no coordination between the applications that needed to read and write data there. Since LabVIEW supports multithreaded execution of VIs, its execution system uses semaphores internally to avoid clashes when VIs need to use shared code. Where will you use semaphores?

Me and You, Rendezvous

Here's another interesting situation that can arise with parallel tasks.

Say that you have two independent data acquisition tasks, perhaps using different hardware, each of which requires an unpredictable amount of time to complete an acquisition. (In other words, they're about as asynchronous as you can get.) But say that you also need to do some joint processing on each batch of data that the two tasks return; perhaps you need to combine their data into a single analysis and display routine. How do you get these things synchronized?

One way to synchronize such tasks is to use the **rendezvous** VIs. Each task that reaches the rendezvous point waits until a specified number of tasks are waiting, at which point all tasks proceed with execution. It's like a trigger that requires a unanimous vote from all the participants; nobody starts until we're all ready.

To arrange a rendezvous in LabVIEW, begin by creating a rendezvous refnum with the **Create Rendezvous VI**. Pass that refnum to all the required tasks, as usual. Use the **Wait At Rendezvous VI** to cause a loop to pause until everyone is ready to proceed. When you're all through, call the **Destroy Rendezvous VI**.

Figure 2.63 shows a solution to our data acquisition problem using rendezvous VIs. We're going to acquire data from two asynchronous sources and then analyze it. In this solution, there are four parallel While Loops that could just as well reside in separate VIs. Leaving everything on one diagram indicates a much simpler way to solve this problem: Just put the two Read Data VIs in a single loop, and wire them up to the analysis section. Problem solved. Always look for the easiest solution!

Starting at the top of the diagram, While Loop A is responsible for stopping the program. When that loop terminates, it destroys the rendezvous, which then stops the other parallel loops. Loops B and C asynchronously acquire data and store their respective measurements in global variables. Loop D reads the two global values and then does the analysis and display. The Wait On Rendezvous VIs suspend execution of all three loops until all are suspended. Note that the Create Rendezvous VI has a size parameter of 3 to arrange this condition. When the rendezvous condition is finally satisfied, loops B and C begin acquisition, while loop D computes the result from the previous acquisition.

There are always other solutions to a given problem. Perhaps you could have the analysis loop trigger the acquisition VIs with occurrences and then have the analysis loop wait for the results via a pair of notifiers. The acquisition loops would fire their respective notifiers when acquisition was complete, and the analysis loop could be arranged so that it waited for both notifiers before doing the computation. And there is no doubt a solution based on queues. Which way you solve your problem is dependent upon your skill and comfort level with each technique and the abilities of anyone else who needs to understand and perhaps modify your program.

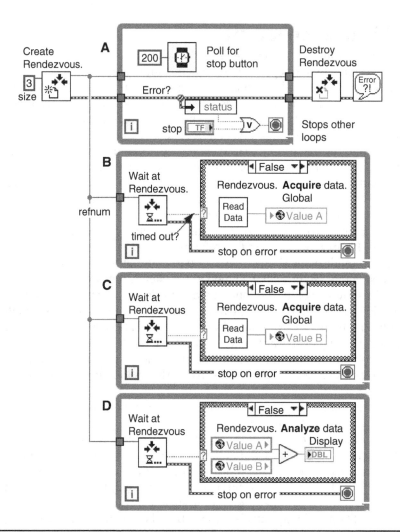

FIG. 2.63 *Rendezvous VIs can synchronize parallel tasks by making them wait for each other.*

Benchmarking Code

Benchmarking code performance and knowing the impact design decisions have on your application are fundamental skills every LabVIEW developer should have. Too often developers only look for performance improvements when there is a problem. Instead, benchmark key sections of your code to get a deeper understanding of how LabVIEW works and how your design choices impact performance. Experienced developers also monitor individual loop execution timing as the application grows to spot problems early.

Most large applications have several parallel processes running at preset intervals. Place a simple **iteration timer** in the top-level loop as a timing indicator to provide a quick peek into a loop's performance. This section of code uses a feedback node to retain the value of the millisecond timer from iteration to iteration. The difference between the new value and the old value gives a nice delta t(ms) for each iteration. Monitor the delta t at each top-level loop throughout development for instant feedback on how code changes impact performance (Fig. 2.64).

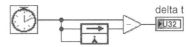

FIG. 2.64 *Use a feedback node and the ms Timer to see how often a section of code runs.*

The Tick Count (ms) primitive is sufficient for most cases, but if you need higher resolution, you can use the High-Resolution Relative Seconds.vi. This VI ships with LabVIEW and returns a relative timestamp with nanosecond resolution on most platforms. It didn't use to be in the palettes. If you use an older version of LabVIEW, you can download the Hidden Gems toolkit via VIPM and it will have this VI in the Hidden Gems palette (Fig. 2.65).

FIG. 2.65 *High Resolution Relative Seconds returns time with ns resolution.*

Iteration timers indicate how often a section of code is executing, but to get an in-depth look at code performance, you need to benchmark the code. This is done by evaluating the time required for the code to run for multiple iterations. Depending on the code complexity, or simplicity, the number of iterations required to get an accurate benchmark may be in the millions.

The following code demonstrates the benchmarking technique for a simple For Loop with a shift register and an addition primitive. Comparing the millisecond timer before and after running the code gives a delta ms that we convert to microseconds per iteration. Reading the front panel control "iterations" is moved outside the timed section to eliminate the impact from any context switching caused by the UI thread. Running this benchmark on a relatively fast i7 for 1 x 10^9 iterations yields 0.00127 μs/iteration. The actual value is not important, but it gives us a baseline for comparison as we make some code changes. At the end of this section is a figure with a comparison table with results for subVIs, classes, and RT targets (Fig. 2.66).

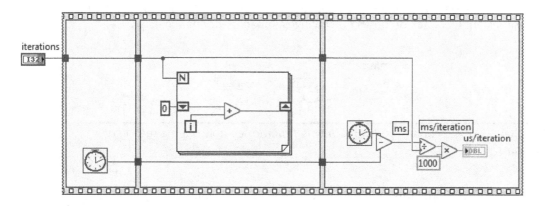

FIG. 2.66 *Benchmark a section of code with the ms Timer.*

The first change is encapsulating the add primitive in a subVI. With no other code changes, the μs/iteration jumps up to 0.068 microsecond. This jump is due to a hidden speed bump lurking in LabVIEW caused by the subVI call overhead. This doesn't mean that you shouldn't use subVIs, but you need to be aware of call overhead in time-critical sections of code.

Fortunately, you can eliminate subVI call overhead by **Inlining** the subVI. Go to the VI properties and check the checkbox on the lower left called "inline subVI into calling VIs." Refer to the shipping example called "VI Execution Properties.vi" for a comparison between the run time performance of an inline subVI against a normal subVI with debugging, a normal subVI without debugging, a subroutine subVI, and the primitive itself. You can find this and other benchmarking examples by going to **Help >> Find Examples...** and searching for the word benchmark. In that example, you will see that the direct function call and inlined call take the same amount of time to execute.

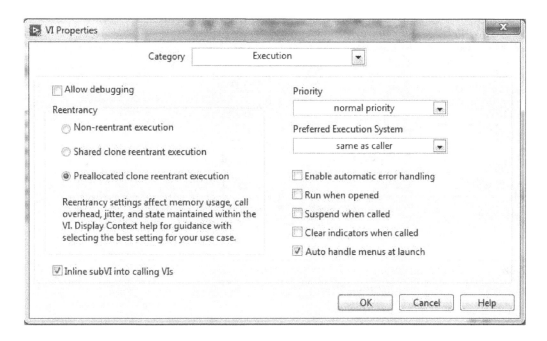

FIG. 2.67 *Check "Inline subVI into Calling VIs" for maximum performance.*

Figure 2.68 has the results for running our simple benchmark on a desktop PC and on a sbRIO 9607 real-time target. We repeated the test for subVIs and classes. A lot of us old-timers shied away from LabVIEW classes because of a performance hit. Classes have improved a lot over the years; based on our quick results, it looks like the hit from classes is about 10 percent vs. a subVI. However, inlining a class method doesn't have the performance increase that you get from inlining a subVI. We're not saying that you shouldn't use subVIs or classes; you need them for clean modular code. But it is important to know that the performance hits are there waiting for the unwary.

The sbRIO-9607 has a 667-MHz dual-core ARM 9. Like all of NI's real-time targets, it has a low-powered, ruggedized CPU, and performance is an order of magnitude slower than a desktop PC. The RT platform has plenty of advantages over a desktop, but speed is not one of them. The overhead of calling a subVI or a class method is almost a millisecond that can be a performance killer if you are trying to run at 1 kHz. In time-critical sections of code, use inlined code when you need performance.

Benchmarks	desktop	sbRIO 9607
Add primitive	0.00127	0.0255
subVI	0.068	0.865
subVI inline	0.001248	0.0258
class	0.079940	0.9814
class inline	0.0091	0.0901

FIG. 2.68 *Benchmarking results on different platforms.*

Another shipping example worth exploring while we are on the topic of benchmarking is "Benchmark Project.lvproj." Again, search the examples for "benchmark." The example code uses a 64-bit timer that is only available on real-time targets. The code keeps an array of all the execution timestamps and has the option to exclude the overhead due to the timestamping operation. Results are returned as a graph of the relative execution times along with jitter statistics.

CHAPTER 3

Data Acquisition

Inputs and Outputs

To automate your lab, one of the first things you will have to tackle is data acquisition—the process of making measurements of physical phenomena and storing them in some coherent fashion. It's a vast technical field with thousands of practitioners, most of whom are hackers as we are. How do most of us learn about data acquisition? By doing it, plain and simple. Having a formal background in engineering or some kind of science is awfully helpful, but schools rarely teach the practical aspects of sensors and signals and so on. It's most important that you get the big picture—learn the pitfalls and common solutions to data acquisition problems—and then try to see where your situation fits into the grand scheme.

This chapter should be of some help because the information presented here is hard-won, practical advice, for the most part. The most unusual feature of this chapter is that it contains very little LabVIEW programming information. There is much more to a LabVIEW system than LabVIEW programming! If you plan to write applications that support any kind of input/output (I/O) interface hardware, read on.

Origins of Signals

Data acquisition deals with the elements shown in Fig. 3.1. The physical phenomenon may be electrical, optical, mechanical, or something else that you need to measure. The sensor changes that phenomenon into a signal that is easier to transmit, record, and analyze—usually a voltage or current. Signal conditioning amplifies and filters the raw signal to prepare it for analog-to-digital conversion (ADC), which transforms the signal into a digital pattern suitable for use by your computer.

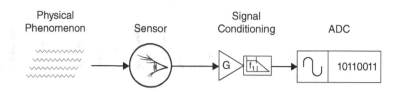

FIG. 3.1 *Elements of the data acquisition process.*

Transducers and Sensors

A transducer converts one physical phenomenon to another; in our case, we're mostly interested in an electric signal as the output. For instance, a thermocouple produces a voltage that is related to temperature. An example of a transducer with a nonelectrical output is a liquid-in-glass thermometer. It converts temperature changes to visible changes in the volume of fluid. An elaboration on a transducer might be called a sensor. It starts with a transducer as the front end, but then adds signal conditioning (such as an amplifier), computations (such as linearization), and a means of transmitting the signal over some distance without degradation. Some industries call this a transmitter. Regardless of what you call it, the added signal conditioning is a great advantage in practical terms, because you don't have to worry so much about noise pickup when dealing with small signals. Of course, this added capability costs money and may add weight and bulk.

Figure 3.2 is a general model of all the world's sensors. If your instrument seems to have only the first couple of blocks, then it's probably a transducer. Table 3.1 contains examples of some sensors. The first one, example A, is a temperature transmitter that uses a thermocouple with some built-in signal conditioning. Many times you can handle thermocouples without a transmitter, but as we'll see later, you always need some form of signal conditioning. The second example, example B, is a pressure transmitter, a slightly more complex instrument. We came across the third example, example C, a magnetic field sensor, while doing some research. It uses an optical principle, called Faraday rotation, in which the polarization of light is affected by magnetic fields.

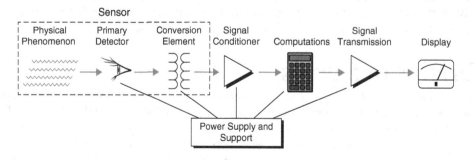

FIG. 3.2 *A completely general sensor model. Many times, your sensor is just a transducer that ends after the conversion element.*

Block	Example A	Example B	Example C
Phenomenon	Temperature	Pressure	Magnetic field
Detector	———	Diaphragm displacement	Faraday rotation
Transducer	Thermocouple	LVDT	Laser and photodiode
Signal conditioning	Cold junction; amplifier	Demodulator	ADC
Computations	Linearize	Linearize; scale	Ratio; log; scale
Transmission	0-10 VDC	4-20 mA	RS-232 serial
Display	Analog meter	Analog meter	Computer system
Support	DC	DC (2-wire current loop)	DC; cooling

TABLE 3.1 *Three practical examples of sensor systems.*

Example A is a temperature transmitter using a thermocouple, with cold-junction compensation, linearization, and analog output.

Example B is a pressure sensor using a linear variable differential transformer (LVDT) to detect diaphragm displacement, with analog output.

Example C is a magnetic field measurement using an optical technique, with direct signal transmission to a computer. This represents a sophisticated state-of-the-art sensor system (3M Specialty Optical Fibers).

A detailed discussion of sensor technology covering all aspects of the physics of transducers and the practical matters of selecting the right sensor is beyond the scope of this book. For the purposes of data acquisition, there are several important things you need to know about each of your sensors:

- The nature of the signal it produces—voltage, amplitude range, frequency response, impedance, accuracy requirement, and so on—determines what kind of signal conditioning, analog-to-digital converter, or other hardware you might need.

- How susceptible is the sensor to noise pickup or loading effects from data acquisition hardware?

- How is the sensor calibrated with respect to the physical phenomenon? In particular, you need to know if it's nonlinear or if it has problems with repeatability, overload, or other aberrant behavior.

- What kind of power or other utilities might it require? This is often overlooked and sometimes becomes a show stopper for complex instruments.

- What happens if you turn off your data acquisition equipment while the sensor still has power applied? Will there be damage to any components?

When you start to set up your system, try to pick sensors and design the data acquisition system in tandem. They are highly interdependent. When monitored by the wrong ADC, the world's greatest sensor is of little value. It is important that you understand the details of how your sensors work. Try them out under known conditions if you have doubts. If something doesn't seem right, investigate. When you call the manufacturer for help, it may well turn out that you know more about the equipment than the designers, at least in your particular application. In a later section, we'll look at a holistic approach to signals and systems.

Modern trends are toward smart sensors containing onboard microprocessors that compensate for many of the errors that plague transducers, such as nonlinearity and drift. Such instruments are well worth the extra cost because they tend to be more accurate and remove much of the burden of error correction from you, the user. For low-frequency measurements of pressure, temperature, flow, and level, the process control industry is rapidly moving in this direction. Companies such as Rosemount, Foxboro, and Honeywell make complete lines of smart sensors.

More complex instruments can also be considered sensors in a broader way. For instance, a digital oscilloscope is a sensor of voltages that vary over time. Your LabVIEW program can interpret this voltage waveform in many different ways, depending upon what the scope is connected to. Note that the interface to a sensor like this is probably TCP/IP or USB serial communications rather than an analog voltage. That's the beauty of using a computer to acquire data: Once you get the hardware hooked up properly, all that is important is the signal itself and how you digitally process it.

Actuators

An actuator, which is the opposite of a sensor, converts a signal (perhaps created by your LabVIEW program) into a physical phenomenon. Examples include electrically actuated valves, heating elements, power supplies, and motion control devices such as servomotors. Actuators are required any time you wish to control something such as temperature, pressure, or position. It turns out that we spend most of our time measuring things (the data acquisition phase) rather than controlling them, at least in the world of research. But control does come up from time to time, and you need to know how to use those analog and digital outputs so conveniently available on your interface boards.

Almost invariably, you will see actuators associated with feedback control loops. The reason is simple. Most actuators produce responses in the physical system that are more than just a little bit nonlinear and are sometimes unpredictable. For example, a valve with an electropneumatic actuator is often used to control fluid flow. The problem is that the flow varies in some nonlinear way with respect to the valve's position. Also, most valves have varying degrees of nonrepeatability. They creak and groan and get stuck—*hysteresis* and *deadband* are formal terms for this behavior. These are real-world problems that simply can't be ignored. Putting feedback around such actuators helps the situation greatly. The principle is simple.

Add a sensor that measures the quantity that you need to control. Compare this measurement with the desired value (the difference is called the *error*), and adjust the actuator in such a way as to minimize the error. The combination of LabVIEW and external loop controllers makes this whole situation easy to manage.

An important consideration for actuators is what sort of voltage or power they require. There are some industrial standards that are fairly easy to meet, such as 0–10 volts dc or 4–20 milliamperes, which are modest voltages and currents. But even these simple ranges can have added requirements, such as isolated grounds, where the signal ground is not the chassis of your computer. If you want to turn on a big heater, you may need large relays or contactors to handle the required current; the same is true for most high-voltage ac loads. Your computer doesn't have that kind of output, nor should it. Running lots of high power or high voltage into the back of your computer is not a pleasant thought. Try to think about these requirements ahead of time.

Categories of Signals

You measure a signal because it contains some type of useful information. Therefore, the first questions you should ask are, What information does the signal contain, and how is it conveyed? Generally, information is conveyed by a signal through one or more of the following signal parameters: state, rate, level, shape, or frequency content. These parameters determine what kind of I/O interface equipment and analysis techniques you will need.

Any signal can generally be classified as analog or digital. A digital, or binary, signal has only two possible discrete levels of interest—an active level and an inactive level. They're typically found in computer logic circuits and in switching devices. An analog signal, on the other hand, contains information in the continuous variation of the signal with respect to time. In general, you can categorize digital signals either as on/off signals, in which the state (on or off) is most important, or as pulse train signals, which contain a time series of pulses. On/off signals are easily acquired with a digital input port, perhaps with some signal conditioning to match the signal level to that of the port. Pulse trains are often applied to digital counters to measure frequency, period, pulse width, or duty cycle. It's important to keep in mind that digital signals are just special cases of analog signals, which leads to an important idea:

TIP
You can use analog techniques to measure and generate digital signals. This is useful when (1) you don't have any digital I/O hardware handy, (2) you need to accurately correlate digital signals and analog signals, or (3) you need to generate a continuous but changing pattern of bits.

Among analog signal types are the dc signal and the ac signal. Analog dc signals either are static or vary slowly with time. The most important characteristic of the dc signal is that information of interest is conveyed in the level, or amplitude, of the signal at a given instant. When measuring a dc signal, you need an instrument that can detect the level of the signal. The timing of the measurement is not difficult as long as the signal varies slowly. Therefore, the fundamental operation of the dc instrument is an ADC, which converts the analog electric signal into a digital number that the computer interprets. Common examples of dc signals include temperature, pressure, battery voltage, strain gauge outputs, flow rate, and level measurements. In each case, the instrument monitors the signal and returns a single value indicating the magnitude of the signal at a given time. Therefore, dc instruments often report the information through devices such as meters, gauges, strip charts, and numerical readouts.

TIP

When you analyze your system, map each sensor to an appropriate LabVIEW indicator type.

Analog **ac time domain** signals are distinguished by the fact that they convey useful information not only in the level of the signal but also in how this level varies with time. When measuring a time domain signal, often referred to as a **waveform**, you are interested in some characteristics of the shape of the waveform, such as slope, locations, and shapes of peaks. You may also be interested in its frequency content.

To measure the shape of a time domain signal with a digital computer, you must take a precisely timed sequence of individual amplitude measurements, or **samples**. These measurements must be taken closely enough together to adequately reproduce those characteristics of the waveform shape that you want to measure. Also, the series of measurements should start and stop at the proper times to guarantee that the useful part of the waveform is acquired. Therefore, the instrument used to measure time domain signals consists of an ADC, a sample clock, and a trigger. A sample clock accurately times the occurrence of each ADC.

Figure 3.3 illustrates the timing relationship among an analog waveform, a sampling clock, and a trigger pulse. To ensure that the desired portion of the waveform is acquired, you can use a trigger to start and/or stop the waveform measurement at the proper time according to some external condition. For instance, you may want to start acquisition when the signal voltage is moving in a positive direction through 0 V. Plug-in boards and oscilloscopes generally have trigger circuits that respond to such conditions.

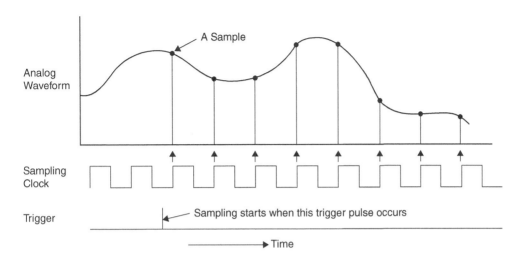

FIG. 3.3 *An illustration of the relationship between an analog waveform and the sampling clock and trigger that synchronize an ADC.*

Another way that you can look at an analog signal is to convert the waveform data to the **frequency domain**. Information extracted from frequency domain analysis is based on the frequency content of the signal, as opposed to the shape, or time-based characteristics of the waveform. Conversion from the time domain to the frequency domain on a digital computer is carried out through the use of a **Fast Fourier Transform (FFT)**, a standard function in the LabVIEW digital signal processing (DSP) function library. The **Inverse Fast Fourier Transform (IFFT)** converts frequency domain information back to the time domain. In the frequency domain, you can use DSP functions to observe the frequencies that make up a signal, the distribution of noise, and many other useful parameters that are otherwise not apparent in the time domain waveform. Digital signal processing can be performed by LabVIEW software routines or by special DSP hardware designed to do the analysis quickly and efficiently.

There is one more way to look at analog signals, and that is in the **joint time-frequency (JTF) domain** (Fig. 3.4) (Qian and Chen, 1996). This is a combination of the two preceding techniques. JTF signals have an interesting frequency spectrum that varies with time. Examples are speech, sonar, and advanced modulation techniques for communication systems. The classic display technique for JTF analysis is the **spectrogram**, a plot of frequency versus time, of which LabVIEW has one of the very best, the Gabor spectrogram algorithm. You can order the LabVIEW Signal Processing Toolkit, which includes a compiled application as well as the necessary VIs to do your own processing with the Gabor spectrogram, short-time FFT, and several others. It works with live data from a data acquisition board or any other source.

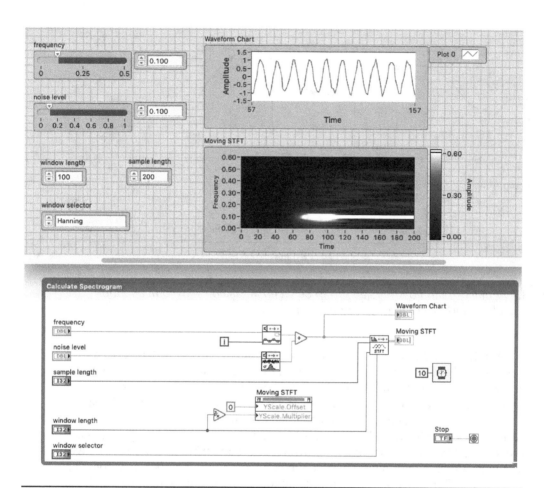

FIG. 3.4 *Joint-time frequency analysis (JTFA) using the STFT Spectrogram PtByPt VI.*

As Fig. 3.5 shows, the signal classifications described in this section are not mutually exclusive. A single signal may convey more than one type of information. In fact, the digital on/off, pulse train, and dc signals are just simpler cases of the analog time domain signals that allow simpler measuring techniques.

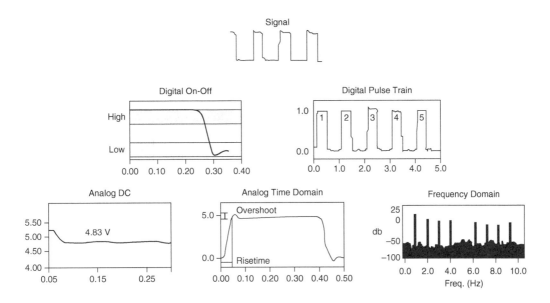

FIG. 3.5 *Five views of a signal. A series of pulses can be classified in several different ways depending on the significance of the signal's time, amplitude, and frequency characteristics.*

The preceding example demonstrates how one signal can belong to many classes. The same signal can be measured with different types of instruments, ranging from a simple digital state detector to a complex frequency analysis instrument. This greatly affects how you choose signal conditioning equipment.

For most signals, you can follow the logical road map in Fig. 3.6 to determine the signal's classification, typical interface hardware, processing requirements, and display techniques. As we'll see in coming sections, you need to characterize each signal to properly determine the kind of I/O hardware you'll need. Next, think about the signal attributes you need to measure or generate with the help of numerical methods. Then there's the user-interface issue—the part where LabVIEW controls and indicators come into play. Finally, you may have data storage requirements. Each of these items is directly affected by the signal characteristics.

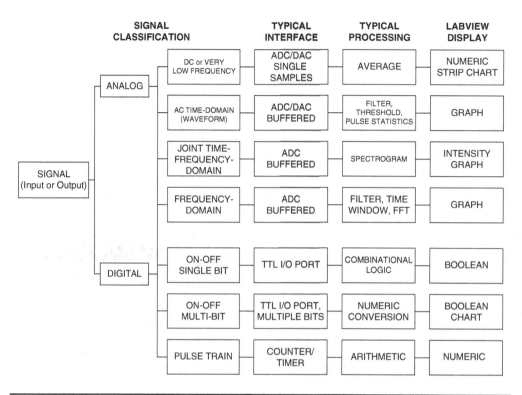

FIG. 3.6 *This road map can help you organize your information about each signal to logically design your LabVIEW system.*

Connections

Prof. John Frisbee is hard at work in his lab, trying to make a pressure measurement with his brand-new computer:

Let's see here.... This pressure transducer says it has a 0- to 10-volts dc output, positive on the red wire, minus on the black. The manual for my data acquisition board says it can handle 0–10 volts dc. That's no problem. Just hook the input up to channel 1 on terminals 5 and 6. Twist a couple of wires together, tighten the screws, run the data acquisition demonstration program, and voila! But what's this? My signal looks like ... junk! My voltmeter says the input is dc, 1.23 volt, and LabVIEW seems to be working OK, but the display is really noisy. What's going on here?

Poor John. He obviously didn't read this chapter. If he had, he would have said, "Aha! I need to put a low-pass filter on this signal and then make sure that everything is properly grounded and shielded." What he needs is **signal conditioning**, and believe me, so do you. This world is chock full of noise sources and complex signals—all of them guaranteed to corrupt your data if you don't take care to separate the good from the bad.

There are several steps in designing the right signal conditioning approach for your application. First, remember that you need to know all about your sensors and what kind of signals they are supposed to produce. Second, you need to consider grounding and shielding. You may also need amplifiers and filters. Third, you can list your specifications and go shopping for the right data acquisition hardware.

Grounding and Shielding

Noise sources are lurking everywhere, but by following some simple principles of grounding and shielding, you can eliminate most noise-related problems right at the source—the wiring. In fact, how you connect a sensor to its associated data acquisition hardware greatly affects the overall performance of the system (Gunn, 1987; Morrison, 1986; Ott, 1988; White, 1986). Another point on the subject: Measurements are inherently inaccurate. A good system design minimizes the distortion of the signal that takes place when you are trying to measure it—noise, nonlinear components, distortion, and so on. The wire between sensor and data acquisition can never improve the signal quality. You can only minimize the negatives with good technique. Without getting too involved in electromagnetic theory, we'll show you some of the recommended practices that instrumentation engineers everywhere use. Let's start with some definitions.

Ground. This is absolutely the most overused, misapplied, and misunderstood term in all electronics. First, there is the most hallowed earth ground that is represented by the electrical potential of the soil underneath your feet. The green wire on every power cord, along with any metal framework or chassis with 120-volts ac main power applied, is required by the National Electrical Code to be connected through a low-resistance path to the aforementioned dirt. There is exactly one reason for its existence: safety. Sadly, somehow we have come to believe that grounding our equipment will magically siphon away all sources of noise, as well as evil spirits. Baloney! Electricity flows only in closed circuits, or loops. You and your equipment sit upon earth ground as a bird sits upon a high-voltage wire. Does the bird know it's living at 34,000 volts? Of course not; there is no complete circuit. This is not to say that connections to earth ground—which we will refer to as **safety ground** from here on—are unnecessary. You should always make sure that there is a reliable path from all your equipment to safety ground as required by code. This prevents accidental connections between power sources and metallic objects from becoming hazards. Such fault currents are shunted away by the safety ground system.

What we really need to know about is a reference potential referred to as **signal common**, or sometimes as a **return** path. Every time you see a signal or measure a voltage, always ask the question, "Voltage—with respect to what reference?" That reference is the signal common, which in most situations is not the same as safety ground. A good example is the negative side of the battery in your car. Everything electrical in the vehicle has a return connection to this common, which is also connected to the chassis. Note that there is no connection whatsoever to earth ground—the car has rubber tires that make fair insulators—and yet, the electrical system works just fine (except if it's very old or British).

In our labs, we run heavy copper braid, welding cable, or copper sheet between all the racks and experimental apparatus according to a grounding plan (Fig. 3.7). A well-designed signal common can even be effective at higher frequencies at which second-order effects such as the **skin effect** and **self-inductance** become important. See Gunn (1987), Morrison (1986), Ott (1988), and White (1986) for instructions on designing a quality grounding system—it's worth its weight in aspirin.

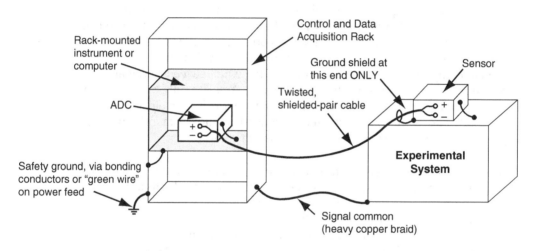

FIG. 3.7 *Taking the system approach to grounding in a laboratory. Note the use of a signal common (in the form of heavy copper braid or cable) to tie everything together. All items are connected to this signal common.*

Electromagnetic Fields. Noise may be injected into your measurement system by electromagnetic fields, which are all around us. Without delving into Maxwell's equations, we present a few simple principles of electromagnetism that you use when connecting your data acquisition system.

Principle 1. Two conductors that are separated by an insulator form a **capacitor**. An electric field exists between the conductors. If the potential (voltage) of one conductor changes with time, a proportional change in potential will appear in the other. This is called *capacitive coupling* and is one way noise is coupled into a circuit. Moving things apart reduces capacitive coupling.

Principle 2. An electric field cannot enter a closed, conductive surface. That's why sitting in your car during a thunderstorm is better than sitting outside. The lightning's field cannot get inside. This kind of enclosure is also called a *Faraday cage*, or **electrostatic shield**, and is commonly implemented by a sheet of metal, screen, or braid surrounding a sensitive circuit. Electrostatic shields reduce capacitive coupling.

Principle 3. A time-varying magnetic field will induce an electric current only when a closed, conductive loop is present. Furthermore, the magnitude of the induced current is proportional to the intensity of the magnetic field and the area of the loop. This property is called *inductive coupling*, or **inductance**. Open the loop, and the current goes away.

Principle 4. Sadly, **magnetic shielding** is not so easy to design for most situations. This is so because the magnetic fields that we are most concerned about (low frequency, 50/60 Hz) are very penetrating and require very thick shields of iron or, even better, magnetic materials, such as expensive mu-metal.

Here are some basic practices you need to follow to block the effects of these electromagnetic noise sources:

- Put sensitive, high-impedance circuitry and connections inside a metallic shield that is connected to the common-mode voltage (usually the low side, or common) of the signal source. This will block capacitive coupling to the circuit (principle 1), as well as the entry of any stray electric fields (principle 2).

- Avoid closed, conductive loops—intentional or unintentional—often known as **ground loops**. Such loops act as pickups for stray magnetic fields (principle 3). If a high current is induced in, for instance, the shield on a piece of coaxial cable, then the resulting voltage drop along the shield will appear in series with the measured voltage.

- Avoid placing sensitive circuits near sources of intense magnetic fields, such as transformers, motors, and power supplies. This will reduce the likelihood of magnetic pickup that you would otherwise have trouble shielding against (principle 4).

Another unsuspected source of interference is your lovely color monitor on your PC. It is among the greatest sources of electrical interference known. Even the interface cards in your computer, for instance, the video adapter and the computer's digital logic circuits, are sources of high-frequency noise. And never trust a fluorescent light fixture (or a politician).

Radio-frequency interference (RFI) is possible when there is a moderately intense RF source nearby. Common sources of RFI are transmitting devices, such as walkie-talkies, cellular phones, commercial broadcast transmitters of all kinds, and RF induction heating equipment. Radio frequencies radiate for great distances through most nonmetallic structures, and really high (microwave) frequencies can even sneak in and out through cracks in metal enclosures. You might not think that a 75-MHz signal would be relevant to a 1-kHz bandwidth data acquisition system, but there is a phenomenon known as *parasitic detection* or *demodulation* that occurs in many places. Any time that an RF signal passes through a diode (a rectifying device), it turns into dc plus any low frequencies that may be riding on (modulating) the RF carrier. Likely parasitic detectors include all the solid-state devices in your system, plus any metal-oxide interfaces (such as dirty connections).

When RFI strikes your data acquisition system, it results in unexplained noise of varying amplitude and frequency. Sometimes, you can observe stray RF signals by connecting a wide-bandwidth oscilloscope to signal lines. If you see more than a few millivolts of high-frequency noise, suspect a problem. Some solutions to RFI are as follows:

- Shield all the cables into and out of your equipment
- Add RF rejection filters on all signal and power leads
- Put equipment in well-shielded enclosures and racks
- Keep known RF sources and cables far away from sensitive equipment
- Keep the person with the walkie-talkie or cellular phone *out of your lab*

Other Error Sources. **Thermojunction** voltages are generated any time two dissimilar metals come in contact with each other in the presence of a temperature gradient. This principle, known as the *Seebeck effect*, is the way in which a thermocouple generates its tiny signal. Problems occur in data acquisition when you attempt to measure dc signals that are in the microvolt to millivolt range, such as those from thermocouples and strain gauges. If you connect your instruments with wires, connectors, and terminal screws made of different metals or alloys, then you run the risk of adding uncontrolled thermojunction voltages to the signals—possibly to the tune of hundreds of microvolts. The most common case of thermojunction error that we have seen occurs when operators hook up thermocouples inside an experimental chamber and bring the wires out through a connector with pins made of copper or stainless steel. The thermocouple alloy wire meets the connector pin and forms a junction. Then the operators turn on a big heater inside the chamber, creating a huge temperature gradient across the connector. Soon afterward, they notice that their temperature readouts are way off. Those parasitic thermojunction voltages inside the connector are added to the data in an unpredictable fashion. Here are some ways to kill the thermojunction bugs:

- Make all connections with the same metallic alloy as the wires they connect
- Keep all connections at the same temperature
- Minimize the number of connections in all low-level signal situations

Differential and Single-Ended Connections. The means by which you connect a signal to your data acquisition system can make a difference in system performance, especially with respect to noise rejection. **Single-ended** connections are the simplest and most obvious way to connect a signal source to an amplifier or other measurement device (Fig. 3.8). The significant hazard of this connection is that it is **highly susceptible to noise pickup**. Noise induced on any of the input wires, including the signal common, is added to the desired signal. Shielding the signal cable and being careful where you make connections to the signal common can improve the situation. Single-ended connections are most often used in wide-bandwidth systems such as oscilloscopes and video, RF, and fast pulse measurements in which low-impedance coaxial cables are the preferred means of transmission. These single-ended connections are also practical and trouble-free with higher-magnitude signals (say, 1 volt or greater) that only need to travel short distances and don't need mV of precision.

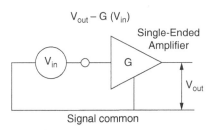

FIG. 3.8 *A single-ended amplifier has no intrinsic noise rejection properties. You need to carefully shield signal cables and make sure that the signal common is noise-free as well.*

Differential connections depend on a pair of conductors where the voltage you want to measure, called the **normal-mode signal**, is the difference between the voltages on the individual conductors. Differential connections are used because noise pickup usually occurs equally on any two conductors that are closely spaced, such as a twisted pair of wires. So when you take the difference between the two voltages, the noise cancels but the difference signal remains. An **instrumentation amplifier** is optimized for use with differential signals, as shown in Fig. 3.9. The output is equal to the gain of the amplifier times the difference between the inputs. If you add another voltage in series with both inputs, called the **common-mode signal**, it cancels just like the noise pickup. The optimum signal source is a **balanced** signal, where the signal voltage swings symmetrically with respect to the common-mode voltage. Wheatstone bridge circuits and transformers are the most common balanced sources.

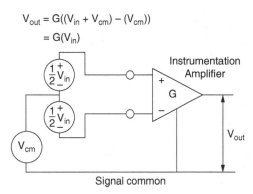

FIG. 3.9 *Instrumentation amplifiers effectively reject common-mode (V_{cm}) signals, such as noise pickup, through cancellation. The normal-mode (differential) signal is amplified.*

The common-mode voltage can't be infinitely large, though, since every amplifier has some kind of maximum input voltage limit, which is usually less than 10 V—watch out for overloads! Also, beware that the common-mode rejection ratio (CMRR) of an amplifier generally decreases with frequency. Don't count on an ordinary amplifier to reject very intense RF signals, for instance. Differential inputs are available on most low-frequency instrumentation such as voltmeters, chart recorders, and plug-in boards, for example, those made by National Instruments (NI). *Rule: Always use differential connections, except when you can't.*

In Fig. 3.10, you can see some typical signal connection schemes. Shielded cable is always recommended to reduce capacitive coupling and electric field pickup. Twisted, shielded pairs are best, but coaxial cable will do in situations where you are careful not to create the dreaded ground loop that appears in the bottom segment of the figure.

Why Use Amplifiers or Other Signal Conditioning?

As you can see, the way you hook up your sensors can affect the overall performance of your data acquisition system. But even if the grounding and shielding are properly done, you should consider **signal conditioning**, which includes **amplifiers** and **filters**, among other things, to reduce the noise relative to the signal level.

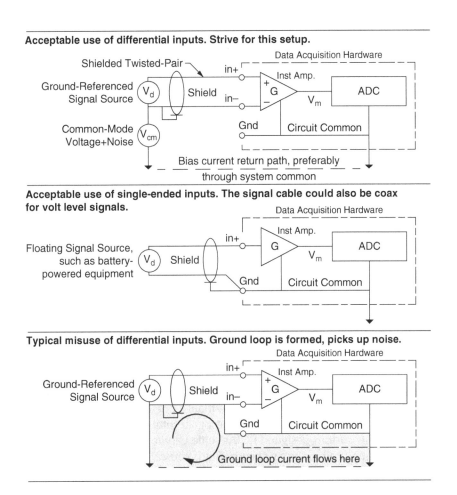

FIG. 3.10 *Proper use of differential and single-ended signal connections is shown in the top two parts of the figure. The bottom part is that all-too-common case in which a ground loop is formed, enhancing the pickup of noise borne by magnetic fields.*

Amplifiers improve the quality of the input signal in several ways. First, they boost the amplitude of smaller signals, improving the resolution of the measurements. Second, they offer increased driving power (lower-output impedance), which keeps such things as ADCs from loading the sensor. Third, they provide differential inputs, a technique known to help reject noise. An improvement on differential inputs, an **isolation amplifier**, requires no signal common at the input whatsoever. Isolation amplifiers are available with common-mode voltage ranges up to thousands of volts.

Amplifiers are usually essential for microvolt signals, such as those from thermocouples and strain gauges. In general, try to amplify your low-level signal as close to the physical phenomenon itself as possible. Doing this will help you increase the **signal-to-noise ratio (SNR)**. The signal-to-noise ratio is defined as

$$\text{SNR} = 20 \log \left(\frac{V_{sig}}{V_{noise}} \right)$$

where V_{sig} is the signal amplitude and V_{noise} is the noise amplitude, both measured in volts rms. The 20 log() operation converts the simple ratio to **decibels (dB)**, a ratiometric system used in electrical engineering, signal processing, and other fields. Decibels are convenient units for gain and loss computations. For instance, an SNR or gain of 20 dB is the same as a ratio of 10 to 1; 40 dB is 100 to 1; and so forth. Note that the zero noise condition results in an infinite SNR. May you one day achieve this.

Most plug-in data acquisition boards include a programmable-gain instrumentation amplifier (PGIA). You change the gain by sending a command to the board, and the gain can be changed at any time—even while you are rapidly scanning through a set of channels. The difference between an onboard amplifier and an external amplifier is that the onboard amplifier must respond quickly to large changes in signal level (and perhaps gain setting) between channels. Errors can arise due to the **settling time**, which is the time it takes for the amplifier output to stabilize within a prescribed error tolerance of the final value. If you scan a set of channels at top speed with the amplifier at high gain, there is the possibility of lost accuracy. But the worst case is scanning a set of channels with different gains—that is, one channel with the lowest gain setting and the next channel with the highest gain setting. NI has gone to great pains to ensure that its custom-designed PGIAs settle very rapidly, and it guarantees full accuracy at all gains and scanning speeds. But if you can afford it, the amplifier-per-channel approach remains the best choice for high-accuracy data acquisition. Also, note that it's not just settling time—on multiplexed-input boards, the input impedance of the amplifier input changes as you change the sampling rate.

Filters are needed to reject undesired signals, such as high-frequency noise, and to provide **antialiasing** (discussed later in this chapter). Most of the time, you need low-pass filters that reject high frequencies while passing low frequencies, including dc. The simplest filters are made from resistors and capacitors (and sometimes inductors). **Active filters** combine resistors and capacitors with operational amplifiers to enhance performance.

Using a signal conditioning system outside of your computer also enhances safety. If a large overload should occur, the signal conditioner will take the hit rather than pass it directly to the backplane of your computer (and maybe all the way to the mouse or keyboard!). A robust amplifier package can easily be protected against severe overloads through the use of transient absorption components (such as varistors, zener diodes, and spark gaps), limiting resistors, and fuses. But your best bet is isolation. Medical systems are covered by federal and international regulations regarding isolation from stray currents. *Never connect electronic instruments to a live subject (human or otherwise) without a properly certified isolation amplifier and correct grounding.*

Special transducers may require **excitation**. Examples are resistance temperature detectors (RTDs), thermistors, potentiometers, and strain gauges, as depicted in Fig. 3.11. All these devices produce an output that is proportional to an excitation voltage or current, as well as the physical phenomenon they are intended to measure. For best results, pick a system that uses ratiometric measurements, where the excitation source (or a signal derived directly from it) is used as the ADC reference. This eliminates most accuracy/noise errors associated with the excitation.

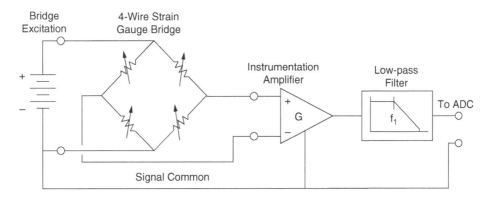

FIG. 3.11 *Schematic of a signal conditioner for a strain gauge, a bridge-based transducer that requires excitation. This signal conditioner has an instrumentation amplifier with high gain, followed by a low-pass filter to reject noise.*

Multiplexing can cause some interesting problems when not properly applied, as we mentioned in regard to the amplifier settling time. Compare the two signal configurations in Fig. 3.12. In the top configuration, there is an amplifier and filter per channel, followed by the multiplexer that ultimately feeds the ADC. This is the preferred arrangement because each filter output faithfully follows its respective input signal without disturbances from switching transients. But in the bottom configuration, the amplifier and filter are located after the multiplexer. This saves money but is undesirable because each time the multiplexer changes channels, the very sluggish low-pass filter must settle to the voltage present at the new channel. As a result, you must scan the channels at a very slow rate unless you disable the low-pass filter, thus giving up its otherwise useful properties.

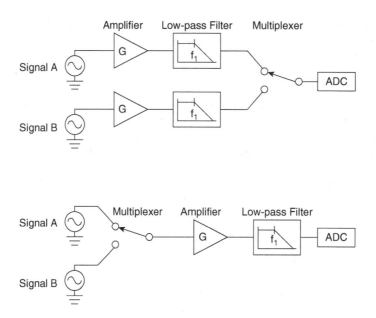

FIG. 3.12 *The preferred amplifier and filter per channel configuration (top); an undesirable, but cheaper, topology (bottom).Watch your step if your system looks like the latter.*

Practical Tips on Connecting Input Signals. Let's look at a few common input signal connection situations that you're likely to come across. We recommend that you refer to the manual for your signal conditioning or plug-in board before making any connections. There may be particular options or requirements that differ from those in this tutorial.

Ground-Referenced Analog Inputs. If one side of your signal has a direct connection to a reliable ground, use a differential input to avoid ground loops and provide common-mode noise rejection.

Floating Analog Inputs. Battery-powered equipment and instruments with isolated outputs do not supply a return connection to signal ground. This is generally good, since a ground loop is easy to avoid. You can connect such a signal to a *referenced single-ended (RSE) input* (Fig. 3.13A), an input configuration available on most data acquisition boards, and called RSE in the NI manuals. Common-mode noise rejection will be acceptable in this case. You can also use differential inputs, but note that they require the addition of a pair of *leak resistors* (Fig. 3.13B and C). Every amplifier injects a small *bias current* from its inputs back into the signal connections. If there is no path to ground, as is the case for a true floating source, then the input voltage at one or both inputs will float to the amplifier's power supply rail voltage. The result is erratic operation because the amplifier is frequently saturated—operating out of its linear range. This is a sneaky problem! Sometimes, the source is floating and you don't even know it. Perhaps the system will function normally for a few minutes after power is turned on and then it will misbehave later. Or touching the leads together or touching them with your fingers may discharge the circuit, leading you to believe that there is an intermittent connection. Yikes!

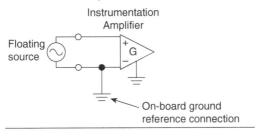

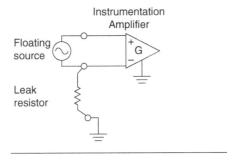

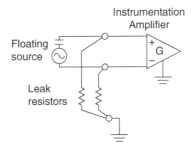

FIG. 3.13 *Floating-source connections: (A) This is the simplest and most economical; (B) this is similar but may have noise rejection advantages; and (C) this is required for ac-coupled sources.*

Rule: Use an ohmmeter to verify the presence of a resistive path to ground on all signal sources.

What is the proper value for a leak resistor? The manual for your input device may have a recommendation. In general, if the value is too low, you lose the advantage of a differential input because the input is tightly coupled to ground through the resistor. You may also overload the source in the case where leak resistors are connected to both inputs. If the value is too high, additional dc error voltages may arise due to *input offset current* drift (the input bias currents on the two inputs differ and may vary with temperature). A safe value is generally in the range of 1–100 kΩ. If the source is truly floating, such as battery-powered equipment, then go ahead and directly ground one side.

Thermocouples. Thermocouples have very low source impedance and low signal voltage, and they may have either floating or grounded junctions. For an excellent tutorial on thermocouples, refer to Omega Engineering's *Temperature Handbook* (2001). For general information on connecting thermocouples to computer I/O, consult NI's website. We generally prefer floating junctions, in which the welded joint between the thermocouple wires is isolated from the surrounding sheath or other nearby metallic items. This removes another chance of a ground loop and results in a floating dc-coupled source as previously described. Always use differential connections to reject noise on this and other low-level signals, and be sure to use a leak resistor—perhaps 1 kΩ or so—on one of the inputs. You must also take care to use extension wire and connectors of the same alloy as that of the thermocouple element. Finally, the connections to the signal conditioning must be *isothermal*; that is, all connections are at the same temperature.

Thermocouples also require a *reference junction* measurement to compensate for the voltage produced at the junction between the thermocouple wires and copper connections. This may be provided by an additional thermocouple junction immersed in a reference temperature bath or, more commonly, by measuring the temperature at the reference temperature block, where the thermocouple alloy transitions to copper, typically near the amplifier. All modern signal conditioners include an isothermal reference temperature feature. The LabVIEW DAQ library includes VIs for making the corrections and performing linearization.

AC Signals. If you want to measure the RMS value of an ac waveform, you have several choices: Use an external AC to DC converter, digitize the waveform at high speed and do some signal processing in LabVIEW, use a plug-in digital multimeter board such as the NI PXI-4065, or use a benchtop DMM with a computer interface. For low-frequency voltage measurements, your best bet is a true RMS (TRMS) converter module, such as those made by Action Instruments and Ohio Semitronics (which also make single- and three-phase ac power transmitters). The input can be any ac waveform up to a few kilohertz and with amplitudes as high as 240 V RMS. To measure low-frequency ac current, use a current transformer and an ac converter module as before. If the frequency is very high, special current transformers made by Pearson Electronics are available and can feed a wideband RMS-to-DC converter. All these transformers and ac converters give you signals that are isolated from the source, making connections to your DAQ board simple. When the shape of the waveform is of interest, you generally connect the signal directly to the input of a sufficiently fast ADC or digital oscilloscope, and then you acquire large buffers of data for analysis.

Digital Inputs. Digital on/off signals may require conditioning, depending upon their source. If the source is an electronic digital device, it probably has TTL-compatible outputs and can be connected directly to the digital I/O port of a plug-in board. Check the manuals for the devices at both ends to ensure that both the levels and current requirements match. Detecting a contact closure (a switch) requires a *pull-up resistor* at the input (Fig. 3.14) to provide the high level when the switch is open. If your source produces a voltage outside of the 0- to 5-VDC range of a typical plug-in board, additional signal conditioning is required, or you can buy NI PXI relay modules that are compatible with high ac and dc voltages.

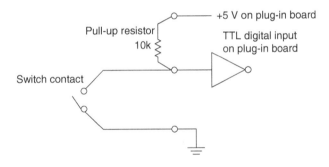

FIG. 3.14 *Contact closures are sensed by a TTL input with the help of an external pull-up resistor.*

Outputs Need Signal Conditioning, Too. To drive actuators of one type or another, you may need signal conditioning. As with input signals, isolation is one of the major considerations, along with requirements for extra drive capability that many simple output devices can't handle directly.

For digital outputs, most plug-in boards offer simple TTL logic drivers, which swing from 0 to about +3 volts, depending on the load (you need to consult the specifications for your hardware to make sure). If you want to drive a solenoid valve, for instance, much more current is needed—perhaps several amperes at 24-volt dc or even 120-volt ac. In such cases, you need a relay of some type. Electromechanical relays are simple and inexpensive and have good contact ratings, but sometimes they require more coil current than a TTL output can supply. Sensitive relays, such as reed relays, are often acceptable. Solid-state relays use silicon-controlled rectifiers (SCRs) or triacs to control heavier loads with minimal control current requirements. Their main limitation is that most units are only usable for ac circuits. If you need to drive other types of logic or require large output voltages and/or currents at high speed, then a special interface circuit may have to be custom-designed. Figure 3.15 shows a couple of simple options for driving light-emitting diodes (LEDs) and heavier dc loads, such as solenoids, using MOSFETs to handle the power.

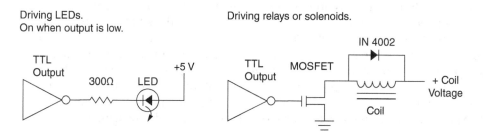

FIG. 3.15 *Simple circuits you can use with digital outputs on MIO and DIO series boards. Drive low-current loads like LEDs directly (left). MOSFETs are available in a wide range of current and voltage ratings for driving heavy dc loads (right).*

One warning is in order regarding digital outputs. When you turn on a digital output device, be it a plug-in board or an external module, which output state will it produce—on, off, or high-impedance (disconnected)? Can it produce a brief but rapid pulse train? We've seen everything, and you will, too, given time. Connect an oscilloscope to each output and see what happens when you turn the system on and off. If the result is unsatisfactory (or unsafe!), you may need a switch or relay to temporarily connect the load to the proper source. This warning applies to *analog* outputs, too.

Most of the analog world gets by with control voltages in the range of ±10 volts or less and control currents of 20 milliamperes or less. These outputs are commonly available from most I/O boards and modular signal conditioners. If you need more current and/or voltage, you can use a power amplifier. Audio amplifiers are a quick and easy solution to some of these problems, although most have a high-pass filter that rolls off the dc response (i.e., they can produce no dc output). Many commercial dc power supplies have analog inputs and are good for low-frequency applications. To control big ac heaters, motors, and other heavy loads, look to triac-based power controllers. They permit an ordinary analog voltage or current input to control tens or even hundreds of kilowatts with reasonable cost and efficiency. One thing to watch out for is the need for isolation when you get near high-power equipment.

An isolation amplifier could save your equipment if a fault occurs; it may also eliminate grounding problems which are so prevalent around high-power systems.

Choosing the Right I/O Subsystem

As soon as your project gets underway, make a list of the sensors you need to monitor along with any actuators you need to drive. Having exact specifications on sensors and actuators will make your job much easier. As an instrumentation engineer, Gary spends a large part of his time on any project just collecting data sheets, calling manufacturers, and peering into little black boxes, trying to elucidate the important details needed for interfacing. Understanding the application is important, too. The fact that a pressure transducer will respond in 1 microsecond doesn't mean that your system will actually have microsecond dynamic conditions to measure. On the other hand, that superfast transducer may surprise you by putting out all kinds of fast pulses because there are some little bubbles whizzing by in the soup. Using the wrong signal conditioner in either case can result in a phenomenon known as *bogus data*.

Since you are a computer expert, fire up your favorite spreadsheet or database application and make up an instrument list. The process control industry goes so far as to standardize on a format known as *instrument data sheets*, which are used to specify, procure, install, and finally document all aspects of each sensor and actuator. Others working on your project, particularly the quality assurance staff, will be very happy to see this kind of documentation. For the purposes of designing an I/O subsystem, your database might include these items:

- Instrument name or identifier
- Location, purpose, and references to other drawings such as wiring and installation details
- Calibration information: engineering units (such as psig) and full-scale range
- Accuracy, resolution, linearity, and noise, if significant
- Signal current, voltage, or frequency range
- Signal bandwidth
- Isolation requirements
- Excitation or power requirements: current and voltage

To choose your I/O subsystem, begin by sorting the instrument list according to the types of signal and other basic requirements. Remember to add plenty of spare channels! Consider the relative importance of each instrument. If you have 99 thermocouples and 1 pressure gauge, your I/O design choice will certainly lean toward accommodating thermocouples. But that pressure signal may be the single most important measurement in the whole system; don't try to adapt its 0- to 10-V output to work with an input channel that is optimized for microvolt thermocouple signals. For each signal, determine the minimum specifications for its associated signal conditioner. Important specifications are as follows:

- Adjustability of zero offset and gain
- Bandwidth—minimum and maximum frequencies to pass
- Filtering—usually antialiasing low-pass filters
- Settling time and phase-shift characteristics
- Accuracy
- Gain and offset drift with time and temperature (very important)
- Excitation—built in or external?
- For thermocouples, cold-junction compensation
- Linearization—may be better performed in software

Next, consider the physical requirements. Exposure to weather, high temperatures, moisture and other contamination, or intense electromagnetic interference may damage unprotected equipment. Will the equipment have to work in a hostile environment? If so, it must be in a suitable enclosure. If the channel count is very high, having many channels per module could save both space and money. Convenience should not be overlooked: Have you ever worked on a tiny module with itty-bitty terminal screws that are deeply recessed into an overstuffed terminal box? This is a practical matter; if you have lots of signals to hook up, talk this over with the people who will do the installation.

If your company already has many installations of a certain type or I/O, that may be an overriding factor, as long as the specifications are met. The bottom line is always cost. Using excess or borrowed equipment should always be considered when money is tight. You can do a cost-per-channel analysis if that makes sense.

Remote and Distributed I/O—Network Everything! Your sensors may not be located close to your computer system, or they may be in an inaccessible area—a hazardous enclosure or an area associated with a high-voltage source. In such cases, **remote I/O** hardware is appropriate. It makes sense to locate the acquisition hardware close to groups of sensors—*remote* from the computer—because it saves much signal wiring that in turn reduces cost and reduces the chance of noise pickup. Many remote I/O systems are commercially available with a variety of communication interfaces. If more than one I/O subsystem can be connected to a common communication line, it is often referred to as **distributed I/O**. Here are a few configurations and technologies to consider.

CompactDAQ. National Instruments' CompactDAQ chassis come in one-, four-, and eight-slot modules with a choice of USB, Ethernet, or Wi-Fi interfaces. You can fill a chassis up with C-series I/O modules and interface easily with DAQMX on a host PC.

FieldDAQ. National Instruments' FieldDAQ series has networked modules for voltage, strain, and temperature. FieldDAQ modules are water and dust resistant to IP-67 with a temperature range of -40°C to 85°C and can sustain 100 g shock and 10 g vibration.

Now you have your signals connected to your computer. Next, we'll take a closer look at how those signals are sampled or digitized for use in LabVIEW.

Sampling Signals

Up until now, we've been discussing the real (mostly analog) world of signals. Now it's time to digitize those signals for use in LabVIEW. By definition, **analog** signals are **continuous-time**, **continuous-value** functions. That means they can take on any possible value and are defined over all possible time resolutions. (By the way, don't think that digital pulses are special; they're just analog signals that happen to be square waves. If you look closely, they have all kinds of ringing, noise, and slew rate limits—all the characteristics of analog signals.)

An **analog-to-digital converter** samples your analog signals on a regular basis and converts the amplitude at each sample time to a digital value with finite resolution. These are termed **discrete-time**, **discrete-value** functions. Unlike their analog counterparts, discrete functions are defined only at times specified by the sample interval and may only have values determined by the resolution of the ADC. In other words, when you digitize an analog signal, *you have to approximate*. How *much* you can throw out depends on your signal and your specifications for data analysis. Is 1 percent resolution acceptable? Or is 0.0001 percent required? And how fine does the temporal resolution need to be? One second? Or 1 nanosecond? Please be realistic. Additional amplitude and temporal resolution can be *expensive*. To answer these questions, we need to look at this business of sampling more closely.

Sampling Theorem

A fundamental rule of sampled data systems is that the input signal must be sampled at a rate greater than twice the highest-frequency component in the signal. This is known as the **Shannon sampling theorem**, and the critical sampling rate is called the **Nyquist rate**. Stated as a formula, it says that $f_s/2 \bullet f_a$, where f_s is the sampling frequency and f_a is the maximum frequency of the signal being sampled. Violating the Nyquist criterion is called **undersampling** and results in **aliasing**. Look at Fig. 3.16 that simulates a sampled data system. I started out with a simple 1-kHz sine wave (dotted lines), and then I sampled it at two different frequencies, 1.2 and 5.5 kHz. At 5.5 kHz, the signal is safely below the Nyquist rate, which would be 2.75 kHz, and the data points look something like the original (with a little bit of information thrown out, of course). But the data with a 1.2-kHz sampling rate is aliased: It looks as if the signal is 200 Hz, not 1 kHz. This effect is also called **frequency foldback**: Everything above $f_s/2$ is folded back into the sub-$f_s/2$ range. If you undersample your signal and get stuck with aliasing in your data, can you undo the aliasing? In most cases, no. As a rule, you should not undersample if you hope to make sense of waveform data. Exceptions do occur in certain *controlled* situations. An example is **equivalent-time sampling** in digital oscilloscopes, in which a repetitive waveform is sampled at a low rate, but with careful control of the sampling delay with respect to a precise trigger.

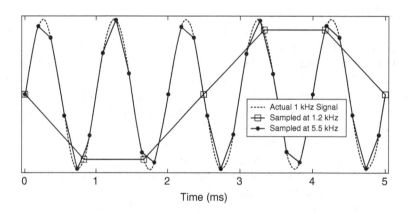

FIG. 3.16 *Graphical display of the effects of sampling rates. When the original 1-kHz sine wave is sampled at 1.2 kHz (too slow), it is totally unrecognizable in the data samples. Sampling at 5.5 kHz yields a much better representation. What would happen if there was a lot of really high-frequency noise?*

Let's go a step further and consider a nice 1-kHz sine wave, but this time we add some high-frequency noise to it. We already know that a 5.5-kHz sample rate will represent the sine wave all right, but any noise that is beyond $f_s/2$ (2.75 kHz) will alias. Is this a disaster? That depends on the **power spectrum** (amplitude squared versus frequency) of the noise or interfering signal. Say that the noise is very, very small in amplitude—much less than the resolution of your ADC. In that case it will be undetectable, even though it violates the Nyquist criterion. The real problems are medium-amplitude noise or spurious signals. Figure 3.17 is a simulated 1-kHz sine wave with low-pass-filtered white noise added to it. If we use a 16-bit ADC, the specifications say its spectral noise floor is about −115 dB below full scale when displayed as a power spectrum. (This corresponds to an RMS signal-to-noise ratio of 98.08 dB for all frequencies up to the Nyquist limit.) Assuming that the signal is sampled at 5.5 kHz as before, the Nyquist limit is 2.75 kHz.

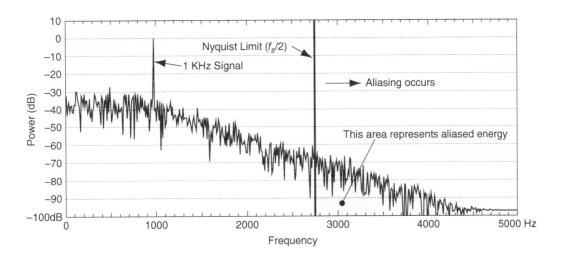

FIG. 3.17 *Power spectrum of a 1-kHz sine wave with low-pass-filtered noise added. If the ADC resolves 16 bits, its spectral noise floor is ≈115 dB. Assuming that we sample at 5.5 kHz, any energy appearing above $f_s/2$ (2.75 kHz) and above ≈115 dB will be aliased.*

Looking at the power spectrum, you can see that some of the noise power is above the floor for the ADC, and there is also noise present at frequencies above 2.75 kHz. The shaded triangle represents aliased energy and gives you a qualitative feel for how much contamination you can expect. Exactly what the contamination will look like is anybody's guess; it depends on the nature of the out-of-band noise. In this case, you can be pretty sure that none of the aliased energy will be above −65 dB. The good news is that this is just plain old uncorrelated noise, so its aliased version will also be plain old uncorrelated noise. It is not a disaster, just another noise source.

Filtering and Averaging

To get rid of aliasing problems, you need to use a **low-pass filter**, known in this case as an **antialiasing filter**. Analog filters are mandatory, regardless of the sampling rate, unless you know the signal's frequency characteristics and can live with the aliased noise. There has to be *something* to limit the bandwidth of the raw signal to $f_s/2$. The analog filter can be in the transducer, in the signal conditioner, on the ADC board, or in all three places. One problem with analog filters is that they can become very complex and expensive. If the desired signal is fairly close to the Nyquist limit, the filter needs to cut off very quickly, implying lots of stages (this is more formally known as the **order** of the filter's **transfer function**).

Digital filters can augment, but not replace, analog filters. Digital filter VIs are included with the LabVIEW analysis library, and they are functionally equivalent to analog filters. The simplest type of digital filter is a **moving averager** (examples of which are available with LabVIEW) that has the advantage of being usable in real time on a sample-by-sample basis. One way to simplify the antialiasing filter problem is to **oversample** the input. If your ADC hardware is fast enough, just turn the sampling rate way up, and then use a digital filter to eliminate the higher frequencies that are of no interest. This makes the analog filtering problem much simpler because the Nyquist frequency has been raised much higher, so the analog filter doesn't have to be so sharp. A compromise is always necessary: You need to sample at a rate high enough to avoid significant aliasing with a modest analog filter; but sampling at too high a rate may not be practical because the hardware is too expensive and/or the flood of extra data may overload your poor CPU.

A potential problem with averaging arises when you handle nonlinear data. The process of averaging is defined to be the summation of several values, divided by the number of values. If your data is, for instance, exponential, then averaging values (a linear operation) will tend to bias the data. (Consider the fact that $e^x + e^y$ is not equal to $e^{(x+y)}$.) One solution is to linearize the data before averaging. In the case of exponential data, you should take the logarithm first. You may also be able to ignore this problem if the values are closely spaced—small pieces of a curve are effectively linear. It's vital that you understand your signals qualitatively and quantitatively before you apply *any* numerical processing, no matter how innocuous it may seem.

If your main concern is the rejection of 60-Hz line frequency interference, an old trick is to average an array of samples over one line period (16.66 ms in the United States). For instance, you could acquire data at 600 Hz and average groups of 10, 20, 30, and so on, up to 600 samples. You should do this for every channel. Using plug-in boards with LabVIEW's data acquisition drivers permits you to adjust the sampling interval with high precision, making this a reasonable option. Set up a simple experiment to acquire and average data from a noisy input. Vary the sampling period, and see if there isn't a null in the noise level at each 16.66-ms multiple.

If you are attempting to average recurrent waveforms to reduce noise, remember that the arrays of data that you acquire must be perfectly in phase. If a phase shift occurs during acquisition, then your waveforms will partially cancel one another or cause distortion. **Triggered** data acquisition (discussed later) is the normal solution because it helps to guarantee that each buffer of data is acquired at the same part of the signal's cycle.

Some other aspects of filtering that may be important for some of your applications are **impulse response** and **phase response**. For ordinary datalogging, these factors are generally ignored. But if you are doing dynamic testing such as in vibration analysis, acoustics, or seismology, then impulse and phase response can be very important. As a rule of thumb, when filters become very complex (high-order), they cut off more sharply, have more radical phase shifts around the cutoff frequency, and, depending on the filter type, exhibit greater ringing on transients. Overall, filtering is a rather complex topic that is best left to the references; the *Active Filter Cookbook* (Lancaster, 1975) is an old favorite.

The best way to analyze your filtering needs is to use a spectrum analyzer. Then you know exactly what signals are present and what has to be filtered out. You can use a dedicated spectrum analyzer instrument (*very* expensive), a digital oscilloscope with FFT capability (or let LabVIEW do the power spectrum), or even a multifunction I/O board running as fast as possible with LabVIEW doing the power spectrum. Spectrum analyzers are included in the data acquisition examples distributed with LabVIEW, and they work quite well.

About ADCs, DACs, and Multiplexers

Important characteristics of an ADC or a digital-to-analog converter (DAC) are resolution, range, speed, and sources of error. (For a detailed look at all these parameters, consult the *Analog-Digital Conversion Handbook,* available from Analog Devices.)

Resolution is the number of bits that the ADC uses to represent the analog signal. The greater the number of bits, the finer the resolution of the converter. Figure 3.18 demonstrates the resolution of a hypothetical 3-bit converter, which can resolve 2^3, or 8, different levels. The sine wave in this figure is not well represented because of the rather coarse **quantization** levels available. Common ADCs have resolutions of 8, 12, and 16 bits, corresponding to 256, 4096, and 65,536 quantization levels. Using high-resolution converters is generally desirable, although they tend to be somewhat slower.

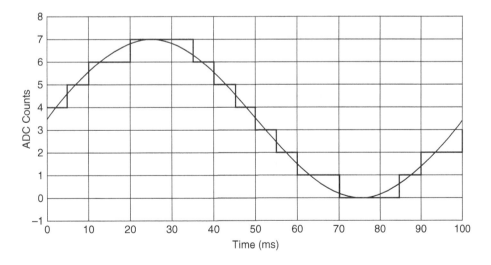

FIG. 3.18 *A sine wave and its representation by a 3-bit ADC sampling every 5 milliseconds.*

Range refers to the maximum and minimum voltage levels that the ADC can quantize. Exceeding the input range results in what is variously termed *clipping, saturation,* or *overflow/underflow,* where the ADC gets stuck at its largest or smallest output code. The **code width** of an ADC is defined as the change of voltage between two adjacent quantization levels or, as a formula,

$$\text{Code width} = \frac{\text{range}}{2^N}$$

where N is the number of bits, and code width and range are measured in volts. A high-resolution converter (lots of bits) has a small code width. The intrinsic range, resolution, and code width of an ADC can be modified by preceding it with an amplifier that adds gain. The code width expression then becomes

$$\text{Code width} = \frac{\text{range}}{\text{gain} \times 2^N}$$

High gain thus narrows the code width and enhances the resolution while reducing the effective range. For instance, a common 12-bit ADC with a range of 0–10 volts has a code width of 2.44 millivolts. By adding a gain of 100, the code width becomes 24.4 microvolts, but the effective range becomes 10/100 = 0.1 volt. It is important to note the trade-off between resolution and range when you change the gain. High gain means that overflow may occur at a much lower voltage.

Conversion speed is determined by the technology used in designing the ADC and associated components, particularly the **sample-and-hold (S/H)** amplifier that freezes the analog signal just long enough to do the conversion. Speed is measured in time per conversion or samples per second. Figure 3.19 compares the typical resolutions and conversion speeds of plug-in ADC boards and traditional instruments. A very common trade-off is resolution versus speed; it simply takes more time or is more costly to precisely determine the exact voltage. In fact, high-speed ADCs, such as flash converters that are often used in digital oscilloscopes, decrease in effective resolution as the conversion rate is increased. Your application determines what conversion speed is required.

There are many sources of **error** in ADCs, some of which are a little hard to quantify; in fact, if you look at the specification sheets for ADCs from different manufacturers, you may not be able to directly compare the error magnitudes because of the varying techniques the manufacturers may have used.

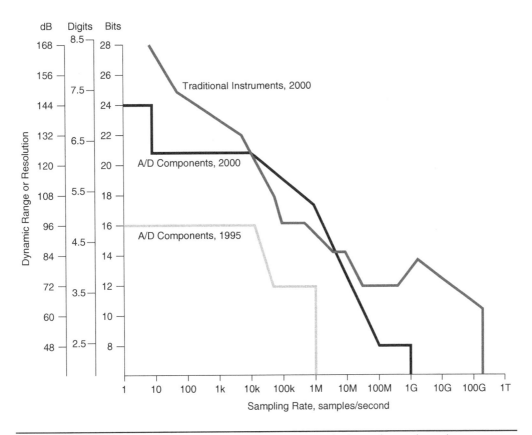

FIG. 3.19 *The performance of A/D components—including plug-in boards—now closely approaches or exceeds the performance of rack-and-stack instruments, and usually at much lower cost. (Courtesy of National Instruments.)*

Simple errors are **gain** and **offset** errors. An ideal ADC follows the equation for a straight line, $y = mx + b$, where y is the input voltage, x is the output code, and m is the code width. Gain errors change the slope m of this equation, which is a change in the code width. Offset errors change the intercept b, which means that 0-volt input doesn't give you zero counts at the output. These errors are easily corrected through calibration. Either you can adjust some trimmer potentiometers so that zero and full-scale match a calibration standard, or you can make measurements of calibration voltages and fix the data through a simple straight-line expression in software. Fancy ADC systems, such as the E-series plug-in boards from NI, include self-calibration right on the board.

Linearity is another problem. Ideally, all the code widths are the same. Real ADCs have some linearity errors that make them deviate from the ideal. One measure of this error is the **differential nonlinearity**, which tells you the worst-case deviation in code width. **Integral nonlinearity** is a measure of the ADC transfer function. It is measured as the worst-case deviation from a straight line drawn through the center of the first and last code widths. An ideal converter would have zero integral nonlinearity. Nonlinearity problems are much more difficult to calibrate out of your system. To do so would mean taking a calibration measurement at each and every quantization level and using that data to correct each value. In practice, you just make sure that the ADC is tightly specified and that the manufacturer delivers the goods as promised.

If a **multiplexer** is used before an ADC to scan many channels, **timing skew** will occur (Fig. 3.20). Since the ADC is being switched between several inputs, it is impossible for it to make all the measurements simultaneously. A delay between the conversion of each channel results, and this is called *skew*. If your measurements depend on critical timing (phase matching) between channels, you need to know exactly how much skew there is in your ADC system.

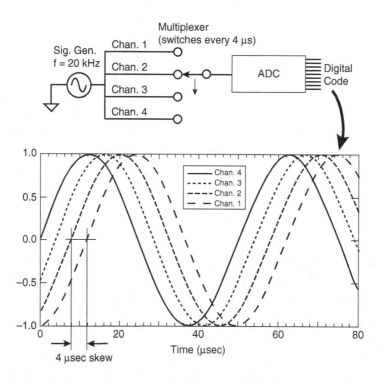

FIG. 3.20 *Demonstration of skew in an ADC with a multiplexer. Ideally, there would be zero switching time between channels on the multiplexer; this one has 4 microseconds. Since the inputs are all the same signal, the plotted data shows an apparent phase shift.*

Multiplexers are used as cost-saving devices, since ADCs tend to be among the most expensive parts in the system, but the skew problem can be intolerable in some applications. Then multiple ADCs or sample-and-hold amplifiers (one per channel) are recommended.

You can remove the timing skew in software as long as you know exactly what the skew is. Most plug-in DAQ boards work as follows: By default, the multiplexer scans through the channel list at the top speed of the ADC. For instance, if you have a 100-kHz ADC, there will be a skew of approximately 10 microseconds between channels. However, the hardware and/or the driver software may increase this value somewhat due to all the extra settling time. For best results, you probably will have to acquire some test data to be sure. You can also choose to adjust the **interchannel delay** (i.e., the skew) with the DAQ VIs in LabVIEW. That feature lets *you* specify the skew, which makes it easier to correct the data's time base. But chances are, even with your adjustments and corrections, the results won't be quite as accurate as the multiple-ADC approach, and that's why we prefer to use an optimum hardware solution.

One misconception about ADCs is that they always do some averaging of the input signal between conversions. For instance, if you are sampling every millisecond, you might expect the converted value to represent the average value over the last millisecond of time. This is true only for certain classes of ADCs, namely, dual-slope integrators and voltage-to-frequency converters. Ubiquitous high-speed ADCs, as found on most multifunction boards, are always preceded by sample-and-hold amplifiers.

The S/H samples the incoming signal for an extremely brief time (the **aperture time**) and stores the voltage on a capacitor for measurement by the ADC. The aperture time depends on the resolution of the ADC and is in the nanosecond range for 16-bit converters. Therefore, if a noise spike should occur during the aperture time, you will get an accurate measurement of the *spike,* not the real waveform. This is one more reason for using a low-pass filter.

When a multiplexer changes from one channel to another, there is a period of time during which the output is in transition, known as the **settling time**. Because of the complex nature of the circuitry in these systems, the transition may not be as clean as you might expect. There may be an overshoot with some damped sinusoidal ringing. Since your objective is to obtain an accurate representation of the input signal, you must wait for these aberrations to decay away. Settling time is the amount of time required for the output voltage to begin tracking the input voltage within a specified error band after a change of channels has occurred. It is clearly specified on all ADC system data sheets. You should not attempt to acquire data faster than the rate determined by the settling time plus the ADC conversion time.

Another unexpected source of input error is sometimes referred to as **charge pump-out**. When a multiplexer switches from one channel to the next, the input capacitance of the multiplexer (and the next circuit element, such as the sample-and-hold) must charge or discharge to match the voltage of the new input signal. The result is a small glitch induced on the input signal, either positive or negative, depending upon the relative magnitude of the voltage on the preceding channel. If the signal lines are long, you may also see ringing. Charge pump-out effects add to the settling time in an unpredictable manner, and they may cause momentary overloading or gross errors in high-impedance sources. This is another reason to use signal conditioning; an input amplifier provides a buffering action to reduce the glitches.

Many systems precede the ADC with a PGIA. Under software control, you can change the gain to suit the amplitude of the signal on each channel. A true instrumentation amplifier with its inherent differential connections is the predominant type. You may have options, through software registers or hardware jumpers, to defeat the differential mode or use various signal grounding schemes, as on the NI multifunction boards. Study the available configurations and find the one best suited to your application. The one downside to having a PGIA in the signal chain is that it invariably adds some error to the acquisition process in the form of offset voltage drift, gain inaccuracy, noise, and bandwidth. These errors are at their worst at high-gain settings, so study the specifications carefully. An old axiom in analog design is that high gain and high speed are difficult to obtain simultaneously.

Digital-to-Analog Converters

DACs perform the reverse action of ADCs: A digital code is scaled to a proportional analog voltage. We use DACs to generate analog stimuli, such as test waveforms and actuator control signals. By and large, they have the same general performance characteristics as do ADCs. Their main limitations, besides the factors we've already discussed, are the **settling time** and **slew rate**. When you make a change in the digital code, the output is expected to change instantaneously to the desired voltage. How fast the change actually occurs (measured in volts per second) is the *slew rate*. Hopefully, a clean, crisp step will be produced. In actuality, the output may overshoot and ring for awhile or may take a more leisurely, underdamped approach to the final value. This represents the *settling time*. If you are generating high-frequency signals (audio or above), you need faster settling times and slew rates. If you are controlling the current delivered to a heating element, these specifications probably aren't much of a concern.

When a DAC is used for waveform generation, it must be followed by a low-pass filter, called a **reconstruction filter**, that performs an anti-aliasing function in reverse. Each time the digital-to-analog (D/A) output is updated (at an interval determined by the time base), a step in output voltage is produced. This step contains a theoretically infinite number of harmonic frequencies. For high-quality waveforms, this out-of-band energy must be filtered out by the reconstruction filter. DACs for audio and dynamic signal applications include such filters and have a spectrally pure output. Ordinary data acquisition boards generally have no such filtering and will produce lots of spurious energy. If spectral purity and transient fidelity are important in your application, be mindful of this fact.

Digital Codes

The pattern of bits—the digital *word*—used to exchange information with an ADC or DAC may have one of several coding schemes, some of which aren't intuitive.

If you ever have to deal directly with the I/O hardware (especially in lower-level driver programs), you will need to study these schemes. If the converter is set up for **unipolar** inputs (all-positive or all-negative analog voltages), the binary coding is straightforward, as in Table 3.2. But to represent both polarities of numbers for a **bipolar** converter, a **sign bit** is needed to indicate the signal's polarity. The bipolar coding schemes shown in Table 3.3 are widely used. Each has advantages, depending on the application.

Decimal Equivalent	Decimal Fraction of Full Scale		Straight Binary
	Positive	Negative	
7	7/8	−7/8	111
6	6/8	−6/8	110
5	5/8	−5/8	101
4	4/8	−4/8	100
3	3/8	−3/8	011
2	2/8	−2/8	010
1	1/8	−1/8	001
0	0/8	−0/8	000

TABLE 3.2 *Straight binary coding scheme for unipolar, 3-bit converter.*

Triggering and Timing

Triggering refers to any method by which you synchronize an ADC or DAC to some event. If there is a regular event that causes each individual ADC, it's called the **time base**, or *clock,* and it is usually generated by a crystal-controlled clock oscillator. For this discussion, we'll define triggering as an event that starts or stops a *series* of conversions that are individually paced by a time base.

When should you bother with triggering? One situation is when you are waiting for a transient event to occur—a single pulse. It would be wasteful (or maybe impossible) to run your ADC for a long time, filling up memory and/or disk space, when all you are interested in is a short burst of data before and/or after the trigger event. Another use for triggering is to force your data acquisition to be in phase with the signal. Signal analysis may be simplified if the waveform always starts with the same polarity and level. Or, you may want to acquire many buffers of data from a recurrent waveform (such as a sine wave) in order to average them, thus reducing the noise. Trigger sources come in three flavors: external, internal, and software generated.

Decimal Equivalent	Fraction of Full Scale	Sign & Magnitude	Two's Complement	Offset Binary
7	7/8	0111	0111	1111
6	6/8	0110	0110	1110
5	5/8	0101	0101	1101
4	4/8	0100	0100	1100
3	3/8	0011	0011	1011
2	2/8	0010	0010	1010
1	1/8	0001	0001	1001
0	0+	0000	0000	1000
0	0−	1000	0000	1000
−1	−1/8	1001	1111	0111
−2	−2/8	1010	1110	0110
−3	−3/8	1011	1101	0101
−4	−4/8	1100	1100	0100
−5	−5/8	1101	1011	0011
−6	−6/8	1110	1010	0010
−7	−7/8	1111	1001	0001
−8	−8/8	not represented	1000	0000

TABLE 3.3 *Some commonly used coding schemes for bipolar converters, a 4-bit example.*

External triggers are digital pulses, usually produced by specialized hardware or a signal coming from the equipment that you are interfacing with. An example is a function generator that has a connector on it called *sync* that produces a TTL pulse every time an output waveform crosses 0 volt in the positive direction. Sometimes, you have to build your own trigger generator. When you are dealing with pulsed light sources (such as some lasers), an optical detector such as a photodiode can be used to trigger a short but high-speed burst of data acquisition. Signal conditioning is generally required for external triggers because most data acquisition hardware demands a clean pulse with a limited amplitude range.

Internal triggering is built into many data acquisition devices, including oscilloscopes, transient recorders, and multifunction boards. It is basically an analog function in which a device called a **comparator** or **discriminator** detects the signal's crossing of a specified level. The slope of the signal may also be part of the triggering criteria. Really sophisticated instruments permit triggering on specified patterns, which is especially useful in digital logic and communications signal analysis. The onboard triggering features of National Instruments' boards are easy to use in LabVIEW, courtesy of the data acquisition VIs. Most boards include an advanced system timing controller (STC) chip with programmable function inputs that solve some of the more difficult triggering problems you may encounter.

Software-generated triggers require a program that evaluates an incoming signal or some other status information and decides when to begin saving data or generating an output. A trivial example is the Run button in LabVIEW that starts up a simple data acquisition program. In that case, the operator is the triggering system. On-the-fly signal analysis is a bit more complex and quickly runs into performance problems if you need to look at fast signals. For instance, you may want to save data from a spectrum analyzer only when the process temperature goes above a certain limit. That should be no problem, since the temperature probably doesn't change very fast. A difficult problem would be to evaluate the distortion of an incoming audio signal and save only the waveforms that are defective. That might require DSP hardware; it might not be practical at all, at least in real time. The NI DAQ driver includes basic software trigger functionality, such as level and slope detection, that works with many plug-in boards and is very easy to use.

A Little Noise Can Be a Good Thing

Performance of an analog-to-digital (A/D) or D/A system can be enhanced by adding a small amount of noise and by averaging (Lipshitz et al., 1992). Any time the input voltage is somewhere between two quantization levels of the A/D system and there is some **dither** noise present, the least-significant bit (LSB) tends to toggle among a few codes. For instance, the duty cycle of this toggling action is exactly 50 percent if the voltage is exactly between the two quantization levels. Duty cycle and input voltage track each other in a nice, proportional manner (except if the converter demonstrates some kind of nonlinear behavior). All you have to do is filter out the noise, which can be accomplished by averaging or other forms of digital filtering.

A source of uncorrelated dither noise, about 1 LSB peak to peak or greater, is required to make this technique work. Some high-performance ADC and DAC systems include dither noise generators; digital audio systems and the NI dynamic signal acquisition boards are examples. High-resolution converters (16 bits and greater) generally have enough thermal noise present to supply the necessary dithering. Incidentally, this resolution enhancement occurs even if you don't apply a filter; filtering simply reduces the noise level.

Figure 3.21 demonstrates the effect of dither noise on the quantization of a slow, low-amplitude ramp signal. To make this realistic, say that the total change in voltage is only about 4 LSBs over a period of 10 seconds (Fig. 3.21A). The vertical axis is scaled in LSBs for clarity. The sampling rate is 20 Hz. In Fig. 3.21B, you can see the coarse quantization steps expected from an ideal noise-free ADC. Much imagination is required to see a smooth ramp in this graph. In Fig. 3.21C, dither noise with an amplitude of 1.5 LSB peak to peak has been added to the analog ramp. Figure 3.21D is the raw digitized version of this noisy ramp. Contrast this with

Fig. 3.21B, the no-noise case with its coarse steps. To eliminate the random noise, we applied a simple *boxcar* filter where every 10 samples (0.5 second worth of data) is averaged into a single value; more elaborate digital filtering might improve the result. Figure 3.21E is the recovered signal. Clearly, this is an improvement over the ideal, noiseless case, and it is very easy to implement in LabVIEW. Later we'll show you how to oversample and average to improve your measurements.

Low-frequency analog signals give you some opportunities to further improve the quality of your acquired data. At first glance, that thermocouple signal with a sub-1-Hz bandwidth and little noise could be adequately sampled at 2 or 3 Hz. But by **oversampling**—sampling at a rate several times higher than the minimum specified by the Nyquist rate—you can enhance resolution and noise rejection.

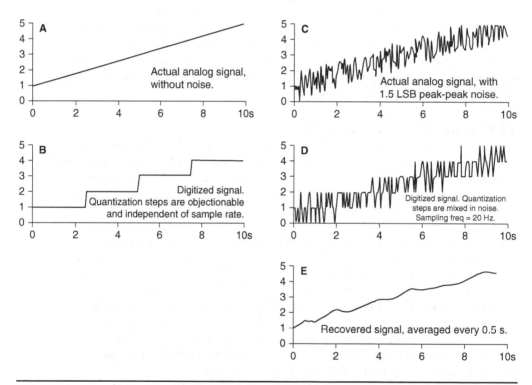

FIG. 3.21 *Graphs in (A) and (B) represent digitization of an ideal analog ramp of 4-LSB amplitude that results in objectionable quantization steps. Adding 1-LSB peak-to-peak dither noise and low-pass filtering [graphs in (C) through (E)] improves results.*

Noise is reduced in proportion to the square root of the number of samples that are averaged. For example, if you average 100 samples, the standard deviation of the average value will be reduced by a factor of 10 when compared to a single measurement. Another way of expressing this result is that you get a 20-dB improvement in signal-to-noise ratio when you average 100 times as many samples. This condition is true as long as the A/D converter has good linearity and a small amount of dither noise. This same improvement occurs with repetitive signals, such as our 1-kHz sine wave with dither noise. If you synchronize your ADC with the waveform by triggering, you can average several waveforms. Since the noise is not correlated with the signal, the noise once again averages to zero according to the square root rule.

Throughput

A final consideration in your choice of converters is the **throughput** of the system, a yardstick for overall performance, usually measured in samples per second. Major factors determining throughput are as follows:

- A/D or D/A conversion speed
- Use of multiplexers and amplifiers, which may add delays between channels
- Disk system performance, if streaming data to or from disk
- Use of direct memory access (DMA), which speeds data transfer
- CPU speed, especially if a lot of data processing is required
- Operating system overhead

The glossy brochure or data sheet you get with your I/O hardware rarely addresses these very real system-oriented limitations. Maximum performance is achieved when the controlling program is written in assembly language, one channel is being sampled with the amplifier gain at minimum, and data is being stored in memory with no analysis or display of any kind. Your application will always be somewhat removed from this particular benchmark.

Double-buffered DMA for acquisition and waveform generation is built into the LabVIEW support for many I/O boards and offers many advantages in speed because the hardware does all the real-time work of transferring data from the I/O to main memory. Your program can perform analysis, display, and archiving tasks while the I/O is in progress (you can't do *too* much processing though). Refer to your LabVIEW data acquisition VI library reference manual for details on this technique.

Performance without DMA is quite limited. Each time you send a command to a plug-in board, the command is processed by the NI DAQ driver, which in turn must get permission from the operating system to perform an I/O operation. There is a great deal of overhead in this process, typically on the order of a few hundred microseconds. That means you can read one sample at a time from an input at a few kilohertz without DMA. Clearly, this technique is not very efficient and should be used only for infrequent I/O operations.

Writing a Data Acquisition Program

We're going to bet that your first LabVIEW application was (or will be) some kind of data acquisition system. We say that because data acquisition is by far the most common LabVIEW application. Every experiment or process has signals to be measured, monitored, analyzed, and logged, and each signal has its own special requirements. Although it's impossible to design a universal data acquisition system to fit every situation, there are plenty of common architectures that you can use, each containing elements that you can incorporate into your own problem-solving toolkit.

Your data acquisition application might include some control (output) functionality as well. Most experiments have some things that need to be manipulated—some valves, a power supply set point, or maybe a motor. That should be no problem as long as you spend some time designing your program with the expectation that you will need some control features.

Plan your application as you would any other, going through the recommended steps. We've added a few items to the procedure that emphasize special considerations for data acquisition: **data analysis**, **throughput**, and **configuration management** requirements. Here are the basic steps.

1. Define and understand the problem; define the signals and determine what the data analysis needs are.

2. Specify the type of I/O hardware you will need, and then determine sample rates and total throughput.

3. Prototype the user interface and decide how to manage configurations.

4. Design and then write the program.

If your system requires extra versatility, such as the ability to quickly change channel assignments or types of I/O hardware, then you will need to include features to manage the system's **configuration**. Users should be able to access a few simple controls rather than having to edit the diagram when a configuration change is needed.

Data **throughput**, the aggregate sampling rate measured in samples per second, plays a dominant role in determining the architecture of your program. High sampling rates can severely limit your ability to do real-time analysis and graphics. Even low-speed systems can be problematic when you require accurate timing. Fortunately, there are plenty of hardware and software solutions available in the LabVIEW world.

The reason for assembling a data acquisition system is to **acquire** data, and the reason for acquiring data is to **analyze** it. Surprisingly, these facts are often overlooked. Planning to include analysis features and appropriate file formats will save you (and the recipients of your data) a lot of grief.

The canonical LabVIEW VI for a simple, yet complete, data acquisition program is shown in Fig. 3.22. It begins with a VI that handles I/O configuration and the opening of any data files. The rest of the program resides in a While Loop that cycles at a rate determined by the sample interval control. The Read Data VI communicates with hardware and returns the raw data, which is then analyzed for display and stored in files. All subVIs that can produce an error condition are linked by an error I/O cluster, and an error handler tells the user what went wrong. Simple as this diagram is, it could actually work, and do some rather sophisticated processing at that. Try simplifying your next data acquisition problem to this level. Add functionality as required, but keep things modular so that it's easy to understand and modify. This chapter is devoted to the building blocks shown in Fig. 3.22.

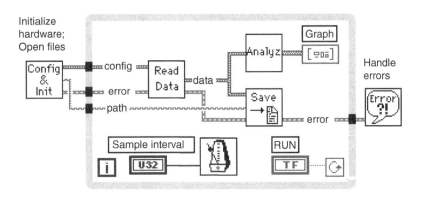

FIG. 3.22 *A generic data acquisition program includes the functions shown here.*

Data Analysis

Data analysis has a different meaning in every application. It depends on the kind of signals you are faced with. For many applications, it means calculating simple statistics (minimum, maximum, mean, standard deviation, etc.) over some period of time. In spectroscopy and chromatography, it means peak detection, curve fitting, and integration. In acoustics and vibration studies, it means Fourier transforms, filtering, and correlation. Each type of analysis affects your LabVIEW program design in some way. For instance, doing a FFT on a large array in real time requires lots of processing power—your system could become so burdened that data collection might be disrupted. Such analysis drives the performance requirements of your VIs. And everyone worries about timing information, both for real-time analysis and for reading data from files. It's obvious that your program has to measure and store time markers reliably and in a format that is useful to the analysis programs.

A technique used in *really* fast diagnostics is to add a timing **fiducial** pulse to one or more data channels. Also known as a *fid*, this pulse occurs at some critical time during the experiment and is recorded on all systems (and maybe on all channels as well). It's much the same as the room full of soldiers and the commander says, "Synchronize watches." For example, when you are testing explosives, a fiducial pulse is distributed to all the diagnostic systems just before detonation. For analog data channels, the fiducial pulse can be coupled to each channel through a small capacitor, creating a small *glitch* in the data at the critical moment. You can even synchronize nonelectronic systems by generating a suitable stimulus, such as flashing a strobe in front of a movie or video camera. Fiducial pulses are worth considering any time you need absolute synchronization among disparate systems.

What you need to avoid is *analysis by accident*. Time and again we've seen LabVIEW programs that grab data from the hardware and stuff it into a file with no thought about compatibility with the analysis program. Then the poor analyst has to grind along, parsing the file into readable pieces, trying to reconstitute important features of the data set. Sometimes, the important information isn't available on disk at all, and you *hope* that it has been written down *somewhere*. Disaster! Gastric distress also occurs when a new real-time analysis need crops up and your program is so inflexible that the new features can't be added without major surgery.

We recommend a preemptive strike. When someone proposes a new data acquisition system, make it a point to force that person to describe, in detail, how the data will be analyzed. Make sure he or she understands the implications of storing the megabytes or gigabytes of data that an automated data acquisition system may collect. If there is a collective shrug of shoulders, ask them point-blank, "... Then why are we collecting data at all?" *Do not write your data acquisition program until you understand the analysis requirements.*

Finally, you can get started. Divide the analysis job into real-time and postrun tasks, and determine how each aspect will affect your program.

Real-Time Analysis and Display

LabVIEW's extensive library of built-in analysis functions makes it easy to process and display your newly acquired data in real time. You are limited by only two things: your system's performance and your imagination. Analysis and presentation are the things that make this whole business of virtual instrumentation useful. The software is the instrument. You can turn a voltmeter into a strip chart recorder, an oscilloscope into a spectrum analyzer, and a multifunction plug-in board into ... just about anything. Here, we'll look at the general problems and approaches to real-time analysis and display.

Once again, we've got the old battle over the precise definition of **real time**. It is wholly dependent upon your application. If 1-minute updates for analysis and display are enough, then *1 minute* is real time. If you need millisecond updates, then *that's* real time, too. What matters is that you understand the fundamental limitations of your computer, I/O hardware, and LabVIEW with regard to performance and response time.

The kind of analysis you need to perform is determined by the nature of your signals and the information you want to extract (Table 3.4). Assuming that you purchase the full version of LabVIEW, there are about 400 analysis functions available. Other functions (and useful combinations of the regular ones) are available from the examples and from others who support LabVIEW through the Alliance Program. If you ever need an analysis function that seems obvious or generally useful, be sure to contact NI to find out if it's already available. NI also takes suggestions—user input is really what makes this palette grow.

Signal Type	Typical Analysis
Analog-DC	Scaling Statistics Curve fitting
Analog-Time Domain	Scaling Statistics Filtering Peak detection and counting Pulse parameters
Analog-Frequency Domain	Filtering Windowing FFT/Power spectrum Convolution/Deconvolution Joint time frequency analysis
Digital On-Off	Logic
Digital Pulse Train	Counting Statistics Time measurement Frequency measurement

TABLE 3.4 *Signal types and analysis examples.*

Continuous versus Single-Shot Data Analysis. Data acquisition may involve either **continuous data** or **single-shot data**. Continuous data generally arrives one sample at a time, like readings from a voltmeter. It is usually displayed on something such as a strip chart, and probably would be stored to disk as a time-dependent history.

Single-shot data arrives as a big buffer or block of samples, like a waveform from an oscilloscope. It is usually displayed on a graph, and each shot would be stored as a complete unit, possibly in its own file. Analysis techniques for these two data types may have some significant differences.

There is a special form of continuous data that we call **block-mode continuous data** where you continuously acquire measurements, but only load them into your LabVIEW program as a block or buffer when some quantity of measurement has accumulated. Multiple buffering or circular buffering can be carried out by any smart instrument, including DAQmx. The advantage of block-mode buffered operation is the reduced I/O overhead: You need to fetch data only when a buffer is half full, rather than fetch each individual sample. The disadvantage is the added latency between acquisition of the oldest data in the buffer and the transfer of that data to the program for processing. For analysis purposes, you may treat this data as either continuous or single-shot, since it has some properties of both.

Here is an example of the difference between processing continuous and single-shot data. Say that your main interest lies in finding the mean and standard deviation of a time-variant analog signal. This is really easy to do, you notice, because LabVIEW just happens to have a statistical function called **Standard Deviation and Variance** that also computes the mean. So far, so good.

Single-Shot Data. A single buffer of data from the desired channel is acquired from a plug-in board. DAQmx returns a numeric array containing a sequence of samples, or waveform, taken at a specified sample rate. To compute the statistics, wire the waveform to the **Standard Deviation and Variance** function and display the results (Fig. 3.23). You might note that there is a conversion function before the input terminal to Standard Deviation.

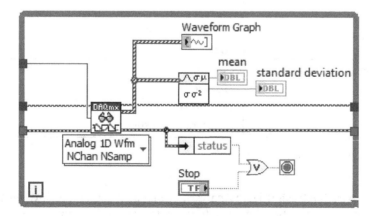

FIG. 3.23 *Statistics are easy to calculate by using the built-in Standard Deviation and Variance function when data is acquired as a single shot (or buffer) with DAQmx.*

Continuous Data. You can collect one sample per cycle of the While Loop by calling **AI Single Scan**, as shown in Fig. 3.24. If you want to use the built-in Standard Deviation function, you have to put all the samples into an array and wait until the While Loop finishes running—not exactly a real-time computation. Or, you could build the array one sample at a time in a Shift Register and call the Standard Deviation function each time. That may seem OK, but the array grows without limit until the loop stops—a waste of memory at best, or you may cause LabVIEW to run out of memory altogether. The best solution is to create a different version of the mean and standard deviation algorithm, one that uses an **incremental** calculation.

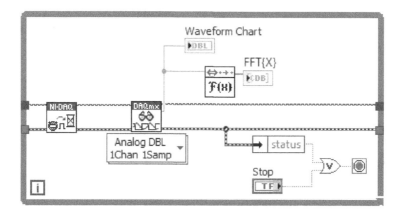

FIG. 3.24 *The Point-by-Point FFT VI operates on a single data point at a time using a moving buffer of data.*

NI introduced another option for single-shot data analysis called the **Point-by-Point** library. It contains nearly all the analysis functions, including signal generation, time and frequency domain, probability and statistics, filters, windows, array operations, and linear algebra. Internally, each function uses the same basic techniques we've just discussed for the Running Mean and Sigma VI. But there's some real numerical sophistication in there, too. Consider what it means to do an FFT calculation on a point-by-point basis. A simple solution would be to just save up a circular buffer of data and then call the regular FFT function on each iteration. Instead, NI implements the mathematical definition of the FFT, which is the integral of a complex exponential. This is, in fact, much faster, but makes sense only when performed on a sliding buffer. As a benchmark, you'll find that a 1024-point FFT runs about 50 percent faster in the point-by-point mode. It's an even larger difference when you're processing a buffer that's not a power of 2.

Faster Analysis and Display. Real-time analysis may involve significant amounts of mathematical computation. Digital signal processing functions, such as the FFT and image processing functions, operate on large arrays of data and may require many seconds, even on the fastest computers. If execution time becomes a problem, you can:

- Make sure that you are using the most efficient computation techniques. Try to simplify mathematical expressions and processes and seek alternative algorithms.

- Avoid excessive array manipulation and duplication.

- Figure out ways to reduce the amount of data used in the calculations. Decimation is a possibility.

- Do the analysis postrun instead of in real time.

- Get a faster computer.

Some of the options you may reject immediately, such as postprocessing, which is of little value when you are trying to do feedback control. On the other hand, if you really *need* that 8192-point power spectrum displayed at 5 kHz, then you had better be using something faster than the average PC. Remember that the LabVIEW user-interface loop runs at 100 Hz, and fast gaming monitors refresh at 120 Hz, so any faster display updates are wasted. Always be sure that the analysis and display activities don't interfere with acquisition and storage of data. One technique for asynchronous display of data is to send buffers of data using a notifier. A notifier, as discussed in Chapter 2 "LabVIEW Fundamentals," is a lossy transfer mechanism. Use a notifier in a high-speed control loop to send data to a slower display loop.

Reducing the Volume of Data. Execution time for most algorithms is roughly proportional to the size of the data arrays. See if you can do something to reduce the size of your arrays, especially when they are to be processed by one of the slower functions. Here are some ideas:

Sample at the minimum rate consistent with the Nyquist criteria and input filtering for your signals. Many times, your data acquisition system will have to sample several channels that have widely different signal bandwidths. You may be able to rewrite the acquisition part of your program so that the low-bandwidth signals are sampled at a slower rate than the high-bandwidth signals. This may be more complex than using a single I/O function that reads all channels at once, but the reduction in array size may be worthwhile.

Process only the meaningful part of the array. Try to develop a technique to locate the interesting part of a long data record, and extract only that part by using the **Split Array** or **Array Subset** functions. Perhaps there is some timing information that points to the start of the important event. Or, you might be able to search for a critical level by using **Search 1D Array**, **Peak Detector**, or a **Histogram** function. These techniques are particularly useful for sparse data, such as that received from a seismometer. In seismology, 99.9 percent of the data is just a noisy baseline containing no useful information. But every so often an interesting event is detected, extracted, and subjected to extensive analysis. This implies a kind of triggering operation. DAQmx has a software triggering feature whereby data is transferred to the LabVIEW data space only if it passes some triggering criteria, including slope and level. This feature works for plug-in boards that don't have similar hardware triggering functionality.

Data decimation is another possible technique. **Decimation** is a process whereby the elements of an array are divided up into output arrays, much as a dealer distributes cards. The **Decimate 1D Array** function can be sized to produce any number of output arrays. Or, you could write a program that averages every *n* incoming values into a smaller output array. Naturally, there is a performance price to pay with these techniques; they involve some amount of computation or memory management. Because the output array(s) is (are) not the same size as the input array, new memory buffers must be allocated, and that takes time. But the payoff comes when you finally pass a smaller data array to those very time-consuming analysis VIs.

Improving Display Performance. All types of data displays—especially graphs and images—tend to bog down your system. Consider using smaller graphs and images, fewer displayed data points, and less frequent updates when performance becomes a problem.

Sampling and Throughput

How much data do you need to acquire, analyze, display, and store in how much time? The answer to this question is a measure of system **throughput**. Every component of your data acquisition system—hardware and software—affects throughput. We've already looked at some analysis and display considerations. Next, we'll consider the input sampling requirements that determine the basic data generation rate.

Modern instrumentation can generate a veritable flood of data. There are digitizers that can sample at gigahertz rates, filling multimegabyte buffers in a fraction of a second. Even the ubiquitous, low-cost, plug-in data acquisition boards can saturate your computer's bus and disk drives when given a chance. But is that flood of data really useful? Sometimes, it depends on the signals you are sampling.

Signal Bandwidth

As we saw in the section on "Sampling Signals," every signal has a minimum **bandwidth** and must be sampled at a rate at least two times this bandwidth, and preferably more, to avoid **aliasing**. Remember to include significant out-of-band signals in your determination of the sampling rate.

If you can't adequately filter out high-frequency components of the signal or interference, then you will have to sample faster. A higher sampling rate may have an impact on throughput because of the larger amount of raw data that is collected. Evaluate every input to your system and determine what sampling rate is really needed to guarantee high signal fidelity.

Sometimes, you find yourself faced with an overwhelming aggregate sampling rate, such as 50 channels at 180 kHz. Then it's time to start asking simple questions such as: Is all this data really useful or necessary? Quite often, there are channels that can be eliminated because of low priority or redundancy. Or, you may be able to significantly reduce the sampling rate for some channels by lowering the cutoff frequency of the analog low-pass filter in the signal conditioner. The fact that *some* channels need to go fast doesn't mean that they *all* do.

Oversampling and Digital Filtering

Low-frequency analog signals give you some opportunities to further improve the quality of your acquired data. At first glance, that thermocouple signal with a sub-1-Hz bandwidth and little noise could be adequately sampled at 2 or 3 Hz. But by **oversampling**—sampling at a rate several times higher than the Nyquist frequency—you can enhance resolution and noise rejection. Noise is reduced in proportion to the square root of the number of samples that are averaged. For example, if you average 100 samples, the standard deviation of the average value will be reduced by a factor of 10 when compared to a single measurement. This topic is discussed in detail in "A Little Noise Can Be a Good Thing" in "Sampling Signals." We use oversampling for nearly all our low-speed DAQ applications because it's easy to do, requires little extra execution time, and is very effective for noise reduction.

Once you have oversampled the incoming data, you can apply a digital low-pass filter to the raw data to remove high-frequency noise. There are a number of ways to do digital filtering in LabVIEW: by using the **Filter** functions in the analysis library or by writing something of your own. Digital filter design is beyond the scope of this book, but at least we can look at a few ordinary examples that might be useful in a data acquisition system.

An excellent resource for filter design is NI's **Digital Filter Design Toolkit**. It is a stand-alone application for designing and testing all types of digital filters. You typically begin with a filter specification—requirements for the filter response in terms of attenuation versus frequency—and use the filter designer to compute the required coefficients that the LabVIEW VIs can use at runtime. The toolkit makes it easy by graphically displaying all results and allowing you to save specifications and coefficients in files for comparison and reuse. Displays include magnitude, phase, impulse, and step response; a z plane plot; and the z transform of the designed filter. Once you have designed a filter and saved its coefficients, an included LabVIEW VI can load those coefficients for use in real time.

FIR Filters. If you are handling single-shot data, the built-in **Finite Impulse Response (FIR)** filter functions are ideal. The concept behind an FIR filter is a **convolution** (multiplication in the frequency domain) of a set of weighting coefficients with the incoming signal. In fact, if you look at the diagram of one of the FIR filters, you will usually see two VIs: one that calculates the filter coefficients based on your filter specifications, and the **Convolution** VI that does the actual computation. The response of an FIR filter depends only on the coefficients and the input signal, and as a result the output quickly dies out when the signal is removed. That's why they call it *finite* impulse response. FIR filters also require no initialization since there is no memory involved in the response.

For most filters, you just supply the sampling frequency (for calibration) and the desired filter characteristics, and the input array will be accurately filtered. You can also use high-pass, bandpass, and bandstop filters, in addition to the usual low-pass, if you know the bandwidth of interest (Fig. 3.25).

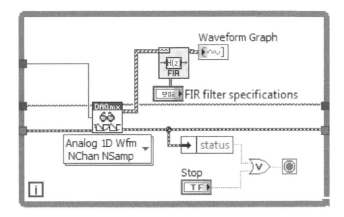

FIG. 3.25 *FIR low-pass filtering applied to an array of waveforms. Bandpass filtering could also be used in this case, if we know the exact frequency of interest.*

The **Digital FIR Filter** VI is set up to perform a low-pass filter operation. The **sampling frequency (fs)** for the filter VI is the same one used by the DAQ VI and is measured in hertz. This calibrates the **low cutoff frequency** control in hertz as well. Since the input signal was 10 Hz, we selected a cutoff of 50 Hz, which is sufficiently high to avoid throwing away any significant part of the signal. Topology is set for the FIR filter by specification, where you supply only the passband and stopband frequencies. Examine the graph and note that the filtered waveform is delayed with respect to the raw data. This is where the Digital Filter Design Toolkit comes in handy. You can choose the optimum cutoff frequency to yield the desired results. Filter design is always a compromise, so don't be disappointed if the final results are in some way less than perfect.

You can compensate for the delay in an FIR filter by routing the filtered signal through the **Array Subset** function and removing the required number of samples from the start of the waveform. That number is approximately equal to the number of taps; being the nonmathematical types, we usually test it out and see what the magic number is. Since the delay is invariant once you set the number of taps, FIR filters can effectively supply zero delay. (Technically, this is because they have *linear phase distortion*.) This is important when you need to maintain an absolute timing reference with respect to a trigger or another signal.

A **time window** generally should be applied to all signals prior to filtering or other frequency domain processing. Time windows reduce **spectral leakage** and other **finite-sequence length** artifacts in spectral estimation calculations such as FFTs and power spectra. Spectral leakage occurs because an FFT has discrete frequency bins. If the signal does not happen to precisely land in one of these bins, the FFT smears the signal energy into adjacent bands. The other problem, due to the finite data record length, shows up as extra energy spread out all over the spectrum. If you think about it, an arbitrary buffer of data probably does not start and end at zero. The instantaneous step from zero to the initial value of the data represents a transient, and transients have energy at all frequencies.

The job of the time window is to gradually force the start and end of your data to zero. It does so by multiplying the data array by some function, typically a cosine raised to some power, which by definition is zero at both ends. This cleans up much of the spectral leakage and finite-sequence length problems, or at least makes them more predictable. One side effect is a change in absolute amplitude in the results. If you use the **Scaled Window** VI from the Signal Processing >> Waveform Conditioning palette, the gain for each type of window is properly compensated. The Digital FIR Filter VI is also properly compensated. There are quite a few window functions available, and each was designed to meet certain needs. You should definitely experiment with various time windows and observe the effects on actual data.

IIR Filters. Infinite Impulse Response (IIR) filters are generally a closer approximation to their analog kin than are FIR filters. The output of an IIR filter depends not only on the *input* signal and a set of **forward coefficients** but also on the *output* of the filter and an additional set of **reverse coefficients**—a kind of feedback. The resulting response takes (theoretically) an infinite amount of time to arrive at its final value—an asymptotic response just like the exponential behavior of analog filters. The advantage of an IIR filter, compared to an FIR filter, is that the feedback permits more rapid cutoff transitions with fewer coefficients and hence fewer computations. However, IIR filters have *nonlinear phase distortion,* making them unsuitable for some phase-sensitive applications. Also, you can choose a set of coefficients that make an IIR filter *unstable*, an unavoidable fact with any feedback system.

IIR filters are best suited to continuous data, but with attention to initialization they are also usable with single-shot data. In the Filter palette, you will find most of the classic analog filter responses in ready-to-use VIs: Butterworth, Chebyshev, Inverse Chebyshev, elliptical, and Bessel. If you are at all familiar with *RC* or active analog filters, you'll be right at home with these implementations. In addition, you can specify arbitrary responses not available using analog techniques. Of course, this gets a bit tricky, and you will probably want the Digital Filter Design Toolkit or some other mathematical tool to obtain valid coefficients.

For classical filters such as the Butterworth filter, controls are quite similar to the design parameters for corresponding analog filters. In particular, you choose the **filter type** (high-pass, low-pass, etc.), and then you choose the **order**. Order, in the analog world, translates to the number of inductors and capacitors in the filter implementation. Higher order results in a sharper response, but as you might expect, the implementation requires additional computations. Sampling frequency and high- and low-cutoff frequencies are calibrated in hertz, as with the FIR filters.

Median Filters. If your data contains **outliers**—also known as *spikes* or *fliers*—consider the **Median Filter** VI. The median filter is based on a statistical, or nonlinear, algorithm. Its only tuning parameter is called **rank**, which determines the number of values in the incoming data that the filter acts upon at one time. For each location i in the incoming array, the filter sorts the values in the range ($i - rank$) to ($i + rank$) and then selects the median (middle) value, which then becomes the output value. The algorithm slides along through the entire data set in this manner. The beauty of the median filter is that it neatly removes outliers while adding no phase distortion. In contrast, regular IIR and FIR filters are nowhere near as effective at removing outliers, even with very high orders. The price you pay is speed: The median filter algorithm tends to be quite slow, especially for higher rank. To see how the median filter works, try out the Median Filtering example VI.

Moving Averagers. Moving Averagers are another appropriate filter for continuous data. The one demonstrated in Fig. 3.26, **Moving Avg Array**, operates on data from a typical multichannel data acquisition system. An array containing samples from each channel is the input, and the same array, but low-pass-filtered, is the output. Any number of samples can be included in the moving average, and the averaging can be turned on and off while running. This is a particular kind of FIR filter where all the coefficients are equal to 1; it is also known as a *boxcar* filter.

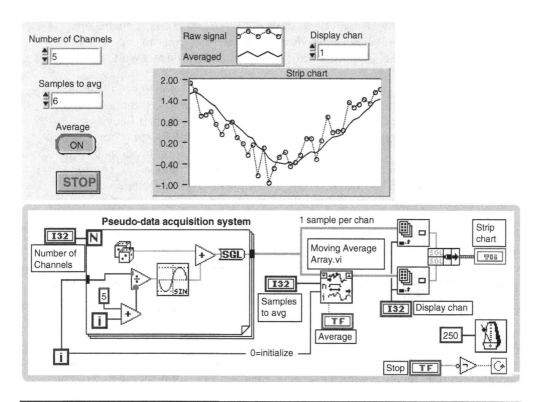

FIG. 3.26 *Demonstration of the Moving Avg Array function, which applies a low-pass filter to an array of independent channels. The For Loop creates several channels of data, like a data acquisition system. Each channel is filtered, and then the selected channel is displayed in both raw and filtered forms.*

Other moving averagers are primarily for use with block-mode continuous data. They use local memory to maintain continuity between adjacent data buffers to faithfully process block-mode data as if it were a true, continuous stream.

After low-pass filtering, you can safely decimate data arrays to reduce the total amount of data. Decimation is in effect a **resampling** of the data at a lower frequency. Therefore, the resultant sampling rate must be at least twice the cutoff frequency of your digital low-pass filter, or else aliasing will occur. For instance, say that you have applied a 1-kHz low-pass filter to your data. To avoid aliasing, the time interval for each sample must be shorter than 0.5 millisecond. If the original data was sampled at 100 kHz (0.01 millisecond per sample), you could safely decimate it by a factor as large as 0.5/0.01, or 50 to 1. Whatever you do, don't decimate without knowledge of the power spectrum of your incoming signal. You can end up with exactly the same result as sampling too slowly in the first place.

Timing Techniques

Using software to control the sampling rate for a data acquisition system can be a bit tricky. Because you are running LabVIEW on a general-purpose computer with lots of graphics, plus all that operating system activity going on in the background, there is bound to be some uncertainty in the timing of events, just as we discussed with regard to timestamps. Somewhere between 1 and 1000 Hz, your system will become an unreliable interval timer. For slower applications, however, a While Loop with a **Wait Until Next ms Multiple** function inside works just fine for timing a data acquisition operation.

The *best* way to pace any sampling operation is with a hardware timer. Most plug-in boards, scanning voltmeters, digitizers, oscilloscopes, and many other instruments have sampling clocks with excellent stability. Use them whenever possible. Your data acquisition program will be simpler and your timing more robust.

If the aggregate sampling rate (total channels per second) is pressing your system's reliable timing limit, be sure to do plenty of testing and/or try to back off on the rate. Otherwise, you may end up with unevenly sampled signals that can be difficult or impossible to analyze. It's better to choose the right I/O system in the first place—one that solves the fast sampling problem for you.

Express VIs provide a simple starting point when you have to get going fast. The DAQ Assistant walks you through configuring your DAQ measurement in just a few simple steps. Figure 3.27 shows the DAQ Assistant's configuration dialog configured to measure 100 samples from each of eight voltage channels. You can interactively test the configuration and change the input signal range if needed. The DAQ Assistant will even show you how to connect the signals for each channel. This example measures simple voltages, but the DAQ Assistant is fully capable of handling thermocouples, RTDs, strain gauges, and just about any measurement you might typically come across in the lab or the field. Once you're done, click OK and the DAQ Assistant will generate the LabVIEW code for your application. You can leave it as an Express VI or right-click and choose to "Generate NI-DAQmx Code." However, once it's converted to a VI, you can no longer use the wizard.

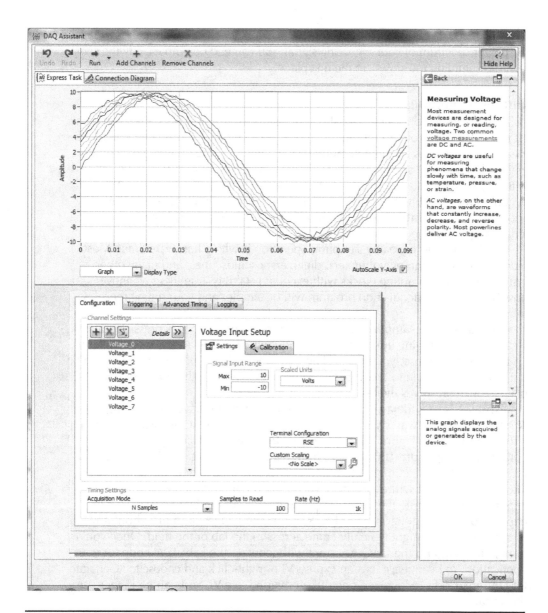

FIG. 3.27 DAQmx Assistant is the easiest way to get started with a DAQ application.

Medium-Speed Acquisition and Processing

Many applications demand that you sample one or more analog channels at moderate speeds, on the order of tens to thousands of samples per second, which we'll call *medium-speed acquisition*. At these speeds, you must have the I/O hardware do more of the time-critical work. The key to any moderate- and high-speed acquisition is **hardware-timed, circular buffered I/O** using **DMA**. Although it sounds complicated, DAQmx sets this up for you transparently. Here's how the technology works:

Timed Buffered I/O. For DAQ applications running at speeds greater than a few tens of hertz, you must turn to hardware-timed buffered I/O. Every DAQ board has a piece of hardware called a **first-in, first-out (FIFO)** memory buffer that is located on the plug-in board. As data is read in, samples from the ADC are temporarily stored in the FIFO pending a bus transfer to your computer's main memory, giving the software extra time to take care of other business. An important piece of information is the size of the available FIFO buffer, which varies among models of plug-in boards. For instance, an E-series board has at least 512 words (a word is 16 bits, or one sample). If you are collecting data from a single channel at 10,000 samples per second, the FIFO buffer will fill up in 51.2 milliseconds. If there is any overhead at all, the driver will not be able to get around to uploading the data before it is overwritten. If your LabVIEW program always had to generate or read each individual sample, there would be little hope of sampling at accurate rates much beyond 10 Hz. With buffered I/O, precision hardware timers control the real-time flow of data between memory buffers and the ADCs and DACs. Buffered operations can be triggered by hardware or software events to start the process, thus synchronizing the process with the physical world. LabVIEW transfers data between the plug-in board and a buffer whenever the buffer is full (for inputs) or empty (for outputs). These operations make maximal use of hardware: the onboard counter/timers and any DMA hardware that your system may have.

DMA. The technique used to get memory off the board and into memory is called direct memory access. DMA is highly desirable because it improves the rate of data transfer from the plug-in board to your computer's main memory. In fact, high-speed operations are often impossible without it. The alternative is interrupt-driven transfers, where the CPU has to laboriously move each and every word of data. Most PCs can handle several thousand interrupts per second, and no more. A DMA controller is built into the motherboard of every PC, and all but the lowest-cost plug-in boards include a DMA interface. The latest boards have multiple DMA channels and *bus mastering* capability. When a board is allowed to take over as bus master (in place of the host CPU), throughput can increase even more because less work has to be done at interrupt time. With DMA, you can (theoretically, at least) acquire data and generate waveforms nearly as fast as your computer's bus can move the data.

Circular Buffered I/O. Circular buffered I/O is a powerful technique because the buffer has an apparently unlimited size, and the input or output operation proceeds *in the background with no software overhead*. A circular buffer is an area of memory that is reused sequentially. Data is streamed from the DAQ hardware to the circular buffer by the DMA controller. When the end of the memory buffer is reached, the buffer wraps around and starts storing at the beginning again. Hopefully you've had time to read the data before it has been overwritten. The only burden on your program is that it must load or unload data from the buffer before it is overwritten. We really like circular buffered I/O because it makes the most effective use of available hardware and allows the software to process data in parallel with the acquisition activity.

Figure 3.28 shows the DAQmx example VI **Voltage—Continuous Input.vi**. This VI follows our canonical DAQ model where we initialize outside the loop, acquire and process (or save) inside the loop, and close out our references after the loop. It's all tied nicely together by chaining the error I/O to enforce dataflow. When we configure our sample clock, DAQmx configures the board for hardware-timed I/O. By enabling continuous samples, DAQmx automatically sets up a circular buffer. Data is streamed to disk inside the loop as a TDMS file. This example can easily stream data from multiple channels at 10,000 samples per second until your hard drive fills up. For more DAQ examples you can modify to fit your application needs, look at the DAQmx examples shipping with LabVIEW.

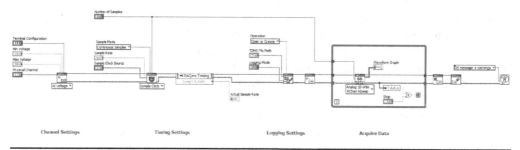

FIG. 3.28 *The Voltage—Continuous Input.vi. This medium-speed example can acquire and save binary data to disk at 10,000 samples per second until you run out of storage.*

Bibliography

Beckwith, T. G., and R. D. Marangoni, *Mechanical Measurements*, Addison-Wesley, Reading, Massachusetts, 1990. (ISBN 0-201-17866-4)

Chugani, M., et al., *LabVIEW Signal Processing*, Prentice-Hall, Englewood Cliffs, New Jersey, 1998.

Gunn, R., "Designing System Grounds and Signal Returns," *Control Engineering*, May, 1987.

Lancaster, D., *Active Filter Cookbook*, Howard W. Sams & Co., Indianapolis, 1975. (ISBN 0-672-21168-8)

Lipshitz, S. P., R. A. Wannamaker, and J. Vanderkooy, "Quantization and Diter: A Theoretical Survey," *J. Audio Eng. Soc.*, 40:(5):355–375 (1992).

Morrison, R., *Grounding and Shielding Techniques in Instrumentation,* Wiley-Interscience, New York, 1986.

Norton, H. R., *Electronic Analysis Instruments,* Prentice-Hall, Englewood Cliffs, New Jersey, 1992. (ISBN 0-13-249426-4)

Omega Engineering, Inc., *Temperature Handbook,* Stamford, Connecticut, 2000. (Available free by calling 203-359-1660 or 800-222-2665.)

Ott, H. W., *Noise Reduction Techniques in Electronic Systems,* John Wiley & Sons, New York, 1988. (ISBN 0-471-85068-3)

Pallas-Areny, R. and J. G. Webster, *Sensors and Signal Conditioning,* John Wiley & Sons, New York, 1991. (ISBN 0-471-54565-1)

Qian, S. and D. Chen, *Joint Time-Frequency Analysis—Methods and Applications,* Prentice-Hall, Englewood Cliffs, New Jersey, 1996. (ISBN 0-13-254384-2. Call Prentice-Hall at 800-947-7700 or 201-767-4990.)

Sheingold, D. H., *Analog-Digital Conversion Handbook,* Prentice-Hall, Englewood Cliffs, New Jersey, 1986. (ISBN 0-13-032848-0)

Steer, R. W., Jr., "Anti-aliasing Filters Reduce Errors in ADC Converters," *EDN,* March 30, 1989.

3M Specialty Optical Fibers, *Fiber Optic Current Sensor Module* (product information and application note), West Haven, Connecticut, (203) 934-7961.

White, D. R. J., *Shielding Design Methodology and Procedures,* Interference Control Technologies, Gainesville, Virginia, 1986. (ISBN 0-932263-26-7)

CHAPTER 4

LabVIEW Object-Oriented Programming

What, Where, When, and Why

One of our challenges when writing this chapter was to answer these four questions. What, where, when, and why LabVIEW object-oriented programming (LVOOP)? It is our hope that this chapter answers all those questions for you. In the end, it is up to you whether you decide to fully adopt LVOOP, go for a mix of LVOOP and non-OOP, or continue to avoid LVOOP.

If you are a LabVIEW programmer who has used LabVIEW for a long time but you are not using LabVIEW classes, this chapter will help you understand what classes are, when to use them, and why. If you are an experienced LVOOP programmer, you may find the Background section interesting as a refresher on how we got here. You can skip the "What?," "Where and When?," "Why?," and maybe even the "How?" sections and jump directly to the SOLID section. If you are a technical leader and are trying to figure out how to bring your team up to speed with LVOOP, you may want to read the whole chapter before you recommend it to the rest of your team.

If you have always been intimidated by LVOOP, don't panic. LVOOP is *not* a completely different language. It is merely a particular way of organizing your G code to provide better scalability and long-term maintainability.

Background

If you want to learn more about how LVOOP came to be, you can read all the details in the History of "History of Object-Oriented Programming for LabVIEW Programmers" section in Chapter 1 of this book.

The question now is why we continue to separate OO from non-OO. Why do we have a whole book about LabVIEW graphical programming and only a section is dedicated to OO? There are a lot of LabVIEW programmers out there who have been using LabVIEW for a long time and still do not use classes. We wanted for them to be able to read this chapter by itself. New LabVIEW programmers probably already learned that an object is just a glorified cluster and have had an easier transition into LVOOP. We believe that some LabVIEW programmers have not integrated classes into their development due to the tax we have to pay when we use OO in LabVIEW. Creating a class requires further design up-front and involves a lot of steps. If you are not careful, you can write code that slows down the IDE or even causes crashes. For real-time applications, there is a performance hit when calling a dynamic dispatch method. Do not get us wrong: the IDE response to classes has only improved with each LabVIEW version, and integrating the GDS toolkit into your development simplifies the number of steps involved in creating a class, gives you a graphical view of your classes, and greatly reduces the technical tax you have to pay to use LVOOP.

With regard to immediacy, experienced LabVIEW programmers are accustomed to opening a block diagram and immediately seeing what the code is doing. Programmers typically spend most of their time on the block diagram, but once a programmer embraces LVOOP, the immediacy is limited. LabVIEW OO programmers spend most of their time on the project. Perhaps one day the project will be graphical, too, and then the immediacy would be there. Or perhaps one day we will have semantic zoom, and the navigation from project overview to code implementation will be seamless. In the meantime, the GDS toolkit with its UML Modeler comes to the rescue by providing a graphical representation of the classes on disk. For NXG, VI Technologies, based in the Netherlands, is working on a UML Class Editor tool that will also help with the visual representation of classes in LabVIEW (for more information visit uml-addon.com).

We believe there is a happy medium. We can keep the benefits of immediacy for parts of the program and take advantage of LVOOP where it is worth it. Where that line between LVOOP and non-OO lies will depend on your style and how comfortable your team is with the OO concepts.

What?

Introduction to Object-Oriented Concepts

If the team you are working with requires you to work using LabVIEW object-oriented syntax or if you have been using it for years but don't quite understand the underlying concepts, we strongly recommend you read the book *The Object-Oriented Thought Process*.[1] In that book, Matt Weisfeld makes an effort to separate the OO concepts from the implementation. Understanding the underlying concepts will make the implementation a lot easier. We will touch on the main concepts in

---

[1] Weisfeld, Matt. *The Object-Oriented Thought Process*—3rd edition: Pearson Education, 2009.

order to make this chapter stand on its own. However, we will not be able to cover every single detail. If you decide to use OO and follow OO design practices, you will find presentations by Stephen Loftus-Mercer, Allen Smith, Dmitry Sagatelyan, and Jon McBee invaluable. They present routinely at CLA Summits and at NI Week. You can find videos of some of these presentations via the Tecnova site.[2]

What Is an Object?

Objects are entities of a system. We already think in terms of objects. When we see an object in real life, we immediately associate attributes and behaviors with the object. For example, if you see an oscilloscope, you may note the brand of the scope and the number of channels it supports. These would be its attributes. You may note that it can amplify or attenuate a signal or trigger a measurement based on a rising edge. These would be its behaviors.

In object-oriented programming, we define objects in our software. Instead of thinking, "Given the requirements of the project, first I have to do X, then I have to do Y, then I have to do Z," you would instead think, "Given the requirements, I will create an object P to do task X and an object R to do tasks Y and Z." You delegate tasks to particular objects in order to make a project more manageable.

How Do We Define Objects?

Every object is of a particular type. The definition of a new type of object is called a **class**. Classes are the primary architecture element of an object-oriented program. We define a class at design time; at runtime, we instantiate one or more objects of each class, assign values to their attributes, and tell them to perform their various behaviors.

What Is a Class in LabVIEW?

A class is a single file that is part of your source code, just like a VI or a typedef. A class file has a ".lvclass" file extension. The file contains the definition of the class attributes (the class private data) and a list of VIs. These VIs define the behaviors of the class. The class has a name, and the VIs that are members of the class use the class name as a prefix to their own file name. For example, if the class is called "Oscilloscope.lvclass" and it has a **member VI** named "Trigger Measurement.vi," then the complete name of the VI is "Oscilloscope.lvclass:Trigger Measurement.vi."[3]

What Is Private Data?

In LabVIEW, you define the attributes of an object using a cluster. In the case of the oscilloscope, the private data would perhaps have two elements: a string (the brand name) and an integer (the number of channels supported). Just like a cluster, the class wire can be bundled and unbundled. But unlike a cluster, where the cluster

---

[2] https://lavag.org/topic/20645-labview-videos-tecnova-download-site/
[3] LabVIEW NXG uses a double colon instead of the single colon for delimited names.

data is public, the class data is private. This means that any VI can bundle and unbundle a cluster and freely access its data, but only members of the class are allowed to bundle and unbundle the class. That privacy means that data stays consistent with the rules established by the class author, so it is far simpler to track down bugs when the data becomes inconsistent because there is a finite number of VIs that can change the data.

What Are Object Behaviors or Methods?
In LabVIEW, you implement all the ways the object can behave via VIs that are part of the class. These member VIs are also called **methods**. In the case of the oscilloscope class, we could have a method to configure the scope, a method to start a triggered measurement, and another method to take a simple snapshot.

What Is the Difference between an Object and a Class?
A class is the library definition that includes the private data and methods for the class. It is the blueprint. In LabVIEW, the class is also the data type definition for the object. Until now, your data types had been numeric, string, cluster, etc. Now you can create your own data types by defining your own classes.

An object is an instantiation of that class. In the oscilloscope example, your `Oscilloscope.lvclass` is the class, but as soon as you initialize your Tektronix scope on your program, the wire is carrying the object that represents the specific Tektronix scope model on your benchtop.

What Are the Differences between LVOOP and Other Languages?
If you have experience with object-oriented programming in other languages, we strongly recommend that you read the *LabVIEW Object-Oriented Programming: The Decisions Behind the Design* white paper at ni.com.[4] We will summarize the main differences here. If you are not familiar with OO, we will explain the foundational concepts in more detail after this table.

C++[5]	C#	LVOOP
A text-based, procedural language		A graphical, dataflow language
Objects passed by value or by reference	Objects exclusively passed by reference	Objects passed by value or by reference through data value references
No ultimate ancestor class	System.Object is the ultimate ancestor	LabVIEW Object is the ultimate ancestor

---

[4] LabVIEW Object-Oriented Programming: The Decisions behind the Design, http://www.ni.com/white-paper/3574/en/
[5] LabVIEW Object-Oriented Programming FAQ, http://www.ni.com/white-paper/3573/en/

C++[5]	C#	LVOOP
Multiple inheritance allowed	Single inheritance with interface support	Single inheritance only, no interface support.[6]
Has constructors		Default constructor only for by value objects; arbitrary constructors for data value references.
Has destructors		No need for destructors on by value objects; destructors available for data value references.
No inherent language support for serialization of data and mutation of data over class revisions	Serialization may be added to classes as needed	Serialization support is fundamental and well-defined for every class. Automatic data mutation history. LabVIEW programmers can retrieve old data, even if the class has been modified.
Has template classes and functions	Has generic classes and generic functions	Template functions only.
Supports pure virtual functions and semi-pure (parent has default implementation but children required to override)	Supports pure virtual functions	Supports semi-pure virtual functions (parent has default implementation but children required to override).

What Is the Unified Modeling Language Standard to Model a Class Diagram?
The Unified Modeling Language (UML) is a family of graphical notations that assist in describing and designing software systems. All software can use UML diagrams; however, software that is object-oriented in nature tends to benefit more from UML diagrams.

UML is an open standard[7] controlled by the Object Management Group (OMG), an open consortium of companies. The UML appeared in 1997 and is the result of

---

[6] During discussions at the CLA events and NIWeek 2019, there were multiple presentations and discussions both about the benefits of having Interfaces in LabVIEW and of possible ways that Interfaces could be added. NI was very interested in these discussions and engaged in debates about how Interfaces could be added to LabVIEW. If these discussions turn into a LabVIEW feature, we will be sure to add examples to our Github repository.

[7] An open standard is a design specification that is publicly available and gives the public various rights to use it. Such standards (including UML) often use a public process for their creation and amendment process.

combining many object-oriented graphical modeling languages from the late 1980s and 1990s.

There are different types of UML diagrams. If you want to get more in depth into UML diagrams, we recommend the book *UML Distilled, Third Edition*, by Martin Fowler. He is one of the authors of the Manifesto for Agile Software Development. In his book, Fowler describes the most commonly used UML diagrams and includes a useful cheat sheet in the front and back cover with typical UML diagram examples.

We have talked with a lot of LabVIEW programmers, and many of them think that the only UML diagram is the class diagram. A class diagram is indeed the most commonly used type of UML diagram. A class diagram describes the types of objects in your application and the relationships between them. The class diagram can also describe the attributes (properties or state data) and behaviors (methods or operations) of a class.

To see an example of an object-oriented project, open the Board Testing example that ships with LabVIEW 8.2 and later. We will use this example throughout the following paragraphs. It is located either in `labview\example\lvoop\BoardTesting\Board Testing.lvproj` or in `labview\example\Object-Oriented Programming\BoardTesting\Board Testing.lvproj`. The example includes a Read_Me.html with more details on both a task-oriented solution and a class-based solution for a hypothetical board inspection system. We will use UML diagrams to describe the classes in that project.

In LVOOP, two classes may have various relationships, but there are three primary kinds of relationships:

1. The Inheritance (aka *"is a"*) relationship
2. The Association (aka *"uses a"*) relationship
3. The Composition (aka *"has a"*) relationship

The following figures describe the different associations seen in the Board Testing example. The main difference between Association and Composition is whether or not class A is passed as a parameter in or out for class B methods but is not stored in B's private data. If it is not in the private data but it is in VI parameters, then it is associated. If it gets stored in the private data, then it is composed. When class A *has a* class B, the owning A class contains the B class or a reference to that class as part of its private data. When class A *uses* class B, class A *uses* class B in some of its methods. In the Board Testing example, the `Get Components.vi` is responsible for determining the Components each Board Design type uses. We created the UML diagrams using the GDS Toolkit that you can download from the LabVIEW Tools Network. You can also download the open-source version via opengds.github.io.

First, let's look at the inheritance relationships for both the Component and the Board Design classes. Both figures include the UML diagram and the LabVIEW class hierarchy view. In these UML diagrams, the inheritance is represented by an arrow that has an empty pointer. When reading the diagram, replace that arrow with the "*is a*" relationship. (See Figs. 4.1 and 4.2.)

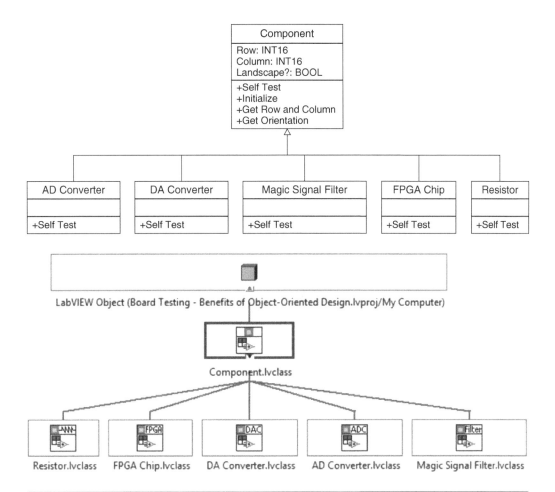

FIG. 4.1 *The UML diagram represents the inheritance diagram for the component used in the LabVIEW Shipping Example "Board Testing - Benefits of Object-Oriented Design." The way to read this diagram out loud is "The AD Converter **is a** Component." The Component has the following attributes: row, column, and landscape. All of the components that inherit from Component have the same attributes. All of the objects can perform a Self Test, Initialize, Get Row and Column, and Get Orientation. The only method that is performed differently for each type of component is Self Test.*

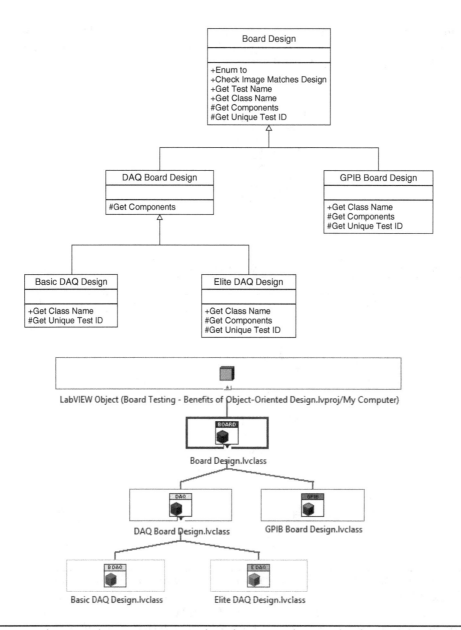

FIG. 4.2 *The UML diagram represents the inheritance diagram for the Board Design used in the LabVIEW Shipping Example "Board Testing - Benefits of Object-Oriented Design." The way to read this diagram out loud is "The Basic DAQ Design **is a** DAQ Board Design." The "DAQ Board Design **is a** Board Design." In this case, none of the classes have any attributes. They all have a unique way to get their components, except for the Basic DAQ Design class that gets its components just as the DAQ Board Design class does.*

Now let's look at the relationship between the Board Designs and the Components. In these UML diagrams, the association is represented by a simple arrow point. When reading the diagram, replace that arrow with the *"uses a"* relationship. (See Fig. 4.3.)

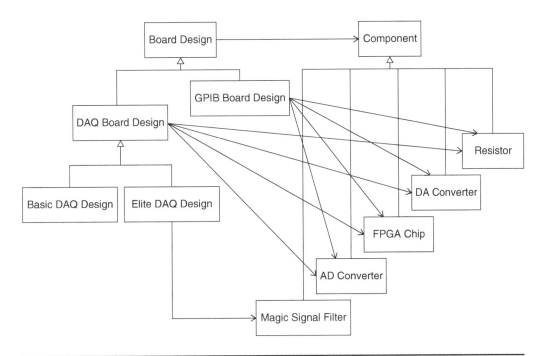

FIG. 4.3 *The UML diagram represents both the inheritance and association relationships for the Board Design and the Component. The way to read this diagram out loud is "The Board Design **uses a** Component." The "DAQ Board Design **uses a** Resistor, DA Converter, FPGA Chip, and AD Converter." The "Basic DAQ Design **uses** the same components as the DAQ Board Design." The "Elite DAQ Design **uses a** Magic Signal Filter and the same components as the DAQ Board Design." To simplify the diagram, we did not include the attributes and methods for each class.*

Note that it is common for programmers to not draw all the "uses" relationships between the child classes. Instead, they only draw them on the parent classes and the exceptions. In our case, the exception would be Elite DAQ Design, which uses Magic Signal Filter.

If you want to investigate where this *"uses a"* is defined in this example, look at the different implementations of the Get Components VI. By the way, these VIs show another benefit of object-oriented programming in LabVIEW, which is the ability to create arrays of objects of the same family. If you remember your early days of learning LabVIEW, an array can only have elements of the same type. Now that you can create your own data types, you can put together all related objects in the same array. (See Fig. 4.4.)

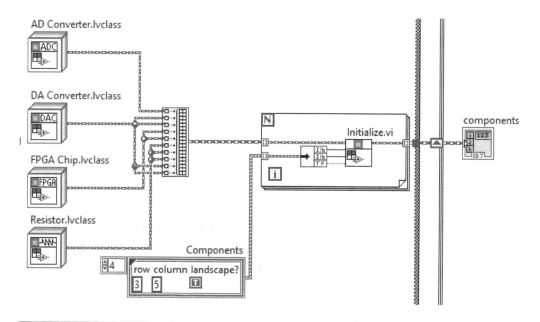

FIG. 4.4 *Detail of the DAQ Board Design.lvclass:Get Components.vi. This code is building an array of different objects of the Component family. AD Converter, DA Converter, FPGA Chip, and Resistor are all Components and can be in the same array.*

What Are the Four Principles of Object-Oriented Programming?

When Stephen Loftus-Mercer and his team at NI were deciding how to implement LVOOP, they first had to decide what object-oriented programming meant for LabVIEW developers. They looked at the four principles of object-oriented programming: encapsulation, abstraction, inheritance, and polymorphism.

What Is Encapsulation? Encapsulation means that object states are protected from outside access and only methods in the class can access them. For LabVIEW, this means the programmers encapsulate class data and tell LabVIEW to allow access to that data only in VIs specified by the programmer. The class data is represented by a special type of cluster. We call it a cluster with superpowers. The main difference between a traditional cluster and a class is that the class cannot be bundled or unbundled by all VIs. Instead of being a .ctl file, when the LabVIEW programmer creates a class in LabVIEW, the .lvclass library includes a .ctl file that can only be accessed via the class library directly in LabVIEW.

In LabVIEW NXG the distinction between a cluster and an object is more subtle. Instead of bundles and unbundles, the data in the cluster is accessed via property nodes, just like we can do now in traditional LabVIEW. NI is moving in the direction of making the transition between traditional LabVIEW programming and LVOOP more straightforward.

What Is Inheritance or the "Is a" Relationship? Inheritance allows you to base class B on class A, where class B has a similar implementation as class A and can reuse most of the code in class A. You would call class A the parent and class B the child. The child is inheriting the parent's behaviors and can have access to the parent's private data via accessors (just like they have access to their parent's wallet in real life!). By default, children implement behaviors as their parent does. Just like in old times the son of a shoemaker would make shoes the same way, but then one day the son decided to innovate. In the code case, in the beginning, the child would have used the parent's "`makes shoes.vi`" but then one day, the child would create their own version of "`makes shoes.vi`." This is called a method override.

What Is The LabVIEW Object?

All classes inherit from the LabVIEW Object. The LabVIEW Object is the ultimate ancestor for all LabVIEW classes. This is what allows LabVIEW programmers to create code that can work for multiple classes, for example, being able to create an array of classes that all have a common ancestor. You can think of the LabVIEW Object as the variant data type, where any data type in LabVIEW can be converted into a variant. Any class can be converted to the more generic LabVIEW Object class. (See Fig. 4.5.)

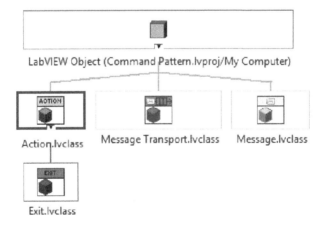

FIG. 4.5 *When you view the class for any class in LabVIEW, you will always see the LabVIEW Object as the ultimate ancestor.*

The "Get LV Class Default Value.vi" and the "Get LV Class Default Value By Name.vi" return the **LabVIEW Object**. These two functions look up a class (one by path and the other one by name) and return an instance of the class with the default values for all private data control fields. The programmer can call these VIs to dynamically load and instantiate any class. This is a common technique you will use in plugin architectures. A common name for this technique is the **factory pattern**. The factory pattern gets its name because it is code that produces different objects depending on the class definition. In other words, it is a factory of objects.

The **LabVIEW Object** does not have any methods defined for it. You will always have to use To More Specific.vi to convert the LabVIEW Object into the class you want to use. (See Fig. 4.6.)

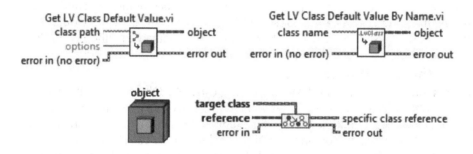

FIG. 4.6 *The LabVIEW Object is the ultimate ancestor. The only VIs that return the LabVIEW Object are Get LV Class Default Value.vi and Get LV Class Default Value By Name.vi. Use the To More Specific Class to convert into the class you want to use.*

What Is Polymorphism versus Dynamic Dispatch? Polymorphism is when your code calls a single function, which can have many different implementations. The function uses the appropriate implementation depending on the types of inputs supplied. While other languages refer to this as polymorphism, LabVIEW already uses that term for polymorphic VIs, so in LVOOP this behavior is called "dynamic dispatch." Dynamic dispatch does the same thing as polymorphic VIs, but at runtime. Remember how the child could decide to make shoes in their own new way? Dynamic dispatch is the mechanism used to call the child's implementation.

What Is Abstraction? According to the Merriam-Webster dictionary, from its roots, abstraction means "something pulled or drawn away." For example, an abstract of a publication is a one-paragraph summary of its contents, with the basic findings "pulled out" of the article. In our case, we take a step back and take a broad view of what the object is trying to accomplish, without going into the implementation details.

You may have heard the term "leaky abstraction." For example, in DQMH we called the DQMH Events **Requests** and **Broadcasts**. We based DQMH on some of the concepts that Justin Goeres presented at CLA Summit 2013. His events were called private events and public events. This is a leaky abstraction because the name indicates how the events are implemented. Private events could only be handled by the event structure inside the module, and the public events could be handled by any event structure that registered to them (and that was outside of the module).

What Are Accessors?   Now that we have an object with private data, we cannot just access this data directly, because it is private. Accessors are special methods, with some optional property node[8] syntactic sugar, that allow you to access the object's private data with bundle and unbundle operations. In LabVIEW, the default is to call accessors with "Read" and "Write" names. We find that confusing when working with hardware. We are used to having a Read DAQmx or Write DAQmx (If you agree with us, at the end of this chapter we show you how we modified LabVIEW to use "Get" and "Set" as the default name).

Where and When?

Fab recommends using LVOOP for things that are objects in real life. Things that you can touch. Also, things that you have multiple versions of or that will have different versions in the future. What does she mean by that?

Creating a class hierarchy for your machine that mixes different chemicals makes sense. You can have a class for a syringe, a class for a linear motor and a rotary motor, and a class for the positioning robotic arm. All of those are things that you can touch.

It is easy to see where we could take advantage of code reuse later. For example, if you decide to buy a different type of linear motor, the functions may be the same but implemented slightly differently by the new vendor. The idea is that you don't have to go find all the places where you call the different functions for the motor, or worse yet, add a Case Structure inside all of the VIs for your current motor so that you can decide what code to execute depending on the actual model in use. Not only does this approach not scale well, now you have to carry around code for devices that you might not need on all your projects!

With LVOOP, all you have to do is create the class for the new linear motor, make it a child of the existing one, override only the methods that are implemented differently, and modify the instantiation of the object to make it possible to choose the new linear motor. That's it. The number of changes is limited, and if you use something like the factory pattern, you don't have to carry around all the classes that you are not using. Better yet, you may even make your code load your classes from

---

[8] In LabVIEW NXG the bundle and unbundle are replaced by property node access for both clusters and classes.

disk, so you could replace the linear motor, create the child, and not even have to recompile your executable! (See Fig. 4.7.)

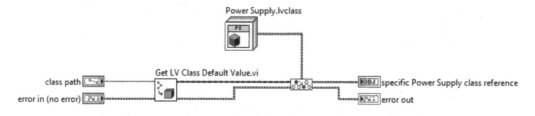

FIG. 4.7 *Factory pattern loads the specific Power Supply class from disk at runtime.*

Now, what does she mean by not recommending the use of LVOOP for things that you cannot touch?

In theory, you can make everything an object. An example of this is the **command pattern**. The command pattern is an abstraction of the Producer–Consumer template.

As a reminder, in the Producer–Consumer[9] you have a loop on the top that determines the actions that the code needs to implement and enqueues them to the loop on the bottom. The loop on the bottom then dequeues and decides what case in the Case Structure within the consumer will execute. (See Fig. 4.8.)

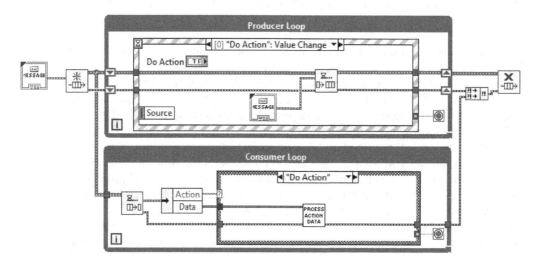

FIG. 4.8 *Producer–Consumer template.*

---

[9] File >> New… >> From Template >> Frameworks >> Design Patterns >> Producer/Consumer Design Pattern (Events)

Chapter 4: LabVIEW Object-Oriented Programming **221**

The command pattern will require a "Command" class, and each of the cases in the Case Structure in the Consumer loop are replaced by the different instantiations of the dynamic dispatch "Execute Command" method. (See Fig. 4.9.)

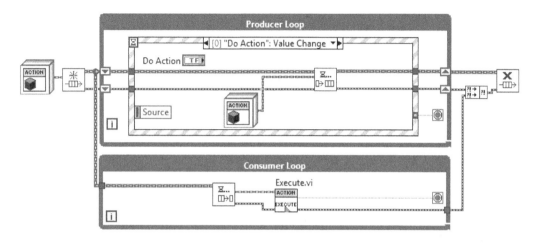

FIG. 4.9 *Command pattern. This example replaces the message data from the Producer–Consumer with an Action or Command object. The Consumer loop replaces the Case Structure that handled the messages with the Execute.vi that can execute different commands thanks to dynamic dispatch.*

You could take the command pattern even further and replace the Queue API with new Message and Message Transport classes. (See Fig. 4.10.)

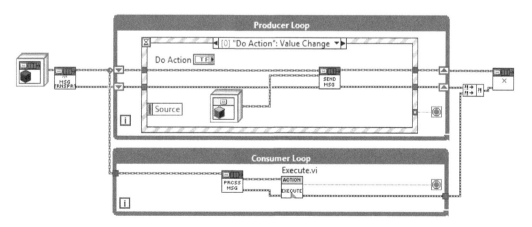

FIG. 4.10 *Command pattern. This example replaces the Queue API with a Message class and a Message Transport class.*

This works well in other programming languages and IDEs. In our opinion, at this level, the benefits of reuse, encapsulation, and inheritance start to diminish when compared with having to manage all the code directly in the project. It is also harder for a programmer not versed in LVOOP to follow the code. Not to mention the difficulties in visualizing something that you cannot touch. This is a clear example of removing the immediacy of the block diagram, because you no longer have a Case Structure to navigate all the possible instances and have to go to the project to understand the code.

Let's look at the DQMH for another example of things that you cannot touch and yet we implemented them as objects. The DQMH includes two objects called "admin" and "DQMH queue." You cannot touch the admin object or the queue. However, we could not envision all the ways people would want to extend their DQMH modules. We were already aware of decisions that might not be popular with everyone (for example, the DQMH dequeue does not expose the time-out terminal). Therefore, even though we couldn't touch it, we decided that it would make our architecture more extensible if we added these two objects. We still limited the number of objects and buried them in the module code to avoid intimidating LabVIEW programmers who are not comfortable with LVOOP.

Let us be clear, LVOOP is very powerful: deciding that everything in your code will be object-oriented might work well for you and your team. We know teams of experienced LabVIEW developers who are very successful using the Actor Framework and pure LVOOP approach on their projects. In general, these teams have a full-time dedicated LabVIEW architect, and everyone on the team is comfortable working with LVOOP. Also, they maintain very good documentation of their class hierarchies, compositions, and any other architecture decisions so that the team members can visualize their classes' relationships.

The following summarizes our view on where LVOOP is appropriate, even though we are well aware that there are always exceptions.

Traditional LabVIEW (no LVOOP) (for example, NI QMH or modules implemented as action engines)	A mix of traditional and LVOOP (for example, DQMH)	Pure LVOOP (for example, Actor Framework)
No need for a full-time architect as long as known design patterns are used.	No need for a full-time architect as long as known frameworks are used.	Dedicated full-time LabVIEW architect.
No one on the team is comfortable with LVOOP.	Different levels of proficiency in the team.	Everyone on the team is comfortable using LVOOP.

Traditional LabVIEW (no LVOOP) (for example, NI QMH or modules implemented as action engines)	A mix of traditional and LVOOP (for example, DQMH)	Pure LVOOP (for example, Actor Framework)
LabVIEW is just another tool; the LabVIEW programmer might go months without touching the code.	Not everyone on the team is dedicated to programming LabVIEW.	Programming in LabVIEW is the main activity for each member on the team; it might even be in their title.
Single programmer. If multiple programmers, everyone needs to understand how every part of the code works.	Only a fraction of the team needs to understand how every part of the code is implemented.	Multiple programmers, not everyone needs to understand how each class is implemented.
Little code reuse or sharing.	Some code reuse and sharing.	There is code that can be shared and reused.
The project will not change often, and you will not keep adding new implementations over time.	The project will change some; the changes are anticipated only in the LVOOP implementation.	The project is anticipated to change often, and you will keep adding new implementations over time.
If the code is not written as spaghetti code, it is very easy to follow. The programmer can identify the Main.VI and follow the block diagram to understand what the code is doing.	The programmer can identify the Main.VI and follow the block diagram to understand what the code is doing. The occasional use of LVOOP is straightforward to follow by double-clicking on the method and navigating through the different implementations.	Hard to follow without an external visualization of the implementation. The programmer starts by studying the class hierarchy and finding the main classes to understand what the code is doing. The programmer works mainly on the project explorer.
Difficult to relate to real-world objects if things are coupled together and there is not a clear delineation of real-world objects.	Real-world objects implemented using LVOOP. The rest is implemented using traditional LabVIEW.	Everything is implemented as an object.

(Continued)

Traditional LabVIEW (no LVOOP) (for example, NI QMH or modules implemented as action engines)	A mix of traditional and LVOOP (for example, DQMH)	Pure LVOOP (for example, Actor Framework)
Modifying one part of the code may result in modifications everywhere in the code.	Modifying one part of the code in the LVOOP is contained. Modifications in the rest of the code may result in modifications everywhere.	Modifying one part of the code is contained and does not result in modifications everywhere in the code.
Cannot be extended.	LVOOP section can be extended.	Very extensible.
Does not require planning ahead (but this is how spaghetti code is born!).	Requires some planning ahead.	Requires extensive planning ahead, but it is worth it.
Chosen for short-term projects.	Chosen for medium-to long-term projects.	Chosen for long-term projects.

LabVIEW programmers might think that just because they use classes, they are wiring an object-oriented program. That is not true. Object-oriented programming is an approach to designing your application, not how you wire it. If you are focusing on state transformations and encapsulated abstractions, you are following object-oriented programming.

An example of this is the LabVIEW component-oriented design described by Jon Conway and Steve Watts in their book *Software Engineering Approach to LabVIEW*.[10] They achieved encapsulation and state transformation by using Components (also known as action engines or functional global variables) with no use of LabVIEW classes at all! Their approach is so clean that even Stephen Loftus-Mercer, the father of LVOOP, said at the CLA Summit in Paris: "Steve Watts is a genius; everyone else, just lay your wires!"

Why?

The promise of benefits of using LVOOP can be summarized by the four pillars of LVOOP: encapsulation, abstraction, inheritance, and polymorphism.

---

[10] *A Software Engineering Approach to LabVIEW*, by Jon Conway and Steve Watts. Prentice Hall, 2013.

Why Do We Want to Encapsulate?

As an architect, you might want to hide information from other developers, or they might not need to have access to that information. Or perhaps, you want to limit the locations where the information can be changed so that when you are troubleshooting, you are only looking at a handful of VIs instead of the entire application. For example, if you are using a queue as the main communication mechanism between your VIs, you might not want just anybody being able to destroy the queue. They should only use the ways and means you offer for them to create and destroy such queues. A way to do this is to make that communication queue reference be the private data of your class.

Also, it is easier to take the encapsulated code and reuse it in another program. Once an object is created, the programmer does not need to understand how it is implemented in order to use it. The original programmer might designate parts of the class as private, and its data/state cannot be modified from the outside. This prevents programmers using the object from tampering with values that they should not touch. The object might also handle its own errors.

Why would you want to limit the number of VIs that have direct access to the object's private data? Well, because it is easier to debug! If anybody can access the private data cluster, then looking for where a bug is introduced implies doing a text search and finding all the bundles and the unbundles where that cluster element is used or modified. If you limit access to the cluster elements, then troubleshooting is limited within the class itself, or we can look for all the places an accessor is called (and yes, this search finds accessors implemented as property nodes too!).

Why Do We Want to Abstract?

Other programmers using the object do not need to understand the implementation as long as they have a general idea of what the end result is. This lets the programmer focus on the problem they are trying to solve and not worry about how a certain method is implemented.

Why Do We Want to Use Polymorphism (Dynamic Dispatch)?

If your environment changes often, you will benefit from dynamic dispatch. For example, the program might use a Tektronix scope today and a Rigol scope tomorrow, or your company may decide to acquire a new scope with higher bandwidth for an upgraded version of the test system.

Why Only Private Data?

Stephen Loftus-Mercer and his team decided that there would not be public, protected, or community data, but only private data. This was to encourage code correctness. One of the goals for LabVIEW is to not require that the programmer be familiar with computer science concepts. They created this bottleneck where the data can only be accessed in a single way, which means there is a single place you need to debug value changes.

Some LabVIEW architects debate that we should not even have the accessor VIs, that programmers should be discouraged from adding code other than reading/writing the private data. Other architects believe that accessor VIs are a great place for adding range checking and other sanity-check code. In our experience, it is a good practice to always use the accessor VIs (even within the class VIs) instead of using bundles and unbundles. If there are changes to the class definition in the future, the programmer only needs to modify the accessor VI as opposed to looking at all the places where the data is being bundled/unbundled. For example, maybe you decided to include the age of a person as private data at an early stage, only to find out later that you probably should have used their birthdate instead. All you have to do is leave the age as private data and add the birthdate as an additional private data, then modify the age accessors to include the conversion to and from birthdate.

An additional advantage of having private data is that you know outside code depends only on the methods and not on the data itself. This means you can even create base executables that load classes dynamically, and you can create new versions of a single class without having to rebuild your entire application.

How?

How to Use UML Class Diagrams

We find that UML models are the best way to share the core concepts of how our application works with other language developers and even noncoding customers. All we have to do is include a legend of what the arrows and other graphical notations mean, and noncoding customers can read the model.

We believe that if reading the model out loud does not make sense, then the model is not correct and it will be very difficult to implement even if those difficulties are not obvious now. For example, we had a customer who had a solid DAQmx class architecture and then decided to add motion control functionality. He wanted to take advantage of the existing architecture and made his motion control channel a channel of DAQmx Analog output. However, the motion controller used the serial port instead of DAQmx. The model looked like Fig. 4.11, and when we read it out loud it says: The Serial Motion Controller Channel **is a** DAQmx Analog Output channel. This is not true, because the Serial Motion Controller channel uses the serial port and not DAQmx. It does not make sense, and if it does not make sense when reading it out loud, it will make less sense in code! We turned out to be correct, and the final solution ended up looking like spaghetti and required lots of hacks to make the rest of the code work. A lot of those headaches could have been prevented by giving up as soon as the UML diagram said it did not make sense.

Remember that it is a lot easier to erase things on a whiteboard or piece of paper than deleting the code you have already committed to. Paper and marker ink is also cheaper! Do your design work in the cheapest medium possible first, then code.

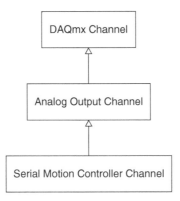

FIG. 4.11 *DAQmx channel hierarchy with out-of-place serial motion controller channel. The way to read this diagram out loud is "The Serial Motion Controller Channel **is a** type of Analog Output Channel. The Analog Output Channel **is a** type of DAQmx Channel." This is not true; the Serial Motion Controller Channel uses the serial port, not DAQmx.*

Traditional LabVIEW Morphs into LVOOP

"For those who choose object designs,
we want the wire and node
to morph naturally
into class and method."[11]

We wanted to help you visualize that transition by describing how your traditional LabVIEW projects would morph into an object-oriented project. Of course, this transition will be a lot easier if you are already writing modularized code. If your code can be described as "The VI" and its block diagram is about 40 screens by 40 screens big, you will need to start breaking it into pieces first.

We will use the Board Testing project example that ships with LabVIEW. It is located either in `labview\example\lvoop\BoardTesting\Board Testing.lvproj` or in `labview\example\Object-Oriented Programming\BoardTesting\Board Testing.lvproj`. The example includes a Read_Me .html with more details on both a task-oriented solution and a class-based solution

---

[11] LabVIEW Object-Oriented Programming: The Decisions Behind the Design, http://www.ni.com/white-paper/3574/en/

for a hypothetical board inspection system. We saved a copy of the project and removed the Object-Oriented Solution so that we could show how to go about converting the Task-Oriented Solution into an Object-Oriented Solution. (See Fig. 4.12.)

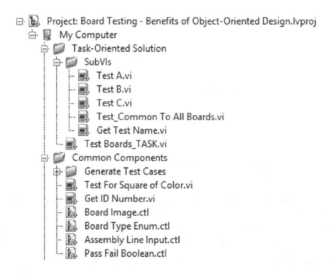

FIG. 4.12 *Board Testing project with only the Task-Oriented Solution.*

The main cluster in this project is the `Assembly Line Input.ctl`. It includes the board image and the enumerator `Board Type Enum.ctl`.

How to Get Started to Refactor a Task-Oriented Project to Object-Oriented

You probably tell beginners that any time they create an enumerator, they should make it a typedef. You tell them that any time you create a cluster, they should make it a typedef.

The LVOOP paradigm is: Should this cluster be a typedef or a class? Should this enum be a standalone typedef, or should it be used to decide which object to instantiate? Should each element of the enumerator represent a child name?

Clusters to Classes—Cluster Morphs into Object

Following the LVOOP paradigm, we are going to convert the `Assembly Line Input.ctl` from a typedef into a class, and the `Board Type Enum.ctl` is what is going to determine the inheritance hierarchy of our board designs. You can follow these steps via the GitHub.com/LGP5/LVOOP repository, under the branch "ConvertTask-OrientedToOO."

First, identify the cluster that you will convert into a class. Right-click on it and select **Convert Contents of Control to Class**. (See Fig. 4.13.)

FIG. 4.13 *Right-click on the control to convert the contents of a cluster saved as a type-defined control into a class.*

This action will replace the contents of the typedef with the newly created class called `Assembly Line Input.lvclass`. Notice that the class kept the name of the original control. You could keep this typedef, but we suggest you get rid of it and replace it with the class directly in any place where the typedef was being called originally. Make sure you replace the argument inside the Queue control and indicators in the code to call the new class. (See Fig. 4.14.)

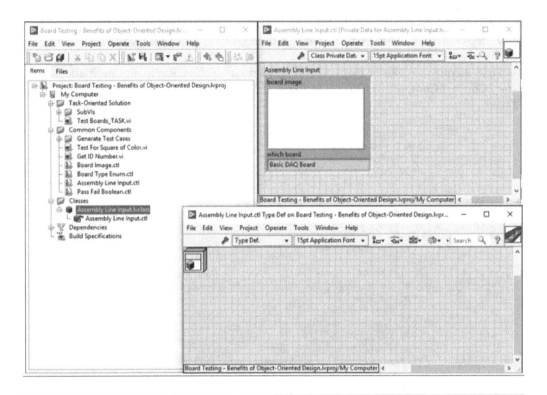

FIG. 4.14 *After converting contents of Assembly Line Input.ctl to class, there is a newly created Assembly Line Input.lvclass.*

You will notice that you now have broken arrows everywhere. Why? Well, remember that one of the pillars of OO is encapsulation, which means you cannot unbundle and bundle the contents of the object in VIs that are not part of the class library. (See Fig. 4.15.)

Let us identify the VIs that access the data inside the Assembly Line Input. You need to decide now if those VIs belong inside the Assembly Line Input class or if you need to create just the accessors and leave those VIs outside the library. (See Fig. 4.16.)

Start by creating the Accessors for the data inside Assembly Line Input. For this, right-click on the Assembly Line Input class on the project and select **New >> VI for Data Member Access…** (See Figs 4.17 and 4.18.)

Chapter 4: LabVIEW Object-Oriented Programming   **231**

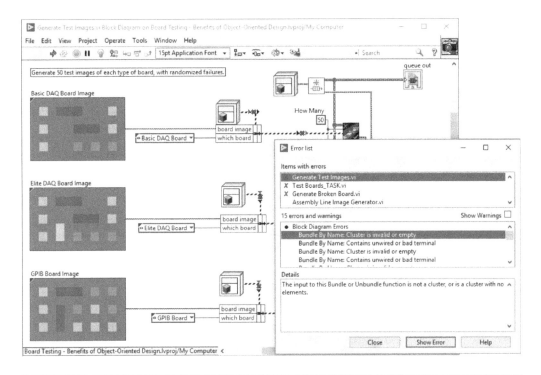

FIG. 4.15 *After converting contents of Assembly Line Input.ctl to class, any VI that unbundles or bundles the elements in the cluster is broken.*

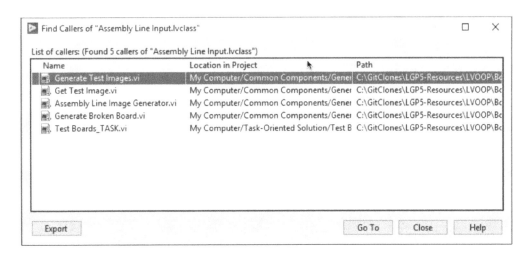

FIG. 4.16 *List of VIs that access the data inside Assembly Line Input.*

232 LabVIEW Graphical Programming

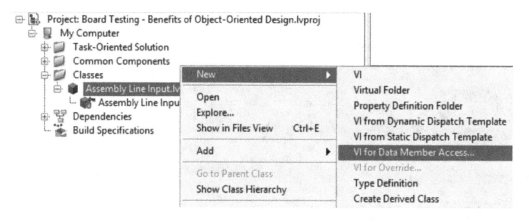

FIG. 4.17 Creating a VI for Data Member Access, also known as Accessors.

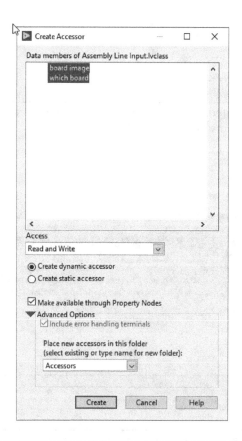

FIG. 4.18 Select to create both Read and Write accessors for both elements in the class.

You selected to make the accessors available through Property Nodes. The name that will show up on the property node will be the localized name you give to the class via its properties. Right-click on the Assembly Line Input class and select its properties. Then in the Documentation category remove the extension `.lvclass` from the localized name. (See Fig. 4.19.)

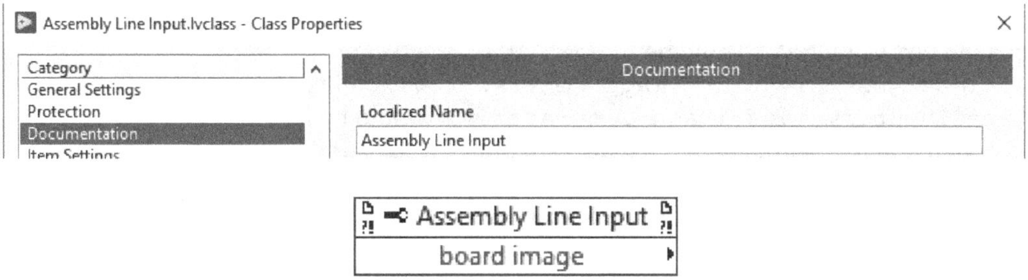

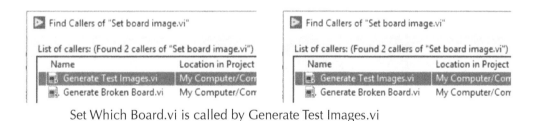

FIG. 4.19 *The Localized Name will be used as the property node name.*

Now that the accessors are available via property nodes, you can go to all the callers of Assembly Line Input and replace the bundles and unbundles with property nodes. A nice side effect of this is that when you try to find the callers for the accessors, now you know exactly what part of the code is setting and getting each element of the Assembly Line Input. No more text searches to find where you are unbundling or bundling an element for that cluster! (See Fig. 4.20.)

Set Which Board.vi is called by Generate Test Images.vi
Get Which Board.vi is called by Test Boards_TASK.vi

FIG. 4.20 *It is easier to identify exactly where the code is setting and getting elements from the Assembly Line Input class.*

At this point, you have running code again and you are using classes, but you are not really applying object-oriented design yet. You need to go back to the list of Assembly Line Input class callers and identify if any of these VIs belongs inside the class. Some considerations are if in the future there would be different ways to implement these VIs. For example, `Generate Test Images.vi` is currently randomly generating the images. Maybe later those images will come from a database. Maybe they will actually be generated via a camera taking pictures directly from the line. This VI is a good candidate for possible extensions. The VI calls `Assembly Line Image Generator.vi`, which, following the same logic, should probably be part of the class as well. You could even rename it to `Image Generator.vi` because Assembly Line is part of the class name and the fully qualified name of the VI would be `Assembly Line Input.lvclass:Assembly Line Image Generator.vi`. Calling it `Assembly Line Input.lvclass:Image Generator.vi` would be less redundant.

What about the `Get Test Image.vi`? `Generate Test Images.vi` puts the images in a queue, while `Get Test Image.vi` dequeues the images from the same queue. Seems like if the creation of the queue and the enqueuing operation are part of the class, then the dequeue operation should also be part of the same class. Also, if in the future we decide to use a different mechanism other than the queue, we would be limiting the change to only the VIs within this class. Following the same logic, you can also move the `Generate Broken Board.vi` and its subVI `Point to Distance on Board.vi` into the class. You can also move all the VIs to be in the same folder as the class in the disk.

Enumerator Morphs into Class Hierarchy

You are using `Board Type Enum.ctl` to determine which board is going through the assembly line. In the code, this enum is determining the case in the Case Structure to execute within `Test Boards_TASK.vi`. (See Fig. 4.21.)

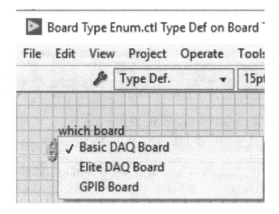

FIG. 4.21 *Board Type Enum.ctl determines which board to test.*

Chapter 4: LabVIEW Object-Oriented Programming **235**

You can use this enumerator as a guide of what class hierarchy you will create. The three items in the enum are Basic DAQ Board, Elite DAQ Board, and GPIB Board. You will create a parent called Board Design, and it will have two children: the GPI Board Design and DAQ Board Design. DAQ Board Design will have two children: the Elite DAQ Design and the Basic DAQ Design. To create the classes, right-click on the Classes virtual folder and select **New >> Class**. To set the new inheritance, right-click on the children classes and select their properties and there, within the inheritance category, set the new inheritance. (See Fig. 4.22.)

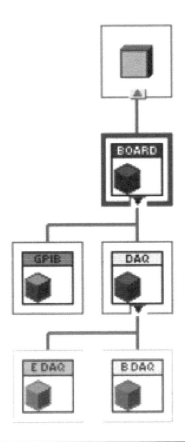

FIG. 4.22 *New Board Design class hierarchy as defined by the Board Type Enum control elements.*

Now you need to create a VI that will select which Board object the code will operate on based on the Board Type Enum. You will call this VI `Board Type Enum to Board Design.vi`. (See Fig. 4.23.)

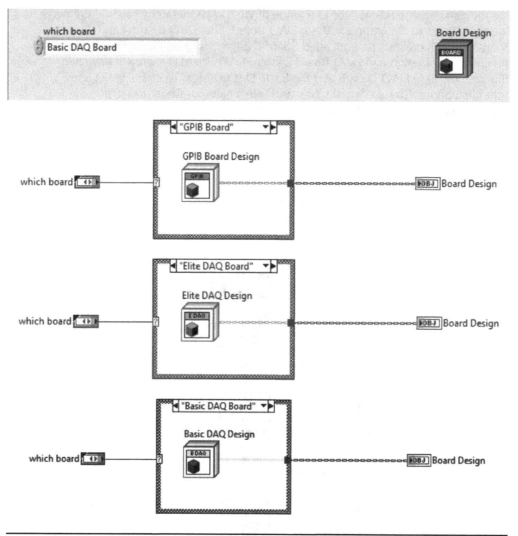

FIG. 4.23 *Board Type Enum to Board Design.vi converts the enumerator to the actual Board Design object type that the rest of the code will use.*

Case Structure Morphs into Dynamic Dispatch

Now that you have your Board Design class hierarchy, you can morph the Case Structure inside the `Test Boards_TASK.vi` into dynamic dispatch. Each case in the Case Structure determines which test to run based on the board type. (See Fig. 4.24.)

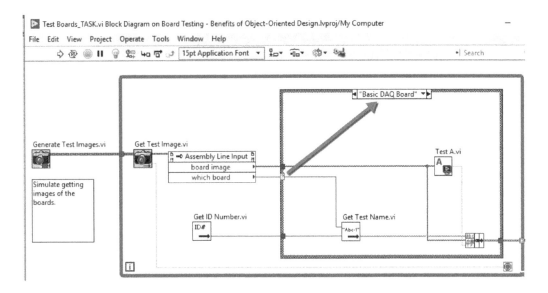

FIG. 4.24 *The Case Structure determines which test to run based on the board type.*

All the test cases call `Get Test Name.vi`. If we assume that the name of the class is the same as the test name, we can use the `Get LV Class Name.vi` to get the test name. Since all the boards seem to require this call, we can make it part of the `Board Design.lvclass:Test.vi` and configure the class to require all its children to call the parent. This way, we put all the common code in that instance of the dynamic dispatch, and we don't have to code the same thing in all the instances of `Test.vi`. (See Fig. 4.25.)

You need to decide which code will go inside the `Test.vi` for each class. Since you already set the `Board Design.lvclass:Test.vi` to be called by all other `Test.vi`, you may want to include the contents of `Test_Common to All Boards.vi` in there, too.

Explore the other tests, and you will notice that `Test B.vi` calls `Test A.vi`, meaning that both the test for the Elite DAQ board and the Basic DAQ board need to call Test A. You will place the contents of `Test A.vi` inside the `DAQ Board Design.lvclass:Test.vi`. The Basic DAQ Design class will not need to create its own Test.vi because it can use the same as the `DAQ Board Design .lvclass:Test.vi`, but you will need to create the `Elite DAQ Design .lvclass:Test.vi` that will include the contents of `Test B.vi`. Finally, create a `Test.vi` for the GPIB Board Design class that will include the contents of `Test C.vi`. (See Fig. 4.26.)

Now you have everything you need to convert the case structure inside the `Test Boards_TASK.vi` to use dynamic dispatch. As you do this, you realize that you forgot the Board ID to be included as part of the name. The good news is that

238 LabVIEW Graphical Programming

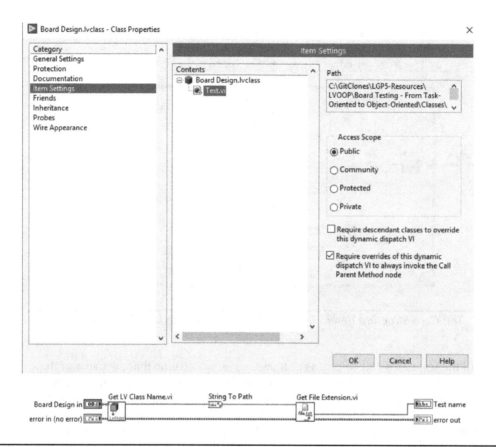

FIG. 4.25 *Include inside Board Design.lvclass:Test.vi a call to get the class name. Make this VI require overrides to call the parent method.*

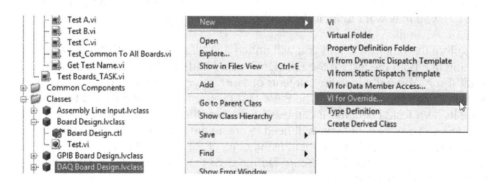

FIG. 4.26 *Right-click on DAQ Board Design.lvclass to create a new VI for Overriding the Test.vi.*

Chapter 4: LabVIEW Object-Oriented Programming **239**

this change has to be done in only the `Board Design.lvclass:Test.vi` because all the board design tests call this VI. You probably should also rename the `Test Boards_TASK.vi` to `Test Boards_Object-Oriented.vi` because this code is no longer task-based. Do you think you are done refactoring this code from task-based to object-oriented? (See Fig. 4.27.)

From:

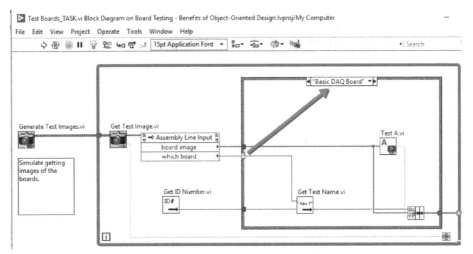

To:

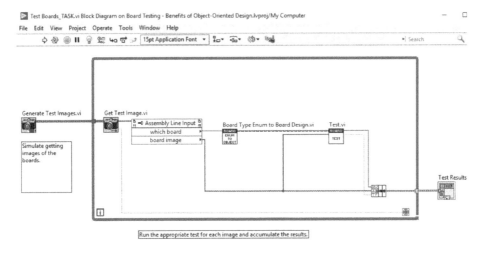

FIG. 4.27 *Case Structure morphs into dynamic dispatch. Each case in the Case Structure is now replaced by different instances of the Test.vi.*

Comparing the refactored code to the way the shipping example did it, you can see that the shipping example left the Assembly Line input as a cluster and had a complete new hierarchy for the components. Based on the task-oriented code alone, we could not define the different colors on the board image. In a real project, you would define what each color represents and refactor this code even further to include a components class hierarchy.

A better approach than refactoring existing code is to create a model of the application and start from scratch with LVOOP. In general, we recommend that you create a model of your solution before you start coding. Based on what the model shows and what the existing code shows, determine what would be the best approach. Then copy only the code that can be repurposed from the task-oriented solution.

LVOOP by Reference

So far in this chapter we have only used **by value** classes. The object is on the wire and if you fork the wire, you create a completely new copy of the object. If different areas of code need to modify the same object, then you need a **reference**. Other languages implement OO by reference. In LabVIEW, you create by reference objects by wrapping your objects with a DVR. Note that you do have to modify the properties of your classes within the inheritance category to determine if you, as the class designer, authorize LabVIEW to allow the developers using this class to wrap the class with a DVR or not. It is your responsibility as a class designer to determine if this class will be used by reference and, if so, whether to provide the methods to wrap the class with a Data Value Reference (DVR) or whether you let future developers wrap the object with a DVR outside of the class. (See Fig. 4.28.)

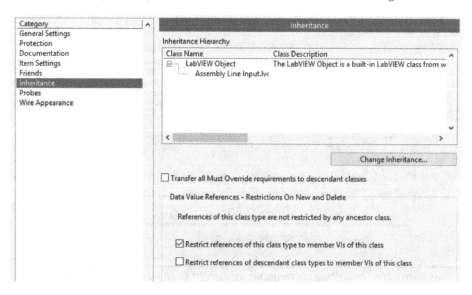

FIG. 4.28 *The lower-right section of the Inheritance class properties determines DVR access to this class.*

A check on **Restrict references of this class type to member VIs of this class** means that only member VIs of the class can create a DVR. This is the default behavior for any new class.

A check on **Restrict references of descendant class** means that only member VIs of this class can create a DVR for descendants of this class.

Again, if different areas of code need to modify the same object, then you need a reference. The reason LabVIEW makes it difficult to create a DVR for a class is that you should avoid handling your objects by reference. Any time you share resources, you have the potential of race conditions and managing the code becomes more difficult. However, there are times where you will not have any other option.

DVR or Single-Element Queues

A single-element queue (SEQ), as the name implies, is a queue of size 1. They were in use long before DVRs were introduced in LabVIEW. The reason they are still used is that they are named and DVRs are not. Another difference between DVRs and SEQs is that DVRs are blocking and SEQs are not. Blocking means that the access to the code wrapped by the reference cannot be accessed in parallel. LabVIEW indicates this blocking mechanism via the In Place Element structure that you use to access the contents of the DVR, although the DVR read in an In Place Element structure got a right-click option to allow parallel reads starting with LabVIEW 2017.

G Object-Oriented Programming Development Suite Toolkit (GDS)

Before there were LabVIEW native classes, there was G Object-Oriented Programming (GOOP). The GOOP Development Suite (GDS) toolkit (formerly known as Endevo or Symbio GOOP toolkit) introduced G Object-Oriented Programming concepts around 1998. It was a joint venture between NI and Endevo. GOOP 1 provided classes but there was no inheritance. GOOP 2 introduced inheritance. Around 2004 Endevo released a version that provided inheritance. LabVIEW 8.20 (August 2006) introduced native classes and also gave GOOP access to a new project provider to include right-click menus and wizards directly from the project explorer. GOOP 3 was the first version to include NI LabVIEW 8.20 LVOOP inheritance.

NI bought the GDS toolkit from Symbio in 2014 and made it freely available. NI opened the source to allow the community to view the internal workings of the tool and even contribute to improving it. Mikael Holmstrom, one of the main Symbio/Endevo GOOP Toolkit developers, has kept an open-source version of the toolkit going via OpenGDS.github.io (to get the source code go to GithHub.com/opengds/OpenGDS). Eventually, the changes in OpenGDS made it to the NI version of the GDS toolkit. One of the reasons NI wanted to make sure that GDS would not go away is that it extends a lot of the features that NI had not included natively.

Differences between GOOP and LVOOP

LVOOP or LabVIEW classes are native classes built into LabVIEW; they are part of the G language, and GOOP is the reference architecture and template support system. GOOP is the IDE enhancement and a set of template classes that allow you to create reference architectures. You can have both working together, so there is no need to select LVOOP or GOOP.

When using GDS to create classes, the wizard asks what type of class to create, allowing you to choose between GOOP 300, GOOP 400, and LVOOP. The difference between these three is mainly how the classes are created:

- GOOP 300 → Used an uninitialized Shift Register with an array of the class attributes; the reference number returned when creating an object corresponded to the index in the array
- GOOP 400 → By reference classes using DVR
- LVOOP → By value classes using native G classes to LabVIEW

GOOP Enhancements
When NI introduced LVOOP native classes, the class creation involved a lot of steps before you could even start working with a class.

- Right-click to create a new class
- Start with an empty class, save it first, and name it
- Create the icon
- Create the wire

Now you can start adding methods.

GDS added a streamlined approach to creating a class, with all of these four steps done in a single step.

- Clone classes.
- Create different icons on the project for your class. This allows you to differentiate your classes with different icons other than the cube icon.

Design Patterns Ability to create entire new design patterns and instantiate all the different pieces for the pattern, and you can add your own design patterns.

Actor Framework under GDS There are a lot of steps involved: you need to remember to add the Actor Framework classes to your project, now add a new actor, and then get some empty spaces to add your functionality.

GDS allows instantiating a new actor in fewer steps.

GDS Templates You can create your own templates.

GDS is class-based, so to add your templates, you only have to override the methods for the class provider for your class.

You create a reader class to be able to interact with the UML so that the UML Modeler can analyze a class created with this template.

GDS UML Modeler UML stands for Unified Modeling Language. UML is a standard language, and programmers use it to model their solutions independently of the programming language they will use to implement it. What is notable about the GDS UML Toolkit is that it provides LabVIEW-centric UML features such as a way to model the friendship relationship, and the difference between a by value and by reference dependency.

Mikael started working on the UML Modeler around 2001. He created it because he needed a way to test the new inheritance feature. UML diagrams were a good fit for this test because every element in the diagram had a spot in the inheritance tree. Once he figured out how to analyze classes and display them in the diagram, he and his team realized how powerful this tool was. LabVIEW is a graphical programming language, and the UML Modeler provided a graphical representation of the class hierarchy tree. What started as a test plan ended up becoming a key part of the product. Once they figured out that they could do reverse engineering (analyzing all the classes in a project and describe their relationships), they started using the UML Modeler as the first step in project creation and generating code from a UML diagram.

Another advantage of the UML Modeler is that it can describe dependencies. The tool can help you identify illegal relationships between classes because you can easily see if there is, for example, a dependency to a typedef.

Circular Dependencies A circular dependency is when VIs inside class A call VIs from class B and VIs inside class B call VIs from class A. These circular dependencies tend to slow down LabVIEW because the development environment has a hard time figuring out the relationships between classes. They are also a headache to separate and defeat the purpose of modularizing code. Most of the time, a circular dependency is a code smell, and you need to figure out where the source of the smell comes from. You don't want code that stinks!

Sometimes, the solution is to create a third class or library that contains the common code. This way VIs inside class A call the VIs from the third class and VIs inside class B call VIs from that third class too. Class A and B no longer depend on each other.

You can find more details on how to address this in the SOLID section of this chapter.

HAL: Hardware Abstraction Layers

A common application of object-oriented programming in LabVIEW is hardware abstraction layers. Earlier in this chapter, we defined what abstraction means, but in this case we are taking a step back and taking a broad view of our hardware and what it is trying to accomplish, without going into the implementation details. Our abstract class will define the elements of our private data and the generic methods that our hardware will implement. For example, if we were to create the HAL for a DC power supply, we would define the following methods:

- Set Current Limit
- Set Voltage Level

- Set Output State
- Measure Current Draw
- Reset Instrument

At this point, we do not care how an NI power supply or an Agilent 3634A power supply implements these commands. We only care about the fact that a power supply can implement these methods.

A word of caution here: Countless times we have seen people taking the manual for one of their devices and then proceeding to implement every single function available in it. This is a recipe for disaster. The likelihood of having every manufacturer in the world implementing the same functions for their version of a given instrument is very low. We recommend that before you start implementing a HAL, only focus on the methods you actually need. This not only means less code for you to write, but it also means that you will minimize the risk of getting to a function that does not exist in all of your devices. If it does not exist in that device, then you probably do not want it in your application anyway!

The main design pattern you need to be aware of when creating a HAL is the factory pattern that we mentioned earlier in this chapter. This will allow you to define at runtime the actual hardware that your HAL will call. This avoids carrying around classes that are not used in a particular code implementation. Check out the Delacor blog for an example of a DMM HAL.[12]

MAL: Measurement Abstraction Layers

We mentioned earlier that one of the challenges with HALs is that not all manufacturers implement the same functions on their devices. We recommended focusing on the functions your application will be using and implementing only those. In the test and measurement world, we have seen that abstracting the measurements instead of the hardware itself forces the developers to only use the hardware functions they need. Also, if one manufacturer implements the same measurement in a completely different sequence of functions, this is transparent to the calling code. We are overriding the measurement itself, not the individual functions.

Actor Framework: The Most Recognizable LVOOP Architecture

See Actor Framework section in Chapter 5 for more information.

---

[12] Simplifying Your Hardware Abstraction Layer HAL with LVOOP and DQMH part 1, Michael Howard. Delacor.com/simplifying-your-hardware-abstraction-layer-hal-with-lvoop-and-dqmh-part-1/

SOLID Principles of Object-Oriented Design

One of the goals of LabVIEW is to give you the power to write large-scale software even if you do not have any formal computer science training. However, there comes a point when you realize that if you knew a little bit more about computer science, you could write your programs in such a way that make it easier to adapt to changes in requirements. Perhaps you do have the computer science background but you struggle to apply the same principles you have used in text-based programming to your graphical programs. Either way, understanding these five principles will help you write better programs.

SOLID principles are a subset of agile software design principles.[13] How many times have you completed close to 80 percent of your project just to have a requirement change the program dramatically, moving that 80 percent completion to 20 percent completion? The only constant in our projects is change. The promise of following SOLID principles is to write code that adapts to change gracefully, even when those changes impact the application in ways we did not think about when we started designing the application. (See Fig. 4.29.)

FIG. 4.29 *The folks at LosTechies.com created a series of Creative Commons–licensed posters that illustrate the SOLID principles. Joe Villa took the posters, cleaned up the typography a little, and posted them under the same Creative Commons license.*[14]

---

[13] https://agilemanifesto.org
[14] The SOLID principles explained with motivational posters. DeVilla Joey. blogs.msdn.microsoft.com/cdndevs/2009/07/15/the-solid-principles-explained-with-motivational-posters/

SOLID stands for

	Stands For	Computer Science Definition[15]	Pragmatic LabVIEW Programmer Definition
S	Single Responsibility Principle (SRP)	A class should have only a single reason to change.	Your VIs, classes, and libraries should have one reason to change.
O	Open-Closed Principle	Software entities should be open for extension but closed for modification.	Aim for packaging your libraries early on. Adding new features should involve creating new code instead of modifying the existing code.
L	Liskov Substitution Principle	Subtypes must be substitutable for their base types.	Your classes' children should keep their parents' promises. A child can overdeliver on those promises but never underdeliver.
I	Interface Segregation Principle	Clients should not be forced to depend on methods that they do not use.	Many client-specific interfaces are better than one general-purpose interface. Provide other LabVIEW programmers who use the code you are developing only the inputs they need.
D	Dependency Inversion Principle	High-level modules should not depend on low-level modules. Both should depend on abstractions. Abstractions should not depend on details. Details should depend on abstractions.	Design your code to use generic implementations (abstractions) instead of depending directly on specific implementations.

SOLID principles are applicable to all software entities[16]:

- Functions (VIs)
- Classes

---

[15] Martin, R. C., *Agile Software Development: Principles, Patterns, and Practices,* Prentice Hall, Upper Saddle River, New Jersey, 2003.
[16] LabVIEW Classes State of the Art, presentation at NI Week 2012 by Stephen Mercer.

- Libraries
- Modules
- Others

We need to emphasize that the lofty goal of SOLID is to make software designs easier to understand, maintain, and extend. However, applying all five principles at the same time on all the code you write is an unattainable goal. Why should you bother learning more about this? So you can understand when you are violating a principle and be able to explain why you are violating it and be fully aware that you are willing to pay the technical debt when the time comes. Do not feel like you need to follow these principles as if they were the law of the land. With time you will learn which principles you need to adhere to and under which circumstances. The promise at the end of the SOLID rainbow is that your code will be more flexible and that when the inevitable change request comes, your code will be able to sustain the changes without falling apart or becoming spaghetti code.

The main promoter of these five OO principles is Robert C. Martin. If you have been to any of Dmitry Satagelyan's presentations, you have seen at least a slide or two on Martin's book *Agile Software Development*.[17] Martin is also known as Uncle Bob and is one of the original Agile Manifesto authors. Uncle Bob's focus was on dependency management. You probably have suffered through poor dependency management if you have been asked to support legacy code, and when you start opening one VI, you realized there are missing toolkits, add-ons, or packages in vi.lib and user.lib. You might not have had a name for it, but you knew that this was an example of poor dependency management. Uncle Bob's main point is that when you manage dependencies well, your code is flexible, robust, and reusable.

We need to repeat that the most important lesson about SOLID principles is that you will never be able to implement them all on the same software entity. You will have to violate one or two principles, but the important thing is to understand why you are violating a principle and whether you are OK with paying the technical debt when the time comes.

SRP: Single Responsibility Principle
"A class should have only one reason to change."

---

[17] Martin, R. C., *Agile Software Development: Principles, Patterns, and Practices*, Prentice Hall, Upper Saddle River, New Jersey, 2003.

FIG. 4.30 *The folks at LosTechies.com created a series of Creative Commons–licensed posters that illustrate the SOLID principles. Joe Villa took the posters, cleaned up the typography a little, and posted them under the same Creative Commons license.*[18]

Basically, entities should have exactly one job. You should be able to describe that job in one sentence. In the case of classes specifically, a class that follows this principle has one, two, or three primary methods; for example[19]

- Do
- Start, Stop
- Setup, Execute, Teardown

These calls may have many properties to support those few methods. If you have a class with tens of methods, then that class might be doing too much and has really more than one job. (See Fig. 4.30.)

This is called cohesion, and it was originally described by Tom DeMarco and Meilir Page-Jones. They defined cohesion as the functional relatedness of the elements of a module. Uncle Bob took that definition as a base and shifted it to relate cohesion to the forces that cause a class to change.

---

[18] The SOLID principles explained with motivational posters. DeVilla Joey. blogs.msdn.microsoft.com/cdndevs/2009/07/15/the-solid-principles-explained-with-motivational-posters/
[19] Computer Science for the G Programmer Year 2, presentation at NI Week 2015 by Stephen Loftus-Mercer and Jon McBee, https://forums.ni.com/t5/Past-NIWeek-Sessions/Computer-Science-for-the-G-Programmer-Part-2/ta-p/3520618

If a class has more than one responsibility, then there will be more than one reason for it to change. A good example of this is when you create a class for the serial instrument you are working with. See example testing for a LabVIEW class section in Chapter 6 Unit Testing. You might be tempted to create a single class called iTacho. If you need to use a simulated version, you would create a child of iTacho. However, the only thing that is changing between the two is the transport mechanism. The physical iTacho uses VISA to send serial commands via the serial port, while the simulated iTacho uses a VI that simulates the serial port. Now, if there is a change in the protocol, you might need to make the changes to both the physical iTacho and the simulated iTacho. There are two reasons these classes can change: due to communication mechanism changes or due to protocol changes. You are violating the single responsibility principle.

A better approach would be to have a class for the iTacho device and a class for the serial port. Now if you need to simulate the iTacho device, you really only need to simulate the serial port. If there is a change in the protocol, you don't have to touch the serial port class; you only modify the iTacho device class. Incidentally, this approach also makes your class easier to test and create unit tests for. Read the Unit Test chapter in this book for more details.

Examples of LabVIEW code that violates the Single Responsibility Principle:

- The Clustesaurus Rex (A cluster so big that it does not fit on the screen and contains all sorts of elements that are not really related to each other. It usually starts as a simple cluster to group a couple of elements but after several weeks of programming, turns into Godzilla!). Others call them Mother clusters.

- VIs with block diagrams that span multiple screens; this is normally a code smell. The code stinks because that VI is trying to do too many things. Stephen Loftus-Mercer discovered that there is a limit to how many structures you can nest within a single block diagram, and it was something like 219! You would not want to be the person who discovered that limit! If all LabVIEW programmers were aiming at satisfying the SRP, nobody would have ever discovered that limit.

- A universal driver that attempts to support all possible devices in a program. For example, a string input that the developer needs to know exactly how to format depending on the instrument she is trying to communicate with. An example of this would be a generic GPIB driver.

Can you go too far? Absolutely, you could end up with several entities that you can describe and understand exactly how each one works but not be able to construct them to work together. Experience will show you when you have not tried to apply the SRP enough and when you went too far.

OCP: The Open-Closed Principle

"Software entities (classes, modules, functions, etc.) should be open for extension, but closed for modification." (See Fig. 4.31.)

FIG. 4.31 *The folks at LosTechies.com created a series of Creative Commons–licensed posters that illustrate the SOLID principles. Joe Villa took the posters, cleaned up the typography a little, and posted them under the same Creative Commons license.*[20]

Bertrand Meyer coined the Open-Closed Principle in 1998 to guide programmers on how to create designs that are stable even when there are changes. If a single change to your program results in changes to several dependent modules, then OCP is advising you to refactor your application so that the next time there is a change, you will not end up in the same situation. If you applied OCP well, then the next time a change like that comes up, you will fix it by adding new code, not by changing the old code that already works.[21]

An example of this is the Options window in LabVIEW. It is a configuration editor that is written in LabVIEW. If you look closely, you will see that the section on the right is a subpanel and the categories on the left select which VI is loading into the subpanel. This means that when NI LabVIEW R&D wants to add a new configuration section, all they have to do is write the new VI that will be loaded, or maybe even only change one of the existing categories. Imagine if all of the categories were on

---

[20] The SOLID principles explained with motivational posters. DeVilla Joey. blogs.msdn.microsoft.com/cdndevs/2009/07/15/the-solid-principles-explained-with-motivational-posters/
[21] Martin, R. C., *Agile Software Development: Principles, Patterns, and Practices*, Prentice Hall, Upper Saddle River, New Jersey, 2003.

the same VI and on a tab control. Now, every time there would be a change, there would be different changes to the VI, and you might even break some of the working code that has nothing to do with the new category. The LabVIEW Options window is open for extension, via the addition of new categories, and it is closed for modification because the LabVIEW programmers who add new categories do not have to modify how the OK and Cancel buttons work. (See Fig. 4.32.)

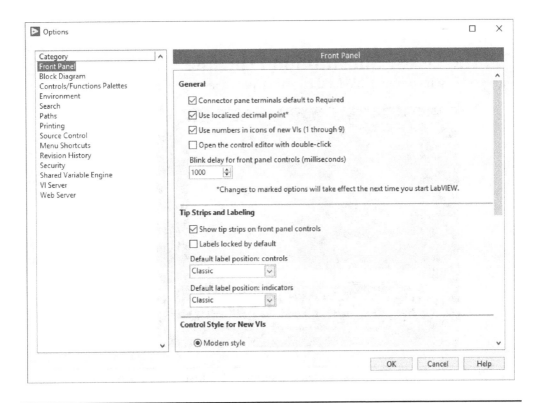

FIG. 4.32 *The LabVIEW Options window.*

Moreover, if the configuration editor window were to be an executable and the categories were loaded at runtime, then the configuration editor is open for an extension because the behavior of the module can be extended by adding new categories. The configuration editor is closed for modification because extending its behavior does not result in changes to the source of the executable. If you make the categories and the editor classes, now the way you can modify them is by creating children of those classes and overriding methods as needed.

One way for you to make LabVIEW remind you if you are violating this principle is to use project libraries (libraries can contain classes) and package them as PPLs

(Packed Project Libraries) as they are ready to be closed for modification. The rest of your code would now be calling the code within the PPL; this means that if the application needs to change, you would violate the OCP principle if you have to change the source code of the PPL. You should be able to extend your PPL by creating children to that class and overriding the methods that need to change. This does not mean that you will never, ever, change that source code again; this is just a reminder to you that you thought you were ready to close that module for modification. If you must change it, make sure you change it in a way that future changes of the same type do not require you to violate the OCP again.

LSP: The Liskov Substitution Principle

"Subtypes must be substitutable for their base types." (See Fig. 4.33.)

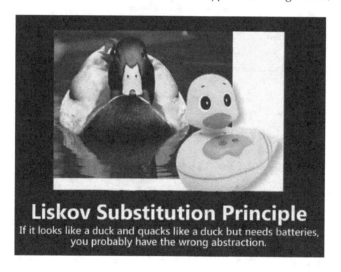

FIG. 4.33 *The folks at LosTechies.com created a series of Creative Commons–licensed posters that illustrate the SOLID principles. Joe Villa took the posters, cleaned up the typography a little, and posted them under the same Creative Commons license.*[22]

Barbara Liskov first wrote this principle in 1988.

"What is wanted here is something like the following substitution property: If for each object o_1 of type S there is an object o_2 of type T such that for all programs P defined in terms of T, the behavior of P is unchanged when o_1 is substituted for o_2 then S is a subtype of T."

---

[22] The SOLID principles explained with motivational posters. DeVilla Joey. blogs.msdn.microsoft.com/cdndevs/2009/07/15/the-solid-principles-explained-with-motivational-posters/

Stephen Loftus-Mercer summarizes this substitution principle as *"Keep your damn promises!"* and he likes to say that this principle answers the question *"Could class X inherit from class Y?"*[23]

Liskov also applies to interface fulfillment. If the interface makes a promise, all the implementing classes have to keep it.

Should Class X Inherit from Class Y?
For X to be a good child class of Y then, if any property is true of all objects of type Y, it must be true for all objects of type X. Moreover, the child should share an identity, state, and behavior with the parent.

Identity
When you give a class a name, you try to name a class based on what it does or what it is. If you name your class a DAQ board, then the child and grandchildren should be some sort of a DAQ board. This goes back to the previous example, where a board that uses the serial port to communicate is not a DAQ board and should not be a child of the DAQ board!

State
The private data of the class usually defines the state of the class. The question you ask here is if every piece of that data is necessary for the function of the child. If not all of the parameters are relevant to the functioning of the child, then you should look to see if inheritance is really the appropriate relationship.

Behavior
All of the methods on the parent class should work on the child. One easy way you can identify if this is the case is to look at how you documented the method. The method documentation should be true for the child method. If the description says "this method will never return errors," then the child method should not return an error.

Like Mother, Like Daughter
Stephen Loftus-Mercer says that classes are female because they use parthenogenetic reproduction, so the LSP can also be summarized as *"Like Mother, Like Daughter."*[24]

If you struggle with fulfilling any of the following conditions, you might want to revisit the SRP and see if your entity can be separated into different entities, since

---

[23] Computer Science for the G Programmer Year 2, presentation at NI Week 2015 by Stephen Loftus-Mercer and Jon McBee https://forums.ni.com/t5/Past-NIWeek-Sessions/Computer-Science-for-the-G-Programmer-Part-2/ta-p/3520618

[24] LabVIEW Classes State of the Art, presentation at NIWeek 2012 by Stephen Mercer.

there may be a hidden responsibility you had not noticed until you tried to add an inheritance.

The Daughter Cannot Demand More of the Callers than the Mother

The parent's preconditions cannot be strengthened in the child. For example, you have a VI and in its documentation, you say that this VI accepts numbers between 0 and 100. The child cannot say that it only accepts 0 to 10. However, the child can be more generous and say that it accepts 0 to 1000.[25]

On the other hand, it would be wrong for the parent method to take any string and for the child override to only take strings that have no numbers in it.

The Daughter May Overdeliver but Not Underdeliver

The parent's post conditions cannot be weakened by the child. For example, if a VI commits to returning all the data within a string and all the information needed is in that string, you cannot have a child that only returns part of the data in the same string.

The Daughter Must Keep Promises Made by the Mother

For example, if the parent promises that a database field needs to increment upon completion of this VI, then the child must also increment that field in the database. One way to avoid this issue is to configure the class to force the children to call the parent and have the parent perform this function. (Do this via right-clicking on the class and selecting Properties, then in the Items Setting section, specify whether to require children to always invoke the Call Parent Method mode. Do this for each method in the class.)

Imagine that you are using a configuration editor similar to the Options window in LabVIEW. As we described earlier, you would extend this configuration editor by adding new categories. You create a new category called "Documentation" that inherits from your category class. The configuration editor knows what to do with category's children, such as how to load them into the subpanel and list them under the category section. Now, imagine that under the hood, this Documentation category modifies other ini tokens that are already handled by some of the other categories, and it does something similar to pressing the OK button. This would cause the entire configuration editor to misbehave because now if the end user presses the Cancel button, whatever changes she already made will be saved, even if she had decided to cancel! Adding this new category causes the configuration editor to misbehave, so the Documentation category violates LSP! It is not keeping its parent's promise to not change ini tokens that do not belong to it.

You would be tempted to add code into the configuration editor so that it knows how to handle the Documentation category. This is now violating the OCP because the configuration editor is no longer closed for modifications! Before going down

---

[25] LabVIEW OOP: Computer Science for the G Programmer. Presentation by Jon McBee and Stephen Loftus-Mercer.

this route, you may want to evaluate if the Documentation category really needs to include code that behaves similarly to pressing the OK button.

The Daughter May Only Return Errors that Are Returned by the Mother (Theoretically)
This is one of the hardest LSP requirements to meet because the parent might not foresee all the possible error conditions for its children. For example, if the configuration was originally read from a database and then the child read the configuration from a file, the child might need to return an error for file not found. The parent might not have foreseen that one of its children would return such an error.

ISP: The Interface Segregation Principle

"Make fine-grained interfaces that are client specific" or "Clients should not be forced to depend on methods that they do not use."

Stephen Loftus-Mercer summarizes ISP informally as *"Give clients only what they will use."* In this context, the client is the target audience for the function you are writing, who is usually the programmer who will use the API that you wrote. The client can also be the hardware target for the code you are writing. The client can also be the end user of the application you are writing. If you give the client more features than they need, they might end up not knowing what to do with the extra features, and they might get in the way of using the features they do want to use.

Stephen Loftus-Mercer also gives a more formal definition as *"Many context-specific APIs are better than one general-purpose API. The goal is to reduce the ambiguity of parameters by having inputs specific to a client's needs. Avoid trying to support the contradictory requirements of two clients."* (See Fig. 4.34.)

The ISP objective is to divide software interfaces into smaller entities in a way where any client only gets the functionality they will actually use.

An example of this would be an action engine that has all the inputs and outputs in its connector pane, but these inputs and outputs are not used in every action. If you create that action engine and give it to one of your peers to use, they will have to open the action engine each time they call one of the actions to see what inputs that are actually used by that case are inside the action engine. One solution to this is to place the action engine in a library, make the action engine and its action enumerator private to the library, and create a public API that wraps every action. These new action VIs would only have the inputs and outputs pertinent to each action VI. In addition, you can mark the inputs as "required," and your peer will never have to open the block diagram of the action engine at all. The action engine itself had a general-purpose API; the public Action VIs have a context-specific API. You can even improve the API further by using the polymorphic approach Steve Watts uses for his components. Wrap all of the individual interfaces with a polymorphic VI, and the selector is the way to decide which interface to use.

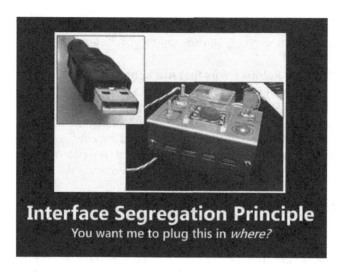

FIG. 4.34 *The folks at LosTechies.com created a series of Creative Commons–licensed posters that illustrate the SOLID principles. Joe Villa took the posters, cleaned up the typography a little, and posted them under the same Creative Commons license.*[26]

Uncle Bob uses an example of a timed door. If you have a regular door, you only care about opening it and closing it. But if you have a timed door that is supposed to sound an alarm if it has been opened more than the target time, now you have a problem. The issue is that the regular door does not have nor needs to know anything about a timer, and none of the code calling a regular door should need to deal with a timer.[27]

First, let us look at a solution that uses native LabVIEW and does not respect ISP. Open the code in the GitHub.com/LGP5/LVOOP repository, under the branch "TimedDoorNoISP." Inspect the code; run some of the testers.

Why do you think this code does not respect ISP? For one, the timed door needs to know about a timer, and any changes done to the timer interface might impact doors. Regular doors should not be affected by changes done to the timer! If you want to see why we say this code does not follow the ISP, try adding a Timer ID as a private data of the timer. This would include providing a new input for `Timer.lvclass:Start.vi` to set the timer id. If you make this change, you will see how all the testers, except for the `Test Alarm API.vi`, break. The culprit is the `Timed Door.lvclass: Unlock.vi` that breaks when it is missing the required Timer ID input to Start.vi.

---

[26] The SOLID principles explained with motivational posters. DeVilla Joey. blogs.msdn.microsoft.com/cdndevs/2009/07/15/the-solid-principles-explained-with-motivational-posters/

[27] Martin, R. C., *Agile Software Development: Principles, Patterns, and Practices*, Prentice Hall, Upper Saddle River, New Jersey, 2003.

However, this also breaks `Test Door API.vi` because although it is meant to only call regular Doors, it is calling `Door.lvclass:Unlock.vi`, and LabVIEW reports an error because one of the dynamic dispatch instances of this VI is broken!

As a possible solution, Uncle Bob suggests using either interfaces or multiple inheritances.[28] Multiple inheritances are currently not possible in LabVIEW, and from the presentations at the different CLA Summits, we strongly believe NI will not add multiple inheritances in future versions. Instead, a lot of discussions have been going on during CLA Summits to bring interfaces into LabVIEW. During discussions at the CLA events and NIWeek 2019, there were multiple presentations and discussions both about the benefits of having interfaces in LabVIEW and of possible ways that interfaces could be added. NI was very interested in these discussions and engaged in debate about how interfaces could be added to LabVIEW. If these discussions turn into a LabVIEW feature, we will be sure to add examples to our GitHub repository. In the meantime, there are some workarounds that we can use in LabVIEW. (See Fig. 4.35.)

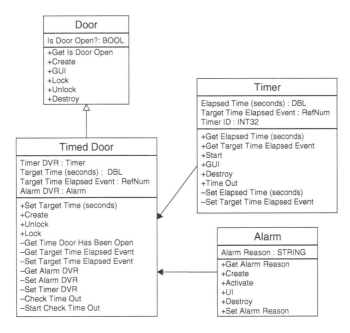

FIG. 4.35 *Timed Door UML diagram. The Timed Door inherits from ("is a") Door. Timed Door owns ("has a") timer and owns ("has an") alarm. This particular solution does not respect ISP.*

---

[28] Martin, R. C., *Agile Software Development: Principles, Patterns, and Practices*, Prentice Hall, Upper Saddle River, New Jersey, 2003.

An interface is a class that does not have private data and defines different behaviors. You would use interfaces, for example, to define `Voltage Measurement Device.lvclass`. This class would not have any private data, and it would have a `Measure Voltage.vi`. Both the `DMM.lvclass` and the `Scope.lvclass` could implement `Measure Voltage.vi` without having to inherit from `Voltage Measurement Device.lvclass`. You do not want to have DMM and Scope inherit from the same class because they are different devices. The Scope can acquire waveforms and the DMM cannot, so it would not make sense for them to be siblings. Having an interface would allow you to reuse code and benefit from dynamic dispatch without making two unrelated classes be members of the same hierarchy.

The Adapter Pattern (Also Known as the Wrapper Pattern)

This is one of the patterns from the Gang of Four book.[29]

The objective of the adapter pattern is to convert the interface of a class into another interface clients expect. The adapter lets classes work together that couldn't work together due to incompatible interfaces.

The idea is to create a wrapper class around the main class. You may have multiple wrappers, each with its own interface, and each wrapper may inherit from different hierarchies. LabVIEW classes are by value, which means that the object can only be one interface at a time. However, with by reference classes, one object can be multiple interfaces at once.

How is the adapter pattern going to help us in the Timed Door code? Open the code in the GitHub.com/LGP5/LVOOP repository, under the branch "TimedDoorISPAdapter" to see the final results of the steps you would have to follow to convert the previous code into one that uses an adapter to segregate the interfaces for the door and the timed door. First, you will notice that you have removed the inheritance association between Timed Door and Door. Timed Door is no longer a Door. Instead, the private data of Timed Door has a Door. (See Fig. 4.36.)

Next, verify that the testers for Timer, Alarm, and Door APIs continue to work. The only testers that are broken are the Timed Door API Tester and the Multiple Doors API Tester. This is expected because now `Timed Door.lvclass:Lock.vi` and `Timed Door.lvclass:Unlock.vi` can no longer call their parent's method: Timed Door no longer has a parent! You will need to adapt the Timed Door into a Door in order to call the Lock and Unlock methods. For this, you created an accessor for the Door and replaced the old parent call with code that adapts the Timed Door to Door, then performs the operation and adapts back to Timed Door. Now, any changes to the timer interface will no longer break the code for `Door.lvclass`. (See Fig. 4.37.)

---

[29] *Design Patterns. Elements of Reusable Object-Oriented Software* by Erich Gamma, Richard Helm, Ralph Johnson, and John Vlissides.

Chapter 4: LabVIEW Object-Oriented Programming 259

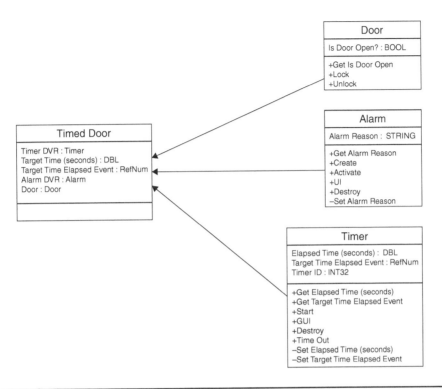

FIG. 4.36 *Timed Door owning Door instead of inheriting from it UML diagram. The Timed Door owns a ("has a") Door. Timed Door owns ("has a") timer and owns ("has an") alarm. In this particular solution, the Door and Timed Door have their own interfaces, and modifications to the Timed Door interface have little to no impact on Door's interface.*

It is important to notice that although you gained the benefit of separating the interfaces for Door and Timed Door, you also lost the benefits of inheritance. Any code that is using Doors that wants to use Timed Doors as well will need to make modifications to their code. For example, you had to change the Multiple Doors API Tester to handle an array of Timed Doors instead of an array of Doors in order to be able to act both on Timed Doors and Doors. Also, you made the code that adapts the Timed Door into a Door (the accessors to get and set the Door) private. If calling code would need to adapt from Timed Door to Door outside of its methods, you would need to make these have a "community" scope and declare the calling code a friend of the Timed Door class. You could also recognize those methods as public methods of a third class. Listing all the options that we could do in a hypothetical case is really hard to do. We are giving you the options we see at this moment, but later we may come up with other options or a better approach.

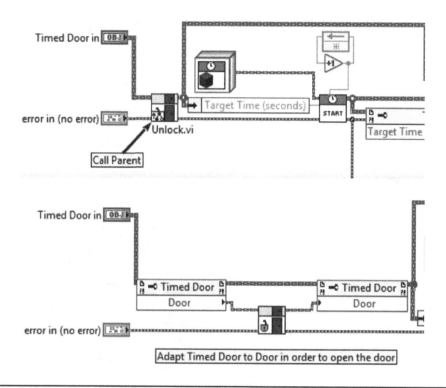

FIG. 4.37 *Timed Door no longer inherits from Door, so it can no longer call the Unlock .vi parent. Instead, Timed Door adapts to Door, calls unlock.vi, and adapts back to Timed Door.*

AZ Interface Toolkit

There have been some workarounds to implementing interfaces; you can check out one of them at www.azinterface.net. Andrei Zagorodni created this toolkit that implements Java-like interface architecture in LabVIEW projects. He presented the first version at the 2018 European CLA Summit in Madrid.

If you visit www.azinterface.net and follow the instructions to install the AZInterface toolkit, you can create an interface class called Door Interface by right-clicking on the My Computer target within the project and selecting **AZ Interfaces >> Create AZ Interface**. Once you create the class, you can right-click on Door.lvclass and select **AZ Interfaces >> Apply Interface**. Do the same for the Timed Door.lvclass. Then you can create the interface methods that you want to implement (right-click the interface class and select **AZ Interfaces >>**

Create Interface Method). In this case, you will create Lock.vi, Unlock.vi and Is Door Open.vi. You need to make sure all of the connector panes are connected to the terminals on the front panels. Select Save All and then you can right-click on Door.lvclass and select **AZ Interface >> Apply Interface** once again, but this time to implement the methods. Do the same for the Timed Door .lvclass. The current version of the AZInterface toolkit (v.2.0.0) does not have automated error prevention features. Pay close attention to the AZInterface toolkit documentation because making a mistake when creating the interface can result in lots of manual modifications. (See Figs. 4.38 and 4.39.)

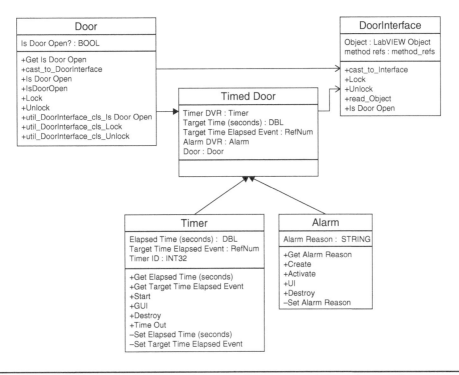

FIG. 4.38 *UML diagram of the Timed Door solution using AZInterface toolkit.*

An example of this code is in the GitHub.com/LGP5/LVOOP repository, under the branch "ISPTimedDoorAZ."

262 LabVIEW Graphical Programming

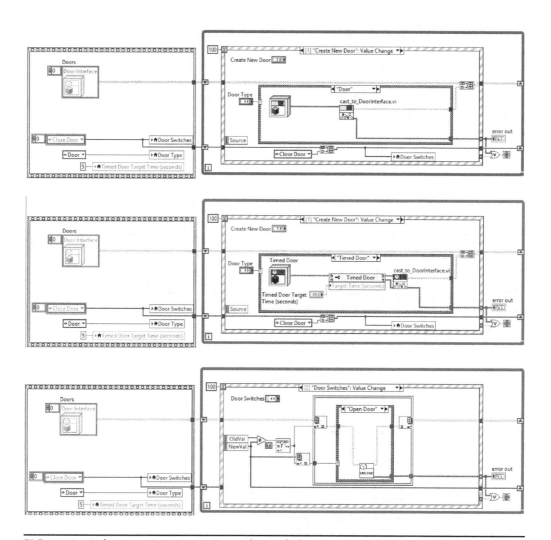

FIG. 4.39 *When creating a new Door, the code has to cast to the Door Interface. When the code executes the code to open or close the Door, each class (Door and Timed Door) is implementing the Door Interface methods.*

DIP: The Dependency Inversion Principle

"Depend on abstract entities, not on concrete entities."

FIG. 4.40 *The folks at LosTechies.com created a series of Creative Commons–licensed posters that illustrate the SOLID principles. Joe Villa took the posters, cleaned up the typography a little, and posted them under the same Creative Commons license.*[30]

Uncle Bob says[31]

a. *High-level modules should not depend on low-level modules. Both should depend on abstractions.*

b. *Abstractions should not depend on details. Details should depend on abstractions.*

Many programmers tend to build their programs where the high-level modules depend on the low-level modules. This means that a change in the low-level modules could end up impacting the behavior of the high-level modules. Uncle Bob believes that this is absurd, because it is the high-level modules that set the business rules, and they should not depend on low-level modules. He uses the word inversion as part of this principle because he is going opposite to what we are accustomed to seeing. (See Fig. 4.40.)

Dependencies

There are two types of dependencies; runtime dependency and source code dependency. You might be thinking that these two things are exactly the same thing, but it turns out they are not.

---

[30] The SOLID principles explained with motivational posters. DeVilla Joey. blogs.msdn.microsoft.com/cdndevs/2009/07/15/the-solid-principles-explained-with-motivational-posters/
[31] Martin, R. C., *Agile Software Development: Principles, Patterns, and Practices*, Prentice Hall, Upper Saddle River, New Jersey, 2003.

Runtime Dependency
The runtime dependency occurs whenever two modules interact at runtime. This is when the code executes.

Source Code Dependency
The source code dependency happens when a method defined in one class is called by methods in another class. Identify the code your entity depends on by placing that entity in a blank project. The dependencies section of the project will list all of your source code dependencies.

Example

Let's look at the simple example that Uncle Bob uses to explain the Dependency Inversion Principle.[32] Open the code in the GitHub.com/LGP5/LVOOP repository, under the branch "DIPButtonLamp," to follow the LabVIEW implementation. (See Fig. 4.41.)

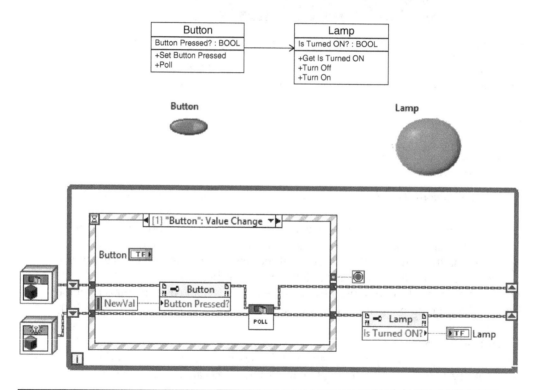

FIG. 4.41 *UML model of a button that "uses a" Lamp and simple code implementation.*

---

[32] Martin, R. C., *Agile Software Development: Principles, Patterns, and Practices*, Prentice Hall, Upper Saddle River, New Jersey, 2003.

You probably already noticed that this button is not very reusable. This button can only turn on lamps that are children of the class Lamp. If we wanted to use that button to turn on a temperature chamber, we wouldn't be able to do so because a temperature chamber is not a lamp. This simple example does not apply the DIP. You have not separated yet the abstractions from the details. Let's look at the source code and runtime dependencies tree for this example. (See Fig. 4.42.)

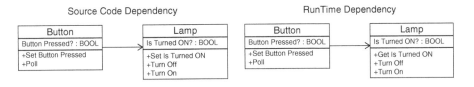

FIG. 4.42 *The Source Code Dependency Tree and the RunTime Dependency Tree are aligned. This code does not implement the DIP.*

In order to be able to invert the dependency, you need to identify what is the abstraction in this case. You had thought about controlling a temperature chamber with the same button, so it seems like you need an abstract class that would be turned on/off via a button. You could call this "Switchable Device," and Lamp could be a child of this class. (See Fig. 4.43.)

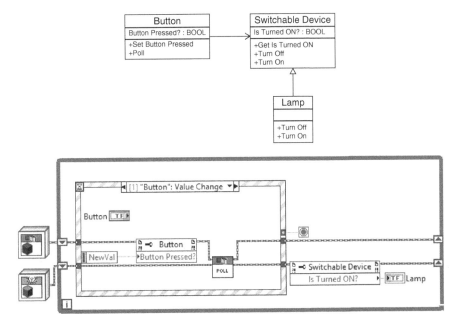

FIG. 4.43 *The button now depends on Switchable Device, and the Lamp is a type of switchable device.*

The code looks the same, but inside `Button.lvclass:Poll.vi` you went from calling the Lamp's code directly to calling the code from the Switchable Device. This means that if you were to add the `Button.lvclass` in a new project, you would only see the Switchable Device dependency. (See Fig. 4.44.)

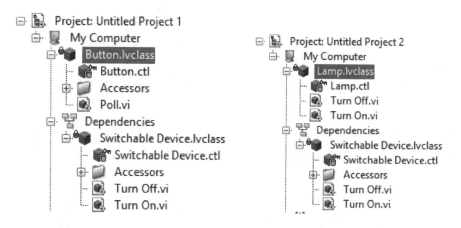

FIG. 4.44 *Source code dependencies: The Button.lvclass only depends on Switchable Device. The Lamp.lvclass only depends on Switchable Device.*

However, at runtime, your button depends on Lamp! (See Fig. 4.45.)

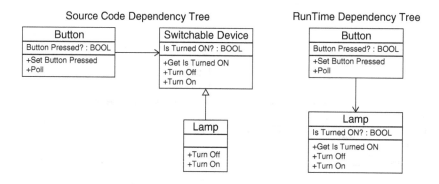

FIG. 4.45 *The Source Code Dependency Tree and the RunTime Dependency Tree are no longer aligned. The dependency inversion occurs at the Lamp level. The Lamp depends on Switchable Device at the Source Code Level. The Button depends on the Lamp at runtime. This code implements the DIP.*

At source code, your `Button.class` depends on the abstraction called "Switchable Device," and Lamp implements the Turn On/ Turn Off abstraction. However, at runtime, the button is calling the `Lamp.lvclass:Turn On.vi` and `Lamp.lvclass:Turn Off.vi` directly. Once you understand that the inversion is happening at the abstract class level, this principle starts making a lot more sense. Implement your code thinking at the abstract level, then at runtime decide which specific implementation your code will use, and applying this principle will become second nature (or so we are told).

If your source code dependency tree and your runtime dependency tree are the same, then you are not applying dependency inversion. And this is not wrong! It is just a different approach. Generally, dependency inversion is a good indicator of object-oriented design; if the dependencies are not inverted, then it is more than likely a procedural design. The goal of this principle is to promote code reuse, create code that is flexible in the presence of change, and improve code testability. This is achieved because the abstraction and the implementation details are separated from each other, and in theory, this should minimize the impact of change.

Interfaces

Don't get too excited yet. As we discussed before, LabVIEW does not support multiple inheritance. Having such a generic abstract class as "Switchable Device" increases the possibility of needing the Lamp class to implement the abstract methods of different classes. We could use a similar approach as the adapter approach described in the DIP; however, the best solution would be to use interfaces, which we don't currently have natively in LabVIEW.

If you visit www.azinterface.net and follow the instructions to install the AZInterface toolkit, you can create an interface class called Switchable Device by right-clicking on the My Computer target within the project and selecting **AZ Interfaces >> Create AZ Interface**. Once you create the class, you can right-click on `Lamp.lvclass` and select **AZ Interfaces >> Apply Interface**. Then you can create the interface methods that you want to implement (right-click the interface class and select **AZ Interfaces >> Create Interface Method**). In this case, you will create `Turn On.vi, Turn Off.vi,` and `Get Device Status.vi`. You need to make sure all of the connector panes are connected to the terminals on the front panels. Save all and then you can right-click on `Lamp.lvclass` and select **AZ Interface >> Apply Interface** once again, but this time to implement the methods. The current version of the AZ Interface toolkit (v.2.0.0) does not have automated error prevention features. Pay close attention to the AZ Interface toolkit documentation because making a mistake when creating the interface can result in lots of manual modifications. (See Fig. 4.46.)

An example of this code is in the GitHub.com/LGP5/LVOOP repository, under the branch "DIPButtonLampAZ."

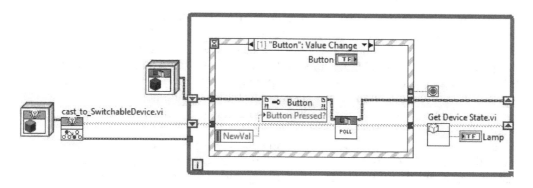

FIG. 4.46 *Using the AZInterface toolkit, the Lamp is cast into the interface called Switchable Device.*

Now we can have `Lamp.lvclass` implement other interfaces if needed, and we can have other classes in the future implement the Turn On and Turn Off methods of the Switchable interface.

Caveats

"The problem with object-oriented languages is they've got all this implicit environment that they carry around with them. You wanted a banana but what you got was a gorilla holding the banana and the entire jungle."

—Joe Armstrong, the creator of Erlang

LabVIEW programmers might think that just because they use classes, they are wiring an object-oriented program. That is not true. Object-oriented programming is an approach to designing your application, not how you wire it. If you are focusing on state transformations and encapsulated abstractions, you are following object-oriented programming.

Object-oriented programming constructs from other languages have not been implemented natively in LabVIEW. As we discussed earlier, we will likely never get multiple inheritances, and we might be better off without them! However, we don't have interfaces, traits, and mixins, which would make it a lot more straightforward to get around situations where multiple inheritances feel like the right solution.

When LabVIEW classes came out, we would routinely get specific requests from customers asking us to not use LabVIEW classes at all. After a while, we realized that this was more due to misunderstanding what classes were, and we don't find that restriction as often as we used to. LabVIEW classes were not included in the LabVIEW Core training courses NI offers, and programmers needed to take the LVOOP class separately. Fortunately, this is changing, and we hope this chapter helps bridge the gap between non-OO programmers and LVOOP enthusiasts.

Accessors Get/Set Instead of Read/Write

As we mentioned earlier, we prefer to use Get/Set instead of Read/Write to avoid confusion with the hardware calls we already use, such as DAQmx Read and DAQmx Write.

At Delacor, the package that every team member installs includes code to change the Read and Write to default to "Get" and "Set." We do this by automating the following steps

1. Close LabVIEW.

2. Make a backup of the following file: `resource\Framework\Providers\LVClassLibrary\NewAccessors\CLSUIP_LocalizedStrings.vi`.

3. Open the original VI and edit the "Write <element>.vi" and "Read <element>.vi" to use "Get" and "Set" instead of "Read" and 'Write."

4. Make these new values the default values.

5. Save the VI.

Classes in LabVIEW NXG

The objective in NXG is to make the transition from non-OO to OO programming much more straightforward than it is today. One of the main changes is to make a cluster and a class be less distinct things and more a common type with different attributes. The cluster type has wide-open data and no methods. Then you can choose to add methods, encapsulate the data, or enable inheritance. If you add all three of those attributes, you have a class type.

In NXG, the bundle and unbundle nodes are gone, replaced with a property node, but that property node can both access member fields and call property methods.

As far as tools are concerned, we have not heard of any plans to migrate the GDS toolkit to NXG. However, we have seen some of the demonstrations of the UML Class Editor that VI Technologies created for NXG (for more information visit uml-addon.com).

References

Many of the presentations from NIWeek, CLA Summits, and other events can be found at the Tecnova site. More details are here: https://lavag.org/topic/20645-labview-videos-tecnova-download-site/

The Object-Oriented Thought Process, Third Edition, by Matt Weisfeld.

UML Distilled: A Brief Guide to the Standard Object Modeling Language, Third Edition, by Martin Fowler.

Agile Software Development: Principles, Patterns, and Practices by Robert C. Martin.
Uncle Bob's articles on principles of OOD, http://butunclebob.com/ArticleS.UncleBob.PrinciplesOfOod
A Software Engineering Approach to LabVIEW, by Jon Conway and Steve Watts.
Design Patterns: Elements of Reusable Object-Oriented Software by Erich Gamma, Richard Helm, Ralph Johnson, and John Vlissides.
Appendix D dedicated to LVOOP in the book *LabVIEW for Everyone: Graphical Programming Made Easy and Fun*, Third Edition, by Jeffrey Travis and Jim Kring.
Decisions behind the Design: Differences between OOP in Text-Based Programming and in LabVIEW: http://www.ni.com/white-paper/3574/en/
LabVIEW Object-Oriented Programming FAQ, http://www.ni.com/white-paper/3573/en/
Applying Common Object-Oriented (OO) Design Patterns to LabVIEW, https://forums.ni.com/t5/Example-Program-Drafts/Applying-Common-Object-Oriented-OO-Design-Patterns-to-LabVIEW/ta-p/3510571
https://blog.ircmaxell.com/2012/07/oop-vs-procedural-code.html
OpenGDS Github repository, https://github.com/opengds/OpenGDS
OpenGDS Github releases, http://opengds.github.io/
GDS support forum, https://forums.ni.com/t5/GDS-Goop-Development-Suite/gp-p/5050
VIshots interview with Mikael Holmstrom and Stephen Loftus-Mercer, https://youtu.be/abXD7M7-Y1Q
LabVIEW OOP: Computer Science for the G Programmer. Presentation by Jon McBee and Stephen Loftus-Mercer. Video from LabVIEW Architects Forum version at https://youtu.be/m3PvWipPDT0
LabVIEW Classes State of the Art, presentation at NIWeek 2012 by Stephen Mercer
Computer Science for the G Programmer Year 2, presentation at NIWeek 2015 by Stephen Loftus-Mercer and Jon McBee. This link also includes their example code from the presentation: https://forums.ni.com/t5/Past-NIWeek-Sessions/Computer-Science-for-the-G-Programmer-Part-2/ta-p/3520618
Jon McBee's presentation about Liskov Substitution Principle, http://www.labviewcraftsmen.com/blog/liskov-substitution-principle
Jon McBee's presentation about Dependency Inversion Principle, http://www.labviewcraftsmen.com/blog/dependency-inversion-principle
Ethan Stern's blog posts on LVOOP,
    http://www.bloomy.com/support/blog/object-oriented-labview-inheritance-part-1-3-part-series
    http://www.bloomy.com/support/blog/object-oriented-labview-encapsulation-part-2-3-part-series
    http://www.bloomy.com/support/blog/object-oriented-labview-polymorphism-part-3-3-part-series
AZInterface toolkit, http://www.azinterface.net/
    https://lostechies.com/derickbailey/2009/02/11/solid-development-principles-in-motivational-pictures/

https://blogs.msdn.microsoft.com/cdndevs/2009/07/15/the-solid-principles-explained-with-motivational-posters/

UML Class Editor Add-on for LabVIEW NXG, http://uml-addon.com

3 Simple Object-Oriented Design Patterns in LabVIEW, http://www.ni.com/newsletter/51506/en/

The Agile Manifesto, https://agilemanifesto.org/

SOLID Actor Programming Presentation at NIWeek 2018 by Dmitry Sagatelyan, https://forums.ni.com/t5/Actor-Framework-Discussions/SOLID-Actor-Programming-Presentation-at-NIWeek-2018/td-p/3794059

Applying Common Object-Oriented (OO) Design Patterns to LabVIEW, https://forums.ni.com/t5/Example-Program-Drafts/Applying-Common-Object-Oriented-OO-Design-Patterns-to-LabVIEW/ta-p/3510571

CHAPTER 5

Why Would You Want to Use a Framework?

Fab likes to tell the story of that time she had lunch with a customer before looking at their code, and when they introduced her to another team member, he welcomed her by saying, "Oh, you are here to help us with The VI." She almost cried when she verified that indeed, the application was "The VI" and it included about 40 by 40 screen sizes, and it was all spaghetti code. Good times!

We have encountered LabVIEW programmers who are content with the state of their code and do not see the need to use a framework. Computer scientists might cringe at this, but from a pragmatic point of view, in many cases this is totally fine. We cannot argue with success, and if your code runs and helps you get your job done, you probably do not need a framework. However, this situation may change when you need to share your code with others, or you need to hire someone to help you with your code. This will be the time when you need to pay the technical debt, and one of the ways to do this will probably be to start using a framework. Alternatively, you can build technical wealth by starting your applications using a framework. If your boss is complaining that you used to add features really fast with LabVIEW and now you are taking longer, you probably need a framework. A framework will establish the rules that you follow on your code. The framework will either encourage other developers to follow those rules or enforce the rules. You know you all follow the same style, and you can share code among each other easily.

For this chapter, we assume that you agree that creating an application using a single VI is no longer an acceptable way to program. We also assume that at this point, after reading this book and working on some other applications, you are not afraid of using project libraries.

More importantly, we assume that we are all in agreement that we want our applications to be modular, where each module has a single task (they are cohesive) and we want to limit the interdependencies between our modules (they are decoupled). We want everyone in our team to follow the same style. We want to make it easier to onboard new team members. We want to create code that is testable and reliable.

In this chapter, first we cover some of the concepts that we will be using. Then we will dive in on two frameworks: Delacor Queued Message Handler (DQMH) and Actor Framework. For each framework, we first cover how to use it and then go into some of the behind-the-scenes details. We recommend you choose one of the existing frameworks, but if you were to decide to create your own framework, the behind-the-scenes sections will guide you through all that is needed to create a good framework. If you are already familiar with framework concepts, skip the first sections in this chapter and go right into the framework you want to learn.

What?
What Is a Process?

For our purposes, a process is anything that runs continuously. We call the different sections of our code modules. In the Advanced Architectures in LabVIEW course[1] by National Instruments (NI), they would be called processes. In the Actor Framework, each actor is a separate process.

A module is a section of the application that can operate as part of the whole application or on its own. It runs continuously and can be started and stopped at different points in the application.

What Is an Abstraction Layer?

According to the *Merriam-Webster* dictionary,[2] from its roots, abstraction means "something pulled or drawn away." For example, an abstract of a publication is a one-paragraph summary of its contents, with the basic findings "pulled out" of the article. In our case, an abstraction layer is one of the layers in our software application. We use it to take a step back and take a broad view of what the application is trying to accomplish, without going into the implementation details for this particular application.

---

[1] http://www.ni.com/training
[2] https://www.merriam-webster.com/dictionary/abstraction

What Is a Framework?

Going back to the *Merriam-Webster* dictionary,[3] one of the definitions of a framework is a frame of reference. In turn, the dictionary defines a frame of reference[4] as a set of ideas, conditions, or assumptions that determine how something will be approached, perceived, or understood.

We see a framework as the frame of reference that provides guidelines on how to develop the code. The framework allows for modifying some of its components and adapts them to a more specific problem. The framework also includes a set of tools and interfaces for integrating the framework into the final solution. A framework guarantees that the interprocess communication works (messages are always delivered), the processes initialize and stop gracefully, and there is a basic error handling system. If the processes in your application stop initializing correctly, stopping gracefully, and the error handling doesn't work, then you know it is due to something you did.

A framework can also be a springboard for any new project you work on. We hear you. Some experienced LabVIEW programmers say that all their projects are unique. We believe you; however, there are a set of features that you keep implementing over and over when you start developing a new application. At the very minimum, a well-designed application will have a group of well-defined modules. We want those modules to start when we tell them to start, stop gracefully when we tell them to stop, and have a bare minimal error handling. This error handling can be as simple as stopping the module on an error.

When Do You Make the Decision to Break Your Application into Parallel Tasks?

There are different considerations. First, what are the size and expected lifetime of the application? If you are in a lab and just want to acquire a set of data, analyze it, and present it, then a monolithic application is probably fine.

If there is the slightest possibility that this application will be part of something larger. For example, now you are in a lab and you are thinking of having a way to compare the acquisition set that you acquired last week with the one you acquired today. Perhaps the system needs to turn a motor or a heating element on or off based on the values acquired. Now we are no longer talking about a simple acquire, analyze, and present type application.

Another consideration is the number of windows the end user will be interacting with. For example, in the lab application, you might want to have a separate

---
[3] https://www.merriam-webster.com/dictionary/framework
[4] https://www.merriam-webster.com/dictionary/frame%20of%20reference

window to configure your acquisition, a different display while the acquisition is taking place, and yet another window for analyzing previous acquisitions. This would also apply to tab controls. Look at your current application. Do you have a tab control there? Tab controls are great to organize the elements on the front panel, but one of their drawbacks is that the terminals on the block diagram are not organized. The more controls and indicators there are on your GUI, the more code your block diagram is implementing. The more code your block diagram is implementing, the more coupled and less maintainable and extensible it becomes. An application that needs several tabs or windows is a good candidate for separating in parallel tasks. A tab control can be replaced by a subpanel.

One consideration that is not code related is the size of the team. If you are working in a team and several people are going to work on the same application, it is a very good idea to ensure the application is modular. This means that different developers can work on their module without stepping on each other's toes. One member can work on the acquisition module while another team member works on the logging module. Finally, one team member works on the GUI and integrates the modules the other two team members worked on. If you work by yourself but there is a chance that other team members might use your code, it will be better if your code is modular enough where you can share only the modules they need. This is much better than handing them a plate of spaghetti code and expecting them to extract the meatballs.

What Are the Design Decisions That Need to Be Nailed Down at the Beginning to Make a Parallel Design Successful?

Fab is a big fan of modeling, or doing a graphical representation of the different modules and what their interactions are going to be. Remember that it is a lot easier to erase things on a whiteboard or piece of paper than deleting the code you have already committed to. It is also cheaper. Do your design work in the cheapest medium possible first, then code.

An example of a model could be bubbles representing each module and arrows going between them representing the data shared between them. (See Fig. 5.1.)

Fab presented a more detailed example of modeling at NIWeek 2018.[5] This presentation resulted in the DQMH version of the Continuous Measurement and Logging Sample project (DQMH CML) that ships with DQMH. Create your own copy of the DQMH CML by clicking Create Project from the Getting Started window in LabVIEW, and then clicking the DQMH CML. (See Fig. 5.2.)

---

[5] http://delacor.com/tips-and-tricks-for-a-successful-dqmh-based-project/

Chapter 5: Why Would You Want to Use a Framework? **277**

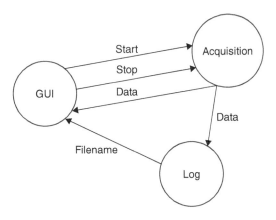

FIG. 5.1 *A simple model for a continuous acquisition application.*

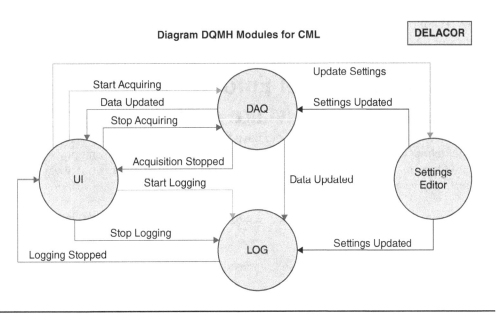

FIG. 5.2 *Diagram of the continuous measurement and logging example using DQMH modules.*

While doing this modeling exercise, you need to determine each module's task. In the continuous acquisition and logging example, you need to decide who is in charge of naming the file. If you are going to name the file based on the timestamp, the Logging module could be in charge of creating the file. If you think that the end user is the one that needs to select the file name based on the test they are running, the GUI needs to send the file name to the Logging module.

You also need to question if you are creating enough modules. One could argue that the Acquisition and Logging modules could be together. If there is a chance that the logging format may change or you think you can use the Acquisition module in other applications that do not need logging, it makes more sense to keep them separated.

The final consideration is how you plan to test each module. A lot of the experienced LabVIEW developers that we encounter are used to creating entire applications and waiting until the entire system is put together to start with serious testing. We prefer to test as we go. Having the Acquisition module separated from the Logging module means that you can test that the Acquisition module works by itself. You do not need anything else to test it. Similarly, you can test the Logging module without having access to the hardware. If you are following test-driven design (TDD) principles, you would even design how you plan to test each module before you start implementing them (see the "What is TDD?" section in Chapter 6).

What Project-Level Enforcement Is Available to Help Make Better Decisions?

The teams we have worked with prefer to have LabVIEW be the enforcer of their style guidelines and good practices. Not a lot of people enjoy having someone else point out the areas where they could have done things better. Also, it is not always possible to have access to the developer who designed the code you are working with.

These are a couple of things that you, a good LabVIEW developer, can do to make LabVIEW the enforcer.

1. Place the VIs in libraries. These libraries can be project libraries (.lvlib) or classes (.lvclass).

2. Configure VI input terminals to be required.[6]

3. If you are using classes:

    a. Configure the overrides for certain methods to require to call the parent.

---

[6] LabVIEW has an INI token; you can also change it via Tools >> Options >> Front Panel >> Connector pane terminals checked, to make all inputs required by default. The token is `reqdTermsByDefault=True`

b. Configure certain methods that must be overridden by a child.

   c. Configure whether you expect other developers to create DVRs for the class or if your class will provide methods to wrap the class into a DVR.

If any developer, even yourself, fails to follow your requirements, LabVIEW will inform them with a broken arrow and a brief explanation of why there is an error. LabVIEW will inform them about the code designer's original intentions. (See Fig. 5.3.)

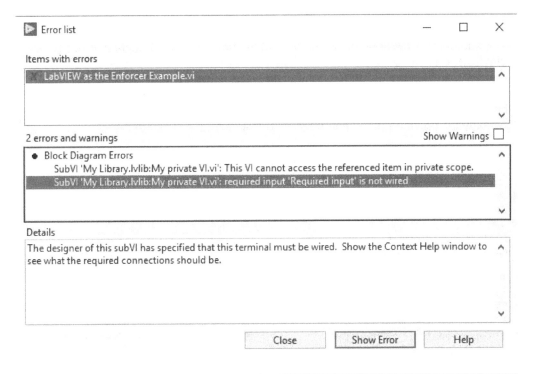

FIG. 5.3 *LabVIEW highlights that a required input is not wired, and in the details section, LabVIEW informs the developer that the designer of this VI had specified that input as required.*

Project Libraries

Libraries have several advantages. All their members share the same icon header. All their members' names become `<library name>:<VI name>.vi`. The library provides the namespace for all the VIs in that library. When a developer uses a library that someone else created, bringing one of the VIs in that library into their project will bring the whole library. Fab likes to joke that it is like in the movie *My Big Fat Greek Wedding*; you invite one member of the family and everyone comes along. This is a good way to tell the developer using the library that there are other VIs related that could be useful to them.

Libraries also offer the advantage of declaring access scope. The developer can decide that certain VIs will be private. This means they do not expect other developers to call them directly. This is a subtle way to encourage the developer to use the public VIs. A good example of this is polymorphic VIs. Fab likes to put the polymorphic instances in a private folder within the library, and only the polymorphic VI itself is public. This means other developers will find out that they cannot call the separate instances and they have to call the polymorphic VI directly. This is another way to show the developer that the polymorphic VI might have other useful instances.

You can also right-click on the library and set its properties to be locked. There is no need for a password. This hides the private VIs from the library, making it clear to others what public VIs they have access to. In the previous example, that would show only the polymorphic VI and none of its instances. (See Fig. 5.4.)

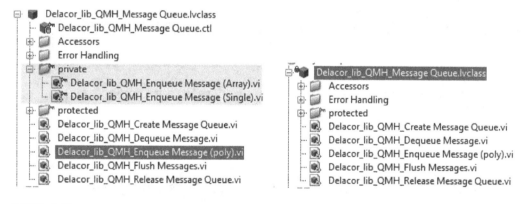

FIG. 5.4 The library on the left has the two implementations of the enqueue polymorphic VI marked as private. After the developer right-clicked on the library and set its protection property to be "Locked (no password)," those private VIs are no longer visible, as shown on the right.

Globals within Project Libraries

Another example where libraries can encourage good behavior is the case of global variables (aka globals). There are a lot of LabVIEW programmers who hate globals. The majority of experienced LabVIEW programmers will agree that if a **global** is of the type **WORM** (Write Once Read Many), then it is not a problem. How can you indicate via the code that your intention is for the global to be WORM? You put the global in a library and make it private. Then only VIs within the library can write to it. You can control that the only VI that is called during initialization writes to the global and then you create public VIs that will access the global private data. It sounds like a paradox to have something be global and private at the same time, but this is the best way we have found to make it clear to all that the global is of the type WORM.

Restricting the access to a global applies as well to functional global variables (FGV), also known as action engines[7] (AE). The FGV itself can be private to the library as well as its enumerator. Then you create public VIs to implement every method (one VI for each option in the enumerator). This means the wrapper VIs make it very clear what inputs are needed for that method only. When you leave an AE without wrappers, other developers using the AE need to open the AE to know if a certain input is needed or not for a particular action.

Some old timers worry about creating too many subVIs. Back in LabVIEW 2 (well, maybe in more recent versions, too), a subVI call took time and made the program less efficient. We do not think this is a concern anymore. However, if you are programming embedded applications or trying to shave off as much time as you can, you can look into **inlining** the public VIs. The compiler will treat an inlined VI as if it was part of the block diagram that calls it. Refer to the example that ships with LabVIEW called "VI Execution Properties.vi,"[8] for a comparison between the runtime performance of an inline subVI against a normal subVI with debugging, a normal subVI without debugging, a subroutine subVI, and the primitive itself.

Frameworks for LabVIEW

There are several frameworks developed by members of the LabVIEW community. In this book, we focus on the Actor Framework because it ships with LabVIEW and the DQMH. However, we encourage you to explore other frameworks like

- Distributed Control and Automation Framework (DCAF)—http://www.ni.com/white-paper/54341/en/

- Top Level Baseline and Top Level Baseline Prime by Norman Kirchner—https://lavag.org/topic/16188-tlb-top-level-baseline-prime-application-template/

- LabVIEW Messenger Library by Dr. James Powell—available via the LabVIEW Tools Network (ni.com/labviewtools)

---

[7] Community Nugget 4/08/2007 Action Engines by Ben Rayner https://forums.ni.com/t5/LabVIEW/Community-Nugget-4-08-2007-Action-Engines/td-p/503801
[8] <LabVIEW>\examples\Performance\VI Properties

- LapDog by Dave Snyder—https://lavag.org/topic/12388-lapdog-an-open-source-mid-level-api-for-lvoop/
- Aloha by S5 Solutions—available via the LabVIEW Tools Network (ni.com/labviewtools)

Note that we are not just choosing DQMH because one of the authors of this book is the lead architect for DQMH. DQMH is one of the most popular LabVIEW frameworks. During NIWeek 2018, Actor Framework and DQMH were the two frameworks that had more dedicated presentations. Also, NI decided to kick off their Frameworks Learning Badges[9] with Actor Framework and DQMH.

To learn more about both frameworks, we recommend creating a project via the Getting Started window in LabVIEW and selecting to create a DQMH project or Actor Framework project. Both projects will include a Documentation virtual folder with very detailed documentation as to how the projects work. You will need to install DQMH via VIPM before you can create the project.

Stephen Loftus-Mercer works for NI. He is considered the father of LabVIEW object-oriented programming, and he designed and implemented Actor Framework. Allen Smith worked very closely with Stephen to develop Actor Framework. Stephen provided the solid computer science theory–based design, while Allen provided the pragmatic approach from an NI customer's point of view. Not a lot of people know this, but Allen started working on Actor Framework before he joined NI, and he has continued to work on Actor Framework even after he left NI.

Advantages of Using a Framework

The main advantage of using a framework is knowing how long it takes to start a project and having a good idea of how long it takes to develop a module, with the knowledge that we are not starting from a blank page and that our project builds as an executable from the very beginning.

A framework has been tested, proven, and used by others. There is no need to reinvent the wheel, or worse, a square tire.

Using a framework enforces good programming practices and promotes the same programming style within a team. The file disk organization is similar. When a team member has to leave because they won the lottery (don't you like that better than being run over by a bus?), other team members can find their way around the code. Everyone in the team speaks the same language and understands the framework nomenclature.

Working on modules makes it easier to have cohesive sections of code that are testable and maintainable on their own. As such, the team leader can distribute the workload among different developers without worrying about having more than one developer working on the same VI.

---

[9] http://ni.com/badges

The team that uses a framework lets team members focus on developing a solution and not worry about architectural decisions like interprocess communication implementation, error handling, and how to start and stop modules.

Both DQMH and Actor Framework come with a set of tools that facilitate the creation of interprocess communication messages (in the case of Actor Framework) and events (in the case of DQMH). Fab has often heard that people would not be using Actor Framework or DQMH if it were not for their tools.

Finally, using a framework improves maintenance tasks and does not necessarily require the original developer to maintain the software. Others familiar with the framework might be able to maintain it.

Disadvantages of Using a Framework

The main disadvantage is the limited programmer's freedom. This could be seen as a good thing by a lot of architects and technical leads in a team, but not necessarily by the developer. The restrictions imposed by the framework forces the developer to use a specific way to start modules, communicate with the modules, and stop them. It adds extra code to every project. Last, there is a learning curve associated with any framework.

Framework versus Design Pattern

In other sections in this book, we talk about the design patterns like FGV, State Machine, Producer Consumer, Client Server, Queued Message Handler, etc. A framework is different than a design pattern. A design pattern is closer to the implementation of an algorithm, while the framework is much more than a design pattern. In LabVIEW, a design pattern is implemented within one VI, where the framework is the collection of multiple VIs working together to assemble an application. For example, in the case of DQMH, the `Main.vi` in a DQMH module is a Queued Message Handler, while the Events are shared via an FGV. Sam Taggart described the design patterns as LEGO pieces, while the framework would be a building built out of hundreds of LEGO pieces.[10]

The Contract between the Framework and the Programmer

The framework provides a specific method of developing code. The framework includes the tools, support code, APIs, templates, etc., to aid in development. The framework guides the flow of control for program execution. The framework guarantees error handling, message delivery, shutdown process, and reference management.

---

[10] Presentation "Choosing a Framework. How to Pick the Right Framework to Meet Your Needs," Samuel Taggart, 2018. http://automatedenver.com/developer-series-presentation/

The developer provides code that is called by the framework to customize the program for a specific application.

In the case of DQMH, DQMH provides the tools to encourage the developer to keep the DQMH API tester up to date every time they create a new event. The developer commits to keeping that DQMH API tester up to date to aid in future troubleshooting and application maintenance.

In the case of Actor Framework, the developer commits to implementing communication between actors by traversing the actor tree.

Why Not Make Your Own Framework?

The main reason is to not reinvent the wheel, or worse yet, ending up with a flat tire! Speaking from experience, developing a framework takes a lot of time and resources, and it is almost impossible to get it right the first time. You can go to VIPM and see that there have been several DQMH versions. With each new release[11] we improve the framework and provide tools to bring your existing DQMH modules up to date. Actor Framework has also gone through several revisions. The very first public version did not ship with LabVIEW. A lot of the concepts and VIs got improved with help from the community before releasing it to a larger audience by shipping it with LabVIEW.

When we talk to other architects who have developed their own frameworks, we quickly realize that we have been through the same anti-patterns and common pitfalls. Another important point is that it is not enough to create the architecture itself. You also need to create the tools that will make the architecture easier to use. We keep hearing that people use DQMH and Actor Framework because they come with the tools to create all the interprocess communication code. DQMH does it via the **Tools >> Delacor >> DQHM Tools** menu, while Actor Framework does it via project providers (right-click menus at the project level). You can also download the Zuehlke Project Explorer menu for DQMH from the Delacor Toolkits Discussions forum. This tool, developed by Jean-Claude Mengisen, provides the right-click menu functionality.[12]

If you decide to create your own framework, consider taking an existing framework and modifying it. Both DQMH and Actor Framework started from the Queued Message Handler. You can see that both teams took very different paths. DQMH is very light on the object-oriented part, while Actor Framework is completely object oriented. DQMH uses events for interprocess communication, while Actor Framework uses queues.

---

[11] DQMH Release notes available at https://delacor.com/dqmh-documentation/
[12] https://forums.ni.com/t5/Delacor-Toolkits-Discussions/Zuehlke-Project-explorer-menu-for-DQMH/m-p/3808053

Criteria to Evaluate Frameworks

Complexity

First, define how complex the problems you are trying to solve are. If all you are doing is acquiring, analyzing, and presenting the data you acquire; if your acquisitions take only a couple of minutes; and if you are the only one using your code, then a State Machine Design Pattern[13] (or template[14]) might be all you need.

If you are trying to control multiple instruments; if you are part of a team of developers tackling the same problem; if there are multiple windows that need to communicate between them, and on top of that, you are connecting to third-party applications, then a state machine will not be enough for your applications.

Another thing to keep in mind is debugging time. Perhaps because she gets called a lot as a firefighter, Fab's focus always tends to be on "how am I going to troubleshoot this?" When she is working on a new project, the first thing she asks the team is "what is the problem that we are trying to solve?" immediately followed by "how are we going to test the individual modules when we are at the customer site and something goes wrong?" This is one of the reasons DQMH comes with a DQMH API tester and the tools encourage the developer to keep that tester up to date. Sam Taggart has created a DQMH-like API tester for Actor Framework to help with this task for that framework.[15]

Learning Curve

Not everyone eats, drinks, and dreams LabVIEW as we do. For a lot of teams, LabVIEW is just one more tool in their tool belt. The team members might not write LabVIEW code all the time. Some of our customers might be heavily involved in LabVIEW a couple of months a year and then not touch it for years. It is completely understandable that when they come back to add a new feature to their project, they might be a little rusty. Another challenge is that the levels of proficiency within the team may be very different.

Each framework will have a different learning curve. With some frameworks, you will be up and running just by watching a couple of videos or going over the documentation. Other frameworks will require more investment in your time and your team's time.

NI offers several LabVIEW training courses.[16] The core courses are called LabVIEW Core 1, Core 2, and Core 3. Additionally, they offer a LabVIEW object-oriented design course. For the best experience with DQMH, you need to have at least a LabVIEW Core 1 and Core 2 level of proficiency. You need to understand what a queue is and be familiar with either the Producer-Consumer or the Queued

---

[13] Application Design Patterns: State Machines. http://www.ni.com/white-paper/3024/en/
[14] Simple State Machine Template Documentation. http://www.ni.com/product-documentation/53321/en/
[15] bit.ly/aftester
[16] http://ni.com/training

Message Handler. Delacor has a series of videos,[17] blog posts,[18] and a forum[19] to get you started. Delacor also provides custom training. For Actor Framework, you need to have at least LabVIEW Core 1, Core 2, Core 3, and LabVIEW object-oriented programming level of proficiency; understand what a queue is; and be very familiar with the Queue Message Handler and how it can be converted into the Command Pattern. NI has an active online Actor Framework community[20] and offers a formal Actor Framework course as well.

Keep in mind the level of proficiency of everyone in your team and who will be maintaining your software (potentially a customer who just knows enough LabVIEW to be dangerous).

Tools Available with the Framework

We have talked so much about the tools that by now you know that they are an important part of a framework. They are like the oil that makes things run smoother. If you compare two frameworks and they look very similar in terms of complexity, learning curve, etc., look at the available tools. You want to make sure the framework includes tools that will do most of the work so you do not have to. You want the tools to take care of tedious and repetitive tasks. You also want tools that encourage the programmer to do the right thing and follow the framework architects' intentions. The tools also ensure that everyone on the team is producing the same code style. You want to evaluate how hard it is to create new modules and new messages, and whether the framework has built-in debugging tools.

Framework Technical Support

When you evaluate a framework, you also need to look at where you will go for support. Visiting the online community for a framework will give you a good idea of how active it is. You can see how fast questions get answered and if there are examples that can help you get started.

To find the Actor Framework support group, just search Google for "Actor Framework read this first" and you will find the online community within NI.

To find the DQMH support group, just search Google for "Delacor Toolkits support," and you will find the online community within NI.

Key Components

After studying the different frameworks available, we have arrived at the conclusion that the key components for any framework are

---

[17] bit.ly/DelacorQMH
[18] delacor.com/blog
[19] bit.ly/DQMH-forum
[20] ni.com/actorframework

- Interprocess communication
- Module initialization
- Stopping processes gracefully
- Error handling strategy

Interprocess Communication

The theme for the 2011 CLA summit was interprocess communication. By the time the summit was over, the list of ways to communicate between parallel processes had grown to more than 20. This was even before **channel wires** were part of the product. The most commonly used methods for interprocess communication within the same target are

- Queues
- User events
- Notifiers

Actor Framework uses queues. DQMH uses queues within the modules as a private communication and user events (called DQMH Events) for the interprocess communication. Of course, there is also the option of communicating between targets. The most common options are

- TCP/IP
- UDP

Module Initialization

If you decide you want to create your own framework, the things to consider for module initialization are

- Do you want one or many instances of your module?
- What communication method will the framework use? In the case of the Actor Framework, the communication method is queues. DQMH uses user events to communicate externally and queues to communicate within the module itself.
- Who owns the references and what is the lifetime of these references? In the case of the Actor Framework, the code that calls the actor owns the

launched actor's queue. This ensures the actor cannot be left running as a zombie. With DQMH the module itself owns its references. DQMH went this route to facilitate communication with TestStand, where references can die between steps.[21] This means DQMH could end up with zombie VIs, but DQMH also offers an API tester, and the developer can use the API tester to stop its associated module if it has become a zombie.

- Do you want to provide initialization information as part of starting your module, or would you create a communication method to initialize the state of your module after you have ensured it launched properly?

Stop Processes Gracefully

As consultants, the number one complaint we get from existing applications is: "My application does not stop gracefully so I have to use Task Manager to kill it." This is not acceptable. At the very minimum a framework should put in place the methodology for the modules to stop gracefully every time.

Stopping gracefully means no last parting errors. For example, if the application already stopped, the end user doesn't care if the device is no longer recognized (it may have been stopped earlier). The code is on the way out, so this is not an error in the application, and the end user should not be distracted by it. Moreover, there should not be any framework-related errors, unless the developer went against one of the framework's guidelines or modified the framework's code.

The stopping code also needs to ensure that any module that needs to do cleanup has enough time to do it.

One thing is certain: stopping gracefully does not mean pressing the Abort Execution button on the toolbar or killing the application via the Task Manager. (See Fig. 5.5.)

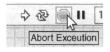

FIG. 5.5 *Abort Execution button on a VI Front Panel toolbar.*

Error Handling Strategy

There are different philosophies toward error handling. We suggest that you try to handle errors as close to the source as possible but defer any user displays or error

---

[21] You can configure TestStand to preserve references in between TestStand steps; the default is to not preserve them.

logs to file to higher levels. It is not a good idea to have the low-level code be able to generate an error when you least expect it. This is one of the reasons Darren Nattinger strongly suggests disabling automatic error handling for your all your VIs.[22] This also why we do not appreciate when some of the low-level VIs that ship with LabVIEW contain a simple error handler that could pop up an error anytime. The team at Delacor even has a VI analyzer test that can identify when any pop-ups are in the low-level VIs.

The next part is to decide if you will have a central error handler module where all your modules report to, or if it will always be the job of the top-level module to handle the errors.

Another important decision is what type of error-related information to log, for example, the system state, last action, or set of actions that happened right before the error.

If you will include a central error handler and you will include the option to log to file, make sure you limit the size of the log file. We once had a customer that had the entire production line stopped because the controller ran out of memory after logging error files for the last 5 years. Chances are nobody needs to look at the errors from so long ago, and in fact, nobody did. This is why the system was taken down by long error log files. Another option is to have a table for all the possible errors on your program and log how many times you have had a particular error number and when was the time of the last occurrence.

Sharing Modules

Once you select a framework, you will be creating independent modules (whether they are DQMH modules, Actor Framework Actors, or other). You could choose to have all your modules directly in the same project. Steve Watts has a couple of articles on his blog regarding project structure and portability. One of those articles advocates[23] even moving things away from vi.lib and user.lib and using folders saved with the rest of the project. This seems to work for projects with low class and low project library count. Check out Watts' other post on project structure[24] (using auto-populating folders and having all dependencies directly in the project as source code) and the Delacor way to structure a project[25] (use only project libraries and it is OK to depend on toolkits, VIs in vi.lib, and other add-ons). Try to understand where both authors are coming from, and based on that, decide what works best for you.

---

[22] An End to Brainless LabVIEW Programming by Darren Nattinger. https://forums.ni.com/t5/Community-Documents/An-End-to-Brainless-LabVIEW-Programming/ta-p/3548039
[23] LabVIEW Life Lessons #1—Project Portability by Steve Watts. https://forums.ni.com/t5/Random-Ramblings-on-LabVIEW/LabVIEW-Life-Lessons-1-Project-Portability/ba-p/3489445
[24] A Tidy Project Is a Happy Project by Steve Watts. https://forums.ni.com/t5/Random-Ramblings-on-LabVIEW/A-Tidy-Project-is-a-Happy-Project/ba-p/3487453
[25] A Method to Project Madness by Fabiola De la Cueva. https://delacor.com/a-method-to-project-madness/

Our way is just a way; it is not "The Way." Just make sure whatever approach you take is based on an informed decision and not just an impulse.

When you start working on large applications, it will be hard to have your entire application source code live within a single project. If you have more than one LabVIEW developer in your team, you will want to have developers working in parallel and using each other's code without stepping on each other's toes. Also, once you start using a framework, the possibilities of reuse increase. You will want to ensure that your modules are portable and that when you share a module with someone else, all the dependencies are clearly defined and come with the module.

Configuring Source Code Control Repository Dependencies

We recommend that you have a repository for each module. If there are a couple of modules that depend on each other, they can be in the same repository. An example of a module that would be in its own repository would be the Serial Port Controller module. An example of a pair of modules in the same repository would be the module that controls the Configuration Editor and the module that will be used to load a category page into the Configuration Editor.

The repository for your application will then configure each of the module repositories as subrepositories. The term is different depending on what source code control you use, but the idea is the same. Your main repository is connected to other repositories. When you update the main repository, it updates all the repositories to the version you configured your main repository to get. These subrepositories go by the names **Git submodules**, **Mercurial subrepositories**, or **SVN Externals**.[26]

You have seen an example of this for the code for this book, the LGP5 GitHub repository. You can either clone the resources repositories independently, or if you clone LGP5, you get all of the repositories in one cloning. Steve Watts also has an article about submodules.[27]

This option requires that all the team members have access to all of the repositories. You also need to figure out who will be in charge of determining what revision of the subrepositories the main repository needs to be linked to and what is the process to consider a subrepository to be in a "released" stage.

Packaging Modules Using VIPM

Assuming that you have a repository for each of your modules (or module groups), you can add a VIPM build specification and build a package from there. The calling

---

[26] How to Organize Your Large LabVIEW Project Using Libraries by Fabiola De la Cueva http://delacor.com/how-to-organize-your-large-labview-project-using-libraries/
[27] I'm Not Being Critical But... (Re-use) Part 3. https://forums.ni.com/t5/Random-Ramblings-on-LabVIEW/I-m-not-being-critical-but-Re-use-Part-3/ba-p/3797642

code will rely on the developers installing the package version that their code needs. One advantage of this approach is that if Jane releases Package 1.2.0.15 and Dana calls it in her module only to find out that it breaks her code, Dana can easily go back to 1.2.0.14 via VIPM, report the issue to Jane, and continue working on her code. Once Jane has a fix, they can repeat the process. You do not need the professional version of VIPM to build packages.

We suggest that if you follow this approach, you create a .vipc file for each repository that lists the packages the repository depends on. When an issue is encountered, like in the example earlier, Dana would update the .vipc to indicate that her code works with version 1.2.0.14. Add to your process to apply the .vipc file every time before opening a project and before building code. If you have a continuous integration process, applying the .vipc file should be part of this process. Note that you do need the professional version of VIPM to create .vipc files, but you can apply .vipc files with the free version.

If your organization is going to go with this option, consider buying the VIPM site license because then you can have package repositories, and developers can load not only the LabVIEW Tools Network packages but the packages for your organization as well. We have seen some workarounds to not having the professional edition where all the packages are saved in a mapped drive or other shared location, and developers know to go there to look for new packages. While this approach works, it is brittle and not as effective as having an organization packages repository recognized by VIPM that automatically refreshes with new packages and new versions.

The main advantage of this approach is that there is a definition of what is "released" code; if it is built into a package, it is released. This also means that you have to apply your building process, for example, running unit tests, before building your package.

Determine the Dependency Tree

We already discussed having a repository for each of your modules. The next step is to determine the dependency tree for all those modules. If your Acquisition module depends on the Configuration Editor module, you need to build the Configuration Editor module first and then have the Acquisition module call the Configuration Editor functions in the Acquisition code that you installed via VIPM. If your modules inherit from code in libraries, you have to build the ancestor package first. (See Figs. 5.6 and 5.7.)

We recommend that you add a prefix at build time to your libraries, so it is easier for developers to identify when they are calling source code (code before building into a package) and when they are calling released code (code installed via VIPM).

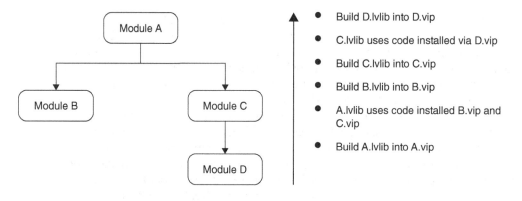

FIG. 5.6 *Building components into packages is a bottom-up process.*

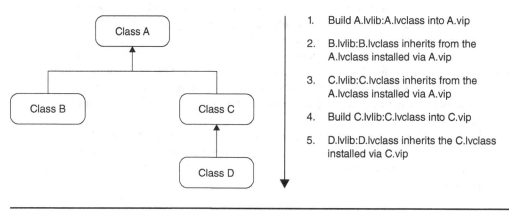

FIG. 5.7 *Building classes into packages is a top-bottom process.*

Packaging Modules in PPLs

When you package a library into a packed project library (PPL), LabVIEW compiles the contents of the library into binary code and into a single file on disk. Matthias Baudot has given excellent presentations on how to use PPLs in your applications, and he describes both the advantages and disadvantages. If you are going to go with this option, we recommend you check out his presentations.[28] He also included an example using DQMH. At the time of the presentation, his application already had about 60 project libraries and he was not done yet. We believe you need to analyze where your pain points are. If your application is so large that it takes a long time to load, going with PPLs may be worth the disadvantages.

---

[28] GDevCon#1: Take Advantage of PPLs and Packages to Enhance LabVIEW Application Performances and Deployment. https://www.studiobods.com/en/gdevcon-1/

Chapter 5: Why Would You Want to Use a Framework?

The way you create a PPL is by creating a new build specification for PPL. Let's look at the differences between a project library and a PPL.

LabVIEW Library	LabVIEW Packed Project Library
• .lvlib	• .lvlibp
• Multiple files	• Single file
• Source code	• Compiled code
• Editable	• Not editable
• Portable across targets and OSes	• Must rebuild for each new target and each OS
• Compatible with LabVIEW version it was created in and newer LabVIEW versions	• Pre-2018—Version specific, must rebuild for newer LabVIEW versions • Post-2018—No longer need to rebuild for new LabVIEW versions; LabVIEW 2018 can load PPLs built-in LabVIEW 2017
• Can take long to load because LabVIEW must compile lvlib before it can use it	• Fast load time because the code is already compiled
• Before building exe, LabVIEW resaves all the files in libraries	• Before building an exe that calls PPLs, LabVIEW copies the PPL into the target directory
• Namespace Library.**lvlib**:VI Name.vi	• Namespace Library.**lvlibp**:VI Name.vi

PPL Advantages	PPL Disadvantages
• Less time to load • Less time to build (copy to target vs. saving all files in the library) • Smaller application exe (the majority of the code is loaded via PPLs) • Smaller updates (only update the PPL that changed) • Single file • Version number • Remove block diagrams (share API while protecting IP) • Ideal for plug-in architectures	• Conditional Disable Structure always executes "RUN_TIME_ENGINE == False" within code built into a PPL (even if the PPL is called from an executable)[29] • The block diagrams can be removed, but the API is still exposed; any LabVIEW developer can call code in your PPL and determine inputs/outputs names and data types for the public functions • Cannot include non-LabVIEW files • When you open a VI in a PPL, all VIs in the PPL are loaded • PPLs do not support mathscript nodes • Replace with a packed library right-click menu is a one-way process only for LabVIEW 2018 and earlier; starting with LabVIEW 2019, it is a two-way process

(*Continued*)

---
[29] This was still the case at the time of writing this book (LabVIEW 2018).

PPL Advantages	PPL Disadvantages
	• Building palettes is not straightforward[30]
• PPLs cannot build if already in memory

Here are some graphs from the benchmarking that Matthias Baudot did for the time it takes to load a VI that calls code in libraries and the time it takes to load an exe that calls code in libraries. It is pretty clear that if load time is your pain point, using PPLs is the way to go. (See Figs. 5.8 and 5.9.)

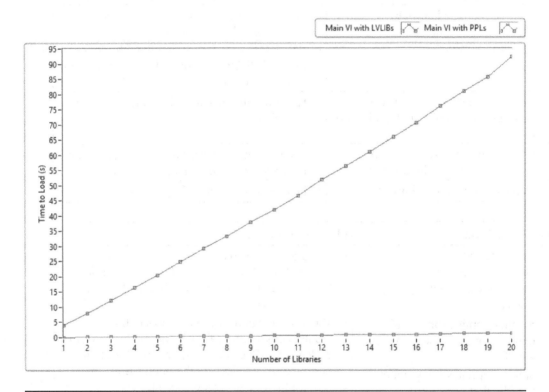

FIG. 5.8 *Time to load the main VI as reported by Matthias Baudot from StudioBods.[31] For this benchmark, Matthias had 50 VIs per library, the main VI calling all VIs in all libraries, and 1 to 20 libraries in the application. Conclusion: It is 65 times faster to load the main VI with 20 libraries using PPLs rather than LVLIBs.*

---

[30] When you build a project library that contains a palette into a PPL, the palette in the PPL still references the LVILIB items. (We were using LabVIEW 2018 at the time we wrote this book.)
[31] GDevCon#1: Take Advantage of PPLs and Packages to Enhance LabVIEW Application Performances and Deployment. https://www.studiobods.com/en/gdevcon-1/

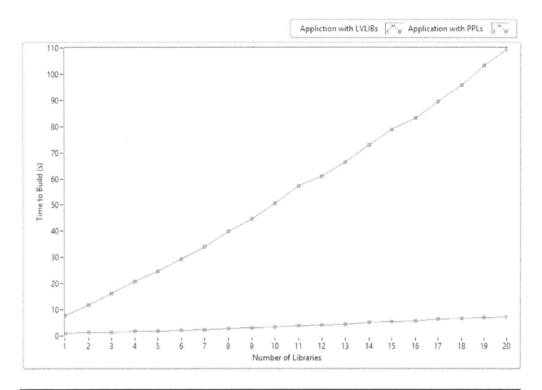

FIG. 5.9 *The time to build the application as reported by Matthias Baudot from StudioBods.[32] For this benchmark, Matthias had 50 VIs per library, main VI calling all VIs in all libraries, and 1 to 20 libraries in the application. Conclusion: It is 12 times faster to build an application with 20 PPLs rather than 20 LVLIBs.*

Determine the Dependency Tree

We already discussed having a repository for each one of your modules. The next step is to determine the dependency tree for all those modules. If your Acquisition module depends on the Configuration Editor module, you need to build the Configuration Editor module first and then have the Acquisition module call the Configuration Editor PPL functions. If your modules inherit from code in libraries, you have to build the ancestor first. (See Figs. 5.10 and 5.11.)

---

[32] GDevCon#1: Take Advantage of PPLs and Packages to Enhance LabVIEW Application Performances and Deployment. https://www.studiobods.com/en/gdevcon-1/

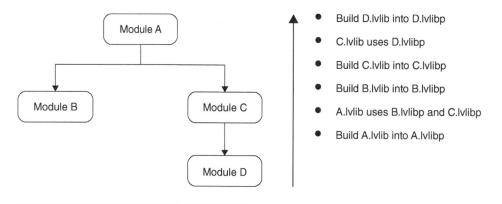

FIG. 5.10 *Building components into PPLs is a bottom-up process.*

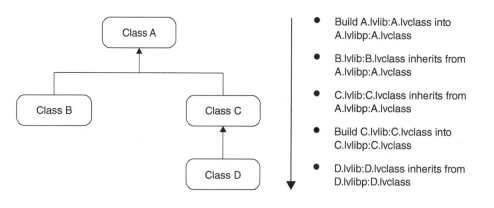

FIG. 5.11 *Building classes into PPLs is a top-bottom process.*

Determine PPL Directory

We strongly suggest that you decide what will be the target location for your team and make sure everyone builds to that location, for example `C:\Company\PPLs\Application Name\Plugins`. If you already read the chapter on working with teams, you already know to build in a separate computer than the one you use for development. This will be very handy when you start building your PPLs because PPLs cannot build if they are already in memory. The build computer will build toward the target `C:\Company\PPLs\Application Name\Plugins`. The developer machines will have the built PPLs in the same target path on their computer.

Build PPL

Add to the custom configuration file for your application the following lines:

```
viSearchPath="C:\Company\PPLs\*"
ShowLoadProgressDialog=False
```

Always exclude dependent packed libraries when building your PPLs. (See Fig. 5.12.)

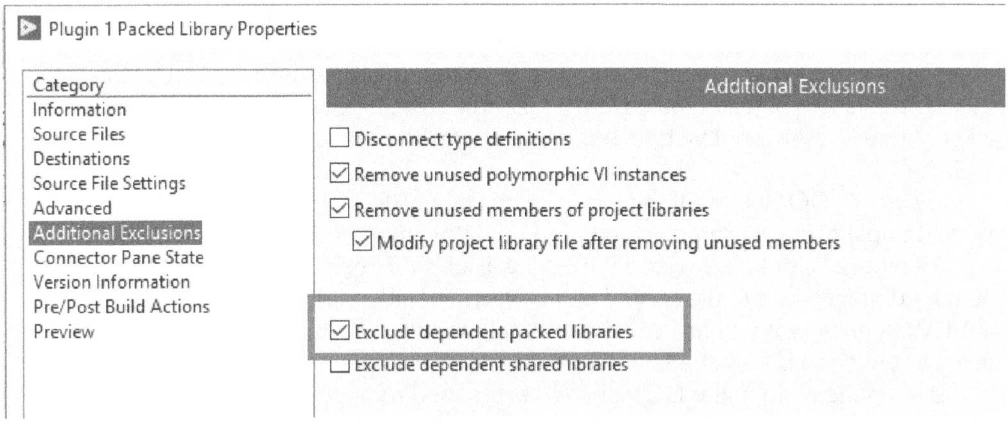

FIG. 5.12 *Always exclude dependent packed libraries.*

Specific Considerations for Frameworks and PPLs

When you package your project library into a PPL, LabVIEW will copy and package any dependency that is not in the Run-Time Engine with the PPL. This means that every DQMH module you build into a PPL and any Actor Framework Actor project library that you build into a PPL will include a copy of the framework libraries in the PPL (DQMH or Actor Framework libraries depending on the framework you are using). You do not want that. Follow these steps to package a DQMH or Actor Framework library:

1. Build your framework library into PPL. Make sure to include all the framework libraries your code depends on.
2. Replace the framework library in your project with the packed version.

3. Build packed versions of your module libraries.[33]
4. Calling code can now use this packed version instead of the code in vi.lib.

DQMH
Use Cases

DQMH modules are useful for applications where multiple tasks occur in parallel, often at different rates. It is also useful when multiple instances of the same front panel are needed. For example, consider an application that controls a single-temperature chamber and needs to log data from multiple units under test (UUTs). If you use TestStand, you can create DQMH modules that TestStand will call, and you can continue to use LabVIEW to debug and troubleshoot the code while it runs in TestStand.

You can use DQMH for all the modules on your code and start with the DQMH Project Template, or you can decide to add a DQMH module to an existing project.

Another use is for teams with different levels of proficiency. The DQMH module could be designed and implemented by one team member, and others with less LabVIEW familiarity could just call the DQMH module API VIs in their code without having to understand how the DQMH module itself is implemented.

It is tempting to go in the DQMH architecture and try to understand all the behind-the-scenes code. We believe that understanding the different DQMH components and how to use the DQMH framework first will make it easier to dive in deeper on the behind-the-scenes section later. Also, keep in mind that you can adapt the DQMH module to fit your needs. So, if you see something you do not agree with, keep on reading because you can create your own DQMH module templates and still benefit from the DQMH tools.[34]

What Is DQMH?

DQMH stands for Delacor Queued Message Handler. The DQMH is a free toolkit available via the LabVIEW Tools Network.[35] DQMH has evolved into a popular framework. Each of the modules you create with DQMH is called a *DQMH module*. There are two components to a DQMH module: the DQMH library and DQMH API Tester.vi. The `API Tester.vi` communicates with the DQMH module via DQMH events. The DQMH toolkit installs tools to create, edit, and validate DQMH components. It also installs a DQMH Project Template and a DQMH Sample Project Template for a CML application (called the DQMH CML sample project).

---

[33] In the case of DQMH modules, you might want to drag the DQMH API Tester into the DQMH library before building the PPL.
[34] https://delacor.com/documentation/dqmh-html/AddingaNewDQMHModulefromaCustomT.html
[35] LabVIEW Tools Network. http://ni.com/labviewtools

DQMH Library

The DQMH library template is based on the NI Queued Message Handler Project Template (NI QMH).[36] The DQMH improves the NI QMH by providing safe, event-based message handling and scripting tools, which makes development easier and by encouraging consistent style among different developers in the same project, which improves efficiency.

DQMH Module Main

Each DQMH module has a `Main.vi`. The full name for the `Main.vi` is `DQMH Library Name:Main.vi`, for example, `Acquisition Module .lvlib:Main.vi`. You can rename the `Main.vi` in your DQMH module.

The DQMH module main repeatedly performs the following steps (see Fig. 5.13):

1. A user interacts with the front panel, causing the Event Structure in the **Event Handling Loop (EHL)** to produce a message.[37] LabVIEW stores the message in a queue.

    a. The `DQMH Module Main.vi` also registers to listen to requests sent by other VIs. Another VI requests that the DQMH module do something, causing the Event Structure in the EHL to produce a message. LabVIEW stores the message in a queue.

2. The **Message Handling Loop (MHL)** dequeues a message from the message queue.

3. The message is a string that matches one of the subdiagrams of the Case Structure in the MHL. Therefore, reading the message causes the corresponding subdiagram of the Case Structure to execute. This subdiagram is called a "message diagram" because it corresponds to a message.

4. Optionally, the message diagram produces another message, storing it in the message queue.

    a. The `DQMH Module Main.vi` can optionally broadcast an event to other VIs that are registered to listen to them.

5. In the NI QMH, the EHL only registers to listen to the Stop message. In the DQMH, the EHL registers to listen to all the **request** events.

---

[36] Queued Message Handler Template documentation. http://www.ni.com/tutorial/53391/en/
[37] Note that EHL not only handles events initiated by interaction with the front panel, code outside the DQMH module can also trigger events that the EHL handles.

300 LabVIEW Graphical Programming

6. The DQMH Module Main.vi has Admin code that manages whether the module is a single instance (**singleton**) or if it can be called multiple times in parallel (**cloneable**).

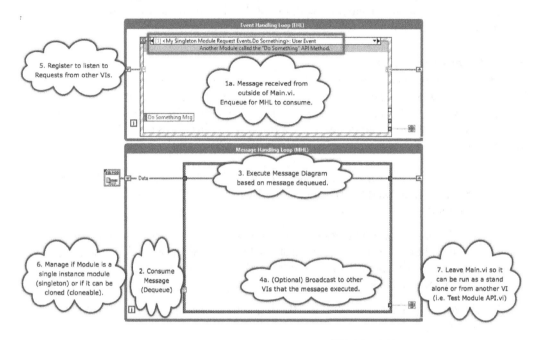

FIG. 5.13 *DQMH Module Main.vi structure.*

The EHL is the producer loop (produces messages), and the MHL is the consumer loop (consumes messages). These loops run in parallel and are connected by the DQMH message queue.

The message queue is a LabVIEW queue that stores messages for consumption. The two-loop architecture allows the EHL to receive external requests or user interactions and send them to this queue as messages while the MHL may be busy processing previous messages. Each message queue belongs to a single MHL. Note that in the DQMH, the VIs that manage the queues are not copied to the project like in the NI QMH. In the DQMH, the queues are wrapped in a class and the managing VIs are stored in vi.lib.

Every iteration of the MHL reads the message at the front of the message queue and then executes the corresponding message diagram in the Case Structure. Although the MHL primarily consumes messages, it also can produce them.

Each loop handles errors using a loop-specific error handler subVI. If the DQMH Module Main.vi encounters an error in the MHL, LabVIEW displays an error message. These VIs are also wrapped in the DQMH queue class and reside in `vi.lib`. If the developer is comfortable with LabVIEW object-oriented programming, she can choose to override some of these VIs and create her own **Message Queue class** as a child of the Delacor Message Queue.[38]

Unlike the NI QMH template that encourages having an MHL for each task, the DQMH encourages having a new DQMH module per task. With the DQMH scripting tools, the developer can select Tools >> Delacor >> DQMH >> Module >> Add New DQMH Module… to quickly add a new module to her project.

Notice the data cluster in the previous diagram. This cluster contains data that each message diagram in the MHL can access and modify. In this template, the cluster is defined as a typedef (`Module Data-cluster.ctl`). Each typedef belongs to a single DQMH module and should only be modified by code in `Main.vi`. We suggest that you never feed the entire cluster to a subVI within the DQMH module. Instead, unbundle only the elements that the subVI needs. Note that you could replace this cluster with an object if you prefer. Also, you can create your own DQMH template[39] with this modification to the data cluster and other modifications. Your template needs to preserve the structure DQMH labels for the scripting tools to continue to work.

DQMH API Tester

The **API tester**, also known as the **tester**, helps troubleshoot, debug, and eavesdrop on the communications between the DQMH module and its calling code. The tester follows the User Interface Event Handler VI template.[40] You commit to keeping this API tester up to date. This VI will aid you to verify that the communication from external code and your DQMH module is working as expected. This also makes this VI the first user of your DQMH API. The DQMH API tester also provides an example as to how you expect other developers to use your DQMH Module. Think of it as running documentation of your DQMH module. Documentation can become stale, but running code never lies.

The DQMH scripting tools will encourage you to edit the API tester every time you create a new **DQMH event** (whether it is a request or a broadcast). This encourages you to think how you are going to test the API calls for your DQMH module even before you implement them.

Note that even if you write an application that has a single DQMH module, the tester has the capability of eavesdropping on the different DQMH events and messages (both in development mode and in an executable).

The `Test Module API.vi` gets added when the developer selects **Tools >> Delacor >> DQMH >> Module >> Add New DQMH Module…** and out of the box it tests the request and broadcast events that come with a DQMH module. (See Fig. 5.14.)

---

[38] <LabVIEW>/vi.lib/Delacor/Delacor QMH/Libraries/Message Queue_class/Delacor_lib_QMH_Message Queue.lvclass
[39] http://delacor.com/documentation/dqmh-html/AddingaNewDQMHModulefromaCustomT.html
[40] File >> New… >> From Template >> Frameworks >> Design Patterns >> User Interface Event Handler.

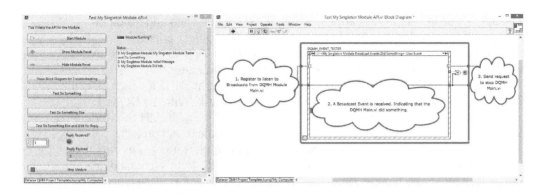

FIG. 5.14 *DQMH API tester structure.*

The tester for singleton DQMH modules registers to listen to the broadcasts from the DQMH Module Main.vi and handles each broadcast event in the Event Structure. When the tester stops, the last request to be sent is to stop the module (in case it was not stopped earlier). It is important to execute that last request to avoid leaving the DQMH Module Main.vi running as a zombie.[41]

One of the advantages of the DQMH Module Main.vi design is that it can run headless. Another advantage is that the tester can be used as a "sniffer" to monitor the status of a DQMH module that is already running.

DQMH Events

The DQMH Events are LabVIEW code that carries custom data throughout the application. Implemented using custom events (also known as user events),[42] these events allow different parts of an application to communicate asynchronously when the events are programmatically generated. An Event Structure handles both types of events: end-user interaction with the front panel (e.g., a value-change event on a control on the front panel) and programmatically generated user events. A DQMH module can send events to multiple Event Structures, and multiple locations in the code can send events to a DQMH module. DQMH events are of the type many-to-many, which means multiple event generators to multiple event receivers.

There are two types of DQMH events: **DQMH request events** and **DQMH broadcast events**.

---

[41] A zombie module is one that is running, but you do not have a direct way to stop it (no user interface). Instead of killing your zombie DQMH modules, just open the API tester, click on Start Module button to establish a connection, and click on Stop Module.

[42] Creating custom events LabVIEW help entry. http://zone.ni.com/reference/en-XX/help/371361N-01/lvhowto/creating_user_events/

DQMH Request Events

A **DQMH request event** is code that fires an event requesting the DQMH module to do something. Multiple locations in the code can send events to the DQMH module. Request events are many-to-one. Requests are asynchronous events. The code that fires the event does not wait for a response from the DQMH module. If you need your calling code to wait for the reply from the DQMH module, you can create a **Request and Wait for Reply DQMH event**.

Name your request events using imperative names, for example, "Start Acquisition."

The core DQMH request events that come with every DQMH module are

- **Show Module Panel:** By default, the module does not show the front panel of its `Main.vi` when it starts. Execute this request to show the front panel of the `DQMH Module Main.vi`. This request cannot be removed via **Tools >> Delacor >> DQMH >> Event >> Remove DQMH Event…**

- **Hide Module Panel:** Execute this request to hide the front panel of the `DQMH Module Main.vi`. This request cannot be removed via **Tools >> Delacor >> DQMH >> Event >> Remove DQMH Event…**

- **Get Module Execution Status:** Execute this request to request the `DQMH Module Main.vi` to report its current execution status via the **Update Module Execution Status broadcast**. This request cannot be removed via **Tools >> Delacor >> DQMH >> Event >> Remove DQMH Event…**

- **Show Diagram:** Execute this request to show the block diagram of the `DQMH Module Main.vi`. This request does not work in executables. This request can be removed via **Tools >> Delacor >> DQMH >> Event >> Remove DQMH Event…**

- **Stop Module:** Execute this request to stop the `DQMH Module Main.vi`. This request cannot be removed via **Tools >> Delacor >> DQMH >> Event >> Remove DQMH Event…**

DQMH Broadcast Events

A DQMH broadcast event is code that fires an event broadcasting that the DQMH module did something. Multiple Event Structures can register to handle the broadcast events. Broadcast events are one-to-many.

Name your broadcast events using the past tense, for example, "Acquisition Started."

The core DQMH broadcast events that come with every DQMH module are

- **Module Did Init:** The `DQMH Module Main.vi` broadcasts that it initialized without issues. This broadcast cannot be removed via **Tools >> Delacor >> DQMH >> Event >> Remove DQMH Event…**

- **Status Updated:** The `DQMH Module Main.vi` broadcasts a new status. This broadcast cannot be removed via **Tools >> Delacor >> DQMH >> Event >> Remove DQMH Event...**

- **Error Reported:** The `DQMH Module Main.vi` broadcasts that an error occurred. This broadcast cannot be removed via **Tools >> Delacor >> DQMH >> Event >> Remove DQMH Event...**

- **Module Did Stop:** The `DQMH Module Main.vi` broadcasts that it did stop. This broadcast cannot be removed via **Tools >> Delacor >> DQMH >> Event >> Remove DQMH Event...**

- **Update Module Execution Status:** The `DQMH Module Main.vi` broadcasts whether it is currently running; this is used when the `Start Module.vi` executes and the module `Main.vi` is already running. This broadcast cannot be removed via **Tools >> Delacor >> DQMH >> Event >> Remove DQMH Event...**

DQMH Helper Loops

Sooner rather than later you will need to do a periodic repetitive task. The temptation is to enqueue the same message over and over and have the MHL run the same case over and over. One disadvantage of this is that messages can get enqueued in between the repetitive task enqueued and the task will not be executed periodically. There would be interrupts or delays. Delacor recommends in its DQMH best practices document[43] using the MHL for one-shot actions that don't take long to complete, and using helper loops for repetitive operations.

Use helper loops for periodic, repetitive tasks. This is a loop that runs in parallel to the EHL and MHL and uses its timeout event case to perform the repetitive task.

The helper loop timing can be controlled by setting the timeout value of the Event Structure or by implementing your own timing code within the timeout case of the Event Structure inside the helper loop.[44] Setting the timeout to –1 (via the shift register in the helper loop) puts the helper loop to sleep, and setting the timeout to a positive value wakes up the helper loop. If you need precise timing, take into account how long the code inside the timeout takes to complete and adjust the next timeout timer accordingly.

Note that it is very important that a new event registration refnum is used for the helper loop. DO NOT FORK THE EVENT REGISTRATION REFNUM WIRE![45] (See Figs. 5.15 to 5.17.)

---

[43] DQMH Best Practices. https://delacor.com/dqmh-documentation/dqmh-best-practices/
[44] DQMH Actors by Joerg Hampel. https://delacor.com/dqmh-actors/
[45] Registering to Broadcast Events from Many Modules. https://forums.ni.com/t5/Delacor-Toolkits-Discussions/Registering-to-broadcast-events-from-many-modules/m-p/3472557

Chapter 5: Why Would You Want to Use a Framework? **305**

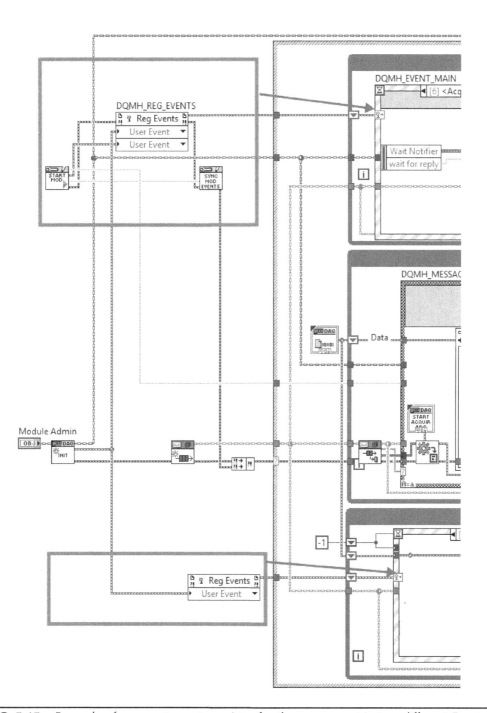

FIG. 5.15 *Example of a correct way to register for the same event in two different Event Structures. One registration for event per Event Structure.*

306 LabVIEW Graphical Programming

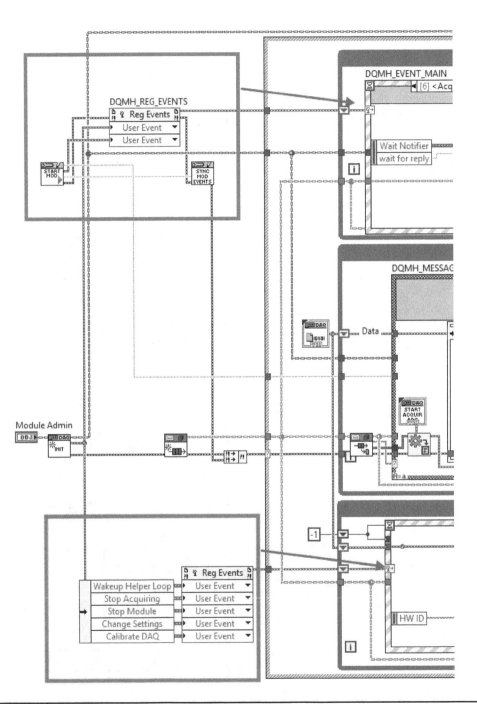

FIG. 5.16 *Another example of a correct way to register for the same event in two different Event Structures. One registration for event per Event Structure.*

Chapter 5: Why Would You Want to Use a Framework?

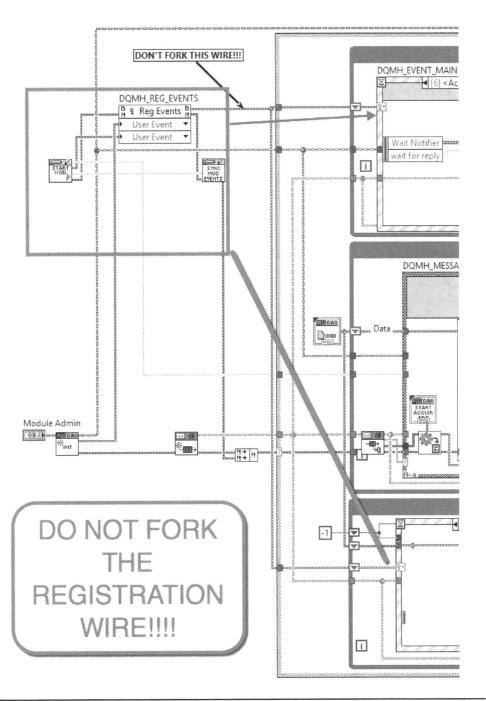

FIG. 5.17 Example of how not to register for the same event in two different Event Structures. DO NOT FORK THE EVENT REGISTRATION WIRE!

A pretty common misunderstanding is that the timeout event in an event case should not be used for critical code or code that you need to reliably execute periodically. This should be rephrased as you should avoid using a timeout event for handling critical events in an Event Structure that handles multiple critical events. In the case of the helper loop, the timeout is the critical event. It is OK if the other events handled interrupt the timeout. We actually want them to interrupt the timeout because those other events are Stop Module and Put Helper Loop to Sleep.[46]

Another misconception is that the timeout case in the Event Structure has bugs. At the CLA summit in 2011, Justin Goeres presented about his public and private events. He showed how you can register for a cluster of events and only handle the events you care about. Then Steen Schmidt pointed out that if the code registered for an event, and the Event Structure did not have a specific event case to handle that event, then the Event Structure timer would still reset when any of the events in the cluster fired. This led half of the room to say this was a bug and the other half to say it was a feature. Norm Kirchner stood up on the table protesting, and this led to a whole discussion about all the things the CLAs wanted to see improved with respect to events. Starting with LabVIEW 2013, this is no longer the case. Only the events handled by the Event Structure (this means the Event Structure is registered to handle them *and* has an event case configured to handle that event) will cause the timer to reset. If the code registers for an event but does not have a specific event case to handle that event, when that event fires, the timer does not reset. Unfortunately, there is no video of that CLA summit episode or pictures of Norm Kirchner standing up on the table and protesting, but there is some audio if you care to listen.[47] (See Fig. 5.18).

Another example of helper loops is using parallel helper loops to handle chatty events.[48] Like we explained before, you can think of the Event Structure as a queue that handles events of different types. If the code that fires the event is firing it at a high rate, other events will have a hard time getting in and showing up on time. If this is a concern, separate your EHL into multiple EHLs (or helper loops). (See Fig. 5.19.)

DQMH Helper Loops in Cloneable DQMH Modules

All the cloneable instances of a DQMH module share the same events. This allows the caller to send the same event at the same time to all the cloneable instances by assigning a -1 to the Module ID input. When you create a helper loop inside a cloneable module `Main.vi`, you need to ensure you are adding a call to `Cloneable DQMH Module.lvlib:Addressed to This Module.vi`;

---

[46] Helper Loops for Active/Repetitive Tasks Discussion on Timeout Event Case. https://forums.ni.com/t5/Delacor-Toolkits-Discussions/Helper-loops-for-active-repetitive-tasks/m-p/3753521#M400

[47] 003 VISP Justin Goeres—CLA Summit. http://vishots.com/interview-justin-goeres-cla-summit/

[48] Tips and Tricks for a Successful DQMH Based Project—Tip 2: Use Multiple Event Handling Loops. https://delacor.com/tips-and-tricks-for-a-successful-dqmh-based-project/

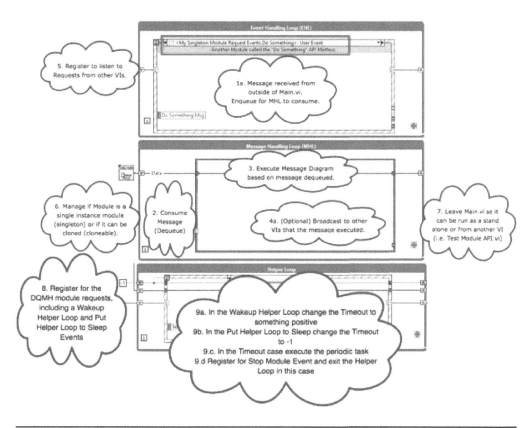

FIG. 5.18 *Helper loop runs in parallel to the EHL and MHL. It registers to "Wake Up Helper Loop," "Put Helper to Sleep," and "Stop Module" request events. In the timeout event case, the periodic task takes place.*

otherwise, all the helper loops for all the cloneable instances will be waking up even when the code sends the request to only one of the events.

Also, if your cloneable instances need precise timing in the helper loop, you will need to use local user events in the DQMH cloneable module as opposed to relying on DQMH private requests. You need to do this because anytime any of the DQMH cloneable instances fire the Put Helper Loop to Sleep event, all of the DQMH cloneable instances will handle that event, even if it is only to verify that the message is not for them. If all your cloneable instances go to sleep at the same time, this is not a problem for you. (See Figs. 5.20 and 5.21.)

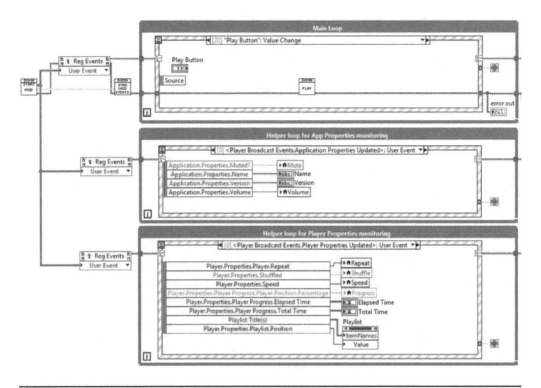

FIG. 5.19 *Example of multiple EHLs handling events that happen at very different rates. The example is a media player, and changes to the elapsed time reporting will always show on time, even when the volume is updated as well.*

DQMH and State Machines

Repeat after us "A DQMH is not a state machine" and "No queue message handler is a state machine." The MHL does look very similar to the state machine template, and you might have enqueued a message to happen after other messages successfully. In our experience though, doing this can lead to issues that are hard to debug. A QMH can have messages enqueued in between the messages you enqueue, and another part of the code may enqueue a high-priority message (enqueue at the other end of the queue) and cause havoc in your sequence. If you do need a series of actions to happen back to back, our suggestion is to create a state machine SubVI for that message case.[49]

---

[49] De la Cueva, Fabiola. Tip 4 at Tips and Tricks for a Successful DQMH-Based Project. delacor.com/tips-and-tricks-for-a-successful-dqmh-based-project

Chapter 5: Why Would You Want to Use a Framework? **311**

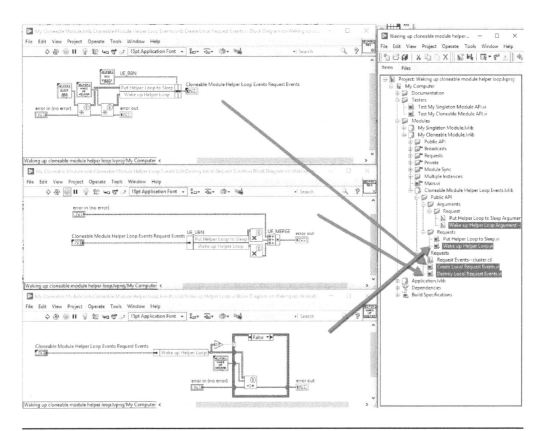

FIG. 5.20 Local DQMH events for a DQMH cloneable module that needs to ensure that other DQMH cloneable instances do not interrupt its cloneable helper loop.

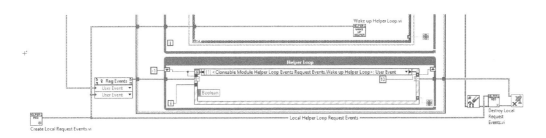

FIG. 5.21 DQMH cloneable Main.vi calling the creation of local request events, the firing of one of them, and the destruction of the local request events.

How to Use DQMH

First, you need to decide if you will be adding calls for a DQMH module in your existing code, if you will be adding modules to an existing project, or if you will be adding modules to a blank project. If you want to see an example of how simple the calling code can be, explore the simple Thermal Chamber Controller examples. If you want to see an example of multiple DQMH modules communicating between each other, and you know your project will have one singleton DQMH module and one cloneable DQMH module, then use the DQMH project template. If you want to see an example closer to a real-life application, start by creating a project template from the **DQMH CML sample project**. If all you need is to add a new DQMH module to your existing project or to a blank project, just add your new DQMH module via the **Tools >> Delacor >> DQMH >> Module >> Add New DQMH Module...** menu option.

Simple Code Calling a DQMH Module

First, let's look at how you would call a DQMH module in your code. You could have created this DQMH module or someone else in your team created it for you. The code that calls the DQMH module can be very simple. Refer to the three simple VIs in the Thermal Chamber Control example[50] for an illustration of this approach. The VIs are `Thermal Chamber Controller.vi`, `Thermal Chamber Controller with DUT.vi`, and `Thermal Chamber Controller with Multiple DUTs.vi`. These VIs illustrate that there is no need to have an Event Structure in the calling code and provides an example of two DQMH modules being used in a traditional "Configure >> Read/Write >> Close" application. (See Fig. 5.22.)

Please note that you do need to call both the `Start Module.vi` and the `Synchronize Module Events.vi` to launch the module even if your code will not have an Event Structure.

The `Thermal Chamber Controller.vi` example implements a thermal chamber controller with a **singleton** DQMH module. The chamber is configured to reach a temperature set point. Once the temperature is reached, the VI stops. (See Fig. 5.23.)

The `Thermal Chamber Controller with DUT.vi` example contains a copy of the `Thermal Chamber Controller.vi` example (see Fig. 5.23). This example uses a single instance of a cloneable DQMH module for the DUT. Once the thermal chamber reaches the temperature set point, the UUT performs a self-test

---

[50] Thermal Chamber Control example available from the Help >> Find Examples... >> Directory Structure >> DelacorDelacor QMH >> DQMH Fundamentals—Thermal Chamber DQMH Fundamentals—Thermal Chamber.lvproj

Chapter 5: Why Would You Want to Use a Framework? **313**

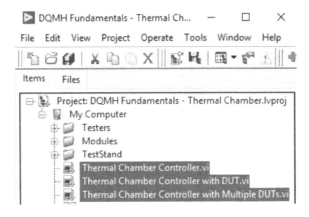

FIG. 5.22 *Thermal Chamber example simple VIs calling DQMH modules API calls.*

and returns the result. The UUT code is configured for an 80 percent pass rate for demonstration purposes. The example also illustrates how to manage a cloneable DQMH module that is running as a singleton (a single instance of the UUT). (See Fig. 5.24.)

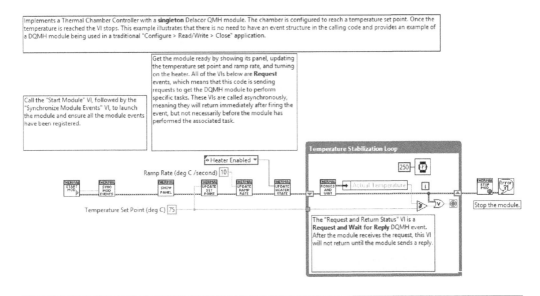

FIG. 5.23 *Thermal Chamber Controller.vi. The code that calls the DQMH module API can be very simple; it does not need to have an Event Structure.*

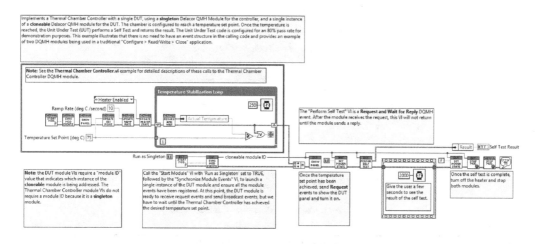

FIG. 5.24 Thermal Chamber Controller with DUT.vi. Notice that the cloneable DQMH module is running as a singleton by wiring a true constant to the Start Module.vi.

The `Thermal Chamber Controller with Multiple DUTs.vi` example contains a copy of the `Thermal Chamber Controller.vi` example (see Fig. 5.24). Unlike `Thermal Chamber Controller with DUT.vi`, this example uses multiple instances of a cloneable DQMH module for the DUT. Once the thermal chamber reaches the temperature set point, each one of the multiple UUTs performs a self-test and returns the result. The example also illustrates how to manage multiple instances of a cloneable DQMH module. (See Fig. 5.25.)

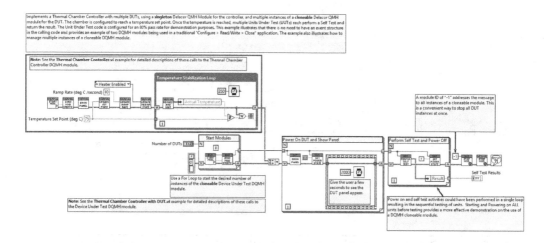

FIG. 5.25 Thermal Chamber Controller with Multiple DUT.vi. Notice that the code uses the Module ID wire to determine which UUT is being called at each For Loop iteration.

Exploring, Debugging, and Troubleshooting Code That Calls the DQMH Module API

Imagine that your team member Jane wrote the `Thermal Chamber Controller_DQMH.lvlib` module, and she shared it with you. Jane is on vacation, and you need to figure out how this thermal chamber code works. One option would be to call Jane and ask her for more details, but you do not want to be that coworker, right?

Your first instinct might be to explore more about how the code is implemented. You will not get too far by using highlighted execution and stepping into the subVIs because the majority of these VIs are just wrappers to generate event calls (see Fig. 5.26). This is one of the drawbacks of DQMH code. You lose the immediacy of the block diagram, but do not despair! You can still figure out what the code is doing, when, and how it is doing it via the API tester.

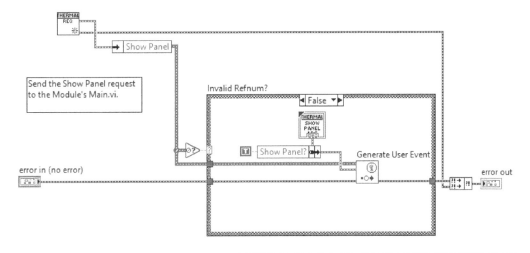

FIG. 5.26 *The DQMH request VIs are wrapping a Generate User Event.vi call. This code is the block diagram for the Show Panel.vi.*

In the `DQMH Fundamentals - Thermal Chamber.lvproj`, open the `Test Thermal Chamber Controller_DQMH API.vi`. (See Fig. 5.27). Run this VI and press the **Start Module** button. You can see how the thermal chamber controller DQMH module works by clicking on the different buttons on the API tester. Notice the **Status** indicator on the lower right gives you a log of what operations the controller has done. One of the functions that come with all DQMH modules is the option to **Show Block Diagram for Troubleshooting**. This comes in handy when you want to use highlighted execution or other debugging techniques directly on the module's code.

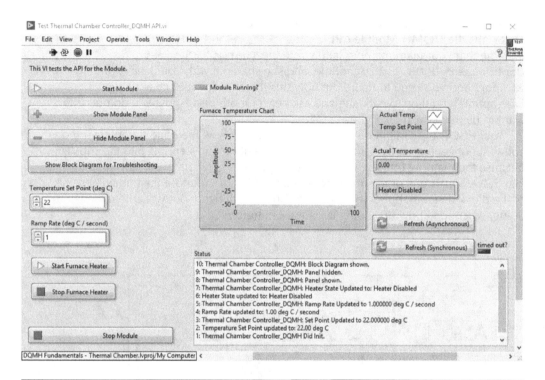

FIG. 5.27 *Test Thermal Chamber Controller DQMH API.vi is the tester for the thermal chamber controller. As any DQMH API tester, it contains the core functions to start the module; show/hide its front panel; show the block diagram for troubleshooting; and to stop module, a status indicator, and a module running? indicator. The developer created the rest of the buttons and indicators as she added new request and broadcast events to the DQMH module.*

You can use the tester as a manual of how Jane developed the thermal chamber controller. If you want to see how she starts the module, you right-click on the **Start Module** button and select **Find >> Find Terminal**. This will show you the code that you are expected to use to start the thermal chamber controller. You can continue to do this for the rest of the buttons.

You can also use the tester as a **sniffer** during execution. For example, run the `Thermal Chamber Controller.vi` and then run the `Test Thermal Chamber Controller_DQMH API.vi`. You will immediately see the **Status** indicator populating with all the information as to what the thermal chamber is doing. You can also inject requests to the module. This is handy, for example, when you want to simulate corner cases.

Troubleshooting a DQMH Request Event Debuging **DQMH Request Events** is easier if you use the following features of DQMH: The framework wraps all the DQMH request events in VIs and stores them within the **Requests** virtual folder; the DQMH `Main.vi` has an **EHL**, and DQMH uses a standard event naming format. You know that all DQMH modules probably follow this format because the DQMH tools use scripting to create the DQMH module and its events. Unless Jane got creative and messed with the scripted code, you can be pretty sure that she used the default DQMH format.

Let's say you are debugging your calling code, and a request seems to not be working as expected. You would find where in your code the VI is not working as expected. You would open this VI and find that the block diagram is a wrapper for the `Generate User Event.vi`. For simplicity, let's say this request is the **Show Panel**. You would open the `Show Panel.vi` and see that is generating an event called Show Panel, and this is one of the requests for the thermal chamber. If you press <Ctrl+Shift+E> while you are inside the `Show Panel.vi`, the project explorer highlights that this VI is within the **Requests** virtual folder of the `Thermal Chamber Controller_DQMH.lvlib`. If you want to know where the DQMH module is handling this event, all you have to do is open the `Thermal Chamber Controller_DQMH.lvlib:Main.vi`, identify the EHL, and look there for the event named <Thermal Chamber Controller_DQMH Request Events. Show Panel>:User Event. (See Fig. 5.28.)

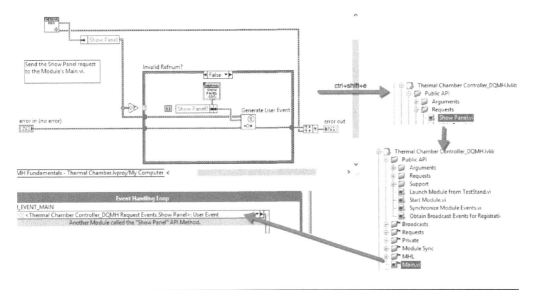

FIG. 5.28 *To debug a DQMH request event, find the request event call in your code, open it, and press <Ctrl+Shift+E> on your keyboard; that takes you to the location in the library. From there, identify the EHL in the DQMH Main.vi and look for <DQMH Module Name>Request Events.<Request Name>:User Event.*

Troubleshooting a DQMH Broadcast Event Debugging **DQMH Broadcast Events** is easier if you use the following features of DQMH: The framework wraps all the DQMH broadcast events in VIs and stores them within the **Broadcasts** virtual folder; the broadcasts can only be fired within the DQMH module code (they are private to the DQMH library) and DQMH uses the standard event naming format. You know that all DQMH modules probably follow this format because the DQMH tools use scripting to create the DQMH module and its events. Again, unless Jane got creative and messed with the scripted code, you can be pretty sure that she used the default DQMH format.

If you identify on your code that you are registered for a broadcast and you want to find where this broadcast is fired within the DQMH module, you would identify the name of the broadcast by looking at the event name <Module Name>Broadcast Events.<Broadcast Name>:User Event. From there, you would go to the DQMH module in the project explorer and look within the **Broadcasts** virtual folder. Let's say you are looking for Thermal Chamber Controller_DQMH Broadcast Events.Status Updated:User Event. You would right-click on Status Updated.vi and select Find >> Callers. In general, the only VI calling broadcasts is the Main.vi. However, there are cases where you would call a broadcast within other VIs in the module. For example, if you right-click on the Status Updated.vi, you would find that several VIs call this code. This makes sense because you use this broadcast to update the status log indicator on the API tester. (See Fig. 5.29.)

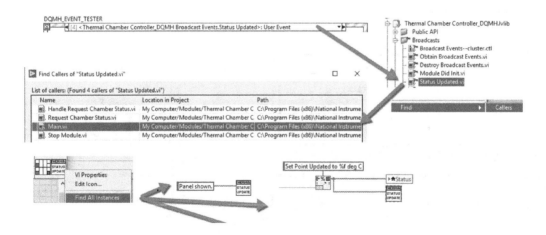

FIG. 5.29 To debug a DQMH broadcast event, identify the broadcast name by looking at the event name of the format <DQMH Module Name>Broadcast Events.<Broadcast Name>:User Event. In the DQMH module library, search for the event within the Broadcasts folder. You can now find all the callers and all the instances where the code fires this broadcast.

DQMH Project Template

The DQMH project template illustrates that the code that calls the DQMH module API does not need to be a DQMH module itself. In the case of the DQMH project template, the calling code is a simple state machine saved in the `Application.lvlib`. This project also shows how you can create your own project templates. There are things like showing the version of the exe and having a specific style for your top-level VI that can be enforced by having a project template for your team.[51] You can explore how we created the DQMH sample project by exploring the `<LabVIEW>\Project Templates\Source\Delacor\Delacor QMH` folder and `README.txt` there.

Create a DQMH project template by clicking on **Create Project** from the **Getting Started window** in LabVIEW and then selecting the DQMH project template. (See Fig. 5.30.)

FIG. 5.30 *Select Delacor >> Delacor QMH to create a Delacor QMH project template from the Create Project menu in the Getting Started window in LabVIEW.*

For instructional purposes, leave the **Include "Do Something" events in new module?** checkbox selected. This will ensure that the modules have a couple of DQMH events besides the core DQMH events. (See Fig. 5.31.)

DQMH Project Structure

Notice that all of the code is contained in project libraries. The only VIs outside of project libraries are the DQMH API testers. This makes sense because the API tester needs to guarantee that it is only calling public code. If they were part of the library, they could accidentally call private code. (See Fig. 5.32.)

The DQMH project template has the following components (see Fig. 5.32):

1. **Testers** that exercise the public API of the DQMH modules in the project. The testers register for the **broadcast** events fired by the `DQMH Module Main.vi` and fire **request events** that will be acted upon in the `DQMH Module Main.vi`.

---

[51] Visit "Using Custom Templates and Sample Projects to Develop LabVIEW Projects" to learn how to create your own project templates. http://www.ni.com/white-paper/14045/en/

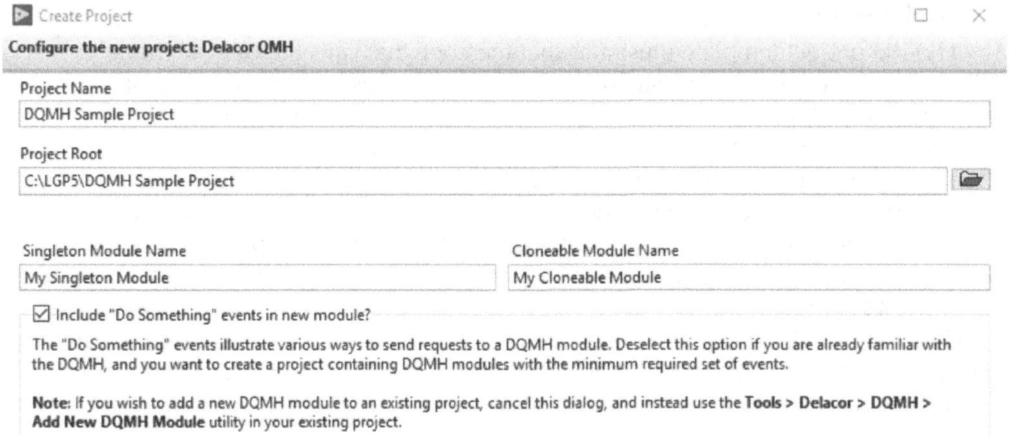

FIG. 5.31 *Create DQMH sample project wizard.*

2. Each DQMH module has its own project library (lvlib). This facilitates having the same icon header for all the members of the library and specifies which VIs are public and which VIs are only used by the library itself.

3. Each DQMH module has a **Public API** virtual folder that lists the VIs that can be called by VIs outside the library.

4. Within the Public API folder, there is a **Requests** virtual folder that lists all the VIs wrapping the firing of the requests that can be sent to the DQMH Module Main.vi.

5. To start a DQMH module, two VIs need to be called: Start Module.vi and Synchronize Module Events.vi. These two VIs take care of launching the DQMH Module Main.vi as well as making sure the DQMH Module Main.vi has created its own **request** and **broadcast** events before the calling VI can fire or register to listen to **broadcast** events.

6. The Obtain Broadcast Events for Registration.vi can be used when the calling VI is not ready to start the DQMH Module Main.vi or it expects the DQMH Module Main.vi to have been started by another module. The Obtain Broadcast Events for Registration.vi will have the **broadcast** events if the DQMH Module Main.vi has already started or it will have empty references, and it can be used to declare the event reference type for registration.

Chapter 5: Why Would You Want to Use a Framework? **321**

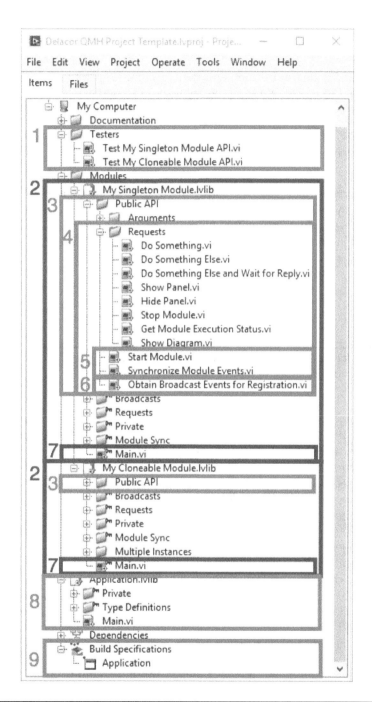

FIG. 5.32 *DQMH project created from the DQMH project template.*

7. The `DQMH Module Main.vi` is a private VI. This encourages developers to use the public API to interact with the DQMH module.

8. The calling code can be another library or simple VIs. In the DQMH project template, a Simple State Machine application is used to call the **singleton** and **cloneable** DQMH modules.

9. The Build Specification for an executable that uses `Application.lvlib:Main.vi` as the startup VI.

Exploring, Debugging, and Troubleshooting the DQMH Project Template

Let's start by verifying that the build specification for the executable works out of the box. In the project explorer, expand on the **Build Specifications** and right-click on the **Application** build specification and select to **Build**. (See Fig. 5.33.)

FIG. 5.33 *Right-click on the Application build specification and select Build to build the executable for this sample project.*

Once the Application Builder is done building your executable, click on the **Explore** button. This will take you to the location in the disk where the executable is. Double-click on `Application.exe` to run the application.

This application does not do much, but it includes things that you probably would want to see in all your applications. On the lower-right corner, it tells the end user what version of the executable they are running, and it has the application name on the window without the VI extension. You can also verify that we can start and stop a singleton DQMH module and start and stop multiple instances of the cloneable modules. To stop the executable, close the window, just as any other Windows application that you use!

Let's start by changing the Application name. To do that, expand on the `Application.lvlib` and open the `Application Name--constant.vi`. (See Fig. 5.34.) Delacor follows the practice of using constant VIs instead of global VIs. A constant VI has a constant on its block diagram connected to an indicator. We do not use globals because there would be a temptation to write to them. Having hard-coded constants means they will not change unless we edit and build a new executable. When determining what should be a constant VI, think about how often it will change. If it will change often, place the value in a configuration file.

Chapter 5: Why Would You Want to Use a Framework? **323**

If you build the exe again and run it, this time the window title will be "LGP5 DQMH sample project application" and the version will be 1.0.0.2.

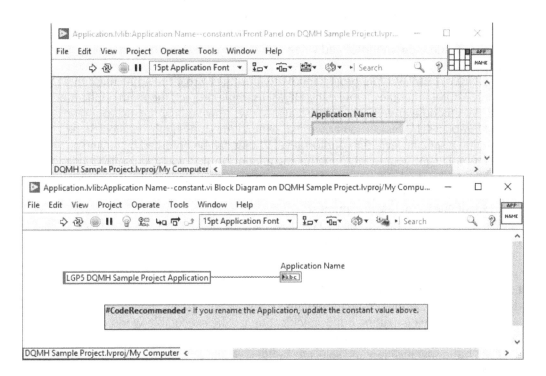

FIG. 5.34 *The exe uses the name inside Application.lvlib:Application Name—constant.vi to label its main window.*

Notice that there is a bookmark called **#CodeRecommended**. If you go on the Project Explorer menu to **View >> Bookmark Manager**, you will see a window with all of the **#CodeNeeded** and **#CodeRecommended** bookmarks. Explore these bookmarks by double-clicking on them. They will guide you through all the changes that you can make to the code. There is also a **#DQMH_HowTo** bookmark to explain how we created parts of this project.

To debug and troubleshoot this project, look for the API testers—in this case, `Test My Singleton Module API.vi` and `Test My Cloneable Module API.vi`. Use the same tips and techniques described in the previous section (simple code calling a DQMH module).

DQMH CML Sample Project Template

CML stands for continuous measurement and logging. It is the DQMH version of the CML sample project that ships with LabVIEW and that uses the NI QMH. Please note that at this level of the application, the DQMH implementation might seem more complicated than the NI QMH implementation. However, the benefits start to show as you add more modules and when you start troubleshooting and debugging the application.

The DQMH CML sample project template illustrates a real-life application that uses only DQMH modules. Unlike the DQMH sample project, the main application is a DQMH module as well. This project also shows how you can create your own project templates. In this particular project, we include showing the version of the exe, having a specific style for your top-level VI that can be enforced by having a project template for your team,[52] and having a launcher that you can use to debug your executable. The launcher was created from making a copy of the Application DQMH API tester. This project also shows using the **command line interface** (CLI) of your executable for debugging purposes.

Create your own copy of the DQMH CML sample project by clicking **Create Project** from the **Getting Started Window** in LabVIEW and then clicking the CML DQMH project template. (See Fig. 5.35.)

Exploring, Debugging, and Troubleshooting the DQMH CML Sample Project Template

Let's start by verifying that the build specification for the executable works out of the box. In the project explorer, expand on the **Build Specifications** and right-click on the **CML** build specification and select to **Build**. (See Fig. 5.36.)

Once the Application Builder is done building your executable, click on the **Explore** button. This will take you to the location in the disk where the executable is. Double-click on CML.exe to run the application.

This application does a lot more than the DQMH sample project. Besides having an application-specific code, in this case acquiring a simulated signal and logging to disk, it also includes some of the things that the DQMH sample project has. On the lower-right corner, it tells the end user what version of the executable they are running, and it has the application name on the window without the VI extension. Test the application by clicking on the **Settings** button, set your preferences, and then click on the **Start** and **Stop** buttons. To stop the executable, close the window, just as any other Windows application that you use.

We mentioned earlier that this exe included the ability to debug. To test this feature, launch a Command Prompt (type CMD in Cortana). Type cd followed by the location where your CML.exe is. You can copy the path in the file explorer, type cd on the CMD window, and then right-click to paste the path. Then type CML.exe --debug; this launches the application in debug mode. (See Fig. 5.37.)

---

[52] Visit "Using Custom Templates and Sample Projects to Develop LabVIEW Projects" to learn how to create your own project templates. http://www.ni.com/white-paper/14045/en/

Chapter 5: Why Would You Want to Use a Framework?

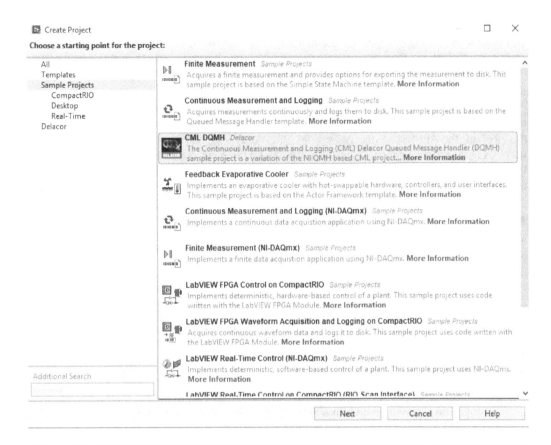

FIG. 5.35 *Select Sample Projects >> CML DQMH to create a DQMH version of the continuous measurement and logging sample project.*

FIG. 5.36 *Right-click on the CML build specification and select Build to build the executable for this sample project.*

FIG. 5.37 *Use the Command Prompt to launch the CML.exe in debug mode.*

This launches an API tester-like UI that you can use to start and stop the module and see the status log of what the application is doing. This allows you to restart your module without having to restart the exe. You can explore the block diagram of `CML Main.vi` to see how we implemented scanning for the CLI arguments and determine if the code is running in debug mode or not.

In the development environment, use the API testers to understand how each module is implemented, and you would call their APIs in a different application. In this project, the API testers are `Test CML UI API.vi`, `Test Acquisition API.vi`, `Test Logger API.vi`, and `Test Settings Editor API.vi`. Use the tips and techniques described in the previous sections to further explore how this application works.

Adding a New DQMH Module

In the previous sections, you learned how to use existing DQMH code and how to explore, debug, and troubleshoot DQMH modules. It is time for you to create your first DQMH module. You need to determine the following:

1. DQMH module name: Pick a name that reflects the DQMH module responsibility. Be wary of names that include the word "and"; this might be an indicator that your module is doing too much. For example, you can name your DQMH module **Acquisition**, but you don't want to name it **Acquisition and Logging**.

2. Determine DQMH module type. By default, DQMH has two options: a **singleton** module or a **cloneable** module. You can also create your own DQMH module templates. If you had already an Acquisition DQMH module template, you could choose that.

3. Define DQMH module API.

    a. Requests: Name them using the imperative form, for example, **Acquire**.

        i. Define if the requests can be asynchronous or if you need to create **Request and Wait for Reply** type events.

        ii. Define the requests' arguments.

    b. Broadcasts: Name them using the past tense, for example, **Acquisition Configured**.

        i. Define the broadcasts' arguments.

You can use the following table to define your module before you start creating code.[53]

---

[53] http://delacor.com/tips-and-tricks-for-a-successful-dqmh-based-project/

Module	Request Event Name	Request Arguments	Wait for Reply?	Reply Arguments	Broadcast Event Name	Broadcast Arguments
Acquisition	Start Acquiring		Yes	HW ID	Acquisition Started	HW ID
Acquisition	Stop Acquiring				Acquisition Stopped	
Acquisition					Data Updated	Data Array

If you had more DMQH modules in your application, you could continue to fill out the rest of the table with the different DQMH modules.

As part of this design process, you will identify if you need a helper loop or not. In this case, you will use the helper loop to ensure the DAQ board is read in a periodic manner. You won't need to create a "Wake Up Helper Loop" event because you can use the Start Acquiring event to wake up the helper loop. You won't need to create a "Put Helper Loop" to sleep either because you can use the Stop Acquiring event for this.

Steps to Add a New DQMH Module

You decided this would be a cloneable DQMH module because you might have more than one DAQ device in your application. If you only have one, you would start it as a singleton and you would be fine.[54] To add a new DQMH module, you will follow the next steps. You can check the DQMH documentation in case these steps have changed after the book was published.[55] (See Fig. 5.38.)

1. Create a blank project (you could also start on an existing project where you wanted to add this module). If you have a virtual folder called **Modules** and one called **Testers**, the DQMH tools will place the DQMH module and its tester on the correct virtual folder.

2. Go to the **Tools >> Delacor >> DQMH >> Module >> Add New DQMH Module...** menu. (See Fig. 5.38.)

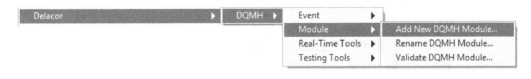

FIG. 5.38 *Delacor Tools menu option to add a new DQMH module to an existing project.*

---

[54] The code for this new DQMH module is at GitHub.com/LGP5/Frameworks repository, under the branch "DQMH-AddNewDQMHModule."
[55] Adding a New DQMH Module. https://delacor.com/documentation/dqmh-html/AddingaNewDQMHModule.html

3. Follow the prompts. Make sure you select **cloneable** for the DQMH type. You know the scripting is done when the module and its tester appear on your project, the **Add New DQMH Module** window disappears, and the cursor goes from busy to the regular cursor. (See Figs. 5.39 and 5.40.)

4. The project now looks like this.

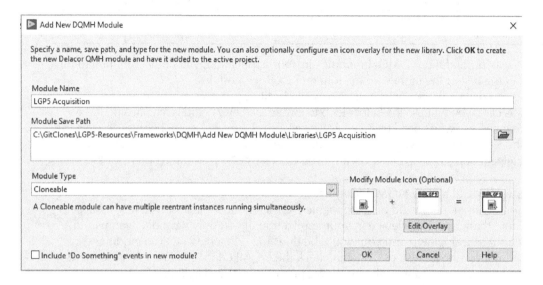

FIG. 5.39 *Enter the module name, the path (by default it saves your module in the Libraries folder within your project folder), select cloneable for the module type, edit the icon overlay, and make sure the "Include 'Do Something' events in a new module?" option is unchecked.*

5. You are ready to add the events you determined during the design section. You will use the **Tools >> Delacor >> DQMH >> Event >> Create New DQMH Event...** menu option to create all the events.[56]

   a. You start by creating the **Start Acquiring** request event. From your table, you can see that this request event is associated with the **Acquisition Started** broadcast event. You can create both events at the same time by selecting the option to create a **Round Trip (Request and Wait for Reply + Broadcast)** event type. (See Fig. 5.41.)

---

[56] https://delacor.com/documentation/dqmh-html/CreatingaNewDQMHEvent.html

Chapter 5: Why Would You Want to Use a Framework? **329**

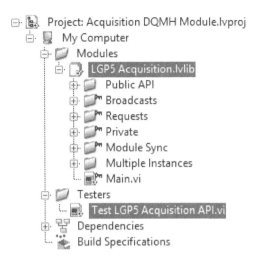

FIG. 5.40 *The DQMH scripting tools added the LGP5 Acquisition library to the Modules virtual folder and the Test LGP5 Acquisition API to the Testers virtual folder.*

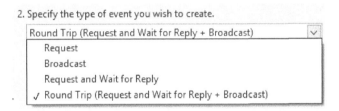

FIG. 5.41 *Selecting the option to create a round-trip event will open two windows for arguments: one for the argument for the request and one for the argument to be used for both the reply and the broadcast.*

    b. Fill out the **Create New DQMH Event Window** (see Fig. 5.42). You don't need an argument for the **Event Arguments Window**, but you will place a **HW ID** string for the **Reply Arguments Window**. You may be prompted about having an empty arguments window; this is just a reminder in case you forgot.

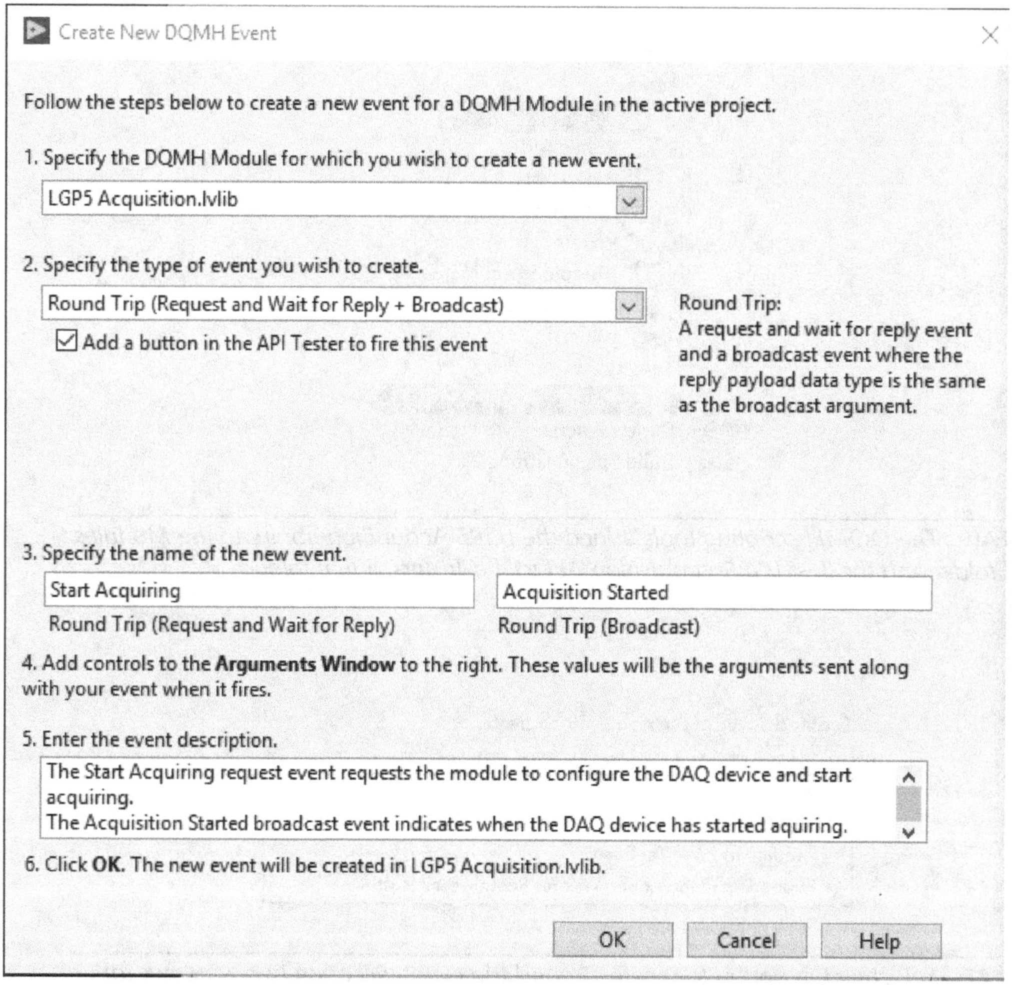

Create New DQMH Event

Follow the steps below to create a new event for a DQMH Module in the active project.

1. Specify the DQMH Module for which you wish to create a new event.

   LGP5 Acquisition.lvlib

2. Specify the type of event you wish to create.

   Round Trip (Request and Wait for Reply + Broadcast)
   ☑ Add a button in the API Tester to fire this event

   Round Trip:
   A request and wait for reply event and a broadcast event where the reply payload data type is the same as the broadcast argument.

3. Specify the name of the new event.

Start Acquiring	Acquisition Started
Round Trip (Request and Wait for Reply)	Round Trip (Broadcast)

4. Add controls to the **Arguments Window** to the right. These values will be the arguments sent along with your event when it fires.

5. Enter the event description.

   The Start Acquiring request event requests the module to configure the DAQ device and start acquiring.
   The Acquisition Started broadcast event indicates when the DAQ device has started aquiring.

6. Click **OK**. The new event will be created in LGP5 Acquisition.lvlib.

   OK   Cancel   Help

FIG. 5.42 *Create New DQMH Event window filled out for the Start Acquiring and Acquisition Started round-trip events.*

    c. The DQMH scripting tools created the **Start Acquiring** and **Acquisition Started** events for you, added them to the API tester, and added the code you need to complete to the `LGP5 Acquisition.lvlib:Main.vi`. The tools will bring the API tester up front as a reminder that you need to start by figuring out how you will test this. For now, put an indicator with the HW ID for when the **Acquisition Started** broadcast arrives and edit the event case that executes when the **Test Start Acquiring** button value changes and wire the **Module ID** (see Fig. 5.43). Once you are done, you will get a running arrow and you will be ready to work on the `Main.vi`.

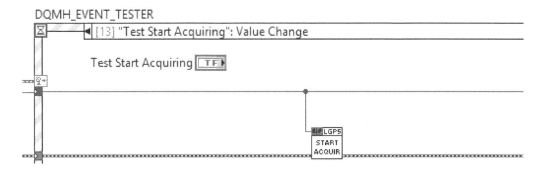

FIG. 5.43 *Wire the Module ID input to the Start Acquiring.vi that the DQMH tools created for you in the API tester.*

d. You will use the `DAQ Device.lvlib` that already has the simulated DAQ functions that the CML DQMH project template uses.[57] You will now edit the MHL case that the DQMH tools created for you. (See Fig. 5.44.)

e. Add the code in the **Start Acquiring** MHL case, and you are ready to run the API tester and verify that your **Start Acquiring** request and your **Acquisition Stopped** work. You can add the **Status Updated** broadcast VI to add more information to the API tester when the action completes. (See Figs. 5.45 and 5.46.)

---

[57] The code for the DAQ Device.lvlib is at GitHub.com/LGP5/Frameworks repository, under the branch "DQMH-AddNewDQMHModule."

332 LabVIEW Graphical Programming

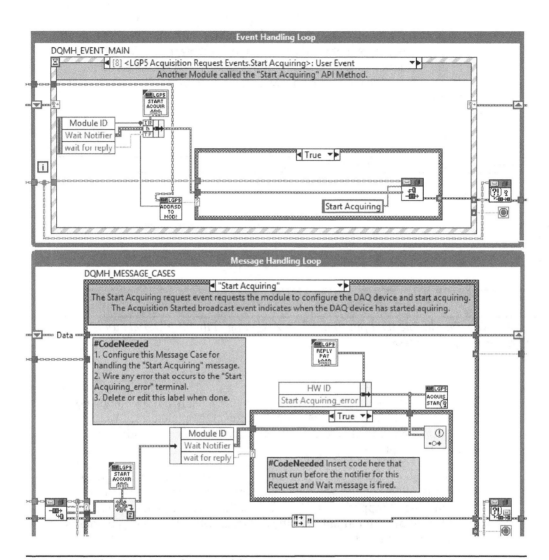

FIG. 5.44 *The DQMH tools created for you the LGP5 Acquisition Request Events.Start Acquiring event case and the Start Acquiring MHL case.*

Chapter 5: Why Would You Want to Use a Framework? **333**

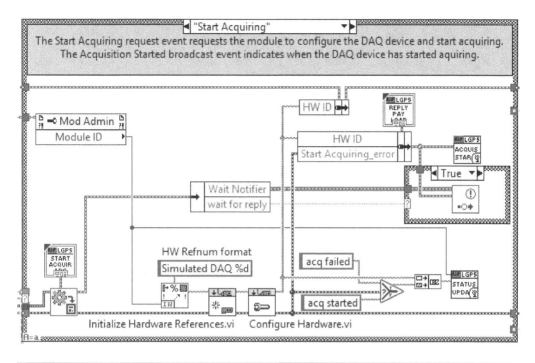

FIG. 5.45 *After you add the DAQ Device calls to initialize and configure the hardware and you add the Status Updated text, you are ready to test.*

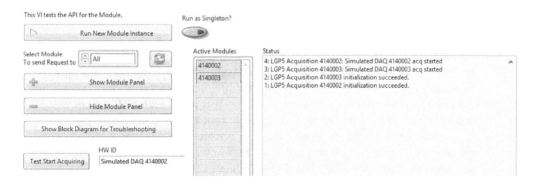

FIG. 5.46 *The API tester after you have pressed the Run New Module Instance button twice and pressed the Test Start Acquiring button. Notice that you get a line for each of the cloneable modules in the status box.*

f. Repeat steps a through c to create a Round Trip event for Stop Acquiring and Acquisition Updated. (See Fig. 5.47.)

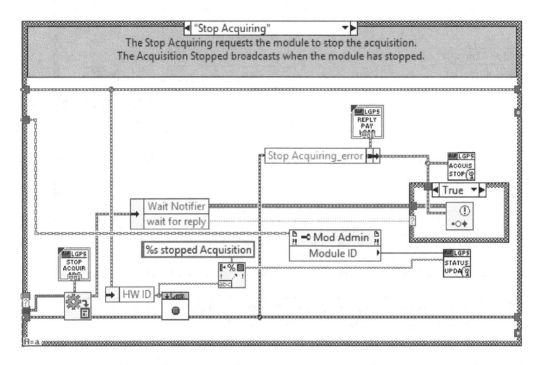

FIG. 5.47 *The DQMH tools create the Stop Acquiring MHL case, and you add the DAQ Device.lvlib:Stop Acquisition.vi and the Status Updated broadcast.*

6. You need to add a helper loop to take care of the acquisition.[58] At this point you realize that you cannot reuse the **Start Acquiring** request to wake up the helper loop because you need to ensure that the acquisition did start and you need to send the **HW ID** as an argument. You will create a **Wake Up Helper Loop** request. This request is a private request, so after you create it, you will move it from the public API to a private virtual folder, and you will not implement the test in the API tester. You will also delete the MHL case and the event handling case that the DQMH tools created for you to

---

[58] There is a feature request for DQMH to include a tool to create helper loops. Maybe by the time you are reading this book, that feature will already be in place. https://forums.ni.com/t5/Delacor-Toolkits-Documents/DQMH-Feature-Requests/ta-p/3537845

Chapter 5: Why Would You Want to Use a Framework? **335**

handle the Wake Up Helper Loop. You don't want code outside the LGP5 Acquisition module to wake up the helper loop. (See Fig. 5.48.)

a. You will register for the Module Request events and make sure you handle the Stop Module request. (See Fig. 5.49.)

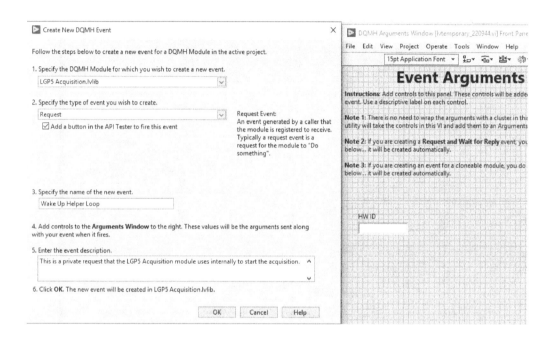

FIG. 5.48 *You created a Wake Up Helper Loop request event with a HW ID argument.*

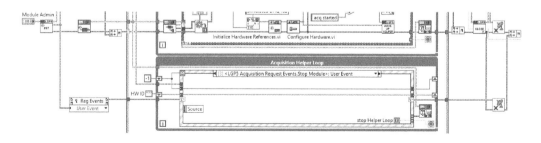

FIG. 5.49 *Add the helper loop and register for the LGP5 Acquisition request events and handle the Stop Module event.*

b. The Start Acquiring MHL case will now send the HW ID via the Wake Up Helper Loop request to the helper loop. (See Figs. 5.50 and 5.51.)
c. The timeout event case in the helper loop is the one doing the acquisition. You are using simulated hardware time, so the timeout of 0 is appropriate. (See Fig. 5.52.)

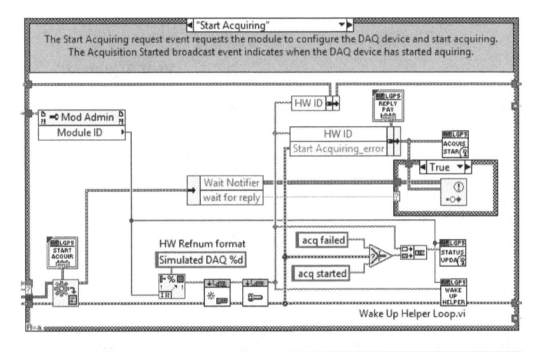

FIG. 5.50 *Start Acquiring MHL case sends the HW ID to the helper loop via the Wake Up Helper Loop request.*

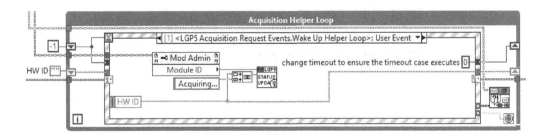

FIG. 5.51 *The helper loop handles the Wake Up Helper Loop request and updates the status with the informative message "<HW ID> Acquiring..."*

Chapter 5: Why Would You Want to Use a Framework? **337**

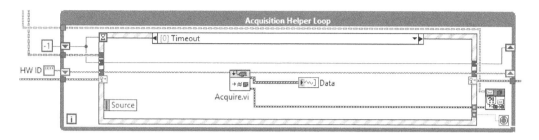

FIG. 5.52 *The timeout case executes the Acquire.vi.*

d. When the helper loop handles the Stop Acquiring event, it sets the timeout to -1, effectively stopping the timeout case from executing. (See Fig. 5.53.)

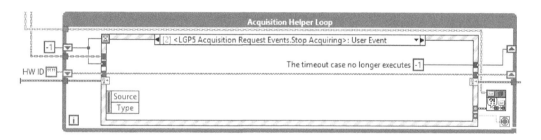

FIG. 5.53 *The helper loop handles the Stop Acquiring event and sets the timeout to -1.*

7. The LGP5 Acquisition module is currently updating the data on its front panel. The only event that you are missing is the **Data Updated**. Use the **Tools >> Delacor >> DQMH >> Event >> Create New DQMH Event...** menu option to create the Data Updated broadcast event. You have to make some decisions on whether you will send the entire data array or whether you will use a DVR to send the data. If your array size is not too big and your data rate is not too high, you can get away with sending the entire data array as is. However, if you see any issues with this approach, wrap the data in a DVR and send the DVR via the broadcast.

   a. It is while implementing this broadcast and using the API tester that you realize that you made a mistake in the helper loop. This is a cloneable module, which means it shares its events with all the other modules. You need to make sure that every event case is calling the LGP5 Acquisition.lvlib:Addressed to This Module.vi that determines if the event case was assigned to this module or not (see Fig. 5.54). An alternative would be to create a local event to this module **Put the Helper Loop to Sleep** and use the main EHL as a bouncer (that is one of the EHL tasks after all) and have it only forward the **Put the Helper Loop to Sleep** when the **Stop Acquiring** is addressed to this module. For this approach, the **Wake Up Helper Loop** would be local to this module as well. We do not show the implementation of the second approach in these images because it is less common.

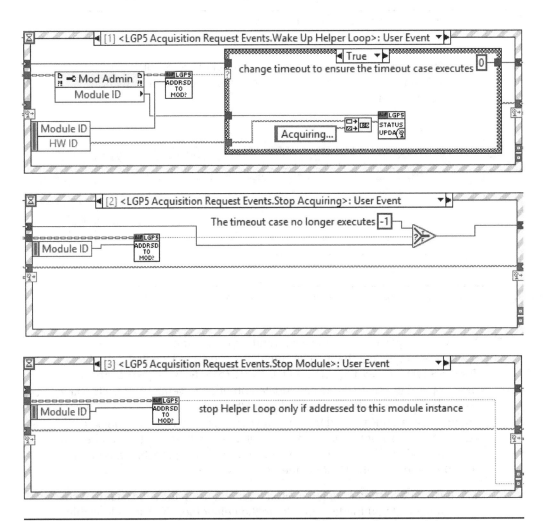

FIG. 5.54 *Fix the helper loop event cases to ensure they are checking if the event is addressed to them.*

Chapter 5: Why Would You Want to Use a Framework? **339**

8. Don't forget to close the hardware references in the Exit MHL case. (See Fig. 5.55.)

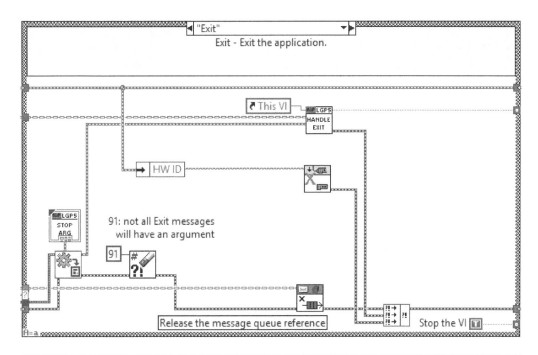

FIG. 5.55 *Close the hardware references in the Exit MHL case.*

Adding a New DQMH Module from a Template

You can create your own DQMH module templates. Create a DQMH module that has the generic Requests, Broadcasts, and modifications that you need for the `Main.vi`. You can modify the core DQMH events to fit your needs; for example, some LabVIEW developers add a timestamp to their **Status Updated** broadcast. You can also include other classes that the DQMH module needs; just make sure your classes and libraries are saved within the DQMH module and in the same folder as the DQMH module in disk. Note that if you need to include a different dependent DQMH template, you will be better off creating a project template instead. (See Fig. 5.56.)

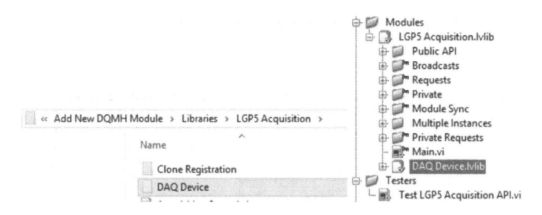

FIG. 5.56 *To create a template from the LGP5 Acquisition module, you need to move the DAQ Device library to be within the LGP5 Acquisition library both in the project and in the disk.*

Once you have created your new DQMH module that you will save as a template, verify that your DQMH module is still recognized as a DQMH module; otherwise, the DQMH tools will not work. To do this, go to the **Tools >> Delacor >> DQMH >> Module >> Validate DQMH Module…** menu option. A prompt will show the available DQMH modules in the current project. Select your DQMH module and follow the prompts.

The Validate DQMH utility looks for specific labels and module contents to determine if a library is a DQMH module or not.[59] If your DQMH module does not appear on the list of available modules, check the DQMH documentation and make sure you did not remove or modify any of the expected components.

You also want to verify that you have addressed all the **#CodeNeeded** and **#CodeRecommended** bookmarks by going to the **View >> Bookmark Manager** menu option and verifying that you have added any bookmarks that will help others use your template.

Once you have validated your DQMH module template, go to the **Tools >> Delacor >> DQMH >> Module >> Save as Template…** menu option and follow the wizard. From now on, when you click on **Tools >> Delacor >> DQMH >> Module >> Add New DQMH Module Template…** menu option, you will have the option to choose the LGP5 Acquisition module. (See Fig. 5.57.)

---

[59] https://delacor.com/documentation/dqmh-html/ValidatinganExistingDQMHModule.html

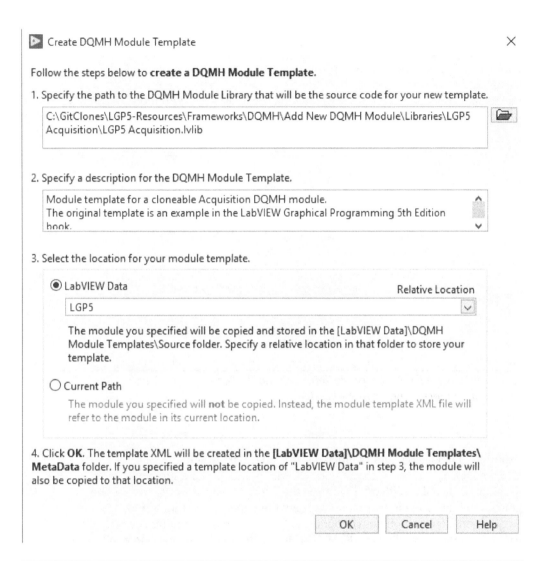

FIG. 5.57 *The Create DQMH Module Template tool allows you to specify the description others will see when they select this as their DQMH type and where you want to save the template. Use the LabVIEW Data option for templates that will be shared across multiple organizations and teams. Use the Current Path if you use the method of subrepositories to link your dependencies to your projects.*

If you want to share this template with others, you can build a VIPM package that would place the metadata and source files in the correct location in their computers. If you are using subrepositories, you can create the module template to use a relative location, and the only files others would have to copy would be the metadata. When the wizard finishes creating the template, it will indicate where it saved the template source and the metadata XML file. (See Fig. 5.58.)

FIG. 5.58 *The Create DQMH Module Template tool indicates where it saved the DQMH module template and its metadata file.*

Exploring, Debugging, and Troubleshooting a DQMH Module
Use the DQMH API tester to run your DQMH module. Make sure you keep that API tester up to date. Even if you do not see the benefit today, trust us, you will be grateful to have done it. Hopefully you saw the benefit even through the LGP5 Acquisition exercise, where we found out things were not working correctly thanks to the API tester. Even if your DQMH module will run as a standalone application and you do not see the need for an API tester, you will still benefit from using the tester as a sniffer.[60] You can even use it as a launcher. See how the DQMH CML project uses a DQMH API tester like VI to launch the application. Use the other tips and tricks described in the previous sections.

---

[60] Using the Tester VI to Troubleshoot an Application While the Application is Running. https://delacor.com/documentation/dqmh-html/UsingtheTesterVItoTroubleshootan.html

DQMH: Behind the Scenes
Interprocess Communication
The queue in a DQMH module is private to that module. This makes troubleshooting and debugging easier because if a state in MHL should not be occurring, you know that the enqueue could only have happened within the module itself. Only private VIs to the DQMH library or an EHL case can enqueue a message.

The communication between outside code and the DQMH module is done via events. This outside code can be a user interaction with the DQMH module's user interface, in which case the event is handled in the EHL inside DQMH `Main.vi`. This outside code could also be code firing a request event.

The communication to outside code from the DQMH module is done via events as well. The DQMH module fires broadcast events.

For a DQMH module, its EHL is the bouncer that decides whether to relay the messages or not to the MHL.

Delacor decided to use events over queues because if a message is enqueued but there is nobody to dequeue, there is a memory leak. However, if a message is generated via an event and nobody is registered to listen to it, there are no memory leaks. The DQMH module does not care if there is code registered or not to its broadcast events. This facilitates having a DQMH module that can be executed as a standalone application or launched by external code. This also makes the API testers useful as sniffers for troubleshooting. They can register to listen to the DQMH Broadcast events, even if the DQMH module was already running before the API tester started.

DQMH Events in Functional Global Variables
DQMH takes advantage of FGVs to pass around the references for the user events without seeing the wires throughout the DQMH `Main.vi`. Delacor opted for this approach for clones too because we wanted to have the option of sending the same message exactly at the same time to all the clones. The DQMH `Main.vi` creates the DQMH events. This ensures the DQMH module owns the references to the events, and if the calling code stops, the DQMH module can continue to work, and any other module registered for the events will continue to communicate with the module.

All the cloneable DQMH instances share the same set of DQMH events. The Delacor scripting tools will add the `DQMH Cloneable Module .lvlib:Addressed to This Module.vi` in the EHL case to ensure the message is sent to this particular instance.

Initialization
A framework must provide a way to initialize the different modules. There are different schools of thought as to whether the arguments should be provided as part of the initialization or at a later call. Our suggestion is to ensure that any initialization of the module that is not part of the DQMH framework is done after

the module has finished initializing. In other words, the DQMH framework already guarantees that the modules will start, so if a module does not start, it is because the developer added extra code during initialization that broke the framework initialization mechanism.

Another discussion is whether the module should own its references or the calling code should create the references. In the case of DQMH, the `DQMH Module Main.vi` creates its request and broadcast events to ensure that even if the calling VI leaves memory, the references are still valid.[61] This feature facilitates calling the DQMH modules via TestStand as well as easily defining the lifetime of a given module's event references.

To call the Start Module, you also need to call the Synchronize for Module Events. You might be wondering why Delacor did not put the **Start Module**, the **Register for Events**, and the **Synchronize for Module Events** in a subVI (see Fig. 5.59). The short answer is Fab does not like to have the Register for Events call inside a subVI. Delacor is also using rendezvous to ensure that the module is ready to receive requests (it has finished registering for the request events), and the calling code is ready to receive broadcasts (it has finished registering for the broadcast events). Keep reading to get the long answer.

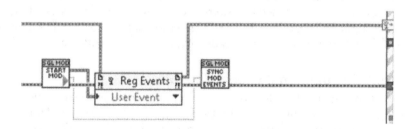

FIG. 5.59 *Combination of Start Module.vi and Synchronize Module Events.vi ensures that the DQMH Module Main.vi is ready to receive requests and that the calling VI is ready to receive the DQMH Module's broadcasts.*

The NI QMH Project Template that ships with LabVIEW used to have the Register for Events inside a subVI. Fab gave Darren Nattinger[62] lots of grief about it. So much so that eventually it was removed from the subVI in LabVIEW 2015. We are sure Fab was not the only one complaining about it, but she might have been one of the loudest.

---

[61] References in LabVIEW belong to the memory space of the calling VI. If you create a reference inside A.vi and then use those references inside B.vi, LabVIEW destroys the references as soon as A.vi goes out of memory, regardless if B.vi is still running.

[62] Darren Nattinger works in LabVIEW R&D. He is known as the fastest LabVIEW programmer and is responsible for the creation of LabVIEW tools such as Quick Drop, VI Analyzer, and NI QMH template, among others.

There are mainly two reasons why we do not like to have the Register for Events inside a subVI:

1. It makes it too easy for a junior developer to branch that wire coming out of that subVI and connect it to more than one Event Structure. This is trouble; if you have attended an Advanced Architectures in LabVIEW[63] course from NI, they have a very good demonstration about it. The Register for Events is creating an event registration engine that handles the events. If nobody is registered to listen to an event, the event doesn't get queued in the event queue. For each event registration, there is one event queue behind the scenes. When the Event Refnum gets forked, it is like having two dequeuers, except it is worse than that! Sometimes both Event Structures will get the event and sometimes only one of them will... DO NOT FORK THE REGISTRATION REFNUM wire! (See Fig. 5.17.)

2. The user event names on the Event Structure are not as descriptive as to when the event registration is on the same VI as where the Event Structure is.

The `Start Module.vi` is in charge of initializing the module. It uses Start Asynchronous Call node to launch the module. The initial versions of DQMH used the VI server Run VI method instead. The reason was to ensure that subsequent calls to the Start Module.vi (which reads the `Exec.State` property of the module main VI in the **"Get Module Main VI information"** VI) would function correctly. We thought that the `Exec.State` property did not return useful values when the module main VI is launched dynamically via a Call By Reference function. Well, it turns out that what we were missing on our initial attempts was to close the reference after the Start Asynchronous Call node! If we close the reference, then subsequent calls to Exec.State property do return the correct information. (See Fig. 5.60.)

In this case, there can only be one instance of the DQMH module, and whenever the `Start Module.vi` is called, it is either launching the module or establishing communication with the same module.

If the developer decides to include data as part of the initialization, they would have to provide a control on the front panel of the DQMH `Main.vi` and connect it to the connector pane. Then modify the VI Reference specifier to use the new connector pane.

The `Start Module.vi` for the DQMH cloneable modules is more complex because it needs to include code to manage multiple instances of the same module. (See Fig. 5.61.)

---

[63] http://ni.com/training

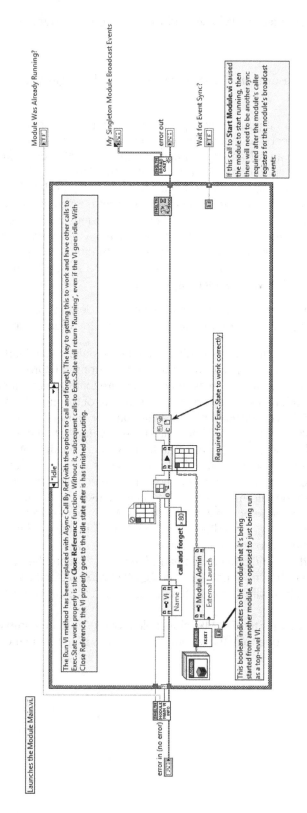

FIG. 5.60 Typical DQMH Start Module.vi block diagram for a singleton DQMH module.

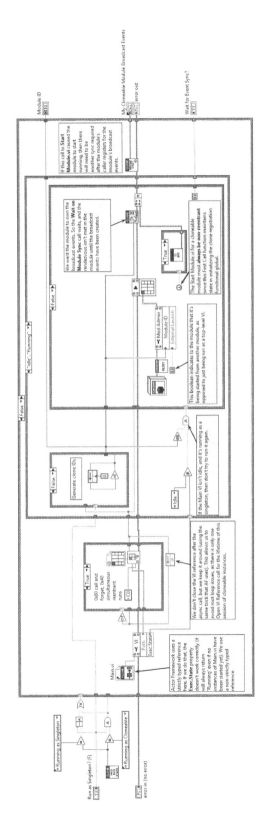

FIG. 5.61 Typical DQMH Start Module.vi block diagram for a cloneable DQMH module.

Stop Module Gracefully

DQMH modules come with a `Stop Module.vi` request. Each module handles this request and stops itself. By default, the Stop Module request is a high-priority request, which means it will be put at the front of the event queue and the message queue. If the module needs to do extra work during exit and handle any pending messages, this VI should be modified for its `Stop Module.vi` to not be high priority. (See Figs. 5.62 to 5.66.)

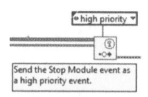

FIG. 5.62 *Inside Stop Module .vi, modify the high priority if this module should take care of any pending events before exiting.*

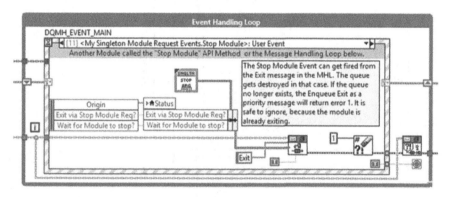

FIG. 5.63 *Event Handling Loop inside the DQMH Main.vi handles the Stop Module request.*

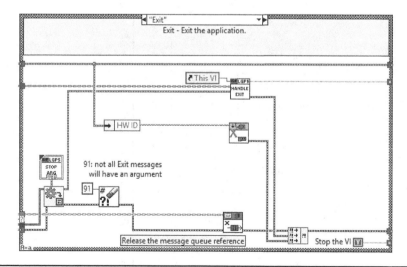

FIG. 5.64 *The Message Handling Loop inside the DQMH Main.vi handles the exit message. The LabVIEW programmer would add any clean-up code needed inside this case.*

Chapter 5: Why Would You Want to Use a Framework? **349**

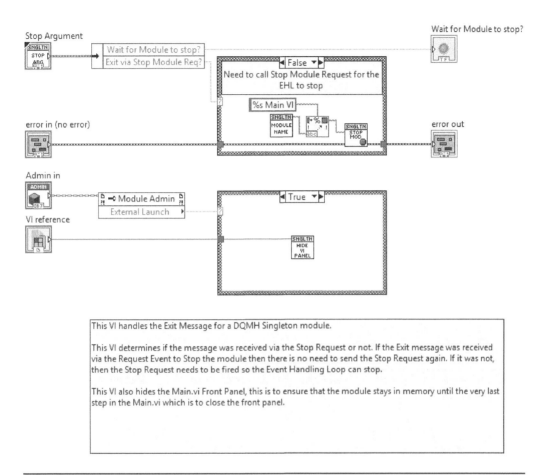

FIG. 5.65 *Within the exit case, the Handle Exit.vi sends the last broadcast—in this case "Module Did Stop."*

The main difference in the cloneable DQMH modules is that the `Close.vi` needs to check if this is the first clone. If it is, it means it owns the references and it cannot completely exit until the rest of the clones have stopped. This module is no longer running in the sense that its EHL and MHL have exited, but it is still alive as a life support mechanism to ensure the other clones still have valid user events to communicate with the rest of the application. (See Fig. 5.67.)

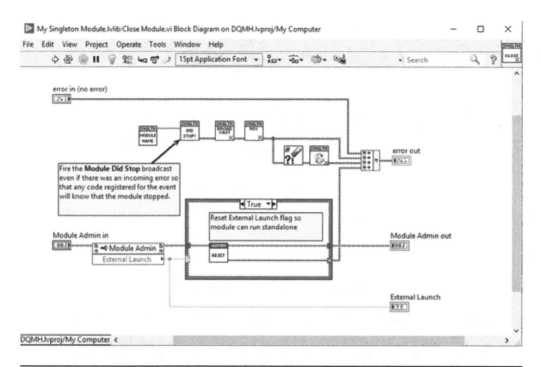

FIG. 5.66 *The very last thing the DQMH Main.vi does is call the Close Module.vi that destroys all the event references.*

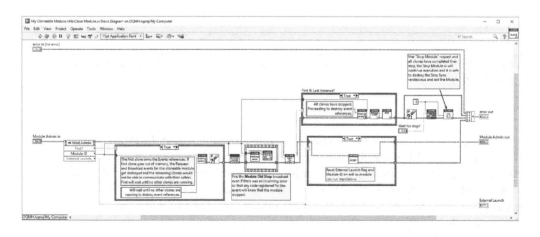

FIG. 5.67 *Cloneable modules need to verify if they are the first clone. If they are, they own the references and cannot exit until the last cloneable module stops.*

Error Handling

The error handling strategy for the DQMH modules is very basic. If a DQMH module is running as a standalone application (you clicked run on the DQMH `Main.vi` itself instead of being called by any other code), the module stops in case of error. If the DQMH module is launched by another VI, such as the DQMH API tester, then it does not stop in case of error. The DQMH module assumes that the calling code is responsible for handling any errors that the DQMH module did not handle directly in the code. You can change this error handling strategy to fit your application's needs.

There are different ways to handle errors in your application. You can choose to have the top-level VI register for the Error Reported broadcast of all the DQMH modules, or you could have a central error handler that can decide what to do with errors and whether to display them, log them, or ignore them. Kabul Maharjan described how Renishaw created a project template that includes a central error handler that not only handles errors but also tracks that the application is running as expected.[64]

Other DQMH Tools

Delacor, the DQMH Trusted Advisors,[65] and other LabVIEW programmers using DQMH are continually coming up with new tools and add-ons. Here are a few of them.

Validating an Existing DQMH Module

With each new version of the toolkit, Delacor makes the commitment to add tests to the DQMH validation tools to upgrade your existing DQMH modules to the latest version. You can also use these tools to ensure the other DQMH tools still consider your modified DQMH module a DQMH module.[66]

DQMH Generic Networking Module

Joerg Hampel from Hampel Software Engineering, a DQMH Trusted Advisor, created a DQMH template to implement DQMH communication between targets. He uses it for remote control of headless applications.[67]

Right-Click DQMH Menu for the Project Explorer

Jean-Claude Mengisen from Zuehlke developed an extension for the project explorer for the DQMH toolkit. Installing his package adds a right-click menu on the project

---

[64] DQMH in Action by Kabul Maharjan. https://delacor.com/dqmh-in-action/
[65] DQMH Trusted Advisors. http://delacor.com/dqmh-trusted-advisors/
[66] DQMH Documentation on Validating an Existing DQMH Module. https://delacor.com/documentation/dqmh-html/ValidatinganExistingDQMHModule.html
[67] DQMH Generic Networking Module by Joerg Hampel. http://delacor.com/dqmh-generic-networking-module/

explorer access to the DQMH tools, in addition to getting to them via the Tools menu.[68]

MGI Panel Manager—DQMH Panels
Derek Trepanier from Moore Good Ideas created the MGI Panel Manager[69] and included two new DQMH templates for using the MGI Panel Manager for the window management instead of the traditional Show Panel and Hide Panel events in a DQMH module.

Sharing Reusable DQMH Modules
For small applications, you can have all the DQMH modules reside in the same project. This would be the case in the DQMH project template and the CML DQMH sample project template. If your application starts to grow into the tens of modules or you want to have multiple developers working in parallel in separate DQMH modules, you would be better off putting those modules in separate projects, or even separate repositories. This has the added advantage of making those modules easier to share with other projects and other areas of your organization.

Sharing DQMH Templates
One way to reuse DQMH modules is to save them as templates and create a unique version of that DQMH module for your new project. If you are going to go down this route, try to make your template as generic as possible. One thing you can do to start building your DQMH templates library is to identify if the DQMH module that you are about to make could be used in other projects. An example of this would be a Serial Port controller; the likelihood of you working on other projects that use a Serial Port controller may be high. If that is the case, first create the DQMH module for a generic serial port controller. This will probably include a connection to the port, port configuration, a helper loop to poll the port, etc. Once you are done, you can save this as a template and now get to work on your project module starting from the template. While you are working on your module, if you discover bugs, you can file a bug tracking case against the project template and eventually fix it in the template. Next time you need a Serial Port controller, you know you can start from that template that has been strengthened throughout several projects. Depending on how your organization works, you could put the source code for that template in its own repository, create its own bug tracking project, and its own product number.

---

[68] Zuehlke Project Explorer Menu for DQMH. https://forums.ni.com/t5/Delacor-Toolkits-Discussions/Zuehlke-Project-explorer-menu-for-DQMH/td-p/3802835/highlight/true

[69] MGI Panel Manager. https://www.mooregoodideas.com/products/panel-manager/index.html

Configuring Source Code Control Repository Dependencies

There are no special considerations for DQMH for this approach. The generic approach we described earlier for frameworks in general applies to DQMH-based projects.

Packaging DQMH Modules Using VIPM

We suggest that when you build VIPM packages, you add a prefix to your project libraries in the package specification. This will help you differentiate source code and code installed via VIPM. Also, make sure to include the DQMH version your packages depend on. This will be useful as Delacor releases new versions of DQMH.

Packaging DQMH Modules in PPLs

Make sure you review all the steps described earlier in this chapter on how to package modules in PPLs. To package the DQMH libraries into PPLs, you will need to do the following:[70]

1. Open the project that has the DQMH module you want to build into a PPL.
2. Create new Library and name it PPL DQMH.lvlib.
3. Move the DQMH classes[71] from vi.lib into the PPL DQMH.lvlib.
4. Save all.
5. Create a blank project named DQMH PPL.lvproj.
6. Add to the project the PPL DQMH.lvlib.
7. Build PPL for PPL DQMH.lvlib by right-clicking on Build Specifications on the project explorer and select to create a New Packed Library. Fill out the different pages, and make sure you set the target build directory to the PPL directory your team uses and build.
8. Go back to the project on step 1 and right click on PPL DQMH.lvlib and select to Replace with a packed library.
9. Save all.
10. Make sure you move back the DQMH classes out of PPL DQMH.lvlib.

---

[70] Video of this process is available at GDevCon#1. Take Advantage of PPLs and Packages to Enhance LabVIEW Application Performances and Deployment. https://www.studiobods.com/en/gdevcon-1/
[71] Delacor_lib_QMH_Cloneable Module Admin.lvclass, Delacor_lib_QMH_Message Queue.lvclass, Delacor_lib_QMH_Module Admin.lvclass

Delacor is working on ways to make this process easier. By the time you read this book, there may be other ways to package DQMH libraries into PPLs. Check the DQMH release notes or the Delacor blog[72] for updates.

Actor Framework
Use Cases

Actor Framework is useful for applications that have multiple modules running independently and need to communicate between them. In Actor Framework, each of these VIs is represented by an actor that is responsible for a task within the application.[73]

It is tempting to go in the Actor Framework architecture and try to understand all the behind-the-scenes code. We believe that understanding what the different AF components are and how to use the AF first will make it easier to dive deeper into the behind-the-scenes code later.

What Is Actor Framework?

The Actor Framework is a software library that Stephen Loftus-Mercer and Allen Smith introduced at NIWeek 2010. The Actor Framework started as a LabVIEW community project and several members of the community contributed to improve it. The library now ships with LabVIEW. It includes a set of classes that implement state and message handling for the application.

The Actor Framework is based on the **Actor Model**, which Wikipedia defines as "…a mathematical model of concurrent computation that treats 'actors' as the universal primitives of concurrent digital computation. In response to a message that it receives, an actor can: make local decisions, create more actors, send more messages, and determine how to respond to the next message received. Actors may modify their own private state, but can only affect each other through messages (avoiding the need for any locks)."[74]

The **Actor Model** has three elements:[75]

1. **Actor:** An encapsulation of data, behaviors, and threads (in this chapter we call them processes).

2. **Message:** A unit of information passed between actors.

3. **Message transport:** A mechanism to move messages to the destination.

---

[72] http://delacor.com/blog/
[73] *Using the Actor Framework in LabVIEW*, Allen C. Smith and Stephen R. Mercer. https://forums.ni.com/t5/Actor-Framework-Documents/Using-the-Actor-Framework-in-LabVIEW-for-framework-version-3-0/ta-p/3504969
[74] Actor Model, Wikipedia. https://en.wikipedia.org/wiki/Actor_model
[75] *Fundamentals of Actor-Oriented Programming*, Dave Snyder, CLA Summit 2014, Austin. https://forums.ni.com/t5/Certified-LabVIEW-Architects/CLA-Summit-2014-Fundamentals-of-Actor-Oriented-Programming/ta-p/3502832

Actor principles[76] or axioms include the following:[77]

1. Actors can receive any message at any time. When an actor receives a message, it can create more actors, it can send messages to actors that it has addresses for, and it can designate what it will do with the next message it receives. An actor has no control over who has its address. Each actor sends messages on its own schedule. Actors must react to unexpected messages gracefully.

2. Actors are self-deterministic. Messages are always requests, never commands. Things are not done to an actor; an actor does things to itself.

"You don't get to chop the chicken's head, the Actors are very powerful chickens, the actor has to agree to chop its own head"—Carl Hewitt, creator of the Actor Model.[78]

Although I prefer Dave Snyder's paraphrased version "If a turkey were an actor, come Thanksgiving we would ask it to chop off its own head."[79]

From Queue Message Handler to Actor Core.vi

`Actor Core.vi` is in essence the MHL in a QMH. (See Fig. 5.68.)

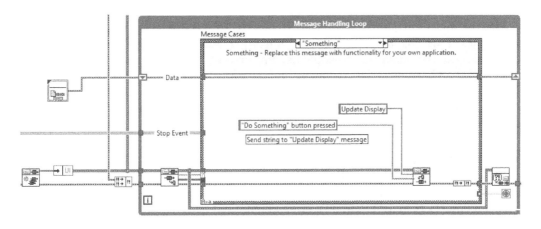

FIG. 5.68 *Message handling loop in the NI QMH Main.vi.*

---

[76] *Fundamentals of Actor-Oriented Programming*, Dave Snyder. CLA Summit 2014, Austin. https://forums.ni.com/t5/Certified-LabVIEW-Architects/CLA-Summit-2014-Fundamentals-of-Actor-Oriented-Programming/ta-p/3502832

[77] Hewitt, Meijer, and Szyperski: The Actor Model (Everything you Wanted to Know…) https://www.youtube.com/watch?v=7erJ1DV_Tlo

[78] Hewitt, Meijer, and Szyperski: The Actor Model (Everything you Wanted to Know…) https://www.youtube.com/watch?v=7erJ1DV_Tlo

[79] *Fundamentals of Actor-Oriented Programming*, Dave Snyder, CLA Summit 2014, Austin. https://forums.ni.com/t5/Certified-LabVIEW-Architects/CLA-Summit-2014-Fundamentals-of-Actor-Oriented-Programming/ta-p/3502832

The main components that you care about in this code are the **Data** cluster, **Queue**, **MHL**, and **Message Cases**. These components will transform into the different Actor Framework components.

QMH Component	AF Component	Notes
Data cluster	Actor class	State information is stored in the actor's private data.
Queue	AF message queue	The message classes wrap a priority queue.
Message handler	Actor core method	Actor.lvclass:Actor Core.vi has the MHL inside it. New messages are added by overriding classes and methods.
Messages	Message class	All messages in Actor Framework inherit from a common class. Each message has its own unique data.

From QMH to Actor Framework table.[80]

In the Object-Oriented chapter (see "Traditional LabVIEW Morphs into LVOOP" in Chapter 4), we talk about how to refactor existing task-based code into object-oriented code. We said a cluster would become an object, the enumerator or case selector would become a class hierarchy and the Case Structure would be replaced by dynamic dispatch. That is exactly what is happening with `Actor Core.vi`. In that chapter, we also talk about the command pattern that you might recognize here. (See Fig. 5.69.)

Actor Framework Library

The Actor Framework library builds on the QMH pattern. The Actor Framework improves the QMH pattern by making each actor reusable and addressing potential timing-related bugs in the communication scheme.

In order to be successful with Actor Framework, you need to have a good understanding of how a QMH[81] works and be comfortable with object-oriented programming (see Chapter 4 dedicated to this topic in this book).

---

[80] *Actor Framework Basics*, Derek E. Trepanier from Moore Good Ideas. https://www.mooregoodideas.com/actor-framework/basics/AF-basics-part-2/
[81] Queued Message Handler Template documentation. http://www.ni.com/tutorial/53391/en/

Chapter 5: Why Would You Want to Use a Framework? 357

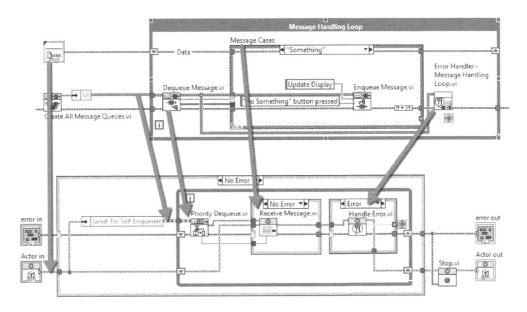

FIG. 5.69 *From QMH MHL to Actor Core.vi. The data cluster became the actor object. The QMH queue became the Send-To-Self Enqueuer queue. The Dequeue Message.vi became the Priority Dequeue.vi, and the message cases structure became the dynamic dispatch for the Do.vi wrapped by Receive Message.vi.*

The library has two parent classes that you will extend through inheritance (see Fig. 5.70):

- Actor: The state and data in a module
- Message: Information passed between actors to initiate changes in the actor's state
    - Stop Msg is a message child that the framework uses to shut down actors
    - Last Ack is a message child that the framework uses to notify when an actor has shut down

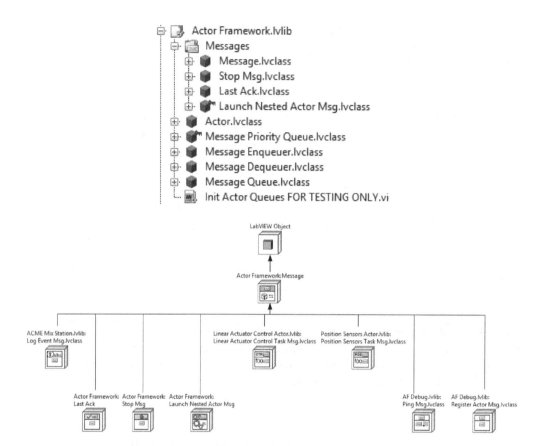

FIG. 5.70 *The Actor Framework Library includes the classes you extend through inheritance, Actor.lvclass and Message.lvclass.*

Actor Class

An actor is a LabVIEW object that represents the state of an independent task in your application. This task is continuously running.

This will be the class that you will use to extend the framework. If you used QMH in the past, you would determine how many message handlers you would have, and you would give each message handler a name and a task. When using Actor Framework, you will determine how many actors you will have, and you will give each one a name and a task. For example, you used to have a message handler to handle the acquisition, and now you will have an actor called acquisition. There are two public methods that the calling code can call: `Launch Actor.vi` (which is deprecated and it is there only for backwards compatibility) and `Launch Root Actor.vi`. You will extend your actor by overriding a couple of protected methods, mainly the `Actor Core.vi`. (See Fig. 5.71.)

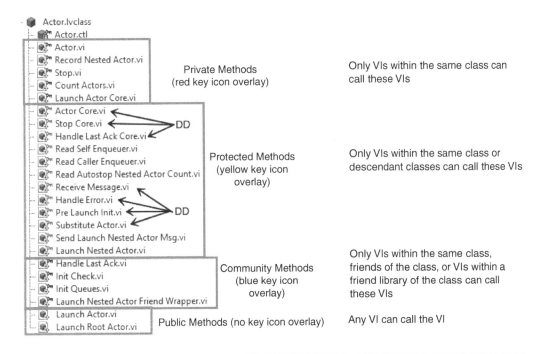

FIG. 5.71 *The Actor.lvclass methods. Notice the access scope for each method inside the Actor class. Your actors will be able to override all the VIs marked as DD in the image (for dynamic dispatch).*

Extending Actors via Inheritance

Since actors are classes, you can extend them by creating new children. In those children you can add methods that will be configured for dynamic dispatch, and their children can choose to override those methods. For example, you can create an actor that is a `Report Generator.lvclass`, and it has a method that is called `Report Generator.lvclass:Generate Report.vi`. This VI takes a string input with the contents for the report. You can then create a child of the Report Generator Actor that is called `HTML Report Generator.lvclass` and overrides `Generate Report.vi`. The new `HTML Report Generator.lvclass:Generate Report.vi` knows how to generate a report using the HTML format. In the future you could extend your application by adding new

Report Generator actors without needing to build your executable again or touching any existing code. If you go back to the object-oriented chapter (see Chapter 4) and look at the *SOLID* principles, you can see how you can take into consideration the Open-Closed and the Liskov Substitution principles when designing your Actor Framework–based application. You can design your parent actors to be open for extension via children but closed for any additional modifications. Also, you can design the actors to make sure their children can keep the promises the parent actors make.

Actor Framework Message Class

A message is an object that represents the information passed between actors to initiate an action by the receiving actor. Every message an actor can receive inherits from the message class. When the actor handles the message in `Actor Core.vi` (similar to how a QMH handles a message in the MHL), the actor will call the `Do.vi` method for that message. When you will create new messages for your actors, you will be overriding that `Do.vi`. In essence, you will be writing the code you used to write inside the MHL case for that message, but now it will be in the `Do.vi` for that message.

Extending Messages via Inheritance

Since messages are classes, you can extend them by creating new children. Notice that the framework already comes with three message children: `Launch Nested Actor Msg.lvclass`, `Stop Msg.lvclass`, and `Last Ack.lvclass`. (See Fig. 5.72.)

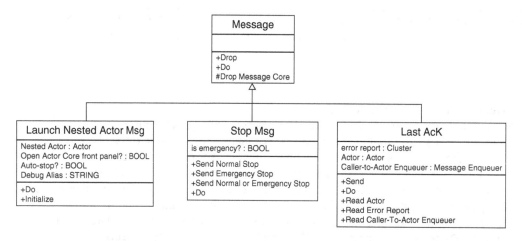

FIG. 5.72 *UML diagram showing the inheritance relationships for the Message class. Launch Nested Actor Msg* **is a** *message, Stop Msg* **is a** *message, and Last Ack* **is a** *message. Message classes are not related to the Actor class. They have an association or* **use a** *relationship.*

Actor Framework Message Queue Class

Actor Framework uses queues for all the communication within the actor itself and between actors. These queues are encapsulated by classes (enqueuer and dequeuer). This ensures that the code calling the Actor API cannot dequeue from the actor's queue and cannot destroy the queue. The calling code can only enqueue new messages. Your actors will be sending their enqueuers to other code that wishes to send messages to the actor. The dequeuer is used by the framework, and you don't need to worry about it, just like when you were writing QMH code and you just knew the dequeuer was there in the MHL and you never copied that dequeue to anywhere else in your code.

The message queue is a priority queue. The messages can have the following priorities:

- Low
- Normal → this is the default and, for the majority of cases, all you need
- High
- Critical → only used by the framework

The Actor Framework Message Maker tool will take care of creating the `Send Message` methods for you, and those will already include the code to enqueue the message. Typically you will not have to worry about using the enqueue method directly.

Actor Framework Helper Loops

By now you understand that the `Actor Core.vi` is in essence the MHL in a QMH. If you need more than just the MHL, for example, if you need an EHL, you will need to add a helper loop that will handle events. Actors are headless by default, which means they do not have a user interface. If you want to have a user interface on your actor, override `Actor Core.vi` and add an EHL. It is your responsibility to provide a method to stop that EHL. You can use the same method that has been used for years to stop the EHL in a QMH; use a local Stop Event. (See Fig. 5.73.)

Oli Wachno wrote about how the DQMH events could be used to provide a DQMH wrapper that would stop the helper loop.[82] More recently, Sam Taggart and Allen Smith, taking inspiration from DQMH templates and DQMH event creation tools, created additional tools to create actors from templates and user events to

---

[82] *DQMH Actor Framework Interface* by Oli Wachno. https://delacor.com/dqmh-actor-framework-interface/

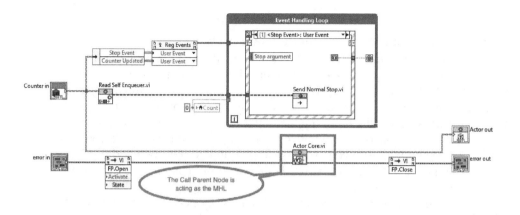

FIG. 5.73 *Counter.lvclass:Actor Core.vi includes an EHL, which is a form of helper loop. This loop handles user interface events as well as the Stop and Counter Updated events that you declared in the Counter.lvclass:Pre Launch Init.vi. The Call Parent Node is acting as the MHL.*

communicate with the actor core UI indicators.[83] Their tool adds an event to the actor's private data, adds the event creation to the `Pre Launch Init.vi`, and adds the code to destroy the event and the event handling case to the EHL in `Actor Core.vi`.

Another use for helper loops in `Actor Core.vi` is to implement repetitive tasks, such as the repetitive reading of a DAQ device. Just like in any QMH, you can also use them to delegate tasks that take a long time, so you don't hold a stage in the MHL hostage. The longer your code takes to process that MHL case, the longer it will get to address any other items in the queue. A good approach is to have the actor receive its message through the regular means and then decide if it is something that needs to be delegated to a helper loop or a separate actor.

The main thing to remember is that you are responsible for deciding when to add a helper loop, starting and stopping the helper loops, and determining how to communicate between the actor loop and the loop you added. You can use `Actor.lvclass:Read Self Enqueuer.vi` to message the actor core from the helper loop (in the example earlier the `Read Self Enqueuer.vi` was used to provide the enqueuer to the `Send Normal Stop.vi` inside the Stop Event User Event case).

---

[83] *Events for UI Actor Indicators* by Sam Taggart. https://automatedenver.com/events-for-ui-actor-indicators/

Launching Actors

Launching an actor means starting up the MHL for that actor. During the launching process, the framework starts the `Actor.vi` and creates the message queue references. Actors are capable of launching other actors. The first actor is the **root-level actor**. Any other actors the code launches from then on are **nested actors**. Nested actors can launch other nested actors. The calling code will get the new actor's enqueuer and will use that to send messages to the actor. (See Figs. 5.74 and 5.75.)

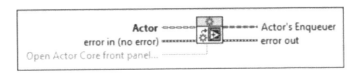

FIG. 5.74 *Actor.lvclass:Launch Root Actor.vi launches an asynchronously running VI that performs tasks and handles messages for the actor. This VI returns a reference to an enqueuer that your code can use to send messages to the newly launched actor.*[84]

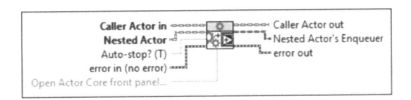

FIG. 5.75 *Actor.lvclass:Launch Nested Actor.vi launches an asynchronously running VI that performs tasks and handles messages for the nested actor. Use this VI to launch actors that are dependent on one or more calling actors. This VI returns a reference to the enqueuer that you can use to send messages to the newly launched actor.*[85]

---

[84] LabVIEW Help for Actor:Launch Root Actor VI. http://zone.ni.com/reference/en-XX/help/371361R-01/lvcomm/af_launch_root_actor/

[85] LabVIEW Help for Actor:Launch Nested Actor VI. http://zone.ni.com/reference/en-XX/help/371361R-01/lvcomm/af_launch_nested_actor/

The calling code can launch an actor at different states in the application, and the code can decide at runtime how many actors of the same type to launch. These means that your application no longer has to wait for all the code to start when the application starts. Your code can start the different actors if and when they are needed.

In general, you will be launching nested actors inside the `Actor Core.vi` of other actors. Do not launch nested actors inside `Pre Launch Init.vi`.

Sending Messages to and between Actors

Your code will use `Message Enqueuer.lvclass:Enqueue.vi` to communicate with the actor it launches. The Actor Framework scripting tools will create all the code you need to send messages.

If your actor needs to send a message to itself, it can reference its own message queue using the `Actor.lvclass:Read Self Enqueuer.vi`. (See Fig. 5.76.)

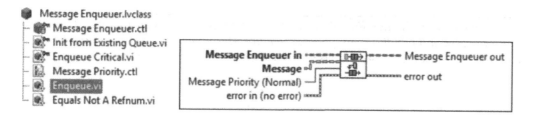

FIG. 5.76 *Message Enqueuer.lvclass:Enqueue.vi sends a message to the actor.*

When an actor launches a **nested actor**, it will communicate with it using the **nested actor's enqueuer**. If the nested actor needs to send a message back up the actor tree, it can get the calling actor's messaging queue using the `Actor.lvclass:Read Caller Enqueuer.vi`.

It is important to note that Actor Framework encourages communication through the actor tree. The root actor sends messages to its nested actors, those nested actors send messages to their nested actors, and so forth. For example, if Actor A launches Actor B and Actor B launches Actor C, then Actor A can only communicate with Actor C by sending a message to B first. Along the same lines Actor C can only communicate with Actor A by sending a message to B first. (See Fig. 5.77.)

Remember we talked about you as a LabVIEW programmer would follow a contract with the framework you chose? You commit to implementing the communication of your actors in a way that the actor tree communication is respected. You will find creative ways to send messages from different branches on the actor tree to other actors. For your own good, don't do it, no matter how tempted you are! If you think you know better and understand the consequences, go ahead, but don't say we didn't warn you!

Chapter 5: Why Would You Want to Use a Framework? 365

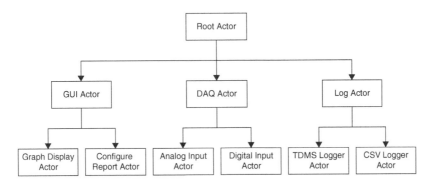

FIG. 5.77 *An example of an actor tree. If the Configure Report Actor needs to send a message to the Log Actor, it should send the message to the GUI Actor, the GUI Actor would relay it to the Root Actor, and Root Actor would relay it to the Log Actor.*

Stopping Actors

The calling code can tell the actor to stop using the `Stop Msg.lvclass:Send Normal Stop.vi`, `Stop Msg.lvclass:Send Emergency Stop.vi`, and `Stop Msg.lvclass:Send Normal or Emergency Stop.vi`. These messages will tell the actor's MHL to stop. Remember that it is your responsibility to stop any helper loops that you started. The `Stop Core.vi` is a good place to stop that helper loop. The actor executes `Stop Core.vi` when it receives the Stop message. (See Fig. 5.78.)

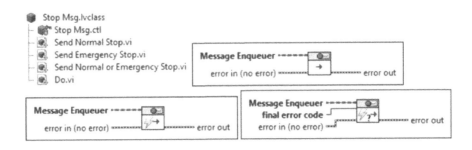

FIG. 5.78 *Use the Stop Msg.lvclass to stop the actor either with a normal stop or an emergency stop. The Emergency Stop message tells the actor to shut down as quickly as possible. This message has critical priority, and the framework will process it before all messages already in the queue. The Normal Stop message tells the actor to run its Stop Core method. This message has normal priority, and the framework will process it after all the high- or normal-priority messages that are already in the queue. The Normal or Emergency Stop message determines if the message is an emergency or not based on the error code.*

Note that the `Launch Nested Actor.vi` has an **Auto-stop?** boolean input. If this input is set to true (which is the default value), then when the calling actor stops, it automatically sends a stop message to the nested actor.

How to Use Actor Framework
Simple Code Launching and Sending Messages to an Actor (actor doesn't communicate back)

We looked around for the simplest example we could find of nonactor code calling an actor. We found the examples from Stefan Lemmens[86] very informative, and even though his most basic example consists of more than one actor, we still recommend you check them out. Stephen Loftus-Mercer published a pair of good examples[87] of how to call an actor tree from non actor applications. These examples use the deprecated `Launch Actor.vi` method and again use more than one actor. As another simple example, we created a Counter actor, just to show how you would start an actor, send messages to it, and stop it. There is no communication back from the actor to the calling code.[88]

The key things to note here are that the calling code is pretty simple. All you need to use this actor are `Launch Root Actor .vi` and the specific `Send Message` VIs. In this case, `Send Increment.vi`, `Send Decrement.vi`, and `Send Reset.vi`. (See Figs. 5.79 to 5.81.)

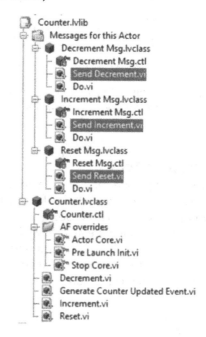

FIG. 5.79 *You only care about the Counter.lvclass and the Send <Message> VIs to communicate with the actor.*

Exploring, Debugging, and Troubleshooting Simple Code Calling One Actor

We have said that our actors are the same as QMHs. In a QMH, you would go to the MHL to find the different cases where your code is being executed. In this case, you know the name of the case being executed because it is part of the VI name

---

[86] Actor Framework from Basic to PPL Plugins by Stefan Lemmens. https://forums.ni.com/t5/Actor-Framework-Documents/Actor-Framework-from-basic-to-PPL-plugins/ta-p/3776945
[87] Calling Actors from Non-Actor Code: Two Ways by AristosQueue. https://forums.ni.com/t5/Actor-Framework-Discussions/Calling-Actors-From-Non-Actor-Code-Two-Ways/td-p/3417363
[88] This example is available at the GitHub.com/LGP5/Frameworks repository, under the branch "AF-ActorCalledByNonActor."

Chapter 5: Why Would You Want to Use a Framework? **367**

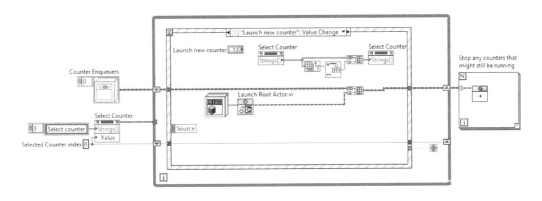

FIG. 5.80 *To launch a new counter, all you have to do is call Launch Root Actor.vi with your Counter actor object as an argument. You keep an array of the counter enqueuers, so you are able to send messages to the different counters.*

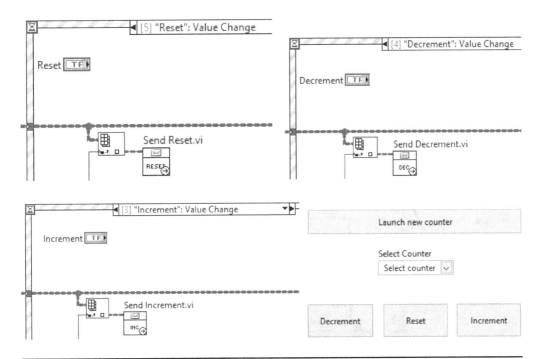

FIG. 5.81 *The different messages have their own Send <message name>.vi, and you call them when you need to perform each operation. The Select Counter ring control is there to facilitate selecting which counter you are communicating with.*

Send <message>.vi. So if you want to explore what the code is doing in the increment case, you would look for the Increment Msg.lvclass:Do.vi. This is equivalent to selecting the case called "Increment" on your MHL Case Structure. (See Fig. 5.82.)

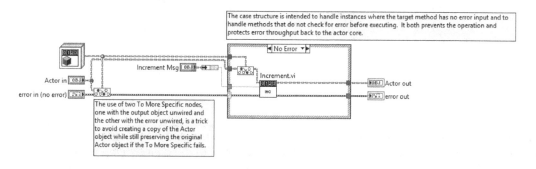

FIG. 5.82 *The code inside the Increment Mgs.lvclass:Do.vi is executing the Increment.vi, which is a method for the counter class.*

If you are already comfortable and familiar with QMHs, then Fig. 5.83 will help you see how the different sections of the counter QMH diagram are now converted into specific components of the counter actor code.[89]

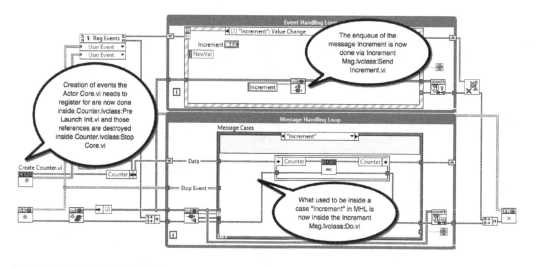

FIG. 5.83 *The block diagram is of the Counter QMH.lvproj >> Main.vi in GitHub. The bubble notes refer to the counter actor example also in GitHub.*

---

[89] A version of the counter QMH code is available at the GitHub.com/LGP5/Frameworks repository, under the branch "AF-ActorCalledByNonActor."

You can use some of the debugging and troubleshooting tools you are already familiar with. If you need to break the code when the message is being handled, put the breakpoint inside the `Do.vi` for the message you are interested in.

Actor Framework Project Template

You probably don't want a single actor and probably want actors that can send information back. The best way to get the full benefit out of actor communication is to have actors call actors and have them communicate among them. When an actor is launched by another actor, it has access to the launcher's queue and can send messages through that queue. This is one of the reasons some LabVIEW programmers call the Actor Framework contagious. Once you create an actor, you are tempted to make the calling code an actor and so on.

The Actor Framework project template illustrates how to create an Actor Framework–based application where multiple tasks need to operate at the same time. It also includes features that you would want in applications, like a splash screen that is fast to launch while the code is loading and a demonstration on how to read from a configuration file.

Create an Actor Framework project template by clicking on **Create Project** from the **Getting Started window** in LabVIEW and then selecting the Actor Framework project template. (See Fig. 5.84.)

FIG. 5.84 *Select Templates >> Actor Framework to create an Actor Framework Template from the Create Project menu in the Getting Started window in LabVIEW.*

Choose a name for the Alpha actor that is descriptive of what that actor will do. In this case, you will choose Linear Actuator Control and select the text you want to see on the VI icons for this actor. Then choose a name for the Beta actor and select the text for its icon—in this case Position Sensors. (See Fig. 5.85.)

Actor Framework Project Structure and Class Hierarchy

When exploring an Actor Framework project for the first time, there are two elements you can use to orient yourself. You need to look at both the project structure and the class hierarchies for both the actors and their messages. Hopefully the project followed the general project structure where actors have a library that contains the `<Actor Name>.lvclass` and the **Messages for <Actor Name>** virtual folder. Find one of those classes in the project, right-click on it, and select **Show Class Hierarchy**. You can expand from there and explore which ones are your actors, which will

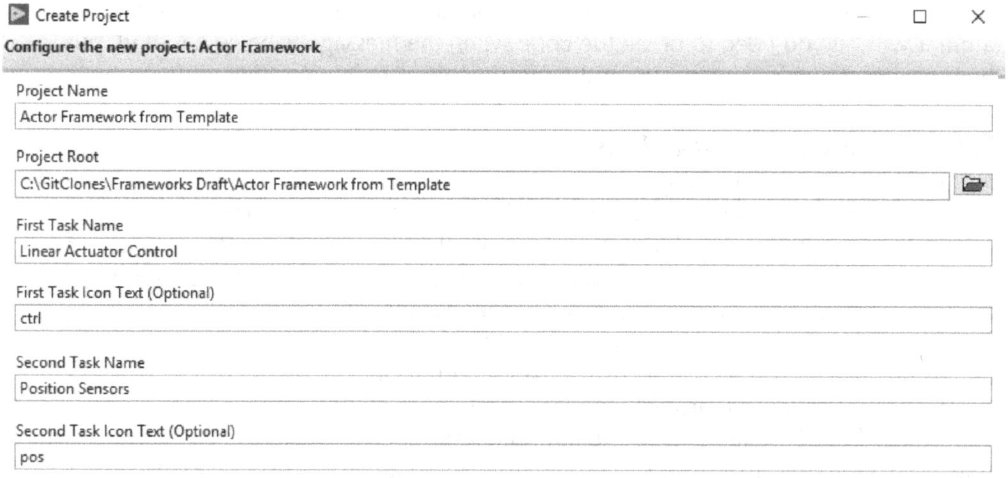

![Create Project dialog screenshot]

Project Name
Actor Framework from Template

Project Root
C:\GitClones\Frameworks Draft\Actor Framework from Template

First Task Name
Linear Actuator Control

First Task Icon Text (Optional)
ctrl

Second Task Name
Position Sensors

Second Task Icon Text (Optional)
pos

FIG. 5.85 *Create Actor Framework project template wizard. You selected to use ACME mix station as the project name and replaced the Alpha actor with Linear Actuator Control and the Beta actor with Position Sensors.*

inherit from the `Actor Framework:Actor`, and which ones are the messages for your actors, which will inherit from `Actor Framework:Message`. One way to find the top-level VI for this application is to open the build specification and look for the **Startup VI**. In this case, it is the `Splash Screen.vi`. (See Figs. 5.86 to 5.88.)

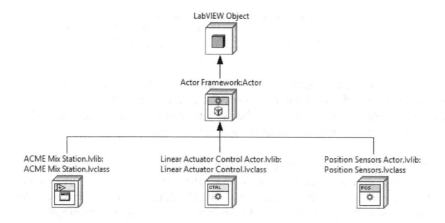

FIG. 5.86 *Actor class hierarchy for the Actor Framework project created from the template. You replaced the project name with ACME Mix Station, Alpha actor with Linear Actuator Control, and Beta actor with Position Sensors Actor.*

Chapter 5: Why Would You Want to Use a Framework? 371

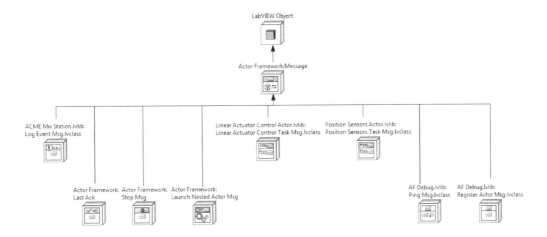

FIG. 5.87 *Actor messages class hierarchy for the Actor Framework project created from the template. There are three messages from the Actor Framework that you use in this project (Last Ack Msg, Stop Msg, and Launch Nested Actor Msg), two Actor Framework debug messages (Ping Msg and Register Actor Msg), and three messages for the actors that the template added for you (Log Event Msg, Linear Actuator Control Task Msg, and Position Sensors Task Msg).*

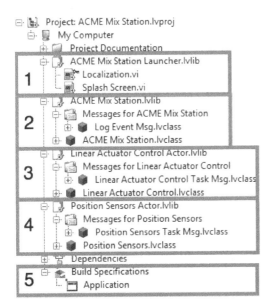

1. Main application library. The Splash Screen.vi is your top-level VI and starts the entire application.

2. ACME Mix Station is one of your actors. It contains a class with the actor's state and methods and one message, Log Event Msg.

3. Linear Actuator Control Actor is one of your actors. It contains a class with the actor's state and methods and one message, Linear Actuator Control Task Msg.

4. Position Sensors Actor is one of your actors. It contains a class with the actor's state and methods and one message, Position Sensors Task Msg.

5. Build specification for an executable.

FIG. 5.88 *Project created from Actor Framework project template where you replaced the ALPHA ACTOR with the LINEAR ACTUATOR ACTOR and the BETA ACTOR with the POSITION SENSORS ACTOR.*

Exploring, Debugging, and Troubleshooting the Actor Framework Project Template
Let's start by verifying that the build specification for the executable works out of the box. In the project explorer, expand on the **Build Specifications** and right-click on the **Application** build specification and select to **Build**. (See Fig. 5.89).

FIG. 5.89 *Right-click on the Application build specification and select Build to build the executable for this sample project.*

Once the Application Builder is done building your executable, click on the **Explore** button. This will take you to the location on disk where the executable is located. Double-click on Application.exe to run the application.

This application doesn't do much, but it includes things that you probably would want to see in all your applications. It contains a splash screen that gives something to the end user to look at while the modules load. The application name on the window matches your application actor's name without the VI extension. You can also verify that you can send individual messages to each actor and send messages to both actors via the Actions menu. To stop the executable, close the window. Just like any other Windows application you use, closing the window stops the application.

Let's explore how you find which actor is your root actor. Granted, most of the time you will know this based on the actor name or knowing what is the first UI you see. One option is to look for all the instances of Actor.lvclass:Launch Root Actor.vi. Another option is to start from the top-level VI. In this case when you open the Splash Screen.vi, you will see that the magic happens at the lower level where the Call By Reference is. This Call By Reference is calling the ACME Mix Station.lvclass:Load App.vi. (See Figs. 5.90 and 5.91.)

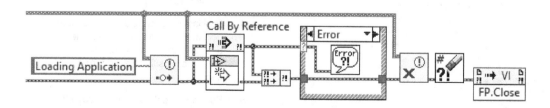

FIG. 5.90 *Explore the Splash Screen.vi. The code that launches the first actor is at the bottom of the screen, where it calls Load App.vi. Double-click on the VI inside the Call By Reference node, and it will take you there.*

Chapter 5: Why Would You Want to Use a Framework? **373**

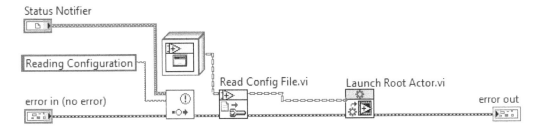

FIG. 5.91 *Sure enough, inside Load App.vi, there it is, Launch Root Actor.vi.*

This will confirm that your root actor is ACME Mix Station.lvclass. You can follow what you know about Actor Framework to explore more. Opening ACME Mix Station.lvclass:Actor Core.vi will show you what else is going on besides the MHL. The first couple of times you have to debug or explore one of these projects, it may be a daunting task, but remember that the Call Parent Actor Core.vi that is in the middle is your MHL, with an EHL above it. It is just a QMH. If you wanted to see what is happening inside each of the cases of the MHL, you would go to the ACME Mix Station.lvlib in the project and expand the ACME Mix Station.lvclass. Look at all its methods. The only method that has a message associated with it is the ACME Mix Station.lvclass:Log Event.vi, and the associated message is Log Event Msg.lvclass. You need to look at both the Log Event Msg.lvclass:Do.vi and the ACME Mix .Station.lvclass:Log Event.vi block diagrams to understand what is going on in the only MHL case for this actor. (See Figs. 5.92 and 5.93.)

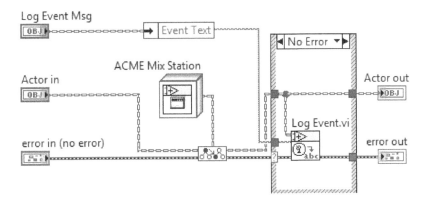

FIG. 5.92 *Log Event Msg.lvclass:Do.vi code calls the Log Event.vi method for the ACME Mix Station actor.*

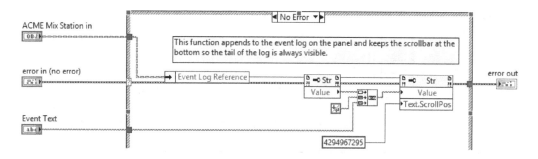

FIG. 5.93 *All the ACME Mix Station.lvclass:Log Event.vi is doing is writing directly a string to the UI of ACME Mix Station.lvclass:Actor Core.vi.*

It is common for Actor Framework code to use the references to indicators on the UI and communicate directly via VI Server property nodes. We do not like this approach because it makes debugging harder later. We know you may think the method name is enough to indicate that the only possible location where the string indicator is changing is the `Log Event.vi`, but another developer could decide to update that same indicator inside another method. We recommend instead using user events and wrapping the generation of the event inside a VI. This is to make it a lot easier to find all the places where a particular indicator is changing. You can use the tool that Sam Taggart and Allen Smith created inspired from DQMH event creation tools.[90]

Follow the same approach to explore the `Actor Core.vi` for your other two actors and the messages they have. Pay special attention as to how both actors accomplish sending a periodic heartbeat message to their caller. One does it by sending the `Log Event` message to the caller via an EHL helper loop, and the other uses the Advanced Actor Framework function `Time-Delayed Send Message.vi`, which is the equivalent to periodically enqueuing in the Actor MHL message. (See Figs. 5.94 and 5.95.)

You can also go to the Project Explorer menu to **View >> Bookmark Manager**. You will see a window with all of the **#CodeNeeded** and **#CodeRecommended** bookmarks. Explore these bookmarks by double-clicking on them. They will guide you through all the changes that you can make to the code.

Actors are supposed to be a simple unit of code that is easy to test, but you need to call a lot of code to set up a tester for an actor. Sam Taggart got tired of this, and based on his experience with DQMH and its API testers, created an actor tester.[91]

---

[90] Events for UI Actor Indicators by Sam Taggart. https://automatedenver.com/events-for-ui-actor-indicators/

[91] DQMH API Tester Like for Actor Framework by Samuel Taggart http://bit.ly/aftester

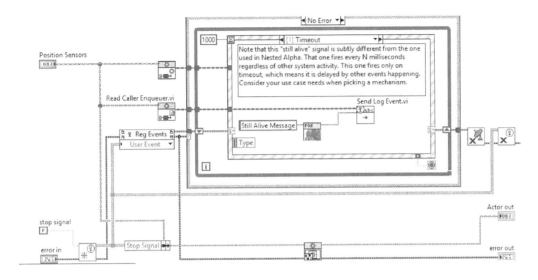

FIG. 5.94 *Example of an Actor Core.vi using a helper loop with an Event Structure sending a periodic message every second via the Timeout Event case, in this case, sending the Log Event message to the ACME Mix Station.*

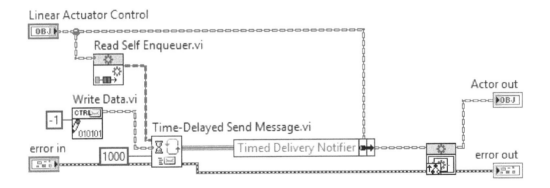

FIG. 5.95 *Example of an Actor Core.vi using Time-Delayed Send Message.vi (this VI is part of the advanced function palette for the Actor Framework). This is equivalent to having a helper loop that is continuously enqueuing the message. In this case the message is assembled via Linear Actuator Control Task Msg.lvclass:Write Data.vi.*

With so many VIs running in parallel and not having a single block diagram to put your breakpoints, debugging and troubleshooting actors can be a real pain. Derek Trepanier, from Moore Good Ideas, created the MGI Monitored Actor Toolkit to aid in this process.[92] Ravi Beniwal and Tim Vargo created a LabVIEW Task Manager that can help you debug and see all the processes that you have running in parallel. This is an open-source community project, so reach out to them if you would like to collaborate.[93] Before leaving NI, Allen Smith created the Actor Framework Debug library that allows monitoring actors via the Desktop Execution Trace Toolkit.[94]

Actor Framework Fundamentals Example (or how to create an AF coffee shop)

Matt Pollock and the Actor Framework LabVIEW community created an example of how asynchronous messaging works using a coffee shop. You can find the background discussions in the forum.[95] Matt based the example on an article that discussed the two-phase commit.[96] Eventually, the example was rolled into LabVIEW, and now you can find it via the shipping examples. Just look for Actor Framework Fundamentals.lvproj.[97]

Exploring, Debugging, and Troubleshooting the Actor Framework Coffee Shop

When you open the project, notice the Readme.html file. It explains not only the actor hierarchy but also how the coffee shop and messages are supposed to work, so be sure to read through it first. By now you know how to identify the top-level VI, so find the one that is calling the Launch Root Actor.vi. In this case this is the Launcher.vi. Run the VI and have fun creating customers, cashiers, and baristas and ordering different coffee drinks. Matt has a great sense of humor, so if you press the stop button before the order is ready, you will get a message informing you that the customer left before the order was ready and the staff is drinking it instead!

Use all the tips and techniques explained earlier in the simple counter and the Actor Framework project template discussion sections of this chapter.

---

[92] MGI Monitored Actor by Moore Good Ideas, Inc. is available via the LabVIEW Tools Network or by searching for Monitored Actor in VIPM
[93] LabVIEW Task Manager by Ravi Beniwal and Tim Vargo https://lavag.org/files/file/245-labview-task-manager-lvtm/
[94] https://forums.ni.com/t5/Actor-Framework-Documents/Desktop-Execution-Trace-Toolkit-Support-for-Actor-Framework/ta-p/3527740
[95] Asynchronous Messaging explained as a coffee shop https://forums.ni.com/t5/Actor-Framework-Discussions/Asynchronous-Messaging-explained-as-a-coffee-shop/td-p/3439197
[96] Your Coffee Shop Doesn't Use Two-Phase Commit by Gregor Hohpe https://www.enterpriseintegrationpatterns.com/docs/IEEE_Software_Design_2PC.pdf
[97] C:\Program Files\National Instruments\LabVIEW 2018\examples\Design Patterns\Actor Framework

Actor Framework Feedback Evaporative Cooler Sample Project Template

You are ready to explore the Evaporative Cooler sample project. If you try it out and start to get lost, try exploring Stefan Lemmens'[98] examples first. Stefan's examples start with two actors and go all the way to building actors into PPLs.

Create your own copy of the Evaporative Cooler Sample project by clicking **Create Project** from the **Getting Started window** in LabVIEW, and then clicking the Evaporative Cooler Project Template. (See Fig. 5.96.)

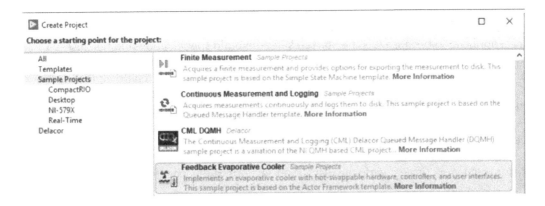

FIG. 5.96 *Select Sample Projects >> Feedback Evaporative Cooler.*

Exploring, Debugging, and Troubleshooting the Evaporative Cooler

Let's start by verifying that the build specification for the executable works out of the box. In the project explorer expand on the Build Specifications and right-click on the **Cooler Application** build specification and select **Build**. (See Fig. 5.97.)

FIG. 5.97 *Right-click on the Cooler Application build specification and select Build to build the executable for this sample project.*

---

[98] Actor Framework from Basic to PPL Plugins by Stefan Lemmens https://forums.ni.com/t5/Actor-Framework-Documents/Actor-Framework-from-basic-to-PPL-plugins/ta-p/3776945

Once the Application Builder is done building your executable, click on the **Explore** button. This will take you to the location in the disk where the executable is. Double-click on `Cooler.exe` to run the application.

This application does a lot more than the project template. It implements an evaporative cooler that has the capability of changing the hardware, controllers, and user interfaces while the application is running. The actors in this project are the user interface, the cooler, the fans, and the water level. When you run the application, you will see that the cooler tries to maintain the desired temperature and that you can change the UI on the bottom.

By now, you should have a general idea of how the application works. Stop the executable and go back to the project in LabVIEW. Run the `Application Launcher.lvlib:Splash Screen.vi` (which you can determine is the top-level application because it is the **Startup VI** in the build specification and because it is the one calling `Launch Root Actor.vi`). Change the **Desired Temperature** and observe how the fans and water pump activate to maintain the temperature. If you want to control the simulation, open the `Simulated Evaporative Cooler.lvclass:Simulated Data_Global.vi` and adjust all the simulation inputs and then watch the reaction on the main application. You can find more details in the project documentation.

Use all the tips and techniques explained earlier in the simple counter and the Actor Framework project template discussion sections of this chapter.

Adding a New Actor

If you used QMHs before, your design and programming steps were probably something like this:

1. Determine how many Queued Message Handlers you need and what each one will have as a task.
2. Determine the contents of the Data cluster.
3. Determine which messages each MHL will be consuming.
4. Determine any specific actions during initialization.
5. Determine any specific actions during the shutdown.

You will be doing something similar with the Actor Framework. The previous list becomes this:

1. Determine how many actors you need and what each one will have as a responsibility.
    a. Each actor will be a child of `Actor.lvclass`.
2. Determine the **state data** for each actor.
    a. This will be the contents of the private data for each actor you create.

Chapter 5: Why Would You Want to Use a Framework? **379**

3. Determine which **behaviors** each actor will have.

    a. These behaviors will be the base for the messages each actor will be consuming.

    b. You will use the Actor Framework scripting tools to create a new class for each message that the actor will use for others to execute the actor's behaviors.

4. Determine any specific actions during initialization.

5. Determine any specific actions during the shutdown.

Let's create an Acquisition actor and a User interface actor that will launch it. Starting from the Acquisition actor, follow the steps. It is always important to design before coding, but with Actor Framework applications, it is absolutely required. If you try the code-and-fix approach, you will quickly find yourself deep into a hard-to-debug mess. Also, know in advance that you will be coding several pieces before you can actually run and test any code. Our recommendation is to get the main messages working from two actors and your launcher set up so you can troubleshoot and test as you go. Also, try Sam Taggard's tool[99] to create DQMH-like API testers for your actors if you think it will help, but first go ahead and write a couple of actors without it, so you can see the full benefit of that tool and understand its use.

Some of the tools you can use during design is the creation of UML diagrams, not only class diagrams to describe your actors with their attributes and methods but also UML sequence diagrams. Sequence diagrams allow you to describe the order of messages and their effect on the actors in your application. (See Fig. 5.98.)

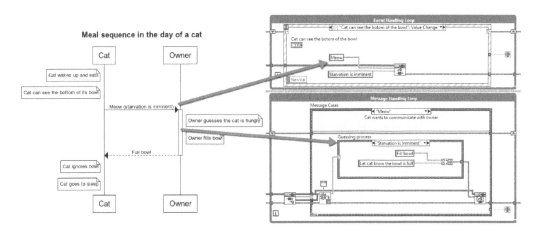

FIG. 5.98 *An UML sequence diagram showing the sequence of actions between a cat and their owner.*

---

[99] bit.ly/aftester

In addition, always keep your actor tree diagram up to date. Remember that actors only communicate with the actor that launched them and with their nested actors. If one actor needs to send a message to a different actor, the message needs to be relayed through the actor tree.

1. You will have two actors:
   a. UI actor
   b. Acquisition actor
2. Private data.
   a. The UI actor doesn't have private data
   b. The Acquisition actor will keep the HW ID as part of its private data
3. Behaviors.
   a. The UI actor will need to
      i. Load the application.
      ii. Display the acquired data.
   b. The Acquisition actor will need to
      i. Configure hardware.
      ii. Initialize hardware references.
      iii. Stop acquisition.
      iv. Close hardware references.
      v. A helper loop to continuously acquire and send the Display message to the UI actor and an option to wake up and put the helper loop to sleep. You could also use a `Timed-Delay Send Message.vi` that functions as a message pump and enqueues the same message over and over at a regular interval. To see an example of this, check out the Actor Framework project template. One of the actor cores for the nested actors uses this approach.
4. Initialization.
   a. The UI actor will need to initialize the UI.
   b. The Acquisition actor will need
      i. An event to wake up the helper loop.
      ii. An event to put the helper loop to sleep.

5. Shutdown.

    a. The Acquisition actor needs to close the hardware references.

Steps to Add a New Actor
To add the Acquisition actor to your project, follow these steps:

1. Create a blank project (you could also start on an existing project where you wanted to add this module).

2. Right-click on My Computer and select **New >> Actor**. (See Fig. 5.99.)

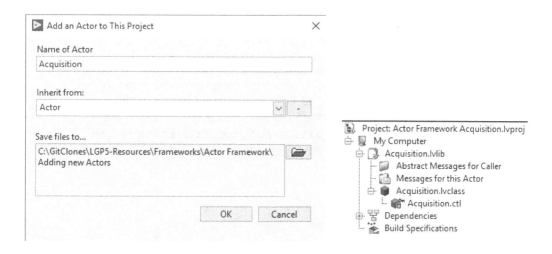

FIG. 5.99 *Right-click on My Computer and select to add a new actor; when the wizard prompts you, name your actor Acquisition. The Actor Framework scripting tools will add an Acquisition.lvlib to your project that will have already the virtual folders for messages and the Acquisition.lvclass that already inherits from Actor.lvclass.*

3. In this step implement the behaviors for the Acquisition actor. For now you can use the functions that we used for the DAQ device in the CML project template. You don't need to get lost into the DAQmx calls and learning how to add an actor at the same time. Also note that at this stage, you could create a quick test VI to ensure all your methods work by themselves. (See Fig. 5.100.)

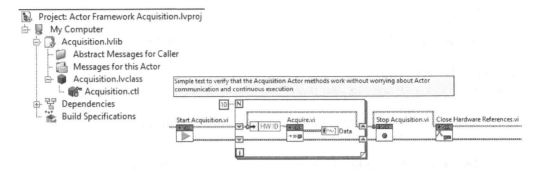

FIG. 5.100 *You added the Acquisition methods to the Acquisition actor and a simple test to verify that those methods work. At this stage, you don't worry about the actor communication or the continuous execution.*

4. You have working methods now, and your actor is defined in terms of state and behaviors. The next step is to create the messages so other actors can communicate with this actor. The Actor Framework scripting tools come to the rescue. The two methods you know you want to call from other actors are the `Start Acquisition` and `Stop Acquisition`. Select both methods, right-click on them, and select **Create Actor Messages**. (See Fig. 5.101.)

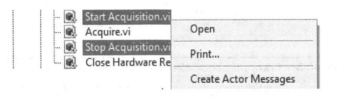

FIG. 5.101 *Right-click on the methods you want to execute by sending a message from another actor and select Create Actor Messages.*

5. You had already decided that you would implement a helper loop. You need to override the `Actor.lvclass:Actor Core.vi` for you to add the helper loop to do the acquisition. You had also decided that you would use events to wake up, put to sleep, and exit the helper loop. You need to override the `Actor.lvclass:Pre Launch Init.vi` and the `Actor.lvclass:Stop Core.vi` for you to add the creation and destruction of the helper loop events. Create a virtual folder called **AF Overrides** (this is you being kind to future you and making it easier to find in this actor class where the Actor Framework Overrides are). (See Fig. 5.103.)

Chapter 5: Why Would You Want to Use a Framework? **383**

FIG. 5.102 *The Actor Framework scripting tools created both the Send and the Do VIs for you. There is nothing else you need to do here. Other actors can now send those messages, and they will automatically tell the actor to execute its own implementation of the Start Acquisition.vi and Stop Acquisition.vi.*

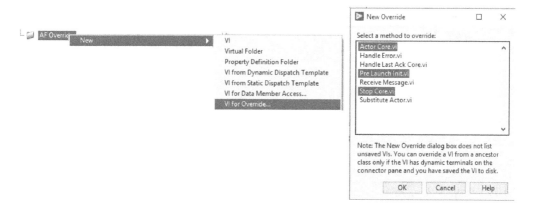

FIG. 5.103 *Right-click on the AF Overrides virtual folder and select to override Actor Core.vi, Pre Launch Init, and Stop Core.vi.*

6. You can wrap the creation, destruction, and firing of the helper loop events in subVIs. It makes it easier to find where these actions are taking place when they are wrapped in subVIs. It is easier to find all the places where a subVI is being called than to do a text search for the event name to find where it is being unbundled. Add the creation of events to `Acquisition .lvclass:Pre Launch Init.vi` (see Fig. 5.104), the destruction of events to `Acquisition.lvclass:Stop Core.vi` (see Fig. 5.105) (also remember to fire the exit helper loop before destroying the event), and the firing of the different events to the `Acquisition.lvclass:Start Acquisition.vi` and the `Acquisition.lvclass:Stop Acquisition.vi`.

FIG. 5.104 *Acquisition.lvclass:Pre Launch Init.vi.*

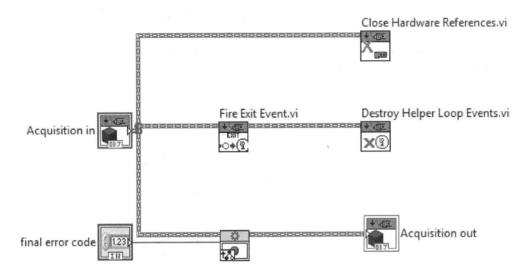

FIG. 5.105 *Acquisition.lvclass:Stop Core.vi.*

7. You are ready to create the new Acquisition UI actor, so you can have another actor to call the Acquisition actor. The Acquisition actor is headless since it does not have a UI. Separating the UI from this actor would make it easier later to create a new UI that would inherit from this Acquisition UI actor and would only override the Actor Core.vi to represent a new view, but the rest of the code would remain the same. Follow step 2 to create the Acquisition UI actor.

8. Add the only method this actor will have, Display.vi. Create your helper loop events, destroy them, and override Actor.lvclass:Pre Launch Init.vi and Actor.lvclass:Stop Core.vi. These two VIs will look very similar to the ones for Acquisition.lvclass. Acquisition UI.lvclass:Actore Core.vi will look a little different; you will be using the display event to update the graph indicator. (See Figs. 5.107 and 5.108.)

Chapter 5: Why Would You Want to Use a Framework? **385**

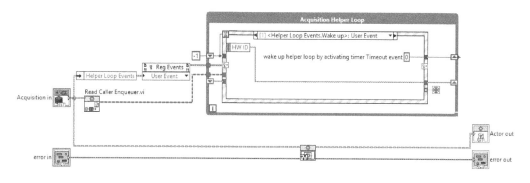

FIG. 5.106 *Acquisition.lvclass:Actor Core.vi.*

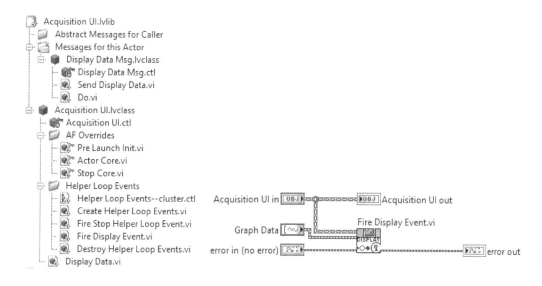

FIG. 5.107 *Acquisition UI actor has a Display Data message. The Display method only fires the display event for the helper loop inside actor core to act on.*

9. You can now add calling the `Acquisition UI.lvlib:Display Data Msg.lvclass:Send Display Data.vi` to `Acquisition .lvclass:Core.vi` when it acquires the latest data. (See Fig. 5.109.)

10. Finally, create the `Launcher.vi`, and you are ready to test your Acquisition—Acquisition UI actors pair. (See Fig. 5.110.)

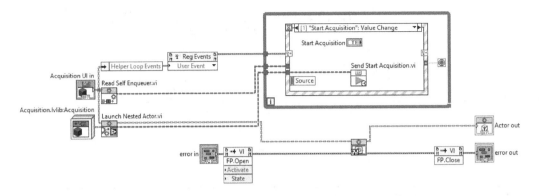

FIG. 5.108 *Acquisition UI.lvclass:Actor Core.vi sends Start Acquisition and Stop Acquisition based on end-user interaction. When it gets the display event, it will update the graph.*

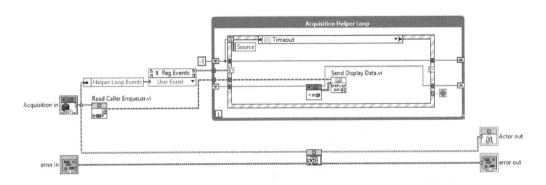

FIG. 5.109 *Acquisition.lvclass:Actor Core.vi calls Display Data Msg.lvclass:Send Display Data.vi with the latest acquired data within the timeout event case.*

FIG. 5.110 *Launcher.vi.*

Exploring, Debugging, and Troubleshooting an Actor

Include some visualization for your actor tree in your documentation and keep it up to date. When you only have two simple actors like the ones here, it might seem useless, but you can appreciate how this can start to get complicated as you add more actors. We cannot emphasize enough how important it is to design your actors before you start coding. Make sure you test your actor methods before you start creating the messages. Organization of your project is very important, and the organization within your actor library too. Following good practices will make it easier for you and others to explore, troubleshoot, and extend your project.

Use all the tips and techniques explained earlier in the simple counter and the Actor Framework project template discussion sections of this chapter.

Actor Framework: Behind the Scenes
Interprocess Communication

Actor Framework uses queues for communication. Each actor has its own queue that drives its MHL and decides what message to execute next. A root actor does not have a caller queue because it is the top-most actor in the actor communication tree. Nested actors have their own queue as well as access to their caller's queue. The Actor Framework provides a full API to use these queues and takes care of managing their creation, destruction, and dequeuing behind the scenes. All that developers care about is determining if they are using the Read Self Enqueuer .vi or the Read Caller Enqueuer.vi to obtain the queue to send a message.

Queues in Actor Framework have priority, and the framework takes care of enqueuing at the appropriate end of the messages depending on the priority chosen. The priorities available to developers are low, normal, and high. There is an internal-use-only critical priority.

The premise of the Actor Framework is that each actor should only worry about its own messages and the messages it sends to its caller. Actor Framework frowns upon actors communicating with other actors that did not launch it or that it did launch them.

You can use other communication methods such as notifiers and user events, but you are responsible of the creation, generation, and destruction of messages. These types of communication are fine when used to communicate with helper loops inside the actor core. Use extreme caution if you decide to use any of these methods of communication to communicate with other actors. This goes against respecting the communication via the actor tree.

Initialization

The Actor Framework template and some of the other examples you can find in the Actor Framework community tend to have a VI called Load App.vi. The Splash Screen.vi calls this VI to ensure the end user knows something is going on while all the libraries load into memory. The main consideration is to decide what will be the root actor for your application. This also makes it easier for other developers to find which actor in a project is the root actor. (See Fig. 5.111.)

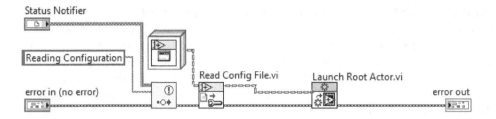

FIG. 5.111 *Load App.vi is in charge of reading the configuration for this application and calling Launch Root Actor.vi.*

Actor Framework provides a way to give initialization information to an actor before launching it. In this case, the `Read Config File.vi` ensures the My Application Actor sent to `Launch Root Actor.vi` has any initialization information as part of its private data. My Application Actor contains, as part of its private data, the actor queues for its nested actors. Note that after the actor launches, its private data can only be modified by the respective message.

The `Launch Root Actor.vi` calls the `Launch Actor Core.vi` but with no enqueuer (see Fig. 5.112). This is because by definition a root actor does not

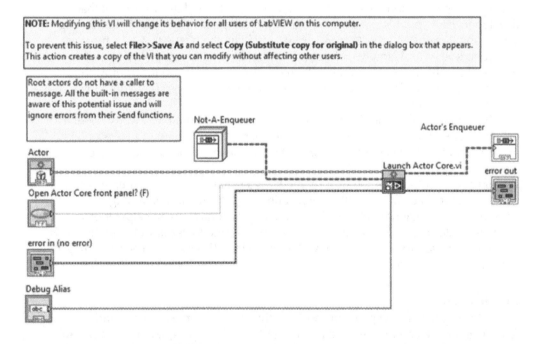

FIG. 5.112 *Launch Root Actor.vi ensures it launches the My Application Actor as the starting point for the task tree. Launch Root Actor.vi calls Launch Actor Core.vi but does not provide a caller enqueuer because the root actor is the top-most actor in the actor communication tree.*

have a caller actor and therefore does not need to communicate with its calling code via an actor queue. This is the main reason we differentiate between the `Launch Root Actor.vi` and `Launch Nested Actor.vi`. (See Fig. 5.113.)

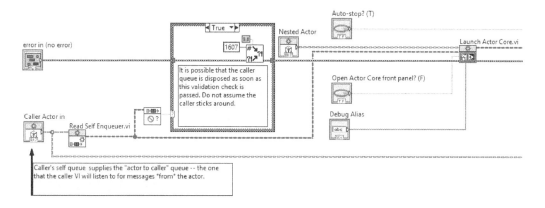

FIG. 5.113 *Detail of Launch Nested Actor.vi where the code obtains the caller actor's queue. This is what allows the nested actors to use the Read Caller Enqueuer.vi later to communicate with their caller.*

`Launch Nested Actor.vi` also calls `Launch Actor Core.vi`, but it provides a caller queue. This is how nested actors get access to their calling actor queue and can use the `Read Caller Enqueuer.vi` to communicate with their caller easily.

Even though we mentioned it earlier, it is worth repeating to not launch nested actors inside `Pre Launch Init.vi`.

Stop Module Gracefully

Actor Framework uses the same principle used in the Producer Consumer template. It destroys the actor queue to generate an error in the dequeue, and this error stops the actor. This works, but we prefer to have an explicit message that asks the modules to stop. (See Figs. 5.114 and 5.115.)

FIG. 5.114 *Actor Framework.lvlib:Stop Msg.lvclass:Send Normal Stop.vi.*

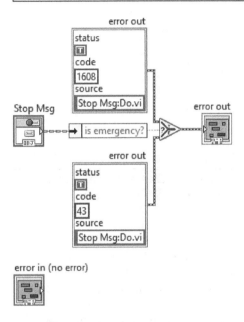

FIG. 5.115 *If you look at the Actor Framework.lvlib:Stop Msg.lvclass:Do.vi, you will confirm that the method to stop the actor is to generate an error.*

Error Handling Strategy

By default, an actor's MHL will produce an error when it receives a message object that it does not recognize. By default, all errors stop the actor. If you need a more graceful and tolerant behavior, you need to override the `Actor.lvclass:Handle error.vi`.

Remember that the Actor Framework uses error 43 to stop an actor. It is important to know this because although it is extremely rare, you could end up with an error 43 that you did not intend to stop your actor! The most common cause is a cancelled file dialog.[100]

---

[100] Delacor has a custom VI analyzer test that detects when the developer is not handling the error output of the File Dialog express VI because error 43 can create lots of issues and it is easy to prevent.

Also, be aware of this issue when debugging with the Desktop Execution Trace Toolkit, where error 43 in the actor module will be displayed upon the intended stop. (See Fig. 5.116.)

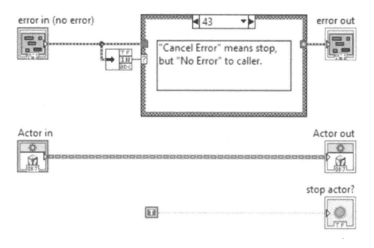

FIG. 5.116 *Actor.lvclass:Handle Error.vi assumes that error 43 means that the actor needs to stop and the caller doesn't need to know about this error. If your actor can produce error 43 and the caller needs to know about it, you need to address this before the actor gets to handle the error (right after the message in MHL is handled). The other case is the default case and just lets the error pass through.*

Other Actor Framework Tools
Debugging Tools for the Actor Framework
We already talked about several tools throughout this chapter. There are several tools to aid with troubleshooting and debugging.

Actor Framework Debug Library for Desktop Execution Trace Toolkit
Actor Framework Debug library that allows the monitoring of actors via the Desktop Execution Trace Toolkit.[101]

MGI Monitored Actor
Derek Trepanier from Moore Good Ideas created the MGI Monitored Actor Toolkit to aid in the troubleshooting and debugging process.[102]

---

[101] https://forums.ni.com/t5/Actor-Framework-Documents/Desktop-Execution-Trace-Toolkit-Support-for-Actor-Framework/ta-p/3527740
[102] MGI Monitored Actor by Moore Good Ideas, Inc., is available via the LabVIEW Tools Network or by searching for Monitored Actor in VIPM.

LabVIEW Task Manager

Ravi Beniwal and Tim Vargo created a LabVIEW Task Manager that can help you debug and see all the processes that you have running in parallel. This is an open-source community project. Reach out to them if you would like to collaborate.[103]

Actor Framework Tester

Sam Taggart developed a tool to create DQMH-like API testers for the Actor Framework.[104]

MGI Panel Manager

Derek Trepanier from Moore Good Ideas created the MGI Panel Manager[105] and included Actor Framework templates for using MGI Panel Manager. MGI Panel Manager itself is built using the Actor Framework.

Advanced Actor Framework

We have only scratched a little beyond the Actor Framework fundamentals in this chapter. We recommend you take the NI Actor Framework training course to learn some of the more advanced concepts. One of the key concepts that you need to be very comfortable with is zero-coupling actors. You need to make sure you understand this concept before you embark in your first large project with Actor Framework. It will save you lots of headaches, and it will make it a lot easier for you to reuse your actors in other projects.

You want to create zero-coupling actors because that makes your children actors interchangeable, encourages you to create actors that are cohesive, and improves the application modularity. The disadvantages of implementing zero coupling are that it takes more effort to implement and you have to be very familiar and comfortable with object-oriented programming and the Actor Framework. If you have read the previous sections and you don't feel comfortable creating a simple CML type of application, don't move on to understanding zero coupling.

The main concept behind zero- or low-coupling messaging is **abstract messages**. These messages determine the API of the message but do nothing. The application decides which actual implementation to use at runtime. The power of this is that you could have created your Acquisition module with an abstract message to Start Acquisition and Stop Acquisition, and it would be at runtime that the Acquisition UI would decide if it would be using the simulated Acquisition module we created here or one that implements DAQmx acquisition, or another actor that implements a third-party acquisition module. The key here is that the Acquisition UI actor could be saved on its own without a fixed dependency to on the Acquisition actor.

---

[103] LabVIEW Task Manager by Ravi Beniwal and Tim Vargo. https://lavag.org/files/file/245-labview-task-manager-lvtm/
[104] DQMH-Like API tester for Actor Framework by Samuel Taggart. http://bit.ly/aftester
[105] MGI Panel Manager. https://www.mooregoodideas.com/products/panel-manager/index.html

Oli Wachno gave a presentation about abstract messaging[106] that uses a simple example with beer. You will go through his example. Imagine you are implementing the control system for filling up a glass of beer. You can either select between bottled or from tap. If you don't drink, that is OK; let's say it is nonalcoholic beer.

You would probably determine both your actor hierarchy and actor messages hierarchy as Oli did for his low-coupling example. (See Fig. 5.117.)

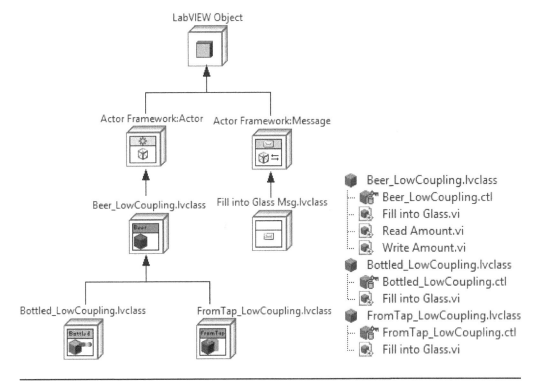

FIG. 5.117 *Class hierarchy for the low-coupling approach for the actors and messaging. In this case you have three actors: Beer, Bottled, and From tap. There is only one message called Fill into Glass Msg. Each actor has their own implementation for the Fill into Glass .vi method.*

A simple VI launching the From Tap and Bottled actors would just launch each actor as the root actor and send the appropriate message to fill into the glass for each actor. (See Figs. 5.118 to 5.120.)

---

[106] Actor Framework—Usage of Abstract Messages by Oliver Wachno, German VIP Days (October 22, 2015). https://forums.ni.com/t5/Actor-Framework-Documents/Abstract-Messaging/ta-p/3529929

394 LabVIEW Graphical Programming

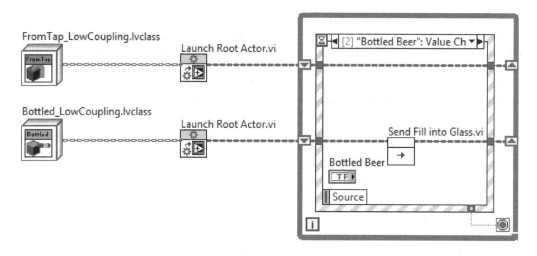

FIG. 5.118 *The calling code launches both actors, and depending on the button pressed, it sends the Fill into Glass message to the appropriate Actor Core.*

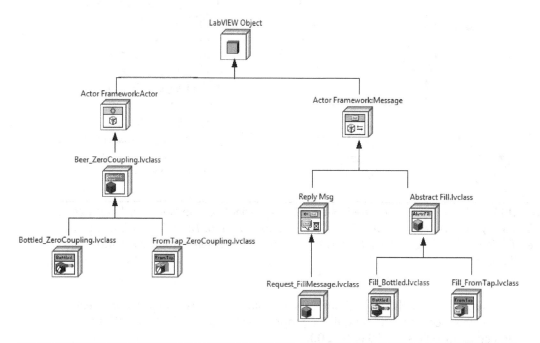

FIG. 5.119 *Class hierarchy for the zero-coupling approach for the actors and messaging. In this case you still have three actors: Beer, Bottled, and From tap. However, now you have more messages. There is a Request Fill message that you will use to define which implementation of the Abstract Fill class you will implement.*

Chapter 5: Why Would You Want to Use a Framework? **395**

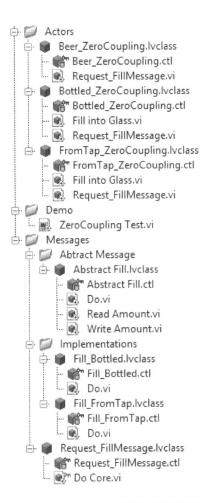

FIG. 5.120 *You had to implement more methods. The parent Beer_ZeroCoupling no longer has implementations of Fill into Glass, Read Amount, or Write Amount. Instead, Beer_ZeroCoupling only has an implementation of Request_FillMessage method, which you will use to define at the children which specific Fill message implementation to use.*

Up to this point, it is all review for you, and this code should be pretty straightforward. Now look at the zero-coupling approach.

The calling code is a little more complex. (See Fig. 5.121 to Fig. 5.124.)

The actual implementation of the Fill message is the typical message that you are already used to.

396 LabVIEW Graphical Programming

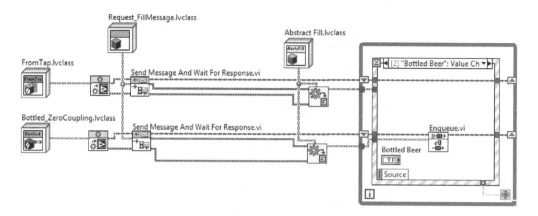

FIG. 5.121 *The calling code launches both actors, but before going into servicing the user interface, it determines what implementation of the Fill method will be used. The Request_FillMessage is serving as the translation step. By using the Send Message and Wait for Response.vi, a synchronous communication between the calling code and the actor is established. Synchronous communication is usually not desired, due to potential issues with blocking the execution. This is why regular messaging between actors is implemented as asynchronous communication. Yet in this use case, synchronous communication can be used without a negative outcome.*

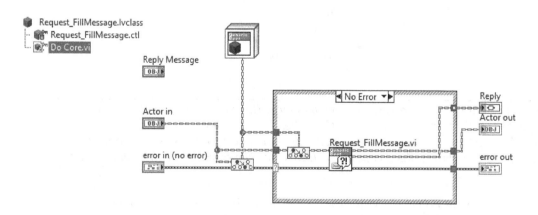

FIG. 5.122 *The code inside Request_FillMessage.lvclass:Do Core.vi is calling the Request_FillMessage.vi method for each actor.*

Chapter 5: Why Would You Want to Use a Framework? **397**

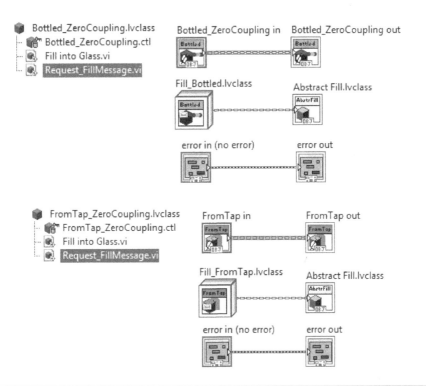

FIG. 5.123 *All the Request_FillMessage.vi method for each actor is doing is translating which implementation of the Fill message each actor should use. Here is where the decoupling happens. You could change your mind on what message implementation the actor would use in the future, and this would be the only VI to change.*

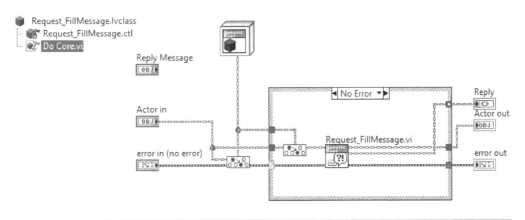

FIG. 5.124 *Fill_Bottled.lvclass:Do.vi is just calling Fill into Glass.vi for the given actor.*

You can find a more detailed tutorial on zero-coupled actors by DMurrayIRL in the Actor Framework community.[107] This chapter should give you all the vocabulary you need and the basis to continue your education in Actor Framework. Visit the online community to expand your knowledge and ask questions.[108]

Other topics to search about are **State Pattern Actor**, which is useful in applications that need actors to act sequentially. If you want to do network communications, take a look at the **End Point Actors**.

Sharing Reusable Actors

For small applications, you can have all the actors reside in the same project. This would be the case in the Actor Framework project template and the Evaporative Cooler Sample project template. If your application starts to grow into the tens of actors or you want to have multiple developers working in parallel in separate actors, you would be better off putting those actors in separate projects, or even separate repositories. This has the added advantage of making those actors easier to share with other projects and other areas of your organization.

Configuring Source Code Control Repository Dependencies

There are no special considerations for the Actor Framework for this approach. The generic approach we described earlier for frameworks in general applies to Actor Framework-based projects.

Packaging Actors Using VIPM

We suggest that when you build VIPM packages, you add a prefix to your project libraries in the package specification. This will help you differentiate source code and code installed via VIPM.

Packaging Actors in PPLs

To package the Actor Framework libraries into PPLs, you will need to do the following:

1. Open a blank project.
2. Right-click on My Computer and select **Add Actor Framework**.
3. Note that only the core Actor Framework classes are included. If your code depends on additional Actor Framework classes such as Reply Msg, Report Error Msg, Self-Addressed Msg, and Batch Msg, you need to include them in the library you will build into PPL.

---

[107] Beginner Tutorial: Zero-Coupled Actor Project by DMurrayIRL. https://forums.ni.com/t5/Actor-Framework-Documents/Beginner-Tutorial-Zero-Coupled-Actor-Project/ta-p/3533889
[108] Online Actor Framework Community. http://ni.com/actorframework

4. Build PPL for `Actor Framework.lvlib` by right-clicking on Build Specifications on the project explorer and select to **Create a New Packed Library**. Fill out the different pages, and make sure you set the target build directory to the PPL directory your team uses and build.

5. Open the project that has the actor you want to package into PPL and right-click on `Actor Framework.lvlib` and select to **Replace with a Packed Library…**.

6. Click Save All.

7. Make sure you move out any classes that you added to the Actor Framework library in step 3.

8. Any new actors that you will package this way should inherit from the PPL packaged actor.

References

http://delacor.com/tips-and-tricks-for-a-successful-dqmh-based-project/
https://www.merriam-webster.com
   https://www.merriam-webster.com/dictionary/abstraction
   https://www.merriam-webster.com/dictionary/frame%20of%20reference
   https://www.merriam-webster.com/dictionary/framework
The course "Design Your Own Framework," Fabiola De la Cueva, Jarobit Pina Saez Bilbao, Spain, February 2018.
Presentation "Choosing a Framework. How to Pick the Right Framework to Meet Your Needs," Samuel Taggart, 2018. http://automatedenver.com/developer-series-presentation/
DQMH API tester like for Actor Framework by Samuel Taggart. http://bit.ly/aftester
Actor Tester code by Samuel Taggart. https://gitlab.com/stagg54/Actor_Tester
https://forums.ni.com/t5/Delacor-Toolkits-Discussions/Zuehlke-Project-explorer-menu-for-DQMH/m-p/3851045/highlight/true#M707
DQMH videos. bit.ly/DelacorQMH
Queued Message Handler Template Documentation. http://www.ni.com/tutorial/53391/en/
Application Design Patterns: Producer/Consumer. http://www.ni.com/white-paper/3023/en/
DQMH Documentation. https://delacor.com/documentation/
DQMH online forum. http://forums.ni.com/t5/Delacor-Toolkits/ct-p/7004
DQMH Getting Started Videos. bit.ly/DelacorQMH
DQMH Best Practices. https://delacor.com/dqmh-documentation/dqmh-best-practices/

Creating Custom Events LabVIEW Help entry. http://zone.ni.com/reference/en-XX/help/371361N-01/lvhowto/creating_user_events/

Community Nugget 4/08/2007 Action Engines by Ben Rayner. https://forums.ni.com/t5/LabVIEW/Community-Nugget-4-08-2007-Action-Engines/td-p/503801

Actor Framework Basics by Derek E. Trepanier from Moore Good Ideas. https://www.mooregoodideas.com/actor-framework/basics/AF-basics-part-1/

https://forums.ni.com/t5/Actor-Framework-Discussions/How-to-build-a-PPL-including-Actor-classes-messages-and-AF-Debug/td-p/3871512

https://forums.ni.com/t5/Actor-Framework-Documents/PPL-Support-in-Actor-Framework-Project-Provider/ta-p/3527818

CHAPTER 6

Unit Testing

Some LabVIEW developers focus on test and measurement applications and are familiar with testing the different components of a larger system, for example, testing all the different parts that make up a phone. Why would it be different for software? We need to ensure that all the different parts that make up our application work well on their own before we put them together. If you spend any time reading about unit testing, you will eventually find the testing pyramid (Fig. 6.1).

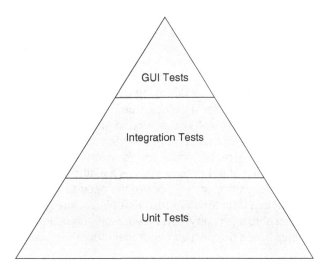

FIG. 6.1 *Testing pyramid.*

The different application modules are tested at the unit level. Then the unit-tested modules are integrated to form the complete application and interact with hardware and other applications. This is called integration testing. Finally, the GUI tests are also known as acceptance tests. This is when the whole system is ready for delivery and the end user verifies the system meets their expectations and has all the requested features.

In this chapter, we look at unit testing in LabVIEW as a practice for creating modular, testable code. LabVIEW makes it easy to test a VI's functionality: just put in some values on the front panel controls, push the run arrow, and check that the outputs match what you expect. For a more complicated VI, you might even write a new VI to test the VI you are developing over a range of inputs. You might even name this VI "`Test X.vi`." Do you ever run those test VIs again? Many times we get to support legacy code, and when we find a folder called "Tests" or "Testers," our excitement goes down as soon as we open the folder and find several stale test VIs. Most of us don't run those testers again unless there is an issue, and by then the test VI may not function because of other changes. Unit testing improves on this by making it easy to run all your tests frequently so they don't go stale. Unit testing is a design practice to help you write robust, testable code that works as designed. Unit tests are also a form of insurance in that when you change the code in the future, you can verify that everything else is still working, or you will know if something else broke (see regression testing defined later in this chapter). Unit tests also serve as working documentation of your code. Code that runs does not lie; stale comments might.

Before we start writing unit tests in LabVIEW, let's go over some concepts and definitions.

What Is Unit Testing?

The common definition of a unit is that it is the smallest testable part of an application. LabVIEW programmers tend to assume this means a VI. Tools like Unit Test Framework from National Instruments (NI) reinforce that assumption. This assumption leads to programmers thinking that all their VIs need to be tested. A more practical definition is to see a unit as a **unit of work**.[1] A unit of work can be a collection of VIs, the public API of a class or a library, or even a collection of libraries. You will then focus on creating unit tests for the critical units of work in your application and not on creating a unit test for every single VI in your code.

Another way of thinking of a unit of work is as a feature. For example, an experienced LabVIEW programmer may not see the need to create a unit test for code that lists the files in a directory. We trust that NI has already tested all the primitives and VIs that ship with LabVIEW. However, if the feature is to create a catalog of instruments based on the files in a folder, now there are more details to

---

[1] R. Osherove, *The Art of Unit Testing*, Second Edition. Shelter Island, NY: Manning Publications Co., 2015.

take into consideration to classify a file as an instrument for that catalog, such as the file name format, file extension, and file contents. All of a sudden creating a unit test makes more sense. LabVIEW knows how to list files in a directory, but it does not know anything about our catalog and its characteristics. We need to test not only that our feature returns all the instrument files in a folder, but also that it knows how to handle files that do not fit the expected format. More importantly, if in the future our definition of an instrument file changes or a new instrument file is added, we already have a test suite to verify that our changes to the code to fit the new instrument do not break the code for the existing instrument types.

Unit testing ensures the unit behaves as expected, and for a given set of inputs, the outputs match the requirements. They focus on software only to keep them fast and simple (if you want to verify you can talk to an instrument, this is considered an integration test). For example, instead of writing a unit test to verify that your device driver talks to an instrument correctly, develop unit tests to verify that each command and response is parsed correctly and all analysis routines are correct.

Unit Testing as Code Documentation

Over the course of a project, you and your team will create a large body of unit tests. These tests make up a working documentation for the API. If you or another developer wants to know how to use a function, she only has to look at the tests for functional examples. This is especially important for those little one-off projects that keep coming back for new features. A project developed with unit tests has predefined test cases to demonstrate the APIs and validate the code, making it much quicker to add a new feature without breaking existing features. Designing with unit tests saves time and money.[2] Fabiola likes to point out that hardware is designed with test points in mind, so it makes sense to do the same for software.

What Makes a Good Unit Test?

Properties of a good unit test:[3]

- Automated and repeatable
- Easy to implement
- Relevant in the future

---

[2] CLA Summit 2014 at CERN, "Save Time and Money with Unit Testing." Video available at https://youtu.be/4TW0dLLioaI

[3] Roy Osherove, *The Art of Unit Testing*, Second Edition. Shelter Island, NY: Manning Publications Co., 2015.

- Easy to read
- Easy to run
- Runs quickly
- Consistent results
- Full control of the unit under test
- Fully isolated from the rest of the code
- Failures should be obvious and easy to debug

Well-written unit tests isolate a unit of work from its dependencies and provide consistent results. A complete set of unit tests should easily exercise and validate every aspect of a unit's behavior. Tests become an act of design. Writing the VI from the point of view of the caller leads to a stronger, easier-to-use, testable API. The unit test becomes the first user of your code, and you get to experience the good/bad API decisions you made and therefore provide a better developer experience to other developers who will use your API. They will have an insight into your state of mind and intentions when the API and unit tests were created.

What Are Assertions?

An assertion, in software engineering, is a statement that evaluates if a given statement is always true. A developer places assertions directly in the code only for use during development, with the idea that assertions are compiled out of the code for deployment. We remove the assertions for an executable to remove the performance impact the assertions could have on the code.

Assertions are used to verify that the code is operating as expected and to document the intended preconditions and post-conditions for a function. When an assertion fails, it should get the developer's attention immediately, and it should emphatically assert that the program is not executing as expected. As such, assertions can cause errors or even terminate the program. Assertions are different than error handling because their purpose is not to gracefully respond to a failure, but just to report it.

The main difference between an assertion and a unit test is that the unit test executes an **isolated section** of the application. The developer has to purposely run the unit tests, where the assertion is always running when in development mode and verifies code compliance during **whole** application execution. (See Table 6.1.)

Assertions	Unit Tests	Error Handling
Use to find hidden bugs during execution.	Use to verify isolated individual code components work as expected.	Handles expected execution failures.
Verify that code preconditions and post-conditions execute as expected. Verify that failure modes never occur.	Test both expected conditions and failure modes. Verify that code works as expected in both cases.	Gracefully handles code failure modes that should not occur but could occur.
Only execute in development mode. Not part of the executable (compiled out).	Execute an isolated part of the application. Not part of the application.	An integral part of the application, available both in development as well as in executable.
Execute when the developer runs code in development mode.	The developer needs to purposely execute unit tests. Can be automated via continuous integration tools.	Always execute when the code runs both in development and executable mode.
Report failures immediately without regard to graceful handling.	Reports failures only when unit tests are executed.	Gracefully handles failures.

TABLE 6.1 *Assertions, unit tests, and error handling comparison table.*

Writing VIs with Testing in Mind

The same good coding practices that we have discussed elsewhere in this book apply here. Specifically, make your VIs perform only one function (highly cohesive code) and avoid unnecessary dependencies with hardware and other modules (loosely coupled code).

The inputs and outputs should be basic data types. Pay extra attention when using clusters, since the more elements there are in a cluster, the more likely that cluster is going to be changing through the life of the project. Please note that we are not saying to never use clusters for your VI inputs; we are saying that if you do have a cluster as an input, this cluster should have only elements that are relevant to that VI. Avoid the "clustersaurus Rex"[4] approach, where a cluster has hundreds of elements and yet the VI only needs a handful of elements from that cluster. This is

---
[4] De la Cueva, Fabiola. *Improving Developer Experience (DX) as a Path to Building Better LabVIEW Applications.* Presented at CLA Summit 2016, Austin, TX. Minute 20:34. https://youtu.be/UZ4d1xuYYvQ?t=1234

important because part of the unit test definition is to define that for a given set of inputs certain outputs are obtained; the more elements in your input, the more elements you have to define.

Test Harness versus Automated Test Frameworks

A test harness is a set of functions and data the developer uses to test a program unit and verify that it works as expected. The automated test framework runs the test harnesses and reports the results. The test framework also provides tools for the developer to create test harnesses.

We used functions instead of VIs for the definition earlier because in LabVIEW we have testing frameworks that implement the test harnesses directly using VIs, such as JKI VI Tester, InstaCoverage, and Caraya. However, the NI Unit Test Framework (UTF) implements the test harnesses by configuring the inputs and expected outputs for those inputs on a table.

What Are the Automated Test Frameworks Available for LabVIEW?

There are several frameworks available to perform unit tests in LabVIEW. JKI has provided the JKI VI Tester and the Caraya Unit Test Framework. You can find these as packages on the LabVIEW Tools Network and the source code on GitHub.com/JKISoftware. More recently, the highly motivated and innovative team at IncQuery Labs has created InstaCoverage, also available on the LabVIEW Tools Network, available for both LabVIEW and LabVIEW NXG. NI UTF is an add-on toolkit included with the professional version of LabVIEW. If you are in a regulated industry, you may want to use NI UTF or InstaCoverage, since both tools are created by ISO 9001–certified companies. Keep in mind that there are no plans for UTF to be available for LabVIEW NXG. The code for JKI Tools is available for review, which may be enough to get through the hurdle of using those tools in a regulated environment. The two tools that provide code coverage metrics are UTF and InstaCoverage. See Tables 6.5 and 6.6 for tool comparisons.

If you are interested in using assertions within your code to verify that it is executing as expected, both Caraya and Peter Horn's LabVIEW Assert API support this approach. The Assert API is available at GitHub.com/PeteHorn/AssertAPI.

We'll demonstrate all of these tools and have examples in our GitHub repository at GitHub.com/LGP5.

What Is the Difference Between Black Box and White Box Testing?

The two types of unit testing are the black box and the white box. When you write a black box unit test, you don't know how the code is written, only that for a certain input, a specific output is expected. Black box unit testing is the starting point for test-driven development. The black box is also commonly used for acceptance tests. When you write a white box unit test, you are aware of how the code is written and the test itself is written to exercise a specific section of code. Code coverage is a reference to how much of the VI runs while the unit tests execute. Code coverage is measured in percentage of source code exercised during tests. The assumption is that the greater percentage of code is exercised during testing, the less opportunity for undetected bugs in the code. Some regulated industries have requirements for minimum code coverage percentages for different application risk levels.

What Is TDD?

For a long time TDD stood for test-driven development. Recently more and more people call it test-driven design. When creating code with testing in mind, you need to put a lot more emphasis in the design. Both interpretations lead to the same result: code that is easy to test.

The unit testing mantra is "test early, test often." This is especially true if requirements change regularly in your projects and/or you work on medium-to-large projects with multiple developers and different skill levels. In **test-driven development (TDD)**, you prepare the unit test before any code is developed and then you or other developers write just enough code to make the unit test pass. TDD coding keeps the team focused on writing modular, purpose-driven code that is cohesive and free of dependencies.

Following TDD tends to produce cleaner code that works, with less opportunity for developers to add extra code that is not in the requirements. As you develop a new feature or module, your team builds up a set of unit tests specific to the feature. Once the feature is complete and passes all its unit tests, merge your code back into the main branch and run the project's complete set of tests to verify bugs were not introduced and inter module dependencies still work. Of course, any bug that turns up is worthy of its own unit test. First, create the unit test that will pass when the bug is fixed but currently fails while the bug is still present. Then proceed to fix the bug and use your new unit test to verify that you have indeed fixed the bug. Enjoy the sense of achievement when you see all those unit tests pass and you get a virtual pat on the back for fixing the bug!

What Is Regression Testing?

Whenever you change your software, even a small modification can have unexpected consequences. These consequences may even affect areas of the code that you have not touched in a long time. When you run your tests to verify that those changes have not broken any existing functionality, you are doing regression testing. You perform regression testing to catch bugs that you may have accidentally introduced to other areas of the code. More importantly, you want to verify that you have not revived an already eliminated bug! When you rerun existing tests that you created when you fixed previous bugs, you ensure that your changes did not result in a regression, hence the name. GUI testing for bugs is slow, but unit tests give you the opportunity to run regression tests multiple times per day

Getting Started with Unit Testing

Ideally, the code you are unit testing has a limited feature set with all inputs and outputs on the connector pane. The code should be complete and self-sufficient and if possible not rely on references, global variables, or action engines. If you do need to rely on external code, use a setup VI to initialize that code before calling the test. The main goal is that for a given set of inputs, you get the same output every time you run the unit test. In the following sections we will explore how to create unit tests using the different tools we have in LabVIEW.

By default the NI UTF stores unit tests in the same folder as the VI. However, we prefer to use a dedicated folder (on disk and as a virtual folder) so that we don't have to include the unit tests in the code we share with others, and they can easily be excluded during a package build. It also makes it easier to programmatically run all of the tests inside the Unit Tests folder. Our suggested project structure includes separate folders for Documentation, Libraries, Unit Tests, and VI Analyzer. (See Fig. 6.2.)

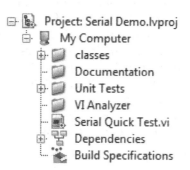

FIG. 6.2 *LabVIEW project structure with unit tests.*

A LabVIEW best practice is to not let the projects get too big and to modularize and package as much as you can. For example, if you are using DQMH, then each module is grouped into a library. Most of the time one project will have one DQMH library and its classes as well as its unit tests. That code then gets packaged via VIPM or a PPL. Having the unit tests in a different folder makes it easier to exclude them in the package/source distribution build.

Follow these steps if you are using UTF and you want to change the default location for your unit tests on your project. Go to the project properties and in the Unit Test Framework category, under the Test Creation section, change the default location for the Test File and Test Vector File Location. This setting will only affect this project. Next time you right-click on a VI and select **Unit Tests >> New Test**, your `.lvtest` file will be created by default in the new location. (See Fig. 6.3.)

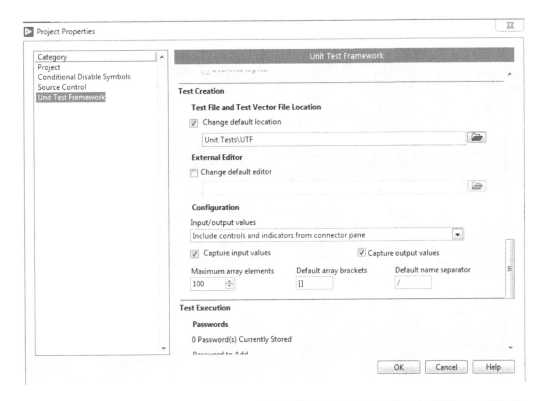

FIG. 6.3 *Configuring the default location for new unit tests and test vectors via Project Properties >> Unit Test Framework >> Test Creation.*

JKI VI Tester, Caraya, and InstaCoverage will not break your tests when you move them. They are, after all, LabVIEW code. Unfortunately, NI UTF may break your tests if they are moved. A way to address this is to edit the .lvtest file directly on a text editor and change the path there before opening the .lvtest with Unit Test Framework again.

Explore the LabVIEW shipping examples for Unit Test Framework to become familiar with the tool. Keep in mind that the units under test for these examples may be primitives. You will never write unit tests to test a LabVIEW primitive, since we trust NI has already done that for us. These shipping examples use primitives so that they can focus on the UTF features instead of explaining what the unit under test does. Later in this chapter, we will explore real-life LabVIEW code and how we test it. (See Fig. 6.4.)

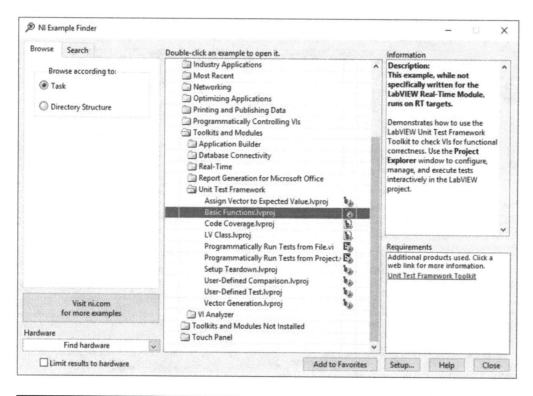

FIG. 6.4 *Unit Test Framework shipping examples.*

Test Coverage

As mentioned in the black versus white testing section earlier, code coverage measures the percentage of source code that executes while the unit tests run. The only two tools that provide this feature are UTF and InstaCoverage. Both tools measure block diagram coverage; they report the percentage of diagrams executed. Before we go over a simple example of code coverage measurement, let's begin by understanding how this coverage calculation works.

How many block diagrams are there in Fig. 6.5?

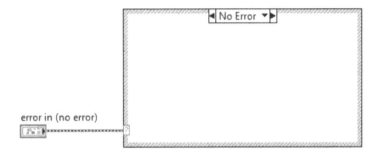

FIG. 6.5 *Simple VI with a case structure with two cases: No Error and Error.*

If you answered three diagrams, you are correct. The block diagram for the VI itself is one diagram, then we have two subdiagrams: one for the **No Error** case and another one for the **Error** case. If you are still not convinced, right-click on the case structure and select **Visible items >> Subdiagram label**. A label is added inside the case structure and that area is called a subdiagram. If our unit tests only test the No Error case, then the unit tests would be covering (VI Diagram + No Error diagram) / 3 diagrams = ⅔ = 66.66%.

Test Coverage Example

The code for this section can be found at https://github.com/LGP5/Unit-Testing.

For this example, we will use a very simple VI that implements a calculator. We are using simple code, so we can focus on how UTF and InstaCoverage measure code coverage and not so much on what the code is doing. Other examples in this chapter go into more details for real-life examples. (See Fig. 6.6.)

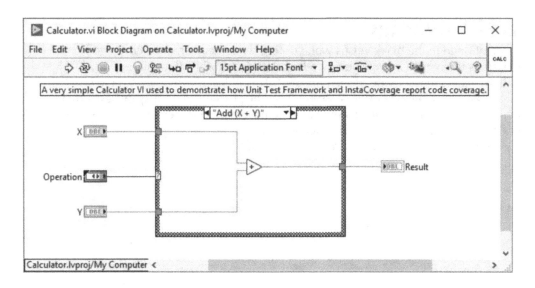

FIG. 6.6 *Simple Calculator.vi.*

Code Coverage Using the Unit Test Framework

First, we configure the project to save the unit tests in the unit tests folder. Otherwise, the tool saves the tests next to the VI under test. (See Fig. 6.7.)

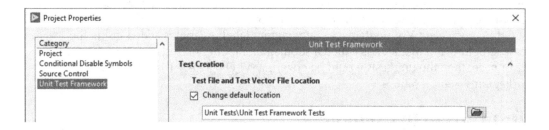

FIG. 6.7 *Right-click on the project and select properties, then in the Unit Test Framework category, change the default location for new tests.*

Next, right-click on the unit under test, `Calculator.vi`, and select **Unit Tests >> New Test**. Next, drag the `.lvtest` created to the Unit Test Framework virtual folder. (See Fig. 6.8.)

First, we will create just the Add test case. (See Fig. 6.9.)

Chapter 6: Unit Testing 413

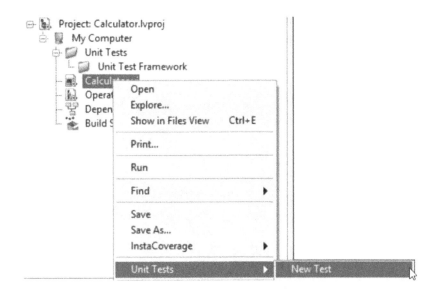

FIG. 6.8 *Right-click on the unit under test and select Unit Tests >> New Test.*

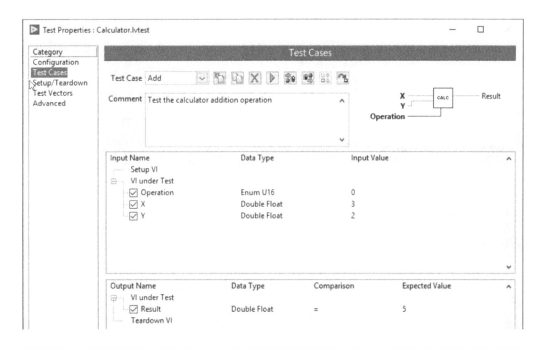

FIG. 6.9 *Calculator.lvtest Add test case.*

Let's run the unit tests by right-clicking on a virtual folder and selecting **Unit Tests >> Run**. We could also click on the green cross on the project explorer window. (See Fig. 6.10.)

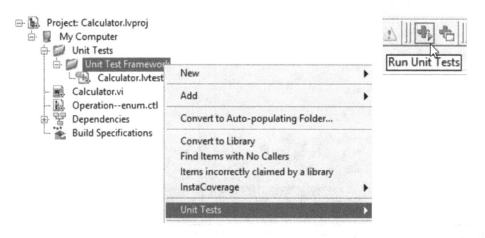

FIG. 6.10 *Run Unit Test Framework tests.*

The results for Code Coverage specify which case structure subdiagrams we have not implemented yet. (See Fig. 6.11.)

If we were to label our case structure as "**Operation?**," we will see the case structure name in the report. This is useful when there are multiple structures in a VI under test. (See Fig. 6.12.)

Code Coverage Using the InstaCoverage

Right-click on the unit under test, `Calculator.vi`, and select **InstaCoverage >> Generate Unit Tests**. (See Fig. 6.13.)

Next, edit the Setup, Teardown, and Test Harness VIs to have the same behavior as the UTF Add Test Case for the `Calculator.lvtest`. (See Fig. 6.14.)

Double-click on `Calculator.instacov` and configure the InstaCoverage test as follows (See Fig. 6.15.)

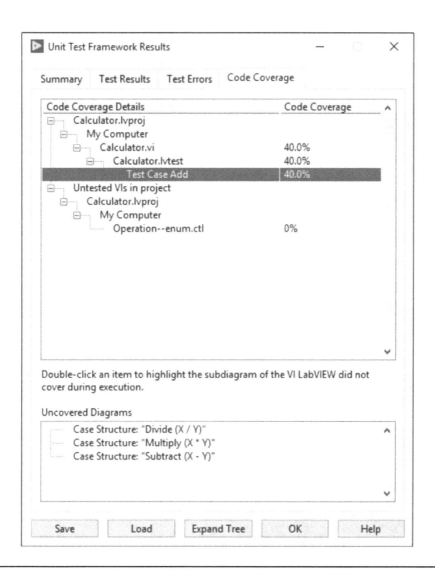

FIG. 6.11 *Unit Test Framework Code Coverage report.*

416 LabVIEW Graphical Programming

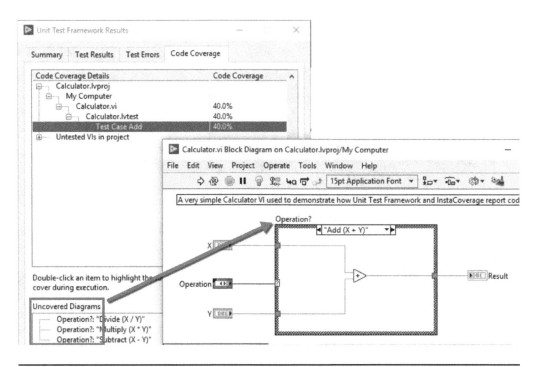

FIG. 6.12 *Unit Test Framework Code Coverage report including the name of the structure.*

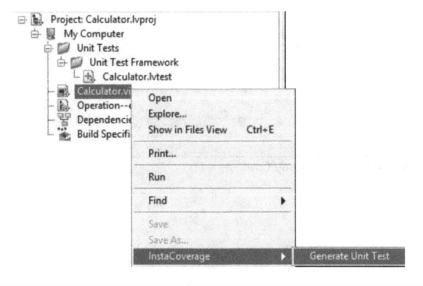

FIG. 6.13 *Right-click on the unit under test and select InstaCoverage >> Generate Unit Test.*

Chapter 6: Unit Testing 417

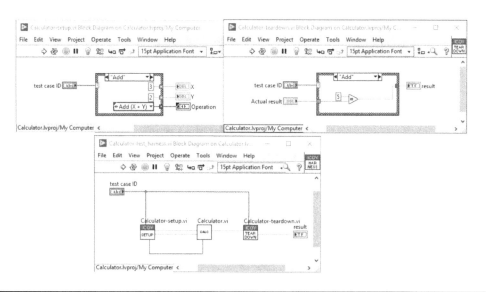

FIG. 6.14 *Add Test Case for Calculator InstaCoverage test.*

FIG. 6.15 *InstaCoverage Config file editor for Calculator.instacov.*

Run the test by clicking on the Run all test cases, and the report shows us which diagrams are not covered. If we double-click on one of the uncovered diagrams, InstaCoverage brings to front the block diagram showing the uncovered diagram. (See Fig. 6.16.)

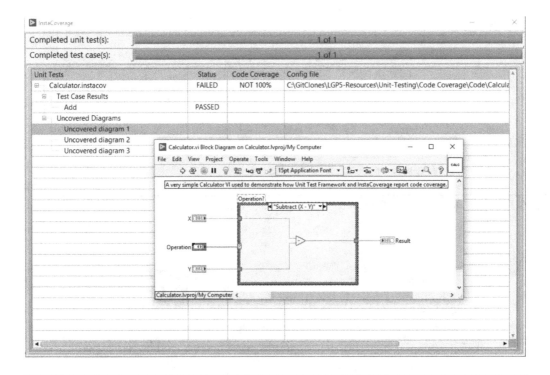

FIG. 6.16 *InstaCoverage test results. We right-clicked on Uncovered diagram 1 and that brought the Calculator.vi block diagram highlighting the uncovered case.*

The team at IncQueryLabs continues to improve InstaCoverage. At the time we wrote this chapter, the tool was reporting only 100 percent or not 100 percent. We hope that eventually the tool will report the actual percentage covered. Also, UTF reports overall project coverage, while InstaCoverage only reports code coverage per `instacov` test. One large advantage of using InstaCoverage is that it runs a lot faster than UTF.

Should You Aim for 100 Percent Coverage?
Requiring 100 percent coverage for a formal certification/verification is not true for all industries nor for all projects. For medical and military projects, there are

different levels of test coverage. It might be intended for specific VIs, but the goals for percentage coverage for overall projects vary. For example, the aviation standard DO-178B requires 100 percent code coverage for safety-critical systems, while the IEC 61508 recommends but does not require 100 percent coverage. It specifies you should explain any uncovered code.

Not all code can be tested; if you aim for 100 percent coverage, the test coverage becomes a goal and not an aid to improve quality. It is important to know what is covered, but it is even more important to know what is not covered and why it is not covered. We have requested IncQueryLabs to add a feature to InstaCoverage to figure out percentage coverage for the entire project. UTF does provide a project coverage metric; however, it is not perfect and sometimes it gives odd results. Also, keep in mind that UTF gets "confused" when the code includes diagram disable structures. Also, UTF only reports the project coverage based on the .lvtests in a project; if there is no .lvtest for a VI, then that VI is not considered for computing the project coverage.

The test coverage of an overall project grows asymptotically; as coverage increases, it is harder to create new tests, and it gets to a point that the return on investment on test coverage starts to diminish and it is better to justify why that section of the code is not being tested than just creating a test to obtain 100 percent. We don't want to fall into the trap of Goodhart's law.

"When a measure becomes a target, it ceases to be a good measure"— Goodhart's Law

Another concern is developers just writing unit tests to increase the percentage covered but not necessarily doing a thorough job in verifying that the code is doing anything useful. Keep in mind that code coverage tools only measure whether the tests execute the code but not the quality of the tests themselves. How good the tests your team creates are will depend on training and on everyone being convinced that we are writing unit tests to make the code better, not just to get a checkmark next to a target coverage goal.

In our experience, team members become convinced of the benefits of unit testing until they get to work on legacy code that has unit tests. One time a LabVIEW programmer told us that the only reason he was willing to add new features to a project with more than 1000 VIs was that there were about 70 unit tests that covered all the public APIs. One hundred percent coverage was not needed to convince him.

Test Vectors

We have met plenty of advanced LabVIEW developers who thought that in order to test a series of inputs, each input needed to be entered individually via a different test case within a Unit Test Framework test. This is not the case. Unit Test Framework supports .lvtest files for the unit test definition and .lvvect files for the test vectors. Unfortunately, the developer experience is not really a good one when going through this method. For one thing, the test vectors cannot be created in the place where you need them, the Unit Test definition window! You have to go to the project, right-click (on a folder or My Computer) and select **New >> Test Vectors**. (See Fig. 6.17.)

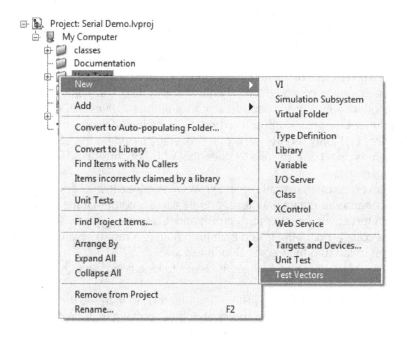

FIG. 6.17 *Creating new test vectors.*

A test vector is an array of values that the developer can give as an input to a unit under test in order to test it. A .lvvect file can contain one or multiple test vectors. The Unit Test Framework has two types of test vectors: sequence and linear. As the name implies, the sequence test vector is an array of values listed sequentially. A linear test vector uses three parameters to define the test vector: start value of the linear test vector, increment step value of the linear test vector, and the total number of values in the linear test vector. (See Figs. 6.18 and 6.19.)

An example of a sequence test vector with three specific values is (3, 6, 15).

An example of a linear test vector definition is (0, 5, 4) (start at 0, increment 5, and have 4 elements); this would result in a test vector with the following values: (0, 5, 10, 15).

FIG. 6.18 *Editing test vector type.*

FIG. 6.19 *Editing test vector values. Tip: Hover over the different cells for the tip strip to display what that cell represents. In this figure, we are editing the number of terms.*

Once you have defined your test vectors, you are ready to use them. On the unit test you want to use the test vectors, go to the Test Vectors category and add your newly created test vectors there. Then on the test case itself, right-click on the input or output you want to define via the test vector and select **Assign Test Vector**. (See Figs. 6.20 and 6.21.)

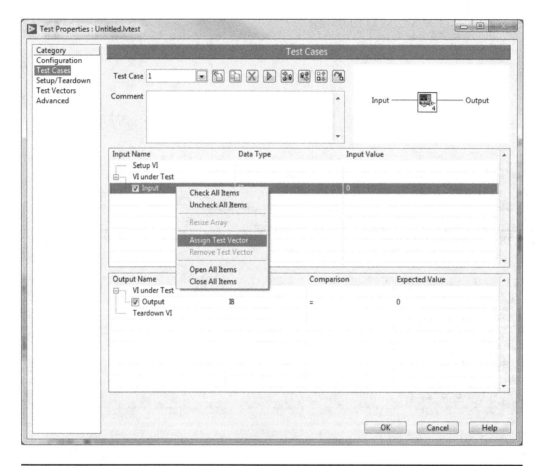

FIG. 6.20 *Right-click on the input to select Assign Test Vector.*

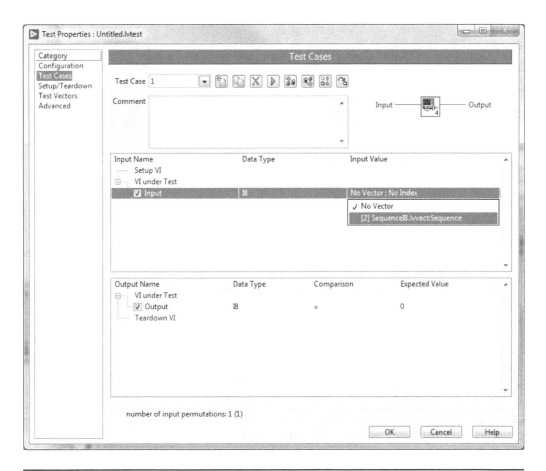

FIG. 6.21 *Use the drop-down box to select from the available test vectors configured via the Test Vectors category on the left.*

You probably saw on the shipping examples that a report was generated as soon as UTF completed the tests. This report was showing up because the example developer right-clicked on the project and in the Unit Test Framework section, she checked the option to generate an HTML report. (See Fig. 6.22.)

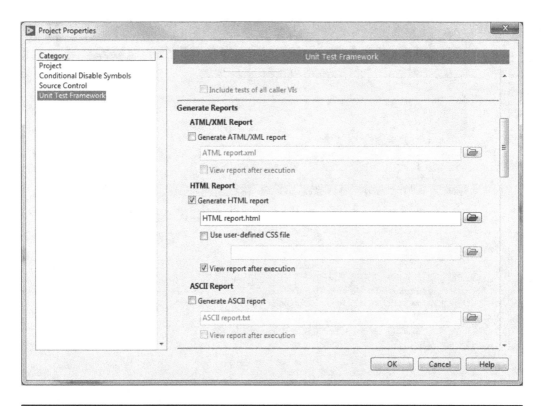

FIG. 6.22 Check the box "Generate HTML report" for UTF to generate a report as soon as it completes running all the tests in the current project.

Please note that both .lvtest and .lvvect files are human-readable and can be edited via Excel or Notepad. You can save time creating your test vectors directly in Excel, but we don't recommend it. However, knowing that these are text files can save you when the Unit Test Framework seems to not recognize your tests after a small change. For example, if you change the name of your inputs/outputs or if you move the test to a different location on disk, you can edit in Notepad the path to the new location and rename your inputs and outputs before opening them in Unit Test Framework and UTF will never know that you sneaked in those changes.

What about Testing Classes? Public versus Private VIs in Libraries

The general idea is to only test the public VIs. We assume that the public VIs will exercise all the private VIs in a class or library. The approach that we favor is to figure out why we want to run a unit test on a private VI. Perhaps this private VI performs some algorithm that we want to make sure works, and we don't want to have any other library calling this method from outside this VI. If this is the case, we should ask ourselves if the VI itself could be a member of a different shared library or class. Perhaps it could be public there and be a dependency for our current code. Alternatively, we can use integration tests to test that private VI. An example of this is the DQMH Unit Testing tools that generate a setup VI charged with starting the DQMH Module and a TearDown VI charged with stopping the DQMH Module. The test VI calls a DQMH request and verifies that the DQMH reports the expected response. In this case, the ultimate goal might be to test a private VI called inside one of the Message Handling Loop cases in the DQMH `Main.vi`, and this VI might be private. In the case of a private method of a class, we might create a wrapper public VI that is a member of the class and its only use is to provide a mean to test that private VI.

Example of Unit Testing for a LabVIEW Class

The code can be found at https://github.com/LGP5/Unit-Testing.

The initial code includes a class for the serial device and a class for the simulated serial device. The UML diagram was created using the NI GDS toolkit. For more information on the GDS toolkit see chapter 4 on object-oriented programming.

This UML diagram represents a very simple set of methods to communicate with a serial device. We are assuming a constant configuration and no extra code to auto-detect the device or anything fancy.

For now, the code only has a "Sample Command" and a "Send Sample Command." The first question we always get is "why not put the command and the sending of the command in the same VI." Well, it makes unit testing a lot easier and we only have to override the Send Sample Command. (See Fig. 6.23.)

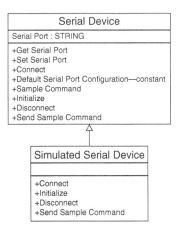

FIG. 6.23 *UML class diagram for the Serial Device class hierarchy.*

Actually, we could even simplify more and change the Send Sample Command to a generic Send Command. This way, we could have multiple commands and we would only have to override the Send Command VI. (See Fig. 6.24.)

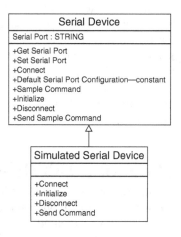

FIG. 6.24 *Simplified UML class diagram for the Serial Device class hierarchy.*

Now that we have defined the general structure of our serial device classes, we will look at the datasheet for the communication protocol that we need to implement and we will be creating a unit test for each command. This way, we can verify that we are implementing the commands correctly. There is no need for us to create unit tests for the Connect, Disconnect, and Send Command methods because they are just wrappers of VISA calls that we trust NI already tested.

The device we are communicating with is called the iTacho. The iTacho is a sensor made up of a noncontact tachometer and an inclinometer.

The iTacho has a very simple communications protocol. In normal operation, the iTacho outputs an output frame every 250 milliseconds. It also accepts a single command to calibrate the inclinometer sensor.

The iTacho appears as a virtual COM port with the following configuration:

- 9600 bps, 8 data bits, no parity, 1 stop bit, no flow control

From the iTacho manual of operation, we gather that the data frame structure is described in Table 6.2.

The manual also provides an example, which will come in very handy for the unit tests. (See Table 6.3.)

Byte 1	Byte 2	Byte 3	Byte 4	Byte 5	Byte 6	Byte 7	Byte 8
Start Code	Length Includes Byte 3 ... n	Tachometer Speed MSB	Tachometer Speed LSB	X gravity LSB	X gravity MSB	Y gravity LSB	Y gravity MSB
0XF1	0x18	0xMM	0xLL	0x00	0x00	0x00	0x00
Byte 9	Byte 10	Byte 11	Byte 12	Byte 13	Byte 14	Byte 15	Byte 16
Z gravity LSB	Z gravity MSB	X Gain Constant MSB	X Gain Constant LSB	X Offset Constant MSB	X Offset Constant LSB	Y Gain Constant MSB	Y Gain Constant LSB
0x00	0x00	0x00	0x00	0x00	0x00	0x00	0x00
Byte 17	Byte 18	Byte 19	Byte 20	Byte 21	Byte 22	Byte 23	Byte 24
Y Offset Constant MSB	Y Offset Constant LSB	Z Gain Constant MSB	Z Gain Constant LSB	Z Offset Constant MSB	Z Offset Constant LSB	Serial Number MSB	Serial Number LSB
0x00	0x00	0x00	0x00	0x00	0x00	0x00	0x00
Byte 25	Byte 26						
Firmware Version	End of Frame						
0x02	0xF5						

TABLE 6.2 *iTacho data frame.*

Byte 1	Byte 2	Byte 3	Byte 4	Byte 5	Byte 6	Byte 7	Byte 8
Start Code	Length Includes Byte 3 ... n	Tachometer Speed MSB	Tachometer Speed LSB	X gravity LSB	X gravity MSB	Y gravity LSB	Y gravity MSB
0XF1	0x18	0xDC	0x31	0x01	0x10	0x06	0x00
Byte 9	Byte 10	Byte 11	Byte 12	Byte 13	Byte 14	Byte 15	Byte 16
Z gravity LSB	Z gravity MSB	X Gain Constant MSB	X Gain Constant LSB	X Offset Constant MSB	X Offset Constant LSB	Y Gain Constant MSB	Y Gain Constant LSB
0x04	0x00	0x7F	0x80	0x00	0x0A	0x83	0xD6
Byte 17	Byte 18	Byte 19	Byte 20	Byte 21	Byte 22	Byte 23	Byte 24
Y Offset Constant MSB	Y Offset Constant LSB	Z Gain Constant MSB	Z Gain Constant LSB	Z Offset Constant MSB	Z Offset Constant LSB	Serial Number MSB	Serial Number LSB
0xFF	0xEB	0x7D	0x70	0xFF	0xF1	0x00	0x01
Byte 25	Byte 26						
Firmware Version	End of Frame						
0x02	0xF5						

TABLE 6.3 *iTacho sample data frame.*

The frame would be interpreted as follows:
Speed: 0x05DC (1500 decimal):26.01 mph (Use Equations 1 and 2)
X Gravity: 0x0101 (257 decimal): **1.00390625 g** (Use Equation 4)
Y Gravity: 0x0006 (6 decimal): **0.0234375 g**
Z Gravity: 0x0004 (4 decimal): **0.015625 g**
X Gain: 0x7F80 (32640 decimal): **0.99612415** (Use Equation 19)
X Offset: 0x000A (10 decimal): **0.01 g** (Use Equation 20)
Y Gain: 0x83D6 (33750 decimal): **1.0299997**
Y Offset: 0xFFEB (-21 decimal): **-0.021 g**
Z Gain: 0x7070 (32112 decimal): **0.98001038**
Z Offset: 0xFFF1 (-15 decimal): **-0.015 g**
Serial Number: 1
Firmware Version 0.2
Where the equations from the manual are as follows:

$$\text{Pulley RPM} = \frac{32768 \, kHz}{\text{Tachometer speed}} \cdot 60 \qquad \text{(Equation 1)}$$

$$\text{Treadmill Speed (MPH)} = \frac{\text{Pulley RPM} \cdot \text{Motor To Pulley Ratio}}{\text{Treadmill Ratio} \left(\frac{RPM}{MPH}\right)} \qquad \text{(Equation 2)}$$

$$X,Y,Z = \frac{\text{Gravity\_Vector\_Read\_From\_iTacho}}{256} \qquad \text{(Equation 4)}$$

$$\text{Gain\_Scaled}_{x,y,z} = \text{Gain}_{x,y,z} \cdot 32767 \qquad \text{(Equation 19)}$$

$$\text{Offset\_Scaled}_{x,y,z} = \frac{\text{Offset}_{x,y,z}(\text{convert to integer})}{1000} g \qquad \text{(Equation 20)}$$

For this example, we will focus on creating the unit tests and functions for two of the VIs that will be used by the iTacho:

1. `Find Complete Frame.vi`
2. `Parse Frame.vi`

Find Complete Frame.vi

This VI will receive a "read buffer in" string, and it will return the first frame found and any leftover bytes. It will use the Start of Frame (0xF1) and End of Frame (0xF5) to identify the frames.

The API for `Find Complete Frame.vi` would look like Fig. 6.25.

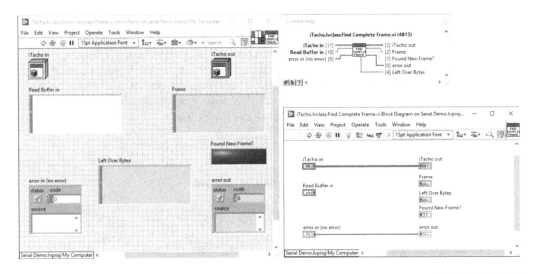

FIG. 6.25 *The API for Find Complete Frame.vi.*

If you do not plan to use classes, the VI would look the same minus the iTacho in and iTacho out objects.

Notice that we have not implemented any code yet. You can update to the early stages of this code in GitHub, when we created the branch called FindCompleteFrame to follow along with this example.

Traditionally, at this point, you probably would start implementing the code for Find Complete Frame.vi. We will follow TDD and we will create the unit test first.

To create the first unit test for Find Complete Frame.vi, we will use the example from the manual.

- Read buffer will receive 0xF115 05DC 0101 0600 0400 7F80 000A 83D6 FFEB 7D70 FFF1 F5

- We expect the output "Found New Frame?" to be true

- Left over bytes will be empty

- Frame will have the exact frame read 0xF115 05DC 0101 0600 0400 7F80 000A 83D6 FFEB 7D70 FFF1 F5

Choose the Unit Testing tool you will be using; the following section shows how we create the tests using a VI, Unit Test Framework, JKI VI Tester, JKI Caraya, and InstaCoverage. By the time this book is published, there might be other options out there.

Unit Testing for a Class—First Step: Create the Unit Test

Unit Testing for a Class—Stand-Alone VI   This is the type of test you might have done in the past to verify if a VI is working. Perhaps you have never done this type of VI and you have relied on changing the front panel controls on the VI and running it until visually you would confirm that the VI was working. We do not advocate implementing these types of tests. However, for demonstration purposes, it will help us better understand what the unit testing tools are doing.

Our stand-alone test would look like Fig. 6.26.

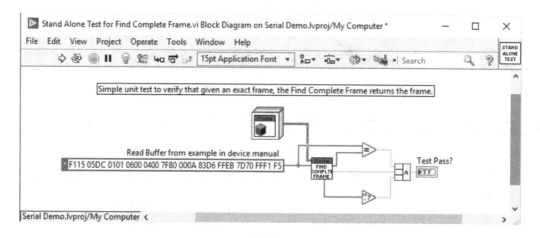

FIG. 6.26   *Stand-alone test VI.*

If we run the test now, it will fail, because we have not implemented the code inside `Find Complete Name.vi`. Following TDD, we would implement the code to make `Stand Alone Test for Find Complete Frame.vi` pass.

Before we move on with the other unit testing tools, let's look at a way to improve this stand-alone test. James McNally, on his blog post "Given-When-Then," goes over what he learned from watching a presentation by Trisha Gee (https://devs.wiresmithtech.com/blog/given-when-then/).

What we want is to describe our test in the following terms:

- **Given:** The preconditions
- **When:** The trigger or action
- **Then:** What the software/system should do in response

And there are general guidelines to know if you are creating your unit test correctly:

- **Given** is small. If it starts to get quite big, this starts to sound more like an integration test and less of a unit test. It should also contain no tests—this is not the subject of the unit test.
- **When** is tiny. This should ONLY be the code you are actually testing.
- **Then** is tiny. This contains your actual tests and assertions. Given a unit test should test one thing there should only be one assertion here or multiple tightly related assertions.

Following this advice, the stand-alone test now looks like Fig. 6.27.

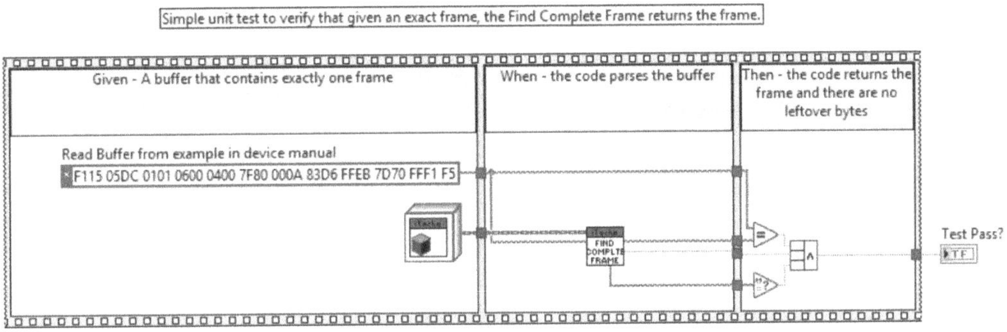

FIG. 6.27 *Stand-alone test VI updated to use Given-When-Then approach.*

This is a lot easier to read and will inform other developers what the test's intention was.

You can download James McNally's VITAC from the LabVIEW Tools Network or at https://github.com/WiresmithTech/VITAC/releases. This package includes templates for the Given-When-Then approach.

Unit Testing for a Class—Unit Test Framework For this example, you will need to have the NI UTF installed.

First, we need to edit the project properties to indicate to the Unit Test Framework that we want to save the new tests in the Unit Tests\UTF folder. Otherwise, the Unit Test Framework saves new unit tests right next to the VI under test. (See Fig. 6.28.)

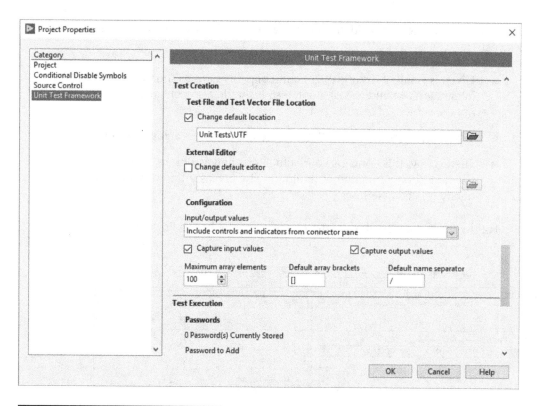

FIG. 6.28 *Configure UTF to save new unit tests under the Unit Tests folder.*

Now, on the project explorer, go to our `Find Complete Frame.vi`, right-click on it, and select **Unit Tests >> New Test**. This creates a `Find Complete Frame.lvtest`. The `.lvtest` file will be saved in the correct location but it is placed as part of the class right next to the VI. We drag it and place it under the `Unit Tests\Unit Test Framework` virtual folder. (See Fig. 6.29.)

Double-click on `Find Complete Frame.lvtest`.

Many LabVIEW developers assume that the test case name has to be a number because the default name is a number, but this is not correct. You can use any text for the test case name. We suggest you use a descriptive name for your test case and include comments. Here, we chose "Find One Frame." Following the **Given-When-Then** approach for your comments will help you and other developers understand the objective of this test case. (See Fig. 6.30.)

Chapter 6: Unit Testing

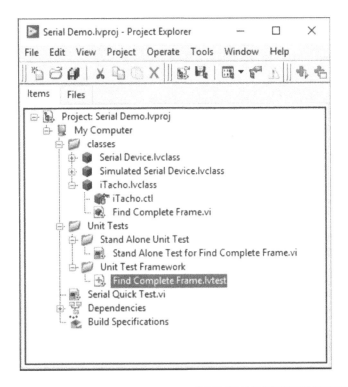

FIG. 6.29 *We dragged Find Complete Frame.lvtest under the Unit Test Framework virtual folder.*

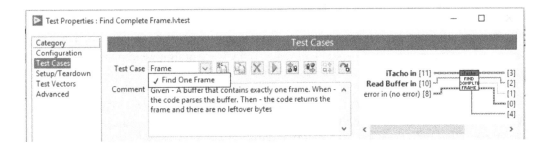

FIG. 6.30 *The test case name can be text.*

There are different ways to enter the inputs and expected results. We could type them directly on the .lvtest definition. However, we have already defined the API for the Find Complete Frame.vi, so let's use the button on the top **Import**

Values from VI to populate the inputs and outputs. For this, we will have to enter the values on the front panel of the `Find Complete Frame.vi` and then press the **Import Values from VI** button. (See Fig. 6.31.)

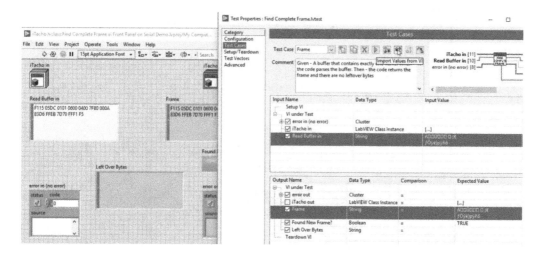

FIG. 6.31 *Use the Import Values from VI button to populate inputs and outputs for the Find One Frame Test Case.*

As you can see, this was the right approach; otherwise, we would have gotten the strings wrong. We wanted to make sure the values were equivalent to the hexadecimal representation for the **Read Buffer In** control and for the **Frame** indicator.

If we run the test now, it will fail, because we have not implemented the code inside `Find Complete Name.vi`. Following TDD, we would now implement the code to make `Find Complete Frame.lvtest` pass.

Unit Testing for a Class—JKI VI Tester For this example, you will need to have JKI VI Tester installed. You can find this package directly via JKI VIPM or you can download the source code via JKI GitHub at https://github.com/JKISoftware/JKI-VI-Tester.

Once JKI VI Tester is installed, your project explorer will have the project toolbar in Fig. 6.32.

FIG. 6.32 *JKI VI Tester project explorer testing bar.*

You can create new tests and run them via this toolbar or via the **Tools >> VI Tester** menu.

We will add a new test case. This will create a new test class that inherits from the JKI VI Tester Test class.

The resulting test will look like Fig. 6.33.

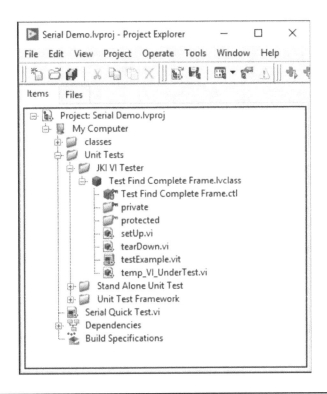

FIG. 6.33 *Serial demo project including JKI VI Tester test.*

For our simple test, we will not use the `setUp.vi` or the `tearDown.vi`. We can right-click on `testExample.vit` and select to create a **New From Template**. This will be our new test case.

Save the new test case as `Test Find Complete Frame.vi` and implement the unit test as described in Fig. 6.34.

The string messages on top of the JKI VI Tester pass/fail tests are useful to indicate where the first failure happens. If we run JKI VI Tester, we get the report in Fig. 6.35.

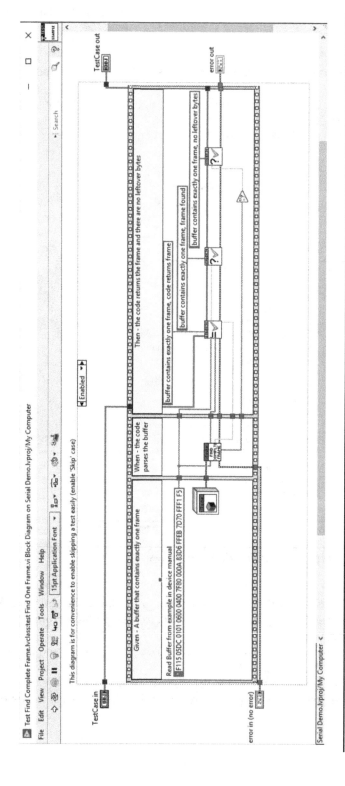

FIG. 6.34　Test Find One Frame.vi JKI VI Tester test.

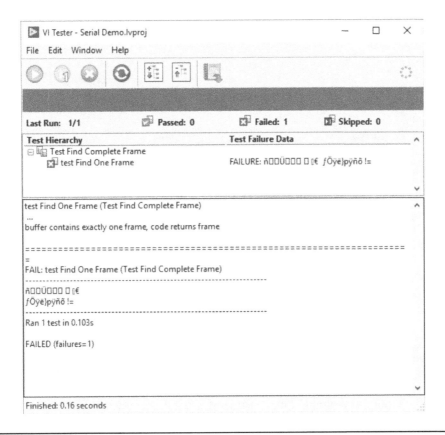

FIG. 6.35 *JKI VI Tester results report.*

We can read the message "buffer contains exactly one frame, code returns frame" and we know this test is failing because we have not implemented the code inside `Find Complete Name.vi`. Following TDD, we would now implement the code to make `Test Find Complete Frame:test Find One Frame.vi` pass.

Unit Testing for a Class—Caraya Unit Test Framework For this example, you will need to have Caraya Unit Test Framework installed. You can find this package directly via JKI VIPM, or you can download the source code via JKI GitHub at https://github.com/JKISoftware/Caraya.

Caraya is both an Assertion Framework and a Unit Test Framework. For this example, we will use the Unit Test Framework features. Caraya gives us the tools to convert our manual test VIs into unit test cases. We will create a copy of the stand-alone test we created earlier and show the steps to convert it into a unit test. We will save the copy as `Caraya Test for Find Complete Frame.vi`. (See Fig. 6.36.)

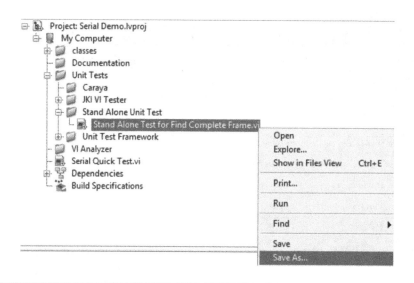

FIG. 6.36 *Save a copy of stand-alone test for Find Complete Frame.vi.*

To convert our stand-alone VI into a unit test, we will add the `Define Test.vi` and `Assert True.vi` from the **JKI Toolkits >> Caraya Unit Test Framework** palette. We need to give the test a name and a label Caraya will use to report the unit test results. In this case, the test name is *Test Find Exactly One Frame* and the label is *Code returns one frame with no leftover bytes*. (See Fig. 6.37.)

Run the `Caraya Test for Find Complete Frame.vi`. (See Fig. 6.38.)

We can read that "Test Find Exactly One Frame" reported as failing: "Code returns one frame with no leftover bytes." We know this test is failing because we have not implemented the code inside `Find Complete Name.vi`. Following TDD, we would now implement the code to make `Caraya Test for Find Complete Frame.vi` pass.

Unit Testing for a Class—InstaCoverage For this example, you will need to have InstaCoverage installed. You can find this package directly via JKI VIPM.

Right-click on the VI under test, `Find Complete Frame.vi`, and select **InstaCoverage >> Generate Unit Test**. (See Fig. 6.39.)

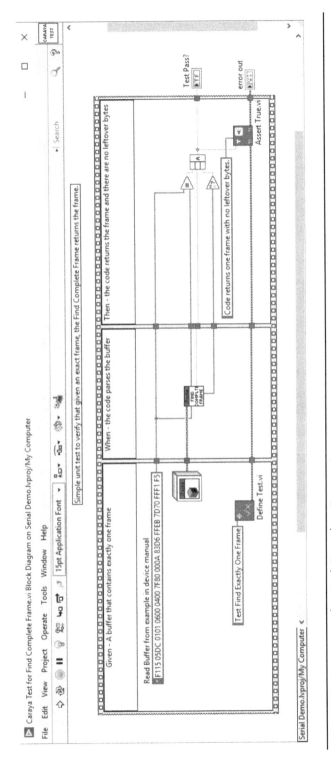

FIG. 6.37 *Caraya test for Find Complete Frame.vi.*

440 LabVIEW Graphical Programming

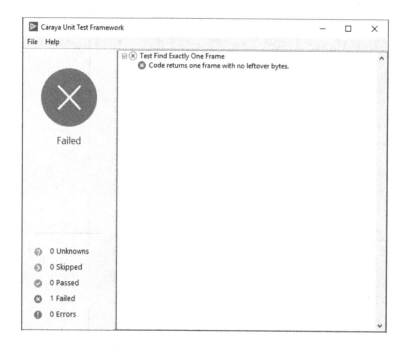

FIG. 6.38 *Caraya reports unit test failure.*

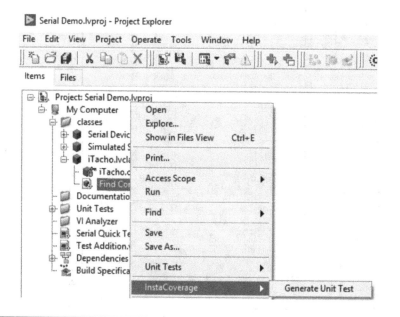

FIG. 6.39 *Create new InstaCoverage unit test by right-clicking on the VI under test.*

InstaCoverage will script the VIs used for the unit test. It will ask you where you want to save the unit tests; we chose to save them under `Unit Tests\InstaCoverage` to follow the same pattern we used for the other examples. We then dragged the resulting virtual folder under our Unit Tests virtual folder. (See Fig. 6.40.)

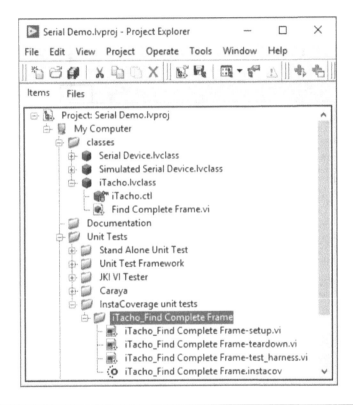

FIG. 6.40 *InstaCoverage scripted the setup, teardown, and test_harness VIs for us.*

Following the same approach of Given-When-Then, we will place the Given within the Setup VI, the When within the test harness VI, and the Then within the Teardown VI.

We begin by modifying the `iTacho_Find Complete Frame-setup.vi`. (See Fig. 6.41.)

We will place the validation code within the `iTacho_Find Complete Frame-teardown.vi`. (See Fig. 6.42.)

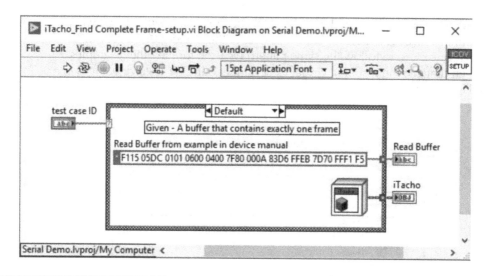

FIG. 6.41 Setup VI with test input conditions, similar to the "Given" section in the stand-alone test.

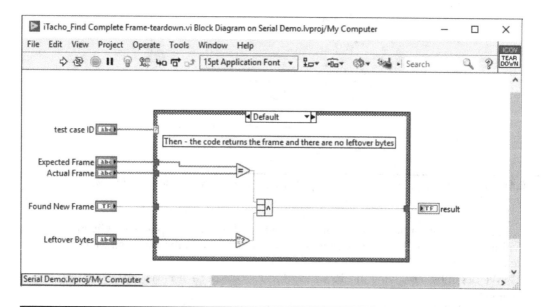

FIG. 6.42 Teardown VI with validation conditions, similar to the "Then" section in the stand-alone test.

Now we can implement the `iTacho_Find Complete Frame-test_harness`. (See Fig. 6.43.)

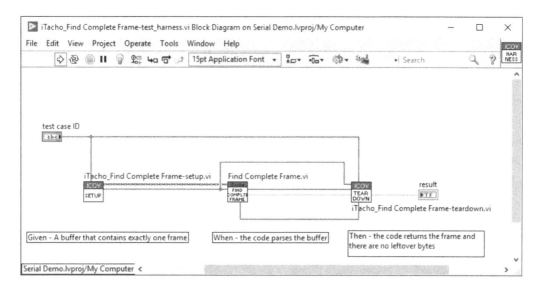

FIG. 6.43 *Test Harness VI with full unit test implementation.*

Note that we could have implemented the whole test directly in the test_harness and leave the setup and teardown VIs empty. You will have to determine what unit test style works best for you and your team.

We now need to configure the test suite. Double-click on the `iTacho_Find Complete Frame.instacov` configuration file. This will open the InstaCoverage: Config file editor. Here, you can add a new test case by clicking on the plus sign. Notice that if we add more test cases in the future, we need to change the default case in our setup/teardown cases to include the test case name *Test Find Exactly One Frame*. Add a comment and requirements and click **Save config** button. (See Fig. 6.44.)

We can read that "Test Find Exactly One Frame" reported as failing (see Fig. 6.45). We know this test is failing because we have not implemented the code inside `Find Complete Name.vi`. Following TDD, we would now implement the code to make `iTacho_Find Complete Frame-test_harness` pass.

FIG. 6.44 *InstaCoverage: Config file editor filled out for the test for Find Complete Frame.vi.*

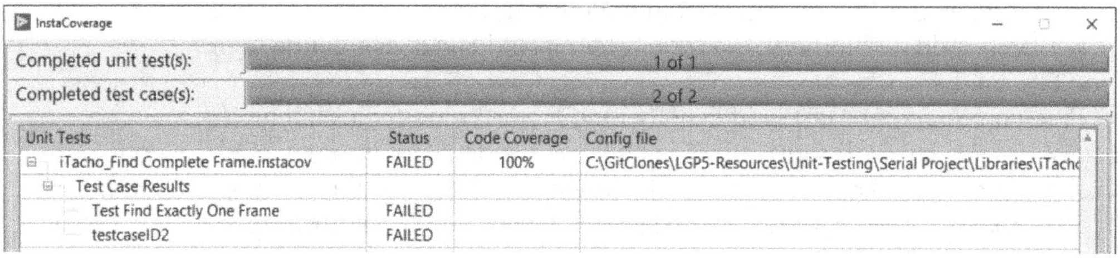

FIG. 6.45 *InstaCoverage test results.*

Unit Testing for a Class—Second Step: Implement Code to Make the Unit Test Pass

We implement the code in Fig. 6.46:

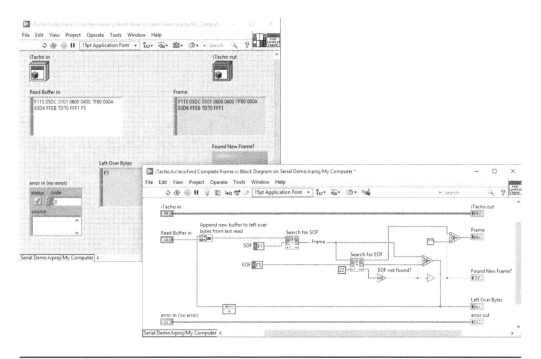

FIG. 6.46 *Find Complete Frame.vi initial implementation.*

However, our unit tests are failing because **Leftover Bytes** is not empty. At this point, we need to evaluate if our assumptions at the beginning were correct or if the unit test definition needs to be modified. We opt for editing the code so we do not include the end of the frame in the leftover bytes. If we had not had unit tests, we might have just left the code as is, and this might have caused other issues down the road, like adding a premature end of the frame to the next frame.

The final implementation looks like Fig. 6.47.

With this, the unit test for one frame passes.

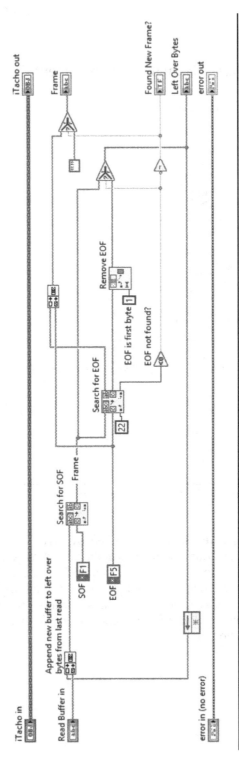

FIG. 6.47 Find Complete Frame.vi final implementation.

Unit Testing for a Class—Third Step: Maintain Unit Tests

We created the bare minimum unit test for this code. We might want to look at creating a test case for an incomplete frame and a test case for multiple frames in the read buffer. We could also wait until that feature is actually needed. At this point, we don't know if we will be reading one frame at a time.

Our new workflow is to always run all existing unit tests before we add a new feature. This is to verify that we are starting out with working code. Then we create the unit test for the new feature and run the tests to verify that the new test fails. Next, write the code to make the new unit test pass, and verify that the rest of the unit tests still work.

If we LabVIEW programmers are going to be taken seriously, then we have to follow good software engineering practices and good programming practices.

When Are Setup and Teardown Required?

Setup and teardown are useful when running a suite of test cases that all require either some code to generate the inputs for the unit under test or call dependencies. In the case of DQMH, for example, we use the setup to launch the DQMH module. The test case would test an individual DQMH request event, and the teardown would stop the module. In our experience, most of the time, when we need to add setup and teardown steps, we are probably doing integration testing and not pure unit testing. For example, if you are testing database calls, the setup would open the connection with the database and the teardown would close the connection. The fact that we are involving the database in our unit tests makes this an integration testing, because we are not isolating our code from external dependencies.

When using JKI VI Tester to test multiple test cases, we suggest that you use the Global setup and Global teardown as opposed to the setup and teardown that comes by default in the test case template. The default setup and teardown executes after each test VI within a JKI VI Tester Test class. The Global setup and Global teardown execute once for the entire JKI VI Tester class. To get to these, right-click on your JKI VI Tester test and select **New >> VI for Override…** and select globalSetUp.vi and globalTearDown.vi. You can now delete from your test the setUp.vi and TearDown.vi from your test class. (See Figs. 6.48 and 6.49.)

448 LabVIEW Graphical Programming

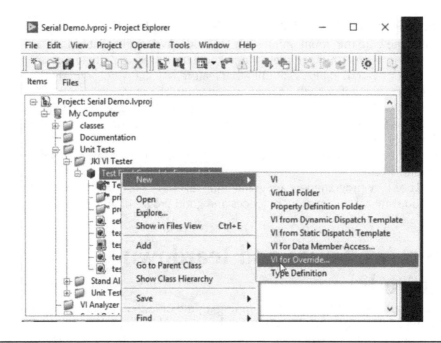

FIG. 6.48 *On the JKI VI Test, right-click to create a new VI for Override...*

FIG. 6.49 *Select globalSetUp.vi and globalTearDown.vi to override.*

Should You Add Test Cases That Are Designed to Fail?

You could include test cases that verify that an error is returned: the presence of a specific error makes the unit test pass. So, the test case is designed to pass when the code fails.

If you are designing tests for a critical part of your code, you want to make sure you are not only testing the success cases but the failures too. This is a challenge for developers new to unit testing. As developers, we are used to creating working code. Our goal is to implement all the features requested, and we don't necessarily think of every corner case. More than once when we create an issue report for developers to fix, we get a reply along the lines of "but the end user is not supposed to do that." Well, if the user can do it, they will do it! A typical example is when the end user cancels the file browser option and it generates error 43 "Operation canceled by user." A good developer makes sure her code can cope with that cancelation. A good unit test developer will make sure she tests both the case where a valid path is given and the case where the operation was canceled by the user or an invalid path was given.

What about Testing When the Expected Output Is an Array of NaN (Not a Number)?

Before LabVIEW 2016 Unit Test Framework Toolkit we could not compare arrays of NaNs (UTF did work when comparing scalars). For those cases, it was necessary to create a user-defined test. NI fixed this bug, and now UTF unit tests can compare arrays of NaNs.

When Would You Create Unit Tests for a DQMH Public API?

The DQMH toolkit installs a tool to create unit tests for a DQMH module. This tool creates a setup VI, a teardown VI, and a VI to test the chosen DQMH Request Event. There are three situations where we would create a unit test for a DQMH module using these tools:

1. We need the DQMH module to be running to test the given request, meaning some initialization in the module itself is needed, some specific state, or we need to call other requests before we test the required request. If you thought "this is integration testing," you are getting the idea. It is integration testing because we cannot isolate the request itself from the DQMH module used to test it.

2. The VI we want to test is private to the DQMH module and it is called within a Message Handling Case in the DQMH module. We can call the DQMH request to test that private VI.

3. GUI tests. If our code calls specific DQMH requests when a button is pressed or other GUI interaction from the user, we can use the DQMH unit test tools to programmatically emulate the end-user GUI interaction by calling the requests as part of the test. And you are right once again: this is definitely integration testing.

Unit Testing for a DQMH Module

The code for this example is in our repository at GitHub.com/LGP5/Unit-Testing. To find all the commits, look for the tag "DQMHTimerTests."

We are going to create a simple DQMH module for a timer. In this case, we want to make sure that our DQMH module is reporting correctly the current time. Our DQMH timer module will be very simple: it will have the functions to start and pause a timer and will broadcast the current time when the timer is running.

We will not use the Message Handling Loop for this example. Everything is happening directly in the Event Handling Loop. Take a look at our Start Timer, Pause Timer, and Unpause Timer; see if you can spot what is missing. (See Figs. 6.50 through 6.53.)

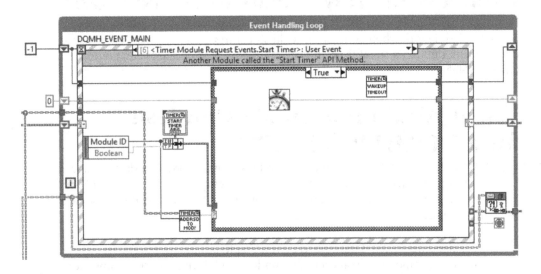

FIG. 6.50 *DQMH Main.vi Event Handling Loop case for Start Timer request.*

Chapter 6: Unit Testing 451

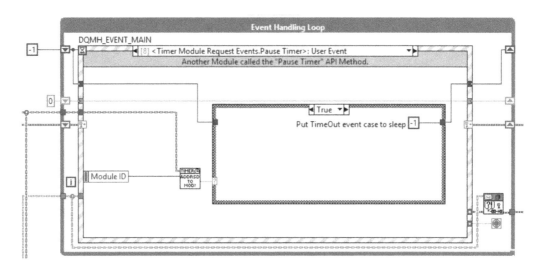

FIG. 6.51 *DQMH Main.vi Event Handling Loop case for Pause Timer request.*

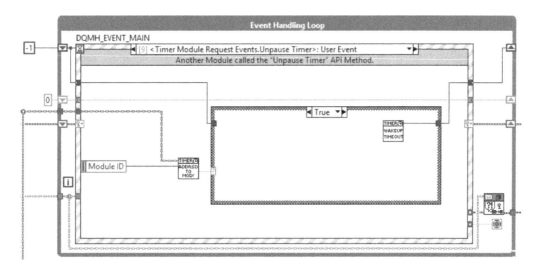

FIG. 6.52 *DQMH Main.vi Event Handling Loop case for Unpause Timer request.*

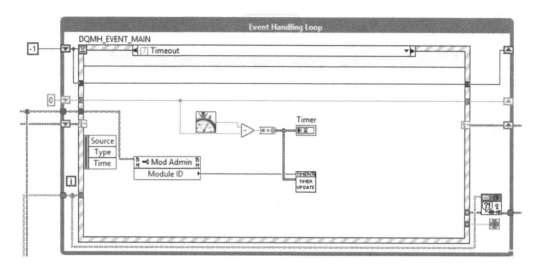

FIG. 6.53 *DQMH Main.vi Event Handling Loop case for Timeout request.*

Did you spot what is missing? No worries if you did not. We will create a unit test to help us find the issue and then when we fix it, the unit test will confirm that we have indeed fixed it.

We will use the DQMH tools to create a new DQMH unit test for the Unpause Timer request. (See Figs. 6.54 and 6.55.)

FIG. 6.54 *Tools >> Delacor >> DQMH >> Testing Tools menu.*

This created a `Timer Module setup.vi`, a `Timer Module teardown.vi`, and a `Test - Timer Module - Unpause Timer1.vi`. The DQMH unit testing tools only create the setup and teardown once. Next time you run the same tool, it will only create a new `Test - Timer Module Unpause Timer2.vi`.

Let's follow the instructions inside the `Test - Timer Module - Unpause Timer 1.vi`. (See Fig. 6.56.)

Chapter 6: Unit Testing 453

FIG. 6.55 *New DQMH Unit Test wizard window.*

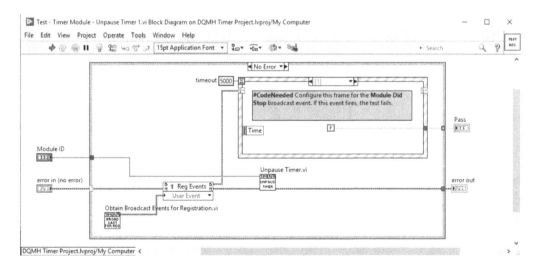

FIG. 6.56 *DQMH scripted a new Test - Timer Module - Unpause Timer 1.vi.*

454 LabVIEW Graphical Programming

We need to configure the event case to handle the Module Did Stop broadcast event. If the DQMH module stops before we get our expected event, we want it to report the test as failing. This has the added benefit of giving us an option to abort the test if needed. The timeout of 5000 milliseconds is also another way of exiting the unit test if no event is received.

We follow the instructions and deleted the #CodeNeeded bookmark when done. (See Fig. 6.57.)

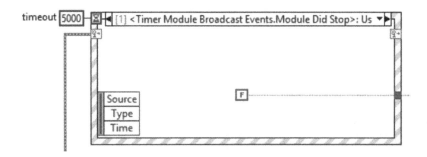

FIG. 6.57 DQMH unit test event handling Module Did Stop broadcast implemented.

If the DQMH module reports an error before the unit test gets the expected response, we want the unit test to report a failure. (See Figs. 6.58 and 6.59.)

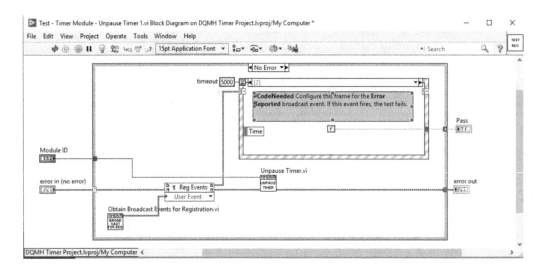

FIG. 6.58 DQMH unit test #CodeNeeded indicating to handle the Error Reported broadcast event.

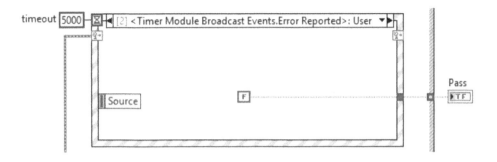

FIG. 6.59 *DQMH unit test event handling Error Reported broadcast implemented.*

Finally, we get to the case where we need to handle the expected event. (See Fig. 6.60.)

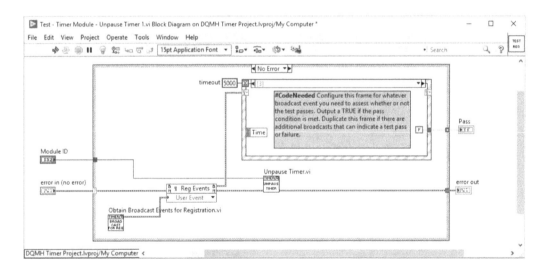

FIG. 6.60 *DQMH unit test #CodeNeeded indicating to handle the broadcast event that we will use to determine if the test passed or failed.*

In order to test the Unpause request, we need to ensure that after we start the timer, let it run, and then pause it for a couple of seconds, we expect that the immediate time broadcasted by the timer will be the same as when we paused. In other words, the timer should freeze; it should not show the value at pause time plus the actual number of seconds the timer was paused.

To do this, we will wrap the Event Structure with a While Loop and we will add an elapsed time to terminate the While Loop. The code will keep track of the Timer Updated broadcast timer and use the final timer to verify that we indeed have a timer showing about 3 seconds. We expect 3 seconds because the initial run is of 2 seconds, then we unpause after another 2 seconds; that means the While Loop has been running for 4 seconds and it will run another 1 second. The total time we expect the timer to run for is about 3 seconds. (See Fig. 6.61.)

Let's run this test as a stand-alone test. (See Fig. 6.62.)

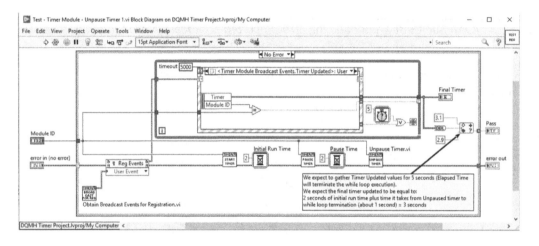

FIG. 6.61 *Initial implementation of Test - Timer Module - Unpause Timer 1.vi.*

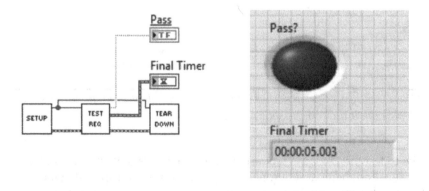

FIG. 6.62 *Stand-alone test vi for testing Unpause Timer request and the test results.*

Not at all what we expected, right? In fact, it looks like the timer never paused. Now you see what is missing: the DQMH timer module is not handling the unpause case appropriately. Let's fix that (see Fig. 6.63).

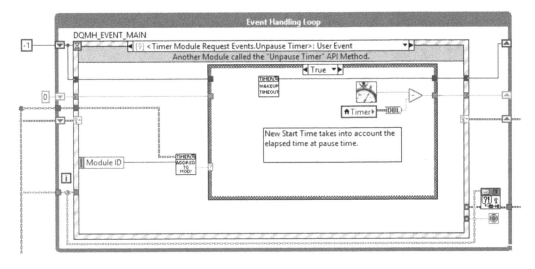

FIG. 6.63 *Final implementation of Test - Timer Module - Unpause Timer 1.vi.*

Now the new start time when the timer unpauses takes into account the elapsed time at pause time. The unit test should pass now. (See Fig. 6.64.)
And it does!

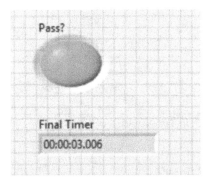

FIG. 6.64 *Test passes now.*

DQMH Unit Testing—JKI VI Tester

Follow the steps outlined earlier to create a new JKI VI Tester case. We need to change the private data of this class to hold the Module ID. (See Fig. 6.65.)

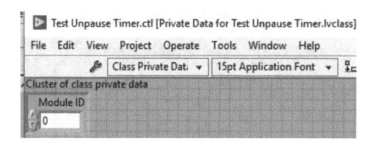

FIG. 6.65 *JKI VI Tester Test Unpause Timer private data.*

Then we modify the `setup.vi` for the class to call the one created by DQMH unit testing tools. (See Fig. 6.66.)

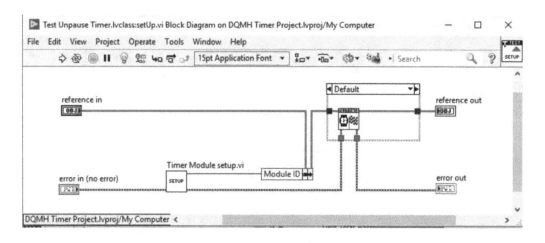

FIG. 6.66 *JKI VI Tester Test Unpause Timer setup.vi.*

Along the same lines, we modify the `teardown.vi` to include the one created by DQMH unit testing tools. (See Fig. 6.67.)

Chapter 6: Unit Testing **459**

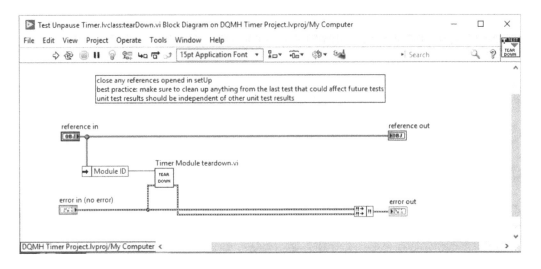

FIG. 6.67 *JKI VI Tester Test Unpause Timer teardown.vi.*

Now we need to call the test we created using the DQMH unit testing tools. (See Fig. 6.68.)

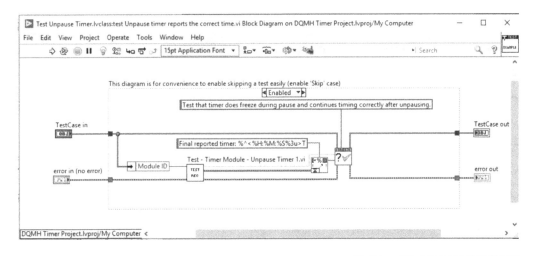

FIG. 6.68 *JKI VI Tester Test Unpause Timer test harness.*

Note that we are adding as a custom message the final reported timer. The JKI VI Tester will report this message in the case where the test fails. See Fig. 6.69 for an example.

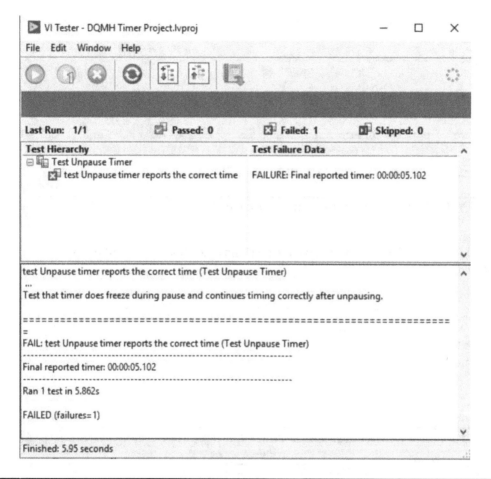

FIG. 6.69 *JKI VI Tester Test Unpause Timer report when the test fails.*

When the test passes, JKI VI Tester only reports the top message. (See Fig. 6.70.)

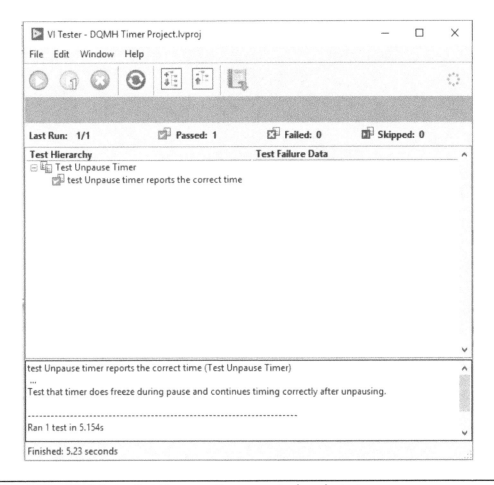

FIG. 6.70 *JKI VI Tester Test Unpause Timer report when the test passes.*

DQMH Unit Testing—Unit Test Framework

We will create the unit test using the test created by the DQMH unit testing tools as our unit under test. (See Fig. 6.71.)

We need to define the Setup and Teardown VIs. Please note that we needed to add a Module ID dup output to our tester so that we could add that to the Teardown VI. We call this page the wire by text page because we are reproducing the code we created in the stand-alone test but without a diagram. (See Fig. 6.72.)

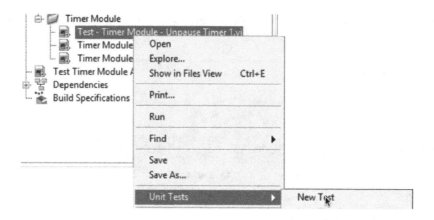

FIG. 6.71 *Right-click on the unit under test to create a new UTF unit test.*

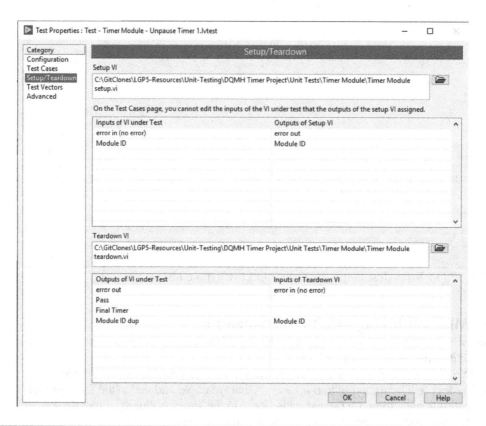

FIG. 6.72 *Test - Timer Module - Unpause Timer 1.lvtest Setup and Teardown VIs definition.*

The test case definition specifies that we expect the test to pass. (See Fig. 6.73.)

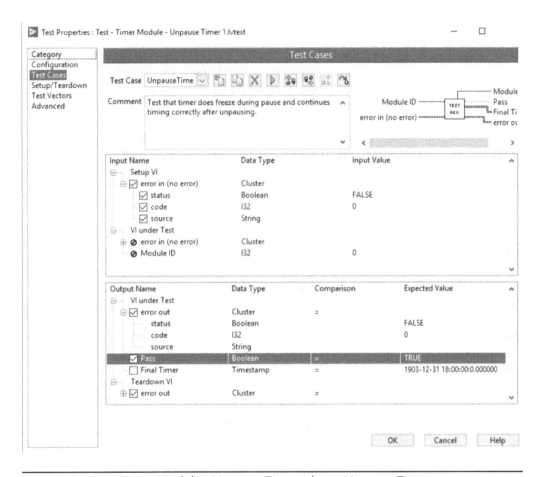

FIG. 6.73 *Test - Timer Module - Unpause Timer 1.lvtest Unpause Timer test case properties.*

DQMH Unit Testing—InstaCoverage

For a step-by-step tutorial on how to integrate InstaCoverage into DQMH Unit Testing, visit https://github.com/IncQueryLabs/LabVIEW-knowledge-base/wiki/Step-by-step-tutorial-for-integrating-InstaCoverage-into-DQMH-Unit-Testing.

What about RT?

Unit Test Framework and InstaCoverage support testing VIs running in the LabVIEW RT target. However, no tool supports code coverage measurement for tests executed on RT targets.

If you will be creating unit tests for your real-time code using UTF, you will need to edit your project properties accordingly. You need to check Enable VI Server on real-time targets. According to UTF help, you must enable the VI Server to execute unit tests on an RT target. LabVIEW disables the VI Server on the RT target when the tests complete.

As mentioned earlier, UTF cannot measure code coverage on real-time targets: the result will always be zero. You can see it for yourself. If you run the unit tests for UTF within the Unit Testing in RT project at GitHub.com/LGP5, the code coverage report will return zero. If you drag the VI and its .lvtest to the desktop and run the unit tests for UTF again, the code coverage report will come back with the correct percentage.

JKI VI Tester cannot execute tests on RT targets.

Example of Unit Testing for LabVIEW RT

Open the Unit Testing RT project in GitHub.com/LGP5. We created a simple VI in the RT target and used JKI VI Tester to create a test on the desktop. This JKI VI Tester runs because we don't have any LabVIEW RT–specific code in the unit under test. However, the tests are not running directly on the target; they are running on the desktop. We do not recommend using JKI VI Tester for testing RT code.

We created a UTF test by right-clicking on the `Simple RT Add.vi` and selecting to create a UTF test. (See Fig. 6.74.)

We also created an InstaCoverage test by right-clicking on the `Simple RT Add.vi` and selecting **InstaCoverage >> Generate Unit Test**. (See Figs. 6.75 and 6.76.)

FIG. 6.74 *Simple RT Add.lvtest UTF unit test does run in the RT target, where the JKI VI Tester test does not.*

Chapter 6: Unit Testing 465

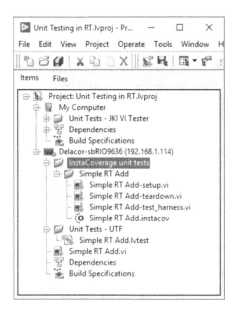

FIG. 6.75 *InstaCoverage created the unit tests under My Computer target, and we dragged them under the RT target.*

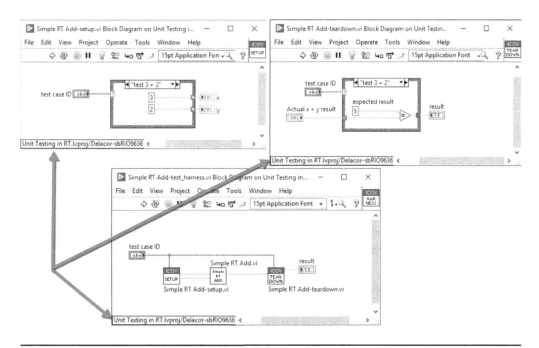

FIG. 6.76 *InstaCoverage unit tests run in the RT target instance.*

What about FPGA VIs?

Testing LabVIEW FPGA code is challenging because there are different layers involved. We had said earlier that we wanted to isolate the code from any hardware or dependency for our unit tests. In the case of LabVIEW FPGA, the code will eventually run on hardware, and we cannot disassociate our code from the hardware. This means there are different levels involved; in the least complex project we will have Host PC code and the FPGA target code. In more complex projects, we will have the Host PC code, the RT code, and then the FPGA target code. We could argue that LabVIEW FPGA testing is always integration testing, since it involves all these layers. Others call it component testing, not quite integration testing, because the code executes either in the Windows PC context or uses simulated I/O via custom VI for FPGA I/O and it does not need the actual FPGA for testing. (See Table 6.4.)

If you do a search at ni.com for Testing and Debugging LabVIEW FPGA Code, you will find a tutorial going into more detail on the different way we can test FPGA code.

Execution Mode	Windows PC[1]	FPGA Simulation Mode	Third-Party Simulation	FPGA Target
Verify Functional Performance	✓	✓	✓	✓
Verify Timing		✓[2]	✓	✓
Verify Third-Party HDL IP		✓[3]	✓	✓
Good for Unit Testing	✓			
Good for Component Testing		✓	✓	✓
Good for System Testing				✓

TABLE 6.4 *Testing and debugging LabVIEW FPGA code. Source: http://www.ni.com/tutorial/51862/en/.*

For testing in Windows PC, also known as testing in the Windows context, we can use Unit Test Framework using the VI in the LabVIEW FPGA section of the code, provided that the VI does not call any LabVIEW FPGA–specific functions. As James

McNally describes on his blog posts, there are a few things to keep in mind. The first obstacle is that the menu option to create a new unit test is not available under FPGA targets. You have to create a .lvtest in an approved target (desktop or RT). Then configure the .lvtest to point to the VI in FPGA that you want to test. See more details in the Examples section at the end of this chapter. (See Fig. 6.77.)

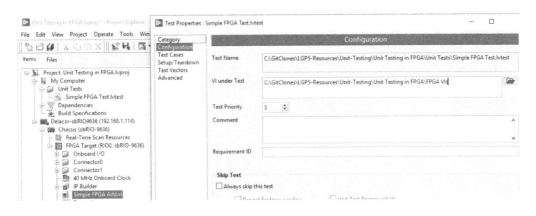

FIG. 6.77 *Unit Test Framework test configuration for FPGA VI.*

In order to have the flexibility of testing FPGA code directly in the Windows context, Ching Hwa Yu creates code to simulate things like Target to Host FIFO and use conditional diagram disable structures to decide if the code will use the FPGA node or the simulation code. This way, all you have to do is drag the VI from the FPGA target to the desktop and the VIs will not break and you can still test them. (See Fig. 6.78.)

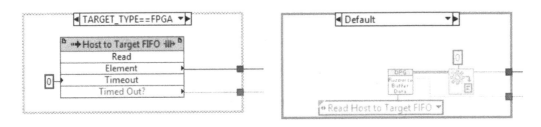

FIG. 6.78 *Use conditional diagram disable and diagram disable structures to facilitate testing either on the desktop, directly on FPGA, or in simulation mode.*

Figure 6.78 is an example of a conditional diagram disabled structure showing the TARGET_TYPE==FPGA case calling the FPGA Host to Target FIFO node. The other case calls a simulated code that does not contain any FPGA nodes. The picture was taken from the Ching Hwa Yu example available via GitHub.com/chinghwayu.

For the FPGA Simulation Mode, you will need to right-click on the FPGA Target and change the Simulated I/O to point to the Simulated I/O for your FPGA VI. (See Fig. 6.79.)

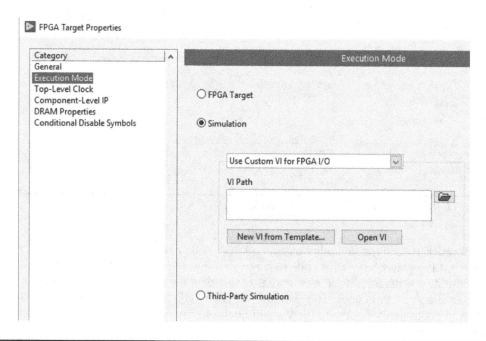

FIG. 6.79 *Configure FPGA Target to run in simulation mode.*

Notice that there is a **New VI from Template...** button; use this button to generate your own simulated IO.

Another important tool when testing FPGA code is the test vectors. FPGA logic builds up through different iterations of the code (pipelining). You can define a test vector that will test the different iterations that will exercise a complete logic unit in the FPGA. However, there is a big caveat: you need to reset the feedback nodes in the FPGA code. James McNally offers the code in Fig. 6.80.

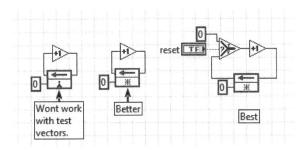

FIG. 6.80 *If you will use UTF and test vectors, make sure you reset the feedback nodes appropriately. Source: https://devs.wiresmithtech.com/blog/4-lessons-learnt-unit-testing-labview-fpga-code/.*

If you need to test FPGA-specific code, you will have to create a Unit Test Framework User Defined Test. The developer writes LabVIEW code that can include an FPGA desktop execution node that will call the FPGA VI. (See Fig. 6.81.)

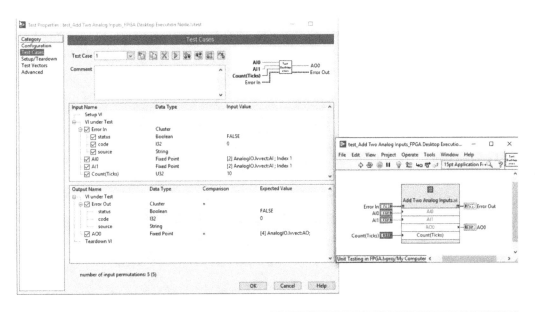

FIG. 6.81 *Use the FPGA desktop execution node in tests that will run on the desktop.*

Example of Unit Testing for LabVIEW FPGA
Testing Using the Unit Test Framework

You have to create a .lvtest in an approved target (desktop or RT). To create this test right-click on **My Computer** and select **New >> Unit Test**. This will create an Untitled.lvtest. Here you will face the next hurdle: you cannot rename a .lvtest file via LabVIEW. You have to close LabVIEW, go to disk, rename it, and go back to LabVIEW... yes, a hassle... (See Fig. 6.82.)

Then configure the .lvtest to point to the VI in FPGA that you want to test. (See Fig. 6.83.)

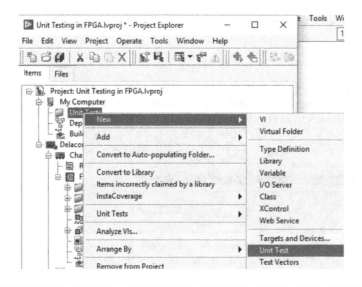

FIG. 6.82 *Create the UTF unit test on MyComputer target.*

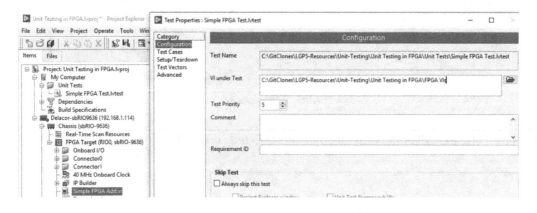

FIG. 6.83 *Point the VI under Test path to the FPGA VI you want to test.*

Testing Using LabVIEW FPGA Simulation Mode

For the FPGA Simulation Mode, we will use the code that Ching Hwa Yu and Jianhua Liu presented at NIWeek 2017 during their "Automated Test of LabVIEW FPGA Code: CI and Jenkins 2 Pipelines" presentation. They use testing to ensure that the commands and parameters that the user interface generates are properly translated to FPGA. They made their code available via GitHub.com/chinghwayu. You will need to install LabVIEW 2016 or later, LabVIEW FPGA, and FlexRIO drivers to see the code. Once you download or clone the repository, you will need to right-click on the FPGA Target and change the Simulated I/O to point to the new location of the Simulated I/O for Digital Pattern Generator.vi. (See Fig. 6.84.)

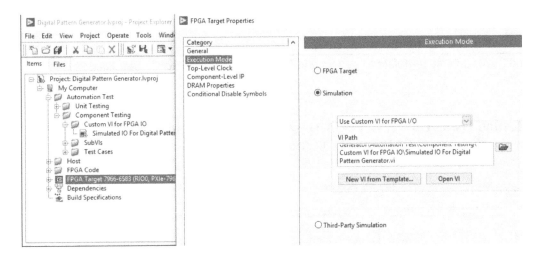

FIG. 6.84 *Verify that the FPGA Target is configured to run in Simulation mode.*

Notice that there is a button called New VI from Template…; that is the VI that Ching Hwa Yu used as a template to generate his own Simulated IO for Digital Pattern Generator.vi. He then added specific simulation code for Reading I/O and Writing I/O. (See Fig. 6.85.)

472 LabVIEW Graphical Programming

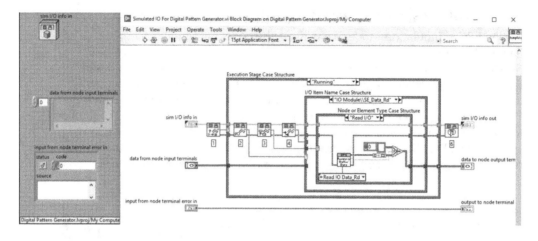

FIG. 6.85 *Modify the template to generate your own simulated IO for Digital Pattern Generator.vi.*

Testing Using LabVIEW FPGA Desktop Execution Node

Now let's test the FPGA VI called Add Two Analog Inputs.vi. (See Fig. 6.86.)

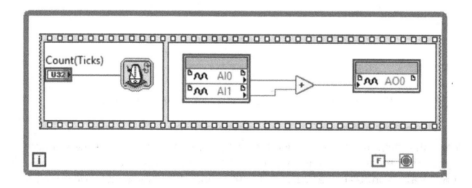

FIG. 6.86 *Add Two Analog Inputs.vi.*

When you place the FPGA Desktop Execution Node on the block diagram of a new VI, a configuration dialog pops up for you to configure which VI to execute. There, you can select the FPGA resources that you want to simulate. (See Fig. 6.87.)

FIG. 6.87 *Configure FPGA Desktop Execution Node to simulate AI0, AI1, AO0, and Count (Ticks).*

Then this new VI can be the unit under test for a Unit Test Framework test. (See Fig. 6.88.)

For this test, the test vectors were defined as a simple sequence for both the AI and AO vectors. (See Fig. 6.89.)

474 LabVIEW Graphical Programming

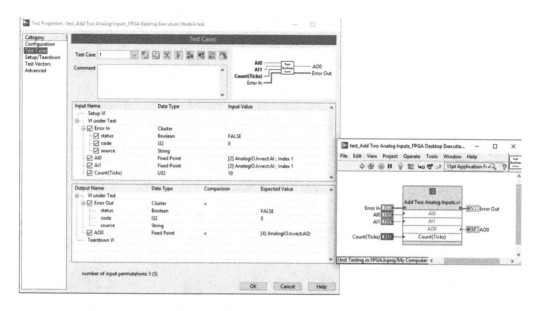

FIG. 6.88 *Configure the UTF test that calls the FPGA Desktop Execution Node.*

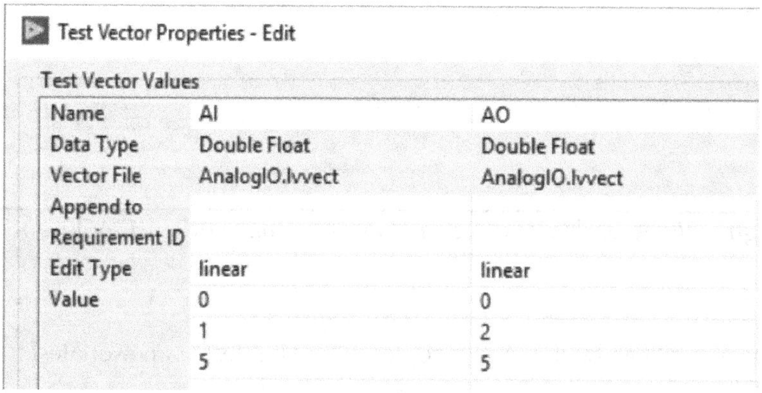

Test Vector Properties - Edit		
Test Vector Values		
Name	AI	AO
Data Type	Double Float	Double Float
Vector File	AnalogIO.lvvect	AnalogIO.lvvect
Append to		
Requirement ID		
Edit Type	linear	linear
Value	0	0
	1	2
	5	5

FIG. 6.89 *AI and AO vectors definition.*

Unit Tests for the GUI?

End-to-end testing is really integration testing, but there may be XControls, user interfaces, etc., that need unit tests. We don't really have good tools for testing the GUI. Other languages have tools that can record actions on the screen. Some of these tools can be used with LabVIEW but require the developer to wrap their code in DLL or .NET. This violates one of the principles of a good unit test: creating a unit test needs to be easy to implement.

This goes back to the discussion about testing private versus public VIs. We can trust that LabVIEW will press a button correctly. We can create a unit under test that would implement the sequence of events that the interaction with the GUI unleashes and then test that new unit under test. The DQMH toolkit includes a scripting tool to create unit tests. If your GUI interactions call DQMH requests in the background, you can use this scripting tool to create your unit tests.

Unit Test Reporting

If you are going to implement continuous integration (CI) and want to run your unit tests automatically via a tool like Jenkins, then you need to generate JUnit reports that can be then sent to the Jenkins' Performance plugin. UTF, JKI VI Tester, and InstaCoverage provide JUnit reports. We recommend looking into Ching Hwa Yu's presentation from NIWeek 2017 where he goes into more detail as to how to integrate LabVIEW unit tests with Jenkins. He has this and other examples on his blog at https://chinghwayu.com/. The IncQueryLabs team also have put together a CI demonstrator for InstaCoverage at https://build.incquerylabs.com/jenkins/job/instacoverage-ci-demonstrator/ with a wiki documentation at https://github.com/IncQueryLabs/instacoverage-ci-demonstrator/wiki.

If you are not working in a CI environment, you can still configure UTF to generate a report at the end of the test. Right-click on your project and in the Unit Test Framework section, configure what type of report you want to generate and where you want to save the reports. You can also use the UTF API to run your UTF unit tests programmatically and use the API call to generate an ASCII, HTML, or XML report.

JKI VI Tester and InstaCoverage also provide an API to run the tests programmatically and output a test report. JKI VI tester uses the Message input to the test results VIs to report text for each test.

Example of Assertions

The code can be found at https://github.com/LGP5/Unit-Testing.

This example is based on a real-life experience. We were working on a demo for a tradeshow. The demo consisted of a 4×4 cart mounted on a motion platform. There was a monitor in front of the cart playing a video game, and the game would transmit position information to the motion controller via UDP. We also had an inclinometer to report what the actual current position was. We couldn't figure out why the cart was always inclined to the extremes. If we had had assertions in the code, we would have found out that a precondition assumption was not met.

According to the datasheet, the voltage-to-angle conversion used the following formula:

$$Angle = \arcsin\left(\frac{V_{out} - Offset}{Sensitivity}\right)$$

Offset = 2.5 V. This is also the inclinometer output at 0°.
Sensitivity = 2 V/g
V_{out} = Inclinometer output.

We also knew that the inclinometer could only measure between −90° and 90°. This means the value inside the arcsin can only be between −1 and 1. These results in the valid inclinometer output could only be between 0.5 V and 4.5 V. When we were testing, we just configured the control to coerce the values. However, the VI that calculates the angle was never meant to be executed on its own or as a top level; we should not have modified the data entry for those controls. If the calling code sends a value out of range, it would get coerced and we were not alerted that something was off. Well, eventually something was off. Maybe it was a miswired sensor or some other hardware issue. We spent days troubleshooting the issue, looking everywhere until we found this VI. Yes, back then we didn't do unit testing or used assertions. (See Fig. 6.90.)

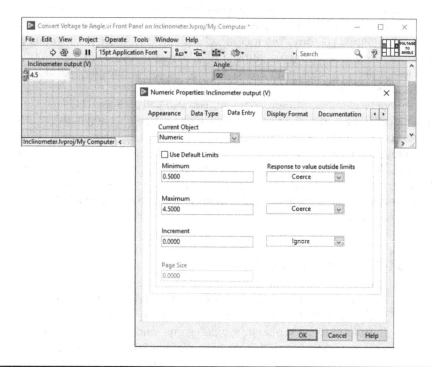

FIG. 6.90 *Convert Voltage to Angle.vi with the control set to coerce values between 0.5 V and 4.5 V.*

Assertions with Caraya

For this example, you will need to have the Caraya Unit Test Framework installed. You can find this package directly via JKI VIPM or you can download the source code via JKI GitHub at https://github.com/JKISoftware/Caraya.

We made a copy of the original VI and called it `Convert Voltage to Angle-with Caraya Assertions.vi`. We removed the data entry coercion for the input and added the Caraya assertion functions for greater or equal and less or equal. Note that these VIs could be built into the executable; following the idea that asserts should be compiled out of deployable code, we put the code inside a conditional diagram structure that is empty for the executable case. (See Fig. 6.91.)

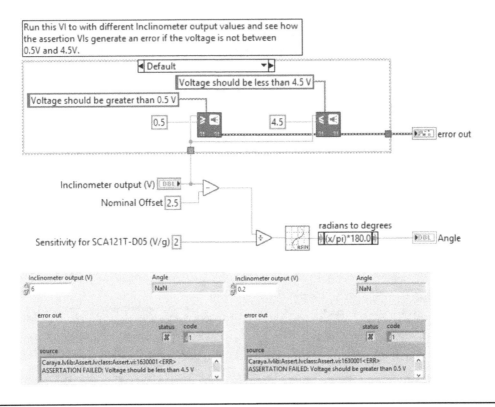

FIG. 6.91 *Convert Voltage to Angle-with Caraya Assertions.vi.*

The problem we have now is that these assertions do not fail loudly enough because our VI is configured to not automatically handle errors. If we want this assert to be loud, we need to configure the VI to automatically handle errors and remove the error output control. This is something that you might not normally do,[5] but this is one of the situations where we would need to enable it. Perhaps in the future, Caraya will have an Assertions monitor, but in the meantime, if you will be using Caraya for implementing assertions, you need to be notified as soon as an assertion fails. (See Figs. 6.92 and 6.93.)

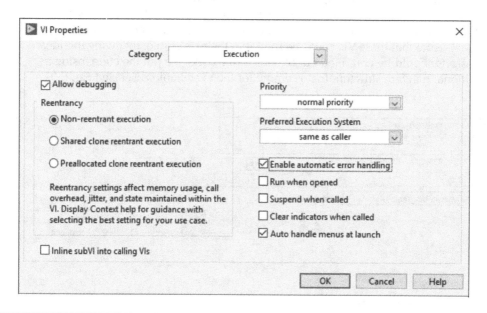

FIG. 6.92 *Configure Enable automatic error handling.*

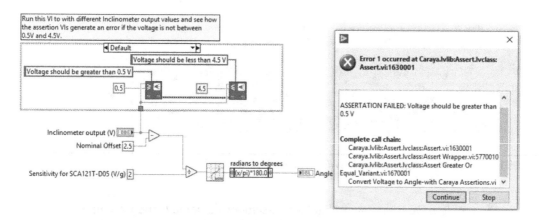

FIG. 6.93 *Now our code asserts as soon as the precondition is not met.*

---

[5] See Darren Nattinger's "End to Brainless Programming" presentation for more details as to why we prefer to have Enable automatic error handling unchecked.

Assertions with AssertAPI

For this example, you will need to have Peter Horn's AssertAPI installed. You can find this package via GitHub.com/PeteHorn/AssertAPI.

We made a copy of the original VI and called it `Convert Voltage to Angle-with AssertAPI Assertions.vi`. We removed the data entry coercion for the input and right-clicked on the Inclinometer output (V) wire and selected Assert API >> Create New Assert. (See Fig. 6.94.)

FIG. 6.94 *With AssertAPI installed, right-click on the wire and select Assert API >> Create New Assert.*

This will bring up a `Build Assert VI.vi` wizard that will create the assert VI following our specifications. We will create two assert VIs: one for the max range and another one for the min range. Please note that we could also have created a single assert VI for "Greater Than Less Than" resulting in a single assert VI. (See Fig. 6.95.)

The options available at the moment we were writing this book were the ones listed in Fig. 6.96.

Peter is continuously improving the tool; when he was reviewing this chapter, he realized that the "None" made it sound like the API will do nothing. He is going to change it to "Display Only," as it will display the results in the main UI.

We create an auto-populating folder for the Assert VIs; this way any future Assert VIs we create will be added automatically to the project. (See Figs. 6.97 and 6.98.)

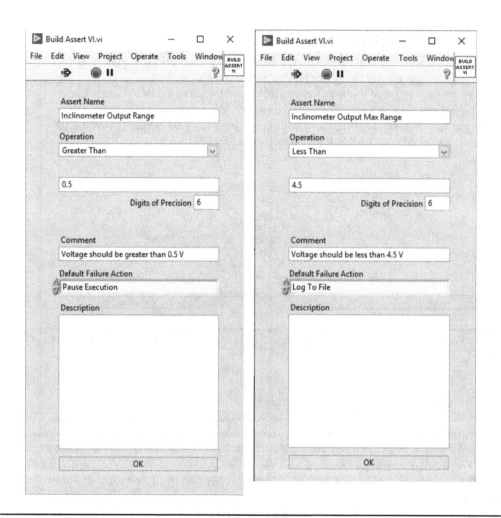

FIG. 6.95 *Build Assert VI.vi wizard configuration for both asserts.*

FIG. 6.96 *Default failure action when the assert fails.*

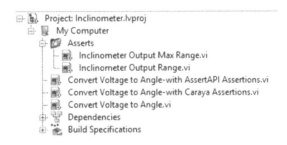

FIG. 6.97 *Asserts auto-populating folder shows the Assert VIs the AssertAPI wizard created for us.*

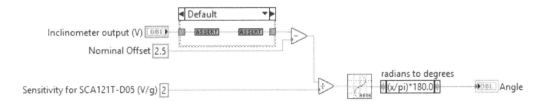

FIG. 6.98 *Compiling assert code out of the executable by wrapping the asserts in a conditional diagram disable structure that does nothing in the executable case.*

Now we have the option to run the VI as is, and it will pause when the voltage is less than 0.5 V and will log to file when the voltage is greater than 4.5 V. The log file will appear on our desktop, and it is a .csv file. This is not a very striking notification. If we want to monitor all of the assertions in our code, we can go to Tools >> Assert API >> Start Assert Display... to monitor the existing Asserts. (See Fig. 6.99.)

You are probably thinking that conditional probes already do this for us. However, probes are not preserved between runs; these assertions are saved with the code. The developer can right-click on each assertion and go directly to where that assertion was reported.[6]

---

[6] We have heard of LabVIEW programmers who save the location of probes to a configuration file so they can load a set of probes back in later when they need to do some debugging.

A. Start Assert Display when the Inclinometer Voltage is 0 V

Assert Name	Status	Failure Action	Data Value	Last Update Timestamp
Inclinometer Output Range	Fail	Pause Execution	0.000000	05:23:37.066 PM 11/02/18

B. Start Assert Display when the Inclinometer Voltage is 6 V

Assert Name	Status	Failure Action	Data Value	Last Update Timestamp
Inclinometer Output Range	Pass	Pause Execution	6.000000	05:24:27.623 PM 11/02/18
Inclinometer Output Max Range	Fail	Log To File	6.000000	05:24:02.624 PM 11/02/18

Contents of Assert logFile

Inclinometer Output Max Range	Inclinometer Output Max Range.vi	Convert Voltage to Angle-with AssertAPI Assertions.vi	Not yet implemented	X < 4.500000	Fail	5:24:27.624 PM	Voltage should be less than 4.5 V

C. Start Assert Display when the Inclinometer Voltage is 3 V

Assert Name	Status	Failure Action	Data Value	Last Update Timestamp
Inclinometer Output Range	Pass	Pause Execution	3.000000	05:25:01.732 PM 11/02/18
Inclinometer Output Max Range	Pass	Log To File	3.000000	05:25:01.732 PM 11/02/18

FIG. 6.99 *As we execute the Convert Voltage to Angle-with AssertAPI Assertions.vi with different values, the Start Assert Display updates with the new data and uses color coding to highlight failures. The developer can right-click on any of the assertions on the table and change the failure action as needed.*

Unit Testing and LabVIEW NXG

From what we have heard at CLA Summits and other NI events, NI is empowering the community to create the unit test tool for NXG. This means that more than likely Unit Test Framework will not exist for LabVIEW NXG. Eventually, JKI might migrate Caraya and JKI VI Tester code for NextGen. InstaCoverage by IncQuery Labs works for LabVIEW and LabVIEW NXG. See Tables 6.5 and 6.6 for tool comparisons.

Tool Comparison

Properties of a Good Test	Stand-Alone Test	Unit Test Framework	JKI VI Tester	InstaCoverage	AssertAPI
Automated	No	Yes	Yes	Yes	Yes
Repeatable	Yes	Yes	Yes	Yes	Yes
Easy to implement	Yes	Yes	There are several steps involved in creating a new test class. Might not be straightforward for people not familiar with LVOOP.	There are several steps involved in creating a new test. InstaCoverage offers wizards and tools to assist in the process.	There are several steps involved in creating a new assertion VIs. AssertAPI offers wizards to assist in the process.
Relevant in the future	Only if the developer is vigilant and keeps the test up to date and continues to run the test	Yes	Yes	Yes	Yes
Easy to run	Yes	Yes	Yes	Yes	Yes
Run quickly	Yes	Depends on the number of VIs already in memory. As the project grows, UTF gets slower.	Takes time to load	Yes (faster to load than UTF and JKI VI Tester)	Yes
Consistent results	Yes	Yes	Yes	Yes	Yes
Test coverage report	No way to know	Yes	No	Yes	No
Fully isolated from the rest of the code	Yes	Yes	No	Yes	No
Failures should be obvious and easy to debug	Yes	Yes	Yes	Yes	Yes

TABLE 6.5 *Tools comparison table.*

Unit Testing Tool Alternatives for LabVIEW

Unit testing tool for LabVIEW[1]	Model[2]	Price	Code Coverage		Real-time target	Source code availability	API (CI support)	NXG
			Measurement	Scalability[5]				
Unit test Framework	xUnit-style	$1,555 or bundled[3]	Yes	No	Yes	None (proprietary)		Not planned
VI Tester	Assertion-based	Free	No	No	No	Yes (BSD-like)		Not known
Caraya			No	No	No	Yes (BSD-like)	Yes[7]	
InstaCoverage	xUnit-style	€279/seat[4]	Yes	Yes	Yes	Inspection[6] (available on request)		Coming soon[8]

1 For "current gen" LabVIEW Windows versions + NXG (3.0). Only Caraya is available for Linux. For external add-ons (like InstaCoverage) licensing is not supported for Linux.
2 Only UTF and InstaCoverage support explicit test configurations (.ivtest and .instacov respectively).
3 Part of LabVIEW Professional edition.
4 Free InstaCoverage Core coming soon (expected in late 2018/early 2019).
5 Speed of test execution as project size increases. For example, InstaCoverage is up to 100x faster than UTF, e.g, 1m vs 1h on average size industry projects.
6 Allows in-house customization.
7 For example, continous integration demonstrator with Jenkins for InstaCoverage: https://build.incquerylabs.com/jenkins/job/instacoverage-ci-demonstrator.
8 InstaCoverage Core for NXG to be released with NXG 3.0 (expected in November 2018).

TABLE 6.6 *Comparison table courtesy of IncqueryLabs, creators of InstaCoverage.*

References

The Unit Testing Group at NI.com https://forums.ni.com/t5/Unit-Testing-Group/gp-p/5182

The Art of Unit Testing Second Edition by Roy Oshevore, August 2009. https://www.manning.com/books/the-art-of-unit-testing-second-edition

James McNally blog series on Unit testing at https://devs.wiresmithtech.com/blog/tag/unit-testing/

Unit Testing in LabVIEW presentation at the LabVIEW Architects Forum, March 1st, 2018. Nate Moehring https://forums.ni.com/t5/Unit-Testing-Group/Unit-Testing-in-LabVIEW/gpm-p/3762773

Fab's presentations on Unit Testing
- CLA Summit 2014 at CERN, "Save Time and Money with Unit Testing"
  - Video available at https://youtu.be/4TW0dLLioaI
- NI Week 2014, Austin, TX, "Save Time and Money with Unit Testing"
- IEEE Spectrum Tech Insiders Webinar Series 'Save Time and Money with Unit Testing"
  - Video available at http://bit.ly/FabIEEE_UnitTesting

JKI VI Tester LabVIEW Tools Network page

JKI VI Tester open source project maintained by JKI at https://github.com/JKISoftware/JKI-VI-Tester

InstaCoverage https://incquerylabs.com/instacoverage/

Step by step tutorial for integrating InstaCoverage into DQMH Unit Testing available at https://github.com/IncQueryLabs/LabVIEW-knowledge-base/wiki/Step-by-step-tutorial-for-integrating-InstaCoverage-into-DQMH-Unit-Testing

https://chinghwayu.com/2017/06/niweek-2017-automated-test-of-labview-fpga-code-ci-and-jenkins-pipelines/

https://smartbear.com/learn/automated-testing/what-is-regression-testing/

Code Complete: A Practical Handbook of Software Construction, Second Edition. Steven C. McConnell. 2004.

LabVIEW Assert API by Peter Horn https://github.com/PeteHorn/AssertAPI

CHAPTER 7

Developing in LabVIEW for Teams

In the beginning it was typical for a LabVIEW programmer to be a lone wolf. Either they had a chance to go to some of the National Instruments (NI) Customer Education trainings or they had to teach themselves LabVIEW. A lucky few had a mentor in their team that introduced them to good programming practices in LabVIEW. The majority of these technicians, scientists, and engineers did not have any programming background and used LabVIEW as a tool to get their job done. With time, the size of LabVIEW teams and the complexity of problems it solves has grown. It is common to visit companies where there are full-time LabVIEW programmers. At the CLA Summits in Austin and in Europe there are equal numbers of presentations and discussions about technical topics as there are about working in teams.

In this chapter, I will guide you through the software engineering practices and tools I recommend for LabVIEW teams to use. The tools I recommend and how I recommend to use them is only a way; I will never claim to know "the way." I have realized that there is no one-size-fits-all and that it is better for the team to decide what tools work best for them. The same goes for software development processes. Agile processes are pretty popular, but even the Agile Alliance[1] will tell you that there is a big focus in the Agile software development community on collaboration and the self-organizing team. Your team needs to find what is best for them. We are no longer talking about code, I can tell you how to best structure your code to get the results you expect. I know how each of the primitives in that LabVIEW code will behave and what I expect from that

---

[1] https://www.agilealliance.org/agile101/

application at the end. But I have no way of knowing the people in your team and how they work better together. Only you and your team know that.

Where Is Your Team At?

Conway's law says that "Organizations which design systems ... are constrained to produce designs which are copies of the communication structures of these organizations."[2] If you want to improve your code, you need to look at your team as well. You might even need to look at your organization!

You probably know that implementing good software engineering practices is going to take some effort, time, and money. We understand that you or your boss will want to know what the return on investment (ROI) is. One way we have measured the ROI is by doing an audit of the process (or lack thereof) and the team's code. We measure how they are doing in different areas such as source code control, deployment, and unit testing. We even look at the human aspect, because we are dealing with people, not code. At the end of the audit, we have a chart that we call "The Delacor Thermometer." (See Figs. 7.1 and 7.2.) This gives a good visual for the managers that are paying for us to help teams implement processes and new tools.

Before working with Delacor									
Human	Design	Process	Tools	Documentation	Efficiency	Testing	Hardware	Code	Overall
9.2	3.4	4.9	1.5	3.8	3.2	3.5	7.0	5.5	4.7
After working with Delacor during the Audit process									
9.2	4.3	6.6	1.8	4.8	4.3	3.5	7.0	7.4	5.4
Where we would like to see The Team in 1 year									
9.3	7.6	8.9	6.3	7.2	5.8	9.2	7.0	9.6	7.9

| 0 | 1 | 2 | 3 | 4 | 5 | 6 | 7 | 8 | 9 | 10 |

FIG. 7.1 *Example of Delacor Thermometers to share with one team where they were before starting to work with Delacor, some of the advances they got just by engaging with Delacor, and the goal for the following year. Notice that the goal was not to get every aspect to pass (above 7). The team chose to focus on their code, testing, and process.*

We emphasize that not every single area needs to be addressed at once and recommend the areas we think will give them the largest ROI. We also give them a target of where we think their thermometer should be in a year. We come back a year later, do another audit, and measure if they made it there or not. We determine the target with the team; the team is really setting the goal based on our assessment,

---

[2] Conway's law en.wikipedia.org/wiki/Conway%27s_law

This document is to gather all the questions/points/controls that need to be addressed when auditing existing code/processes.
Every point needs to be answered, but the actual questions/points/answers are not shared with the customer. Instead we would show a "heat map" (Delacor Thermometer) of the areas they need to work on.

Area	Control	How are we going to assess control?	Guidelines and Instructions for auditor					Auditor fills in these columns		
			Survey	Interview	Review code	Physical evidence	Observations/evidence of whether the control is followed or not	Grade (0 to 10). 10 being excellent compliance.	Pass/Fail	Reasons for Pass/Fail and grade
Pre Interview	Set one hour aside for interview followed by 30 minutes for them to produce any examples/documents they need to send you.									
	In the INTERVIEW make it clear that we will not share individual comments with names, we will share aggregated results with your supervisors. We are recording your answers only for our use to create the report. We are not here to judge your past nor your present, we are here to help your team figure out what is the best approach in the future.			Yes						
Interview intro	Describe a typical day in your job? How much LabVIEW you do?			Yes						
Design	Modeling and planning before coding starts	Does the team have tools defined to model/plan before coding starts? Examples of tools: Visio, websequencediagrams, GOOP UML modeler	Yes				10 = All the architects, software leads in the team are aware of the tools available and enforce modeling before start coding. 5 = Some modeling is done, there are diagrams on whiteboards around the office, but they are not stored anywhere. 0 = No tools, code and fix approach.			
Process	Style Guidelines are verified via VI Analyzer tests	There are clear guidelines as to how to determine the cfg to use, where to find it and when to run the VI Analyzer.	Yes			Yes	10 = Everyone in the team knows when to run the VI Analyzer tests and runs them. 5 = VI Analyzer tests are only executed when someone asks for the results or after bugs are discovered. 0 = No VI Analyzer tests			
Tools	Team uses Source Code Control tools	Ask team members what source code control tool use and if they are using it only to store releases or if it is used frequently.	Yes				10 = all the team trained in scc, everybody commits/pushes often 7 = not all the code in scc 5 = scc only for releases 2 = zip files 0 = no SCC			
Documentation	Style Guidelines exist	Ask team member if document exists	Yes			Yes	10 = document exists 0 = no document			
Efficiency	Build server is used exclusively for building and never/ever used for testing or editing code	Ask team members if they have performed fixes directly in the build server or tested their code there.		Yes			10 = Build server used exclusively to build 5 = Developers regularly fix/test code there 0 = No dedicated build machine			
Testing	Unit Testing used/ran before every build	Ask team member if they run unit tests before every build	Yes			Yes	10 = Yes 0 = No			
Hardware	Manufacturing tests limits come from Theory operation and not from characterization only	Ask team members how they determine their test limits		Yes			10 = From theory of operation 5 = From characterization 0 = Made up			
Code	VI Analyzer results	Run Delacor internal cfg (or other cfg agreed before audit)			Yes		Grade correlates to percentage pass. (for example 95% pass = 9.5 grade)			
Human	Joy/satisfaction at work	Do you enjoy going to work? What is it that you enjoy? What you don't like? Are there people you are afraid to talk to? What is your biggest pain at your job?		Yes						

FIG. 7.2 *Example of the Audit tool Delacor uses to measure where a team is at. The tool Delacor uses has a lot of questions for each section.*

we coach them toward setting realistic goals. The Delacor Thermometer for the team in the figure presented earlier (see Fig. 7.1) was one of the most challenging we ever had because the human column was so high! These people adored their jobs; they believe in the product they are making and do not care about the long commute to get to work. We are used to working with teams that are in disarray, and the state of their code is just a reflection of other problems in their team; most of the time anything we do will be better than where they were before. The challenge here was that no matter what we did, we better not make those people hate their job!

Steve Watts presented his Project Risk Registers on his blog[3] as another way to measure the ROI on following processes and selecting the right kind of tools for your team. You could measure how you are doing today with the things you are already tracking and set a goal where you want to be in 6 months or a year. Then put that goal aside and continue to track your progress. Then come back later and see if you made it where you wanted to be at. It is hard to show ROI if your team is not already tracking things. One way is to ask the team where their tasks and team goals fall within the company's goals. Ask the managers what their biggest issue is. If they say: the time it takes from requirements to having a finished product, ask if they are already measuring that. From there, you can project how much time you think they will save if they follow the processes you are proposing. You don't have to go for 50 percent reduction; maybe all they want is to have an accurate estimate of how long things take. Just having a measurement on how long things take might be enough ROI for that team.

There is no one-size-fits-all; the ROI will be unique for each team.

It is very hard to get teams to move more into thinking on investing in the future and moving away from the bad habits of focusing on technical debt and dealing with the task in front of them. A typical excuse is: we don't have time for that! I can only think of the image of the guys using square wheels, telling the guy with the round wheels that they don't have time for that.

What Is the Problem You Are Trying to Solve?

Before you embark on any project, whether it is starting to code or solving how best to work with your team, the very first question you have to answer is "what is the problem that we are trying to solve?" I like to do an exercise before we start any project called the Agile Inception Deck.[4] I read about it in the Agile

---

[3] Watts, Steve. What's the ROI of all this Process Stuff? forums.ni.com/t5/Random-Ramblings-on-LabVIEW/What-s-the-ROI-of-all-this-Process-Stuff/ba-p/3705742

[4] Rasmusson, Jonathan. "The Agile Inception Deck." *The Agile Warrior*, 6 Nov. 2010, agilewarrior.wordpress.com/2010/11/06/the-agile-inception-deck

Samurai book,[5] but you can find a summary and a template slide deck at the Agile Warrior blog. It guides you through answering ten questions at the start of your project to ensure everyone is indeed on the same page. We even did this exercise before writing the fifth edition of this book. I have successfully used it to get that first quote to customers; we get together with the customer and we go through the exercise to figure out what that they want versus what they actually need.

Another important part of the Inception Deck is the page with the Trade-off sliders, where you define what you are willing to trade off when things in your project get difficult. There are four main aspects that you need to keep in mind: quality, scope, time, and budget. You cannot meet them all in every single project. Understanding the one that your team can cut corners on is important. For example, I know that any application that I am releasing to my customers will have the highest quality; that is one that I am not willing to trade off. Which one of the other three we can trade off will depend on the customer I am working with. I have customers that have no problem increasing the budget, whereas there are others where the release date cannot move. Find out those trade-offs first and then find out four other sliders for your team, print them out, and have them in a visible spot.

I know that some teams that use LabVIEW don't see themselves as software engineers. Until they come to the realization that, indeed, all those wires and graphics are software programming! The next realization is that there might be a better way to do things and probably somebody else already figured it out. Brian Powell says that software engineering is an engineering problem.[6] Once you figure out that when your team is programming in LabVIEW, it is a software team, you can move on to solving the problem of working in LabVIEW as a team as efficiently as possible.

You already picked your development environment, LabVIEW. You might even already done the exercise in the previous section of determining where you are at and how well are you doing in the different areas. We will cover different tools and processes in the following sections; read the ones where you know you need to improve.

What Is Technical Wealth?

I got my inspiration from an article by Andrea Goulet[7] about forgetting technical debt and starting to focus on technical wealth instead. I gave two presentations on the topic: one at CLA Summit[8] and another one at NIWeek.[9] Based on those

---

[5] Rasmusson, Jonathan. *The Agile Samurai: How Agile Masters Deliver Great Software*. The Pragmatic Bookshelf, 2012.
[6] Powell, Brian H. "My Continuously Evolving Thoughts on Software Engineering." GDevCon#1
[7] Goulet, Andrea. Forget Technical Debt - Here's How to Build Technical Wealth firstround.com/review/forget-technical-debt-heres-how-to-build-technical-wealth
[8] De la Cueva, Fabiola. First Thoughts About Technical Wealth. delacor.com/first-thoughts-about-technical-wealth/
[9] De la Cueva, Fabiola. Saving Money by Investing in Technical Wealth. youtu.be/RPL94qCoJZQ

presentations and on Steve Watts's blog post[10] on the same topic, here are the definitions we came up with.

Technical debt refers to suboptimal design decisions based on extenuating circumstances. You accrue technical debt when you choose a suboptimal approach that may look like less effort up front but ends up costing you later. The cost could be a missed deadline, more money invested in the project, or just a headache down the line. All of these costs are far worse than the extra time and effort it would have taken to take the better approach in the beginning.

A typical example of technical debt is not designing up front and starting to code right away. The debt increases even more as you continue to code and fix, patch here, patch there until you get to an application that looks like spaghetti code; it is hard to read and hard to maintain.

Technical investments is the time and money spent building your tool set, processes, and skills. The reward of this investment is the creation of technical assets.

Technical assets include project templates, reusable libraries, software quality tools like VI Analyzer custom tests, style guidelines, a unit testing policy, API testers (like the DQMH Testers), and a chosen framework understood by all members of the team. These technical assets take time to create or take money to buy. Even if you get a free technical asset like DQMH, you still have to invest the time into learning how to use it. When you read technical books or attend conferences, it always seems like these technical assets are just there and don't require any investment.

Technical tax. Every design decision will have a cost associated with it. In the financial world you would be paying your taxes to a government and eventually get to see some benefit from them. I couldn't find what the analogy to the government would be, so we will stick just with the pain of paying taxes and not getting to enjoy part of what you earned.

When you sacrifice ease of use for functionality, you are paying a technical tax. When you sacrifice flexibility for simplicity, you are paying a technical tax. When it takes many steps to create a class, or you have to do certain repetitive tasks in your code every time, you are paying a technical tax. Every decision has a positive and a negative result. You need to pick the technical assets that benefit you at the lowest rate of technical tax.

Technical wealth is the collection of technical assets that eliminate or reduce suboptimal design decisions.

Technical insurance. We pay for insurance in advance so that when there is an accident, we are covered and we don't go bankrupt.

In software projects, our technical insurance includes the unit tests, creating testable code from the beginning, comments that describe intent, and bookmarks (when used as breadcrumbs to indicate "if you change things here, you might want to check out all the places where this bookmark is because chances are you need to

---

[10] Watts, Steve Let's Talk Technical Accounting forums.ni.com/t5/Random-Ramblings-on-LabVIEW/Let-s-Talk-Technical-Accounting/ba-p/3871788

change that too"). Anything that is there so if something not planned happens (e.g., a nasty bug), we are covered and don't go bankrupt!

Other examples of technical insurance are regression testing, source code control, design reviews, and building executables early and often (so you discover early things that work differently in the source code and in the executable).

Technical depreciation is that, unfortunately, your technical assets might become less valuable over time. Maybe there are better or more suitable tools and methodologies available. Your team needs to invest in new technical assets ahead of current ones becoming worthless. Come back at the end of a project and verify what needs improvement and what technical assets are still working for you. You might need to start investing in the next set of tools sooner than you expected. I have helped customers who after all these years are still maintaining code that started in LabVIEW 4.0, and although the code has been upgraded all the way to the latest version of LabVIEW, the coding practices are still back in LabVIEW 4.0 (no user events, no libraries, sometimes not even projects!). These teams let their technical assets completely depreciate!

Technical bankruptcy is when your team needs to throw all your technical assets away and salvage whatever they can.

One of your team's goals could be "Stop accruing technical debt and start building technical wealth instead."

We can push further the analogy. I was teaching Advanced Architectures in LabVIEW, and we started talking about reusable code. Students said they went back at the end of a project and identified reusable code; of course, they rarely have the time to do that.

I suggested instead to identify reusable code as they start coding. For example, we ended up with the Delacor Configuration Editor because we identified we would probably need that again. We created first a project template and that first project was the first "customer" for the project template.

By focusing on investing in our reusable code, we have the wealth for a down payment for a complete project that includes the editor and other features. The number of times we need to take on technical debt are fewer. We also treat our technical assets as another customer. They get their own repository, their own project in our issue tracking system, and we review how much time we are investing every month on those technical assets. We also keep track of how many hours we saved on billable projects. We package some of these technical assets and sell them to our customers. Sometimes as a single tool, sometimes as a line item in a quote for a larger project. We keep track of how much money each package is receiving too. Even if you don't have external customers, you can do something similar; keep track of how much time you are saving. Remind your team and your boss of your success. You can let them know that project template that you created just saved you 25 hours.

We still have to take on technical debt from time to time. For example, a startup that was in a hurry to show something to investors might need us to take some shortcuts. That is OK, as long as they understand we will have to throw away most of that code.

Focus on technical wealth!

From Model to Code

If you are looking for ways to improve the quality of your code, a good place to start is modeling. Are you taking the time to figure out what problem you are trying to solve before you start coding? Do you do any planning ahead of time? Do you stick to the plan?

What Is a Model?

English not being my first language has its advantages. I tend to spend more time figuring out what a word means. I became interested in the meaning of the word model when I realized that in Spanish, we use a different word for a three-dimensional representation of a house. An architect would build a "maqueta." And we use a different word for graphical representations of a problem solution. An engineer would draw a "modelo." When I realized that in English, we use the same word for both representations, it became more clear the intent of a model.

A model is a simplified representation that describes behavior and interactions. These interactions could be with the outside world or within the system.

We have heard Jeff Kodosky for years, almost at every NIWeek, talk about models of computation. When I understood what a model really meant, it made a lot more sense. It became apparent that when Jeff is talking about models of computation, he is talking about simplified representations of calculations.

Wasn't LabVIEW Supposed to Remove the Need for Modeling?

Jeff envisioned LabVIEW as a way to convert those simplified representations of the problem's solution directly into a programming language. There should not be an intermediate step where the developer designs a model and then has to translate it into code. However, our applications in LabVIEW have grown, and the days of having a handful of VIs to only acquire, analyze, and present our data are gone. The drawback of the current approach is that several LabVIEW developers would just jump directly onto solving the problem in LabVIEW without doing that extra step of modeling the solution first. For now this model step is done outside of LabVIEW. Some tools are now available (like the GDS toolkit and the state machine diagram), but for the most part the modeling step has to be done outside of LabVIEW. Remember to do your design in the cheapest medium possible; most of the time this is paper or a whiteboard. I have been to some offices where this step is done on the office windows![11]

Kodosky and his team continue to look for ways to bring more of that modeling to be done directly in LabVIEW. An addition to the LabVIEW language, G, are the channel wires; they were introduced in LabVIEW 2016. The channel wires simplify the representation of communication between loops. I am still wrapping my head

---

[11] I am talking about real windows and offices here, not Microsoft Windows, not Microsoft Office

around them and do not use them very much. However, I agree that a diagram with channel wires represents in a more direct way how the different modules communicate. If you want to learn more about channel wires, check out the section of the LabVIEW Help called "Communicating Data Between Parallel Sections of Code Using Channel Wires."[12] The "Channel Basics.lvproj" shipping example for channel wires is organized in lessons with very detailed documentation.

If you look at the traditional way to use queues to communicate between two loops, it looks like Fig. 7.3.

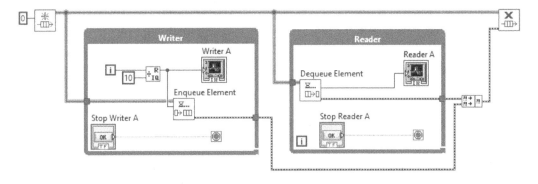

FIG. 7.3 *Image from upper part of block diagram of Channel Basics Lesson 1.vi from the Channel Basics.lvproj.*

Using channel wires, it looks like Fig. 7.4.

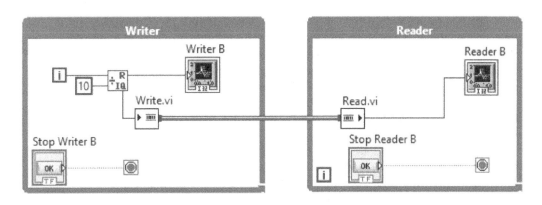

FIG. 7.4 *Image from lower part of block diagram of Channel Basics Lesson 1.vi from the Channel Basics.lvproj*

---

[12] http://zone.ni.com/reference/en-XX/help/371361N-01/lvconcepts/channel_wires_intro/

Both diagrams are doing the same thing, except in the first diagram we need to add the code to create and destroy the queue reference.

This is creating what is called a higher level of abstraction, where you are more focused on the action of communicating data between two loops and you worry less about the communication mechanism.

There is, however, one higher level of abstraction where you just have a writer and a reader and an arrow just mentioning that data is going through (Fig. 7.5).

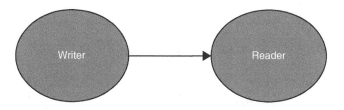

FIG. 7.5 *Simple image of two modules communicating.*

There is hope that one day, the modeling step will also be done in LabVIEW and will keep up with the program as it evolves. During Jeff Kodosky's NIWeek 2018 keynote,[13] he gave us a preview of one of the next things he is working on. Google "The Big Picture by Jeff Kodosky" for a video of this presentation.

In that presentation, Jeff talks about diagrams that display the entire application, from desktop to cRIO to LabVIEW FPGA. Although we are excited about this tool, we couldn't help but cringe at the thought of inter-target spaghetti code!

The other preview was of a more distant tool; this would be a modeling tool that is directly connected to LabVIEW and synchronized with the code. LabVIEW would preserve the storyboard of our high-level sketch as part of the project. Imagine taking a picture of the whiteboard sketch you currently have of your system and being able to click on each part of the diagram and seeing the actual code behind it. Better yet, adding wires to the high-level sketch and seeing them appear in the lower code implementing it. It would be a way to travel through the different levels of abstraction in your application. Double-click on the "writer" bubble; it takes you to the diagram where the writer is generating the data.

For now, we don't have all-in-one modeling to implement code tools. If you are doing a simple acquire-analyze-present program, you can continue to jump directly into LabVIEW and program your application. As the problems you are solving become more complex, the need for that extra step of modeling is back. Time and time again we see that stepping back and creating a model of the solution to our problem is better. It is a lot easier to erase a whiteboard or throw away a

---

[13] Kodosky, Jeff. The Big Picture. youtu.be/D7-ej-cqVqI

piece of paper with our scribbles than it is to delete a program that we have already dedicated time creating.

Why and When to Use Models

Models are a good way to discuss the code with a customer who might not be familiar with LabVIEW. They don't have to be familiar with UML either; all I have to do is provide a legend of what each item in the diagram means and an example of how to read it. Models are a good intermediate step between geek-speak and English. They are also useful when different members of the team will be using different programming languages. Perhaps one of the team members is writing the firmware for a microcontroller we will be communicating with in LabVIEW and another team member is writing an application in C# to communicate with another part of the system. Having a high-level view of the system where everyone speaks the same language regardless of the implementation helps the entire team.

Another advantage of models is that it is easier to erase and move items around on a model than it is to do it on the final software.

Making the model clear for everyone in the team helps with architecture reviews. Again, it doesn't matter if everyone in the team is programming using the same language. We find that reading the diagram or explaining the diagram out loud to someone else gives us a clear idea of whether we are on the right path or not. If it doesn't sound right in our spoken language (e.g., English), it will not work out in software. For example: We wrote a DAQmx class-based project for a customer. Later on, they acquired a motor controller. LabVIEW communicated with the motor controller via the serial port. The customer wanted to take advantage of all the configuration tools in place for the DAQmx class hierarchy. It was easy for him to declare: "A motor controller serial channel is a DAQmx channel." It is clear that is not correct. Sure enough, the code ended up having hacks all around to try to make it work! They could have saved a lot of time by just accepting that "A motor controller serial channel IS NOT A DAQmx channel." This was clear by adding the motor controller serial channel to the application model.

Types of Models

Models can be a simple back-of-a-napkin sketch or they can be a full-blown model that follows a standard such as Unified Modeling Language (UML).

Custom Model

A custom model does not follow any standard. For example, a quick drawing on the whiteboard using bubbles to represent each module in the system and colored arrows to show interactions between bubbles.

UML = Unified Modeling Language

UML is a relatively open standard. It is controlled by the Object Management Group (OMG), an open consortium of companies. We recommend reading *UML Distilled*, Third Edition, by Martin Fowler. This book covers the most used parts of UML.

State Diagrams State diagrams were mentioned on Chapter 3 page 97 of *LabVIEW Graphical Programming*, Fourth Edition. The State Diagram Editor is now a free add-on from NI and can be downloaded via the LabVIEW Tools Network.[14] As a side anecdote, one of the LabVIEW Champions loved this tool and was very disappointed when it was no longer shipping with LabVIEW. He brought it up during one of the LabVIEW Champions meeting, and NI agreed to make it available for free via the LabVIEW Tools Network; so if you like this tool, make sure to thank Ben Rayner when you see him around at NI events.

State diagrams describe a system by defining its states and what would cause the system to transition to a different state. They have been used in LabVIEW because it is straightforward to translate a state diagram into a VI. The states become the cases in the Case Structure; the transitions decide the next case to be executed. (See Figs. 7.6 and 7.7.)

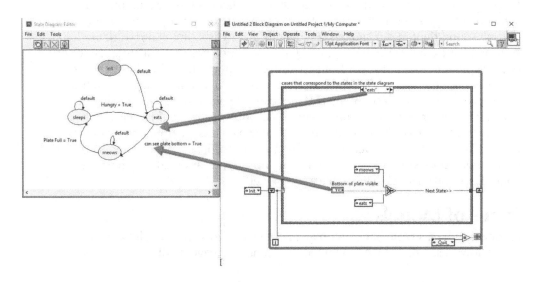

FIG. 7.6 *This is a traditional state machine in LabVIEW. Each frame in the Case Structure represents a state (circle) from the state diagram. This LabVIEW code is not attached to the LabVIEW State Diagram Editor.*

---

[14] Search for LabVIEW State Diagram Toolkit in ni.com/labviewtools.

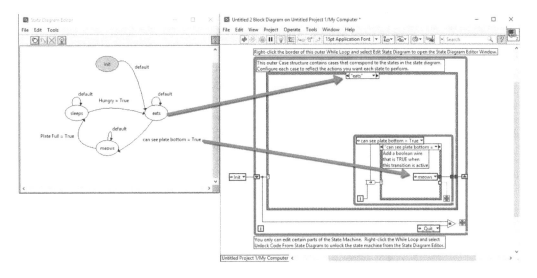

FIG. 7.7 *This LabVIEW code is generated and maintained via the State Diagram Editor. There is an additional inner While Loop to poll for the next state.*

UML Class Diagrams These are the most common UML diagrams out there. When we ask people if they have heard about UML diagrams, they think there is only one type: the class diagram. UML class diagrams represent a class by a box divided into three sections: the class name at the top, the properties, and finally the methods.

There are different arrows to express different relationships between classes. (See Fig. 7.8.)

- Inheritance, also known as an "**is a**" type relationship
- Aggregation, also known as a "**has a**" type relationship
- Composition, also known as a "**part of**" type relationship

UML Sequence Diagrams The UML sequence diagrams represent the sequence of actions within a single scenario. (See Fig. 7.9.)

The loop on the top starts the sequence via the "Cat can see the bottom of the bowl" change in value. It sends a message to the lower loop to go to the Meow state. The lower loop represents the cat's owner. In the Meow state, he sends the messages to himself to fill the bowl and then let the cat know that the bowl is full.

There are many more UML diagrams, but the ones we mentioned here are the most used in the LabVIEW community.

500 LabVIEW Graphical Programming

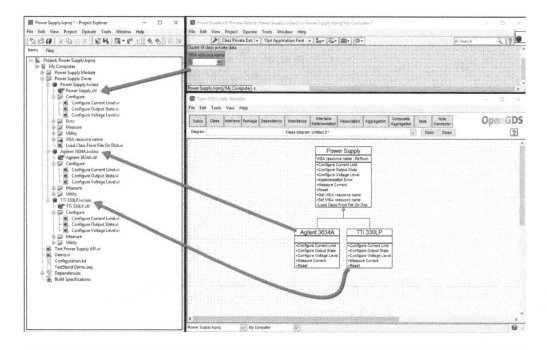

FIG. 7.8 *Open GDS toolkit UML Modeler displays the UML class diagram for the Power Supply class. The Private data for the Agilent and TTi are empty. The way to read this diagram is: The Agilent 3634A is a Power Supply and the TTi330LP is a Power Supply too.*

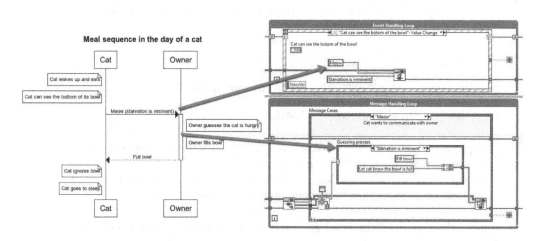

FIG. 7.9 *An UML sequence diagram showing the sequence of actions between a cat and their owner.*

Modelling Tools

Some of the modelling tools that I recommend checking out to see if they fit your teams' needs are

- The LabVIEW State Diagram Toolkit.[15] Seen earlier in this chapter.
- GDS UML Diagram.[16] Seen earlier in this chapter.
    - Round trip (model into code, code into the model). There are some examples of this in the LabVIEW object-oriented programming chapter in this book.
    - Can model Actor Framework.
    - Simple state machine diagram, sequence diagrams, and class diagrams.
- Websequence diagrams is an external tool that does not integrate with LabVIEW yet, but it is really easy to use for UML sequence diagrams.[17]
- Channel wires.
    - In Language features to represent interprocess communication. Available in LabVIEW since 2016. The shipping examples and the feature in general were greatly improved in LabVIEW 2018.

Source Code Control—The Developer's Time Machine

If you have been to any of my presentations on source code control or follow the Delacor blog posts on this subject,[18] you probably have already seen this cartoon (see Fig. 7.10). I still find people who say "I don't use source code control because I am a single developer"... so here it goes again, in print so it is around for a long time...

In the LabVIEW community, we tend to call it source code control. However, the rest of the world calls it version control systems (VCS), source code management (SCM), or revision control system (RCS). I think it is important to share these terms to help you in your search for more information.

One thing I have learned while working with different teams is that every project, regardless of size, benefits from source code control. A good way to figure out if you should use source code control or not is to think about how many hours

---

[15] Search for LabVIEW State Diagram Toolkit in ni.com/labviewtools
[16] Search for GDS in ni.com/labviewtools
[17] www.websequencediagrams.com
[18] delacor.com/category/scc/

502 LabVIEW Graphical Programming

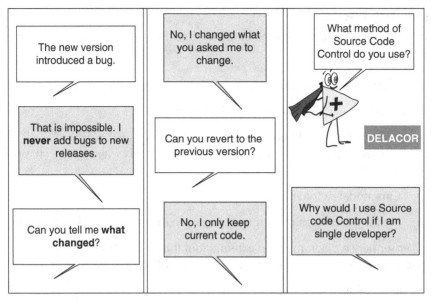

Good programming practices have nothing to do with marital status

FIG. 7.10 *A cartoon describing a real-life situation. The last line came courtesy of Steve Watts.*

of your day you are willing to lose if your project gets corrupted. Make this your trigger to decide whether or not to use source code control. I like to tell the story of that time my husband came into the room asking me to recover his LabVIEW project; it was corrupted. I told him to just revert to the previous version that worked and go from there. He then had to admit that he was not using source code control; it was a small and quick project, why would he. I asked him then how long had he been working on that project; his answer was 6 hours. I am not that mean; I did help him and now he always uses source code control; he even has his schematics in source code control.

What Is Source Code Control?

I don't think I exaggerate when I say that the most important technical asset for any software development team is their source code control tool. It keeps track of every modification done to the code. Developers can "travel back in time" to see what the code looked like in the past and revert changes or determine when a bug was introduced. No more bad names for your files (see Fig. 7.11).

There are two types of source code control tools, centralized and distributed.

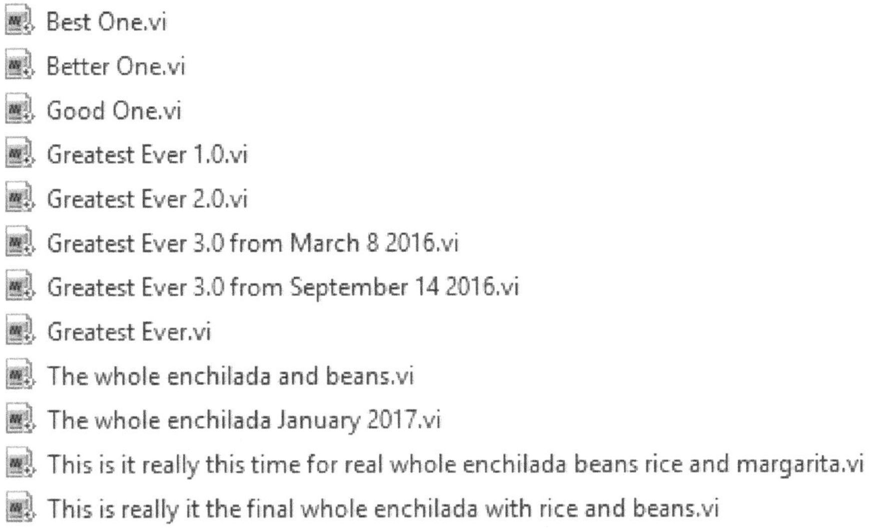

- Best One.vi
- Better One.vi
- Good One.vi
- Greatest Ever 1.0.vi
- Greatest Ever 2.0.vi
- Greatest Ever 3.0 from March 8 2016.vi
- Greatest Ever 3.0 from September 14 2016.vi
- Greatest Ever.vi
- The whole enchilada and beans.vi
- The whole enchilada January 2017.vi
- This is it really this time for real whole enchilada beans rice and margarita.vi
- This is really it the final whole enchilada with rice and beans.vi

FIG. 7.11 *These are actual names we have encountered in the wild; we promise no developer was harmed while gathering these names.*

Centralized Source Code Control

A central copy of the project is in a repository (typically a server). Developers "commit" their changes to that central location. Some examples of this type of tools are CVS, SVN, and Perforce.

The workflow you would use in a centralized source code control system would be

- Step 1: Check out working copy (see Fig. 7.12)
- Step 2: Developer makes changes and commits (see Fig. 7.13)
- Step 3: Other developers get the change by updating their copies (see Fig. 7.14)
- Step 4: If two developers make a change, whoever gets second needs to update first and resolve any conflicts (see Fig. 7.15)

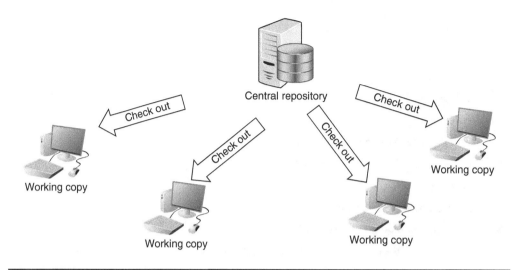

FIG. 7.12 *Step 1: Check out working copy from the central repository to your computer.*

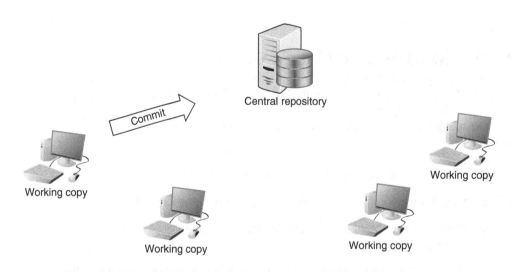

FIG. 7.13 *Step 2: Developer makes changes directly on their computer and commits the changes to the central repository.*

Chapter 7: Developing in LabVIEW for Teams

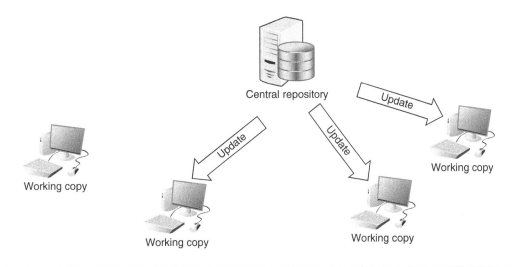

FIG. 7.14 *Step 3: Other developers get the change by updating their working copies to get the latest version in the central repository.*

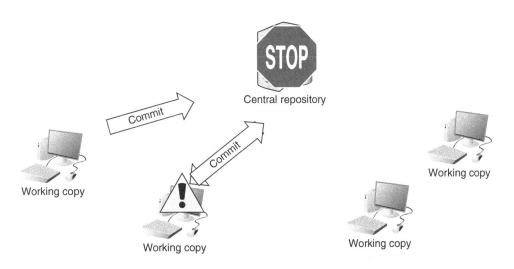

FIG. 7.15 *Step 4: If two developers make a change, whoever gets to do a commit second needs to update first and resolve any conflicts.*

Centralized Source Code Control: Disadvantages

I have seen some disadvantages to using centralized source code control. First the programmer requires a connection to the repository in order to commit, update, or read the log. I don't like that it doesn't respect the programmer's workflow; she has to address any conflicts right then and there, or she doesn't get to commit her changes. This leads to developers not committing as often because they want to put off dealing with conflicts or they fear they are going to break others' code.

Teams that use centralized source code control tools tend to have a single repository for all their projects. They can get away with this because the tools allow them to check out only the section of the repository they are actively working on. Having the entire code base in a single repository means that if something happens to the repository, the entire code base for the entire team is affected. Another disadvantage of this approach is that managing the history of a project is difficult because it is combined with the history of other projects.

Of course, there are ways to address some of these limitations. You can set a policy for your team that you will have multiple repositories. To address the issue of programmers not committing often, provide them with their own branch or a sandbox repository. The one item that is hard to address is the fact that programmers do need a continuous connection to the central repository. This alone is the reason a lot of consultants and contractors use distributed source code control, because they know that at certain customers' sites, they won't have access to the Internet.

Distributed Source Code Control

A central copy of the project is in a repository (typically a server). Each developer has a clone of the entire repository. Developers "commit" their changes to a local repository and "push" their changes to a central location. Some examples of this type of tool are Git and Hg (Mercurial).

The workflow you would use in a distributed source code control system would be

- Step 1: Clone repository (see Fig. 7.16)

- Step 2: Developer makes changes and commits as many times as needed to the local repository (see Fig. 7.17)

- Step 2a: When ready, she pushes all of her commits

- Step 3: Other developers get the change by pulling the changes into their local repository (see Fig. 7.18)

- Step 3a: When ready, they update

- Step 4: If two developers make a change, whoever gets second pulls and can choose to update, resolve any conflicts, and push when convenient (see Fig. 7.19)

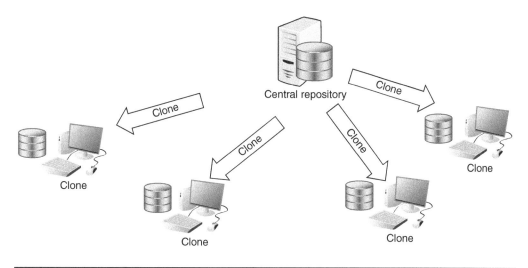

FIG. 7.16 *Step 1: Clone the entire repository to your computer.*

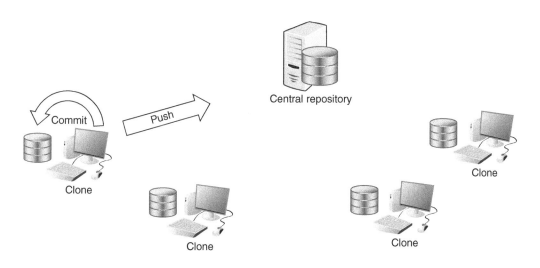

FIG. 7.17 *Step 2: Programmer makes changes to code and commits locally as many times as necessary. When ready, she pushes all of her commits from her local repository to the central repository.*

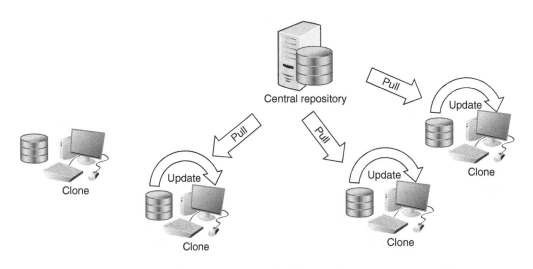

FIG. 7.18 *Step 3: Other developers get the change by pulling the changes from the central repository into their local repository. When they are ready, they can update their working copy to the latest changes. In the meantime, they can continue to work on their own copy.*

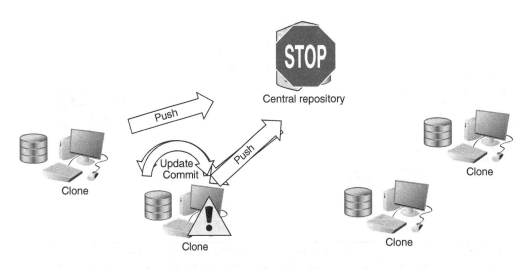

FIG. 7.19 *Step 4. If two developers make a change, whoever gets to push second pulls and can choose when to update, when to resolve any conflicts, and when to push all her commits.*

Branch Merge versus Code Merge

You will hear these two different concepts when working with a source code control tool. **Branch merge** is when the files of a project have been modified in two separate branches and they get combined into a single branch. Different developers can work on different branches. This may or may not involve merging code. **Code merge** is when the same code file has been modified in two separate locations and both modifications get combined into one final version of the file. (See Fig. 7.20.)

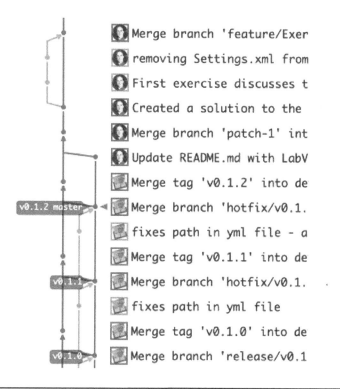

FIG. 7.20 *Branch merge.*

LabVIEW Compare

In LabVIEW 2019, NI started distributing LabVIEW Compare as a command-line executable so that it could be called when a source code control tool wants to compare two different files. LabVIEW Compare is also known as LabVIEW Diff, and the executable is called `LVCompare.exe`. Calling this executable is equivalent to launching the "Compare VIs dialog box" found in **Tools >> Compare >> Compare VIs**. It is available only in the LabVIEW Professional Development System. This tool works like those "Find the differences" games we did as kids. (See Figs. 7.21 to 7.23.)

510 LabVIEW Graphical Programming

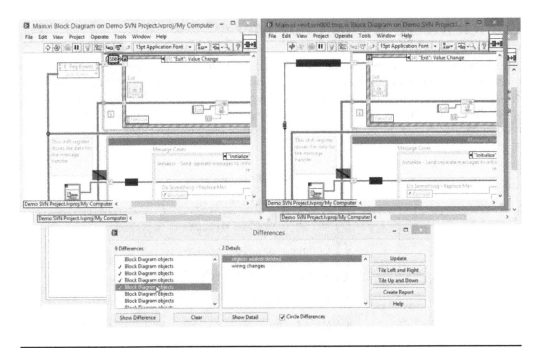

FIG. 7.21 *Code merge.*

FIG. 7.22 *Find the differences.*

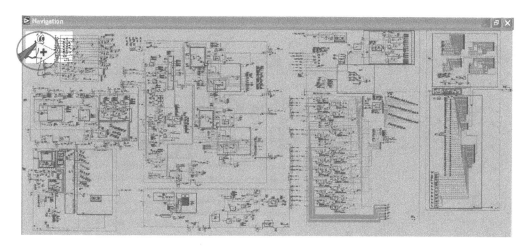

FIG. 7.23 *Find Waldo.*[19]

Configure the Source Code Control Tool to Use LabVIEW Compare

You can find the details on how to configure source code control with LabVIEW Compare in the LabVIEW Help.[20] While the information is there on the Help, there are some caveats to consider when configuring third-party source code control tools.

To configure SVN to use LabVIEW Compare, you can download the Viewpoint Systems SVN Toolkit[21] from the LabVIEW Tools Network, and it will take care of configuring both LVCompare and LVMerge for you.[22]

If you cannot use the toolkit, you can open the TortoiseSVN settings by right-clicking on any folder on your Windows Explorer. In the TortoiseSVN settings configure the Diff Viewer to use LabVIEW Compare for .ctl, .ctt, .vi, and .vit files. You will enter each file type at a time, and in the External program, you will point to LabVIEW Compare. (See Fig. 7.24.)

The files that can be compared/merged with these tools are

```
.vi, .vit, .ctl, and .ctt
.vit = VI Template
.ctt = Control Template
```

---

[19] Code image taken from Rube-Goldberg-Code post in the NI forums. forums.ni.com/t5/BreakPoint/Rube-Goldberg-Code/m-p/649645
[20] Configure Source Code Control with LVCompare. zone.ni.com/reference/en-XX/help/371361R-01/lvhowto/configlvcomp_thirdparty
[21] Go to ni.com/labviewtools and search for SVN
[22] There is a known issue with the Viewpoint Systems SVN Toolkit, in that when Auto-Refresh is on, it might crash LabVIEW. Go to Tools >> Viewpoint TSVN Toolkit >> settings and uncheck Auto-Refresh. You will need to use the refresh button on the toolbar to refresh the icon overlays.

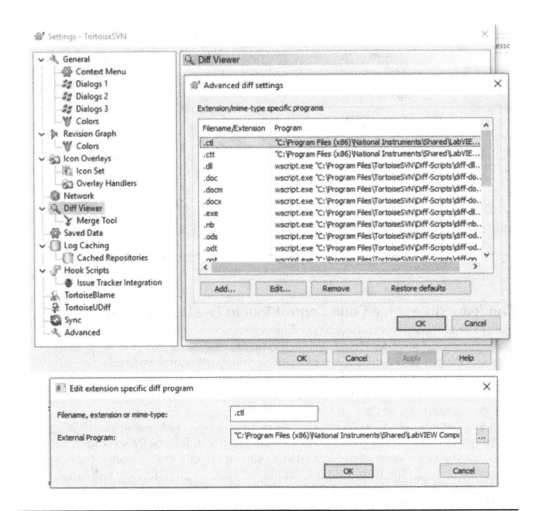

FIG. 7.24 Set the Tortoise SVN Diff Viewer to use LabVIEW compare for LabVIEW files (.ctl, .ctt, .vi, .vit).

The External program is defined as follows:

```
"C:\Program Files (x86)\National Instruments\Shared\LabVIEW Compare\
LVCompare.exe" %mine %base -nobdcosm -nobdpos
```

Where `%mine` refers to the modified file, `%base` refers to the base file, `-nobdcosm` means no block diagram cosmetic changes, and `-nobdpos` means not checking for block diagram position changes. There are other modifiers that you could use, and they are described in the LabVIEW Help.

The team from Endigit also has a blog post on how to configure LVCompare and LVMerge for SourceTree.[23]

LabVIEW Compare requires some LabVIEW wrapping to make it work with TortoiseHg and TortoiseGit. To find the tool, go to the Mercurial user group at NI Forums.[24] Andreas S, system engineer at NI, created a tool that wraps LVCompare so it can be called from Mercurial. James McNally made some changes, and both tools are available via Bitbucket. There is another version of the tool that parses the relative paths to the absolute paths that LabVIEW Compare and LabVIEW Merge need to function. The code is available via GitHub.[25]

To configure TortoiseGit, go to its settings and configure the Diff tool to use LabVIEW Compare and set the modifiers:

```
"C:\Program Files(x86)\National Instruments\Shared\LabVIEW Compare\
LVCompare.exe" %mine %base -nobdcosm -nobdpos
```

LabVIEW Merge

In LabVIEW 2019, NI started distributing LabVIEW Merge as a command-line executable, so it could be called when a source code control tool wants to merge two different files; the executable is called `LVMerge.exe`. Calling this executable is equivalent to launching the "Merge VIs dialog box" found in **Tools >> Merge >> Merge VIs**. It is available only in the LabVIEW Professional Development System. Just like LabVIEW Compare, it works better with small changes on small block diagrams.

Configure Source Code Control Tool to Use LabVIEW Merge

You can find the details on how to configure source code control with LabVIEW Compare in the LabVIEW Help.[26] While the information is there on the Help, there are some caveats to consider when configuring third-party source code control tools.

To configure TortoiseHg to use LabVIEW Merge, edit the Settings file to include

```
[merge-tools]
LVMerge.args = $base $other $local $output
LVMerge.executable = C:\Program Files (x86)\National Instruments\
Shared\LabVIEW Merge\LVMerge.exe
LVMerge.gui = True
LVMerge.binary = True
```

---

[23] Vowles, Marty. "Using LabVIEW's Diff Tool with SourceTree." endigit.com/2017/11/using-labviews-diff-tool-sourcetree
[24] http://zone.ni.com/reference/en-XX/help/371361P-01/lvhowto/configmerge_thirdparty/
[25] SmithEdVIcompare. https://github.com/smithed/vicompare
[26] Configure Source Code Control with LVCompare. zone.ni.com/reference/en-XX/help/371361P-01/lvhowto/configlvcomp_thirdparty

```
[merge-patterns]
**.vi = LVMerge
**.vit = LVMerge
**.ctl = LVMerge
**.ctt = LVMerge
```

To configure TortoiseGit to use LVMerge to edit conflicts, include in the settings:

```
"C:\Program Files(x86)\National Instruments\Shared\LabVIEW Compare\
LVMerge.exe" %base %theirs %mine %merged
```

```
%merged: the conflicted file, where to save
%theirs: the file as it is in the repository
%mine: your own file, with your changes
%base: the original file without your changes
```

- Distributed source code control does not require a connection to the central repository all the time
- Every team member has a complete copy of the repository, including the entire metadata
- It is still advisable to have backups of the repository, but they are not as crucial as with centralized source code control
- Avoid working on the same VI on different branches
- Overuse of LabVIEW Compare and LabVIEW Merge is a symptom of an architecture that is not modular enough and/or a process that has multiple developers working on the same code at the same time

Good Practices for Source Code Control

Source code control tools are great, but they can only serve you well if you use them correctly.

Separate Source Code from Compiled Code

Make sure that everyone in your team has configured their LabVIEW and the project they are working on to separate source code from compiled code. This will minimize the number of files that get modified on disk when a change happens.[27]

Process and Architecture Smells

Throughout the years, we have encountered the insistence on having LVCompare and LVMerge configured in the source code control tools, but this is a process/architecture smell. By smell, we mean that it smells bad, that it is an indication that something is

---

[27] Facilitating Source Control by Separating Compiled Code from VIs and Other File Types. zone.ni.com/reference/en-XX/help/371361K-01/lvconcepts/saving_vis_compiled_code/

not right with the process. It should be very rare that a developer needs to use either tool. Really, the only valid frequent use for LabVIEW Compare is in a regulatory environment, the Q&A person requests to see a redline comparison of what changed. Don't get us wrong; we still value these tools, and it is definitely better to have the source code control tool configured to work with them.

Architectures where the VIs are coupled and not cohesive lend themselves more to developers needing to work on the same VI as another developer for implementing different features.

The library files `.lvlib`, `.lvclass`, and `.lvproj` are XML files; you would think they are text files, except when they are not completely text files. These files do include some binary information. Regular text compare tools can be used to identify some changes, but merge tools should be avoided, as they can corrupt the files.

At Delacor, we assign an owner for the library files. If two developers are working on the same library (but in different VIs) and they need to make a change to the library, they request the owner to make the change or check with the owner if it is OK for them to make the change. Good source code control tools are not a replacement for good-old communication. We have worked with teams that are not collocated, and sometimes we don't even work in the same time zone; in those cases, we use tools like Skype, Slack, e-mail, or whatever works for that team to stay in constant communication.

Commit Notes

One of my customers is a psychologist who is a very proficient LabVIEW programmer too. The first time I explained to him that when he was ready, he needed to commit his code, he told me, "Who said software engineers were afraid of commitment or couldn't commit?" Then we moved on to talking about resolving conflicts and he said, "Who said software engineers were not good at conflict resolution?" (See Fig. 7.25.)

FIG. 7.25 *Image by xkcd.*[28]

---

[28] Image from https://xkcd.com/1296/
Image licensed under Creative Commons Attribution, author says it could be included in books as long as we give credit. https://xkcd.com/license.html

Code that runs doesn't lie, but it cannot describe intent. Discuss with your team how long your commit/push descriptions should be. There are some suggestions that you can use to start the discussion, like the seven rules of a great Git commit message.[29]

1. Separate subject from body with a blank line.
2. Limit the subject line to 50 characters.
3. Capitalize the subject line.
4. Do not end the subject line with a period.
5. Use the imperative mood in the subject line.
6. Wrap the body at 72 characters.[30]
7. Use the body to explain *what* and *why* vs how.

I would add a line on which issue tracker cases this commit solves. I prefer this in line with the subject, but others prefer it at the bottom of the commit message.

One of my customers used to give me grief about how detailed my commit notes were. When I work with him, we do pair programming, we share screens, and we are both coding together. I explain my thought process while I code, and he can do the same. When we are ready to commit, his tendency is to add a commit note that says "made some changes." I never let him. He stopped giving me grief after a couple of times that I was able to use the commit notes to find our way to when, how, and why we had introduced a bug.

Be kind to future you and leave enough information so if you ever need to go back to that revision of the code, you will be able to find out what changed, why it changed, and how you changed it. Sometimes I add bookmarks to my commit notes to pinpoint exact locations in the code that changed.

With some of the teams I work, the programmers are not in the same room; sometimes they are not even in the same time zone. If a bug tracker case needs to be assigned to another programmer, leaving clear notes for when they need to take over is also nice for them. I like it when I walk in the morning to my computer and I find clear commit notes from the person in Australia who was working on the code the night before.

Pull or Merge Requests

Bitbucket and GitHub choose the name pull request because the first action from the team member would be to pull the feature branch. GitLab and Gitorious chose the name **merge request** because that is the final goal, to merge the changes pushed

---

[29] Beams, Chris. The Seven Rules of a Great Git Commit Message. https://chris.beams.io/posts/git-commit/#seven-rules

[30] You can configure some tools to add a redline to indicate where the 72 characters are at.

to the base branch.[31] Check with your Git or Mercurial host tool to see what they call it and how they go about implementing this process.

When working in teams where all the team members are working on the same application, it is important to have a process in place where someone else reviews the code before integrating it with the rest of the code. There is another section in this chapter where we talk about a formal code review; some of the practices described in that section can also be used during the discussion section of a **pull request.**[32,33] Formal code reviews are not always possible, especially when the team is not colocated. Pull requests are available via the distributed source code control hosting tools. This allows you to tell others about changes that you have done to the code and that you have already pushed to the branch in the repository. Once you open a pull request, you can communicate with others in the team and discuss and review the changes with collaborators. Others in your team may request that you do further changes, and everyone can see the subsequent commits; once everybody agrees, you can merge the changes into the main branch. There needs to be an official approval before the changes can be merged with the repository. The developer can add a summary of their changes. Once all the changes are approved, the developer can perform a **Merge a pull request**.

If you are using Bitbucket or GitHub, there are ways to configure branches in your repositories to force a pull request. This will force members of the team to go through the pull request process before pushing. Tools like GitLab don't have a pull request, but they do have a **Merge Request.**[34]

If you decide to collaborate in an open-source project, you more than likely will be using pull requests. By the way, if you want to find out LabVIEW projects where you could contribute or use, do a search in the most popular open-source project hosts (github.com, bitbucket.org, and gitlab.cm) for LabVIEW projects.[35]

Ignore Files

There are certain files that you don't need to keep in source code control. If you are using git, edit your `.gitignore` to ignore the following files; same thing for `.hgignore`. You can also configure SVN to ignore the same files.

```
# Metadata
*.aliases
*.lvlps
```

LabVIEW creates these two files every time you open and work on your LabVIEW project, and LabVIEW can re-create them without any problem. You don't need to keep track of them in source code control.

---

[31] https://about.gitlab.com/2014/09/29/gitlab-flow/
[32] https://www.atlassian.com/git/tutorials/making-a-pull-request
[33] https://help.github.com/articles/about-pull-requests/
[34] https://docs.gitlab.com/ee/user/project/merge_requests/
[35] McNally, James. Open Source LabVIEW—How to Contribute. https://devs.wiresmithtech.com/blog/open-source-labview-how-to-contribute/

A more complete `.gitignore` file includes all of the following content:

```
# Libraries
*.lvlibp
*.llb

# Shared objects (inc. Windows DLLs)
*.dll
*.so
*.so.*
*.dylib

# Executables
*.exe

# Metadata
*.aliases
*.lvlps
```

GitHub already creates this file when you create a repository and specify that the language you are using is LabVIEW.

Establish a Source Code Workflow

In my experience, for the most part, teams that use SVN (or other centralized source code control tools) avoid branching the main code. They tend to work on the trunk and only use branches as sand boxes; they avoid merging back to the trunk. This is because SVN is not really friendly toward merging.

Distributed source code control tools are a lot better at merging. In fact, as two developers start working on the same folder, the tool automatically creates a branch of the code for each developer, albeit it is not a named branch. I have seen both approaches: teams working always on the trunk and only using branches as sandboxes, never merging them back, and approaches that use named branches.

If working on the trunk works for your team, go that route. If you have enough features that need to be implemented in parallel and a large team, or working on the trunk is not working for you and you want to explore a feature-driven approach, then try out **Git-flow** or **Hg-flow**. If that is not your cup of tea, then check out **GitHub flow**, and if you prefer an approach that differentiates only between production and development, use GitLab flow. Of course, you are also welcome to come up with your own workflow.

Git-flow

Git-flow is a workflow design that was first published by Vincent Driessen.[36] It is a branching model that has been widely adopted by text-based programmers. It works very well when there are multiple developers working on the same project, and it scales well.

---

[36] Driessen, Vincent. A Successful Git Branching Model. nvie.com/posts/a-successful-git-branching-model/

It allows for parallel development and isolates the finished work from the work in progress. Each new feature is separated from the rest, and it is relatively easy to go from one feature branch to another branch (Fig. 7.26).

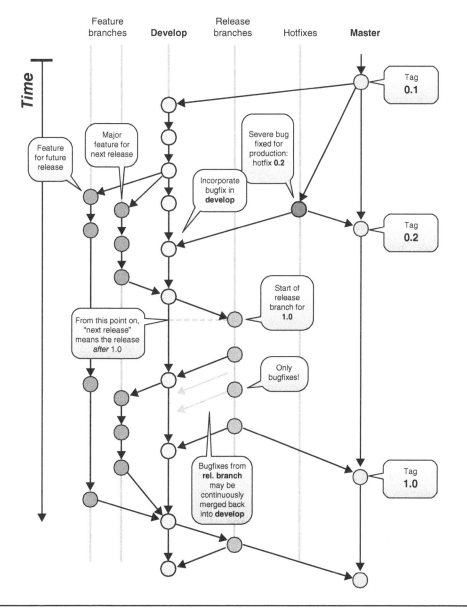

FIG. 7.26 *Git-flow model.*[37]

---

[37] Image source Driessen, Vincent. A Successful Git Branching Model. nvie.com/posts/a-successful-git-branching-model

A similar approach can be done using Mercurial. We assisted one of our customers, Precision Acoustics in the UK, to establish the process their team would follow. Each team is different, and we end up using different tools depending on the team.

For this particular team, we settled on using Hg (Mercurial). A diagram of their Hg-flow looks like Fig. 7.27.

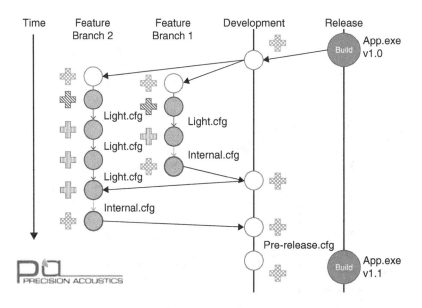

FIG. 7.27 *Hg-Flow model.[38] Image includes markers for where the unit tests need to pass, where to run VI Analyzer configuration files (.cfg), and where to have code reviews (thicker circle right before merging back to Development branch). Code reviews can be formal code reviews or conducted via pull request.*

We included in the model not only when a branch starts but also when the LabVIEW programmer is expected to run VI Analyzer tests, which type of VI Analyzer configuration to run, when to run unit tests, and when to conduct formal code reviews. You can see that in the Development branch the unit tests are always passing. The developer branches the code and verifies that the tests still pass; if they fail, then it could be that the unit tests are workstation dependent and that should not be the case. Before the developer can start working on the feature, she needs to figure out why the unit tests are failing. You want to always start with code working. Then she proceeds to create the unit tests that she will use to test that the feature she is working on works. As she commits her code, she runs the light version of

---

[38] Morris, Paul. Lone Wolf to Dream Team: A Journey into Team-Based Development. learn.ni.com/center-of-excellence/resources/1142/lone-wolf-to-dream-team-a-journey-into-team-based-development

the VI Analyzer configuration tests. Once all the unit tests pass, she knows she is done and she is ready for a code review. Once the code review is done, if everyone agrees, the programmer merges her branch back into the Development branch.

If you are using SourceTree as your client for Git or Hg, it already comes with a Git-flow button that will create the branches for you.

GitHub-Flow

GitHub-flow[39] is branch-based workflow. It works better for projects that are deploying new releases regularly.

The main rule for GitHub is that the team considers anything in the `master` branch to be ready for deployment. Following this idea means that anytime you plan to touch the code for any reason (adding new features or fixing issues), you create a branch first (Fig. 7.28).

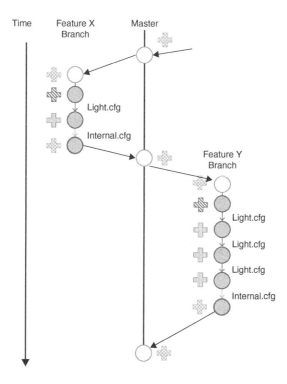

FIG. 7.28 *GitHub-flow model. Image includes markers for where the unit tests need to pass, where to run VI Analyzer configuration files (.cfg), and where to have code reviews (thicker circle right before merging back to Development branch). Code reviews can be formal code reviews or conducted via pull request.*

---

[39] Understanding the GitHub flow. https://guides.github.com/introduction/flow/index.html

Create the branch for the new feature or bug fix you will work on. You verify first that the unit tests pass. Then make any commits to your branch while you are modifying code to make those unit tests pass. Once all your new unit tests pass, you verify that all the other unit tests that existed before you started working still pass. You are ready to merge back into the `master` branch; you either conduct a formal code review or open a **pull request**.

During the pull request, a discussion and code review begins, and this conversation is captured in the tool you are using. Any subsequent push you do will show up as part of the same pull request because the tool assumes you are reacting to comments from the pull request discussion. Depending on the tool, some of them allow you to include images on your comments, something that comes in very handy to us graphical programmers. Once everyone is happy, you can deploy and test directly in production. If this causes issues, you can roll back your changes by deploying the existing `master` branch.

Once you verify that the changes work in production, you can merge your code into the `master` branch. Depending on the tool you are using and whether your tool is linked to your issue tracking system, you can even close cases as part of your merge by including text like `Closes #DQMH-42`, where the text after the hashtag is the issue ID in your issue tracking system that is linked to the code changes you just merged.

Other Source Code Control Workflows

There might be other approaches by the time this book is in your hands. I know of other flows like GitLab-flow[40] that is very similar to GitHub-flow without the assumption that you are able to deploy to production every time you merge a feature branch. GitLab-flow then suggests that you create a branch with the explicit name of `production`. Atlassian's website is a good resource for other workflows[41] available.

How to Select the Source Code Control Tool and the Workflow for Your Team

We have worked with different teams of different sizes and levels of proficiency. The majority were not computer scientists and had no defined workflow before we started working with them. Some of them might not even have been using source code control or were using it just as a backup service (only commit right at delivery time!); this made me sad. Over the years we have seen that for individual developers centralized source code control is enough. The only thing that pushes a single-person shop to move to distributed is if they find themselves working at other locations where they might not have access to their repository.

---

[40] Introduction to GitLab Flow https://docs.gitlab.com/ee/workflow/gitlab_flow.html
[41] Paolucci. Nicola Simple Git workflow is simple https://www.atlassian.com/blog/archives/simple-git-workflow-simple

For teams of two or more people where more than one person will be working on the same application, distributed source code control is the way to go. In our experience Hg is easier to learn than Git, especially if you already have some experience with SVN. TortoiseHg is very similar to TortoiseSVN. I worked once with a team where everyone had already accepted the move to Hg, but one of the team members was reluctant and kept complaining about the two-step process (commit–push and pull–update). I helped that team member configure TortoiseHg to treat commit–push in a single step and the same for pull–update. To me that defeats the purpose of being able to commit to my local repository and only push when I am ready, but that worked for that team. After a while our reluctant team member saw how well it was working for everyone else and eventually removed that setting.

Git is a lot more popular and powerful, but with power comes responsibility. There are a lot more things that you can do to your repository with Git; just rest assured that it is very hard to destroy a repository and that even when you might think you lost it all, chances are all of your commits can be recovered from your clone repository. The main tip with Git is always double-check that you are working on a branch; if you are using SourceTree, look at the left column and verify that your branch is bolded. Also, remember that Git is a three-step process: first you stage your files, then you commit, and finally you push.

At the time we wrote this book some of the most popular options in the cloud are Github (Git), Atlassian's Bitbucket (Git and Hg), and Gitlab (Git). If you need an option inside your firewall, then you can look at GitBlit, Atlassian's Bitbucket (only Git).

As far as the workflow, we have seen both teams where they all continuously merge their changes to the `master` branch or `trunk` and teams that go for one of the -flow approaches we described earlier. The main difference tends to be on the size of the application and whether the team needs to continue to support released versions of the code while they continue to work on the next version. For example, if we release *Our Awesome App 2.1*, we start new branches for features for 2.2. Then an end user finds a bug that we need to release. If we had been working on the `master` branch, we would not be able to release without releasing some of the new features for 2.2. If we are using Git-flow, we can fix the bug in a separate branch and merge it with the main branch without including any of the 2.2 changes. We release *Our Awesome App 2.1.1* without issue. If your team has a single customer within the same company and you don't care about releasing new features when a patch is needed then this might not be as important to you. GitHub-flow and GitLab-flow are simpler and work well if you don't have to support old releases of your software. As always, a total valid option is to find what is the best <your team name here> -flow.

Workstations

Have you ever been in a situation where someone says: "it doesn't work" and somebody else says: "well, it does work on **my** machine." Another example is when you come back to a project that you have not touched in months and all of a sudden there are lots of broken arrows. More than likely someone forgot to push a dependency to the repository. You didn't notice before because you are used to building the executable on the same machine where you develop. I don't know about you, but I prefer to find out about these issues right after I made the mistake and not three years later, when either my mind doesn't have any recollection of the project or when the original developer and machines are long gone!

Your objective is to improve the deployment process. You want to discover any missing dependencies as soon as possible. You want to find any problems with relative paths as soon as you add a new path to your code. You want to make sure that your installation works on a clean machine and that any secondary installers or patches work as well. You want your deployment process to be repetitive and reliable. In an ideal world any member of your team should be able to distribute a new executable just by pressing one button. But not having an automated build process is not an excuse to not have a build server! You can have a build server, where you manually launch your builds.

Another problem that text-based programmers and LabVIEW FPGA developers suffer more often than regular LabVIEW developers is waiting for the code to compile and build (see Fig. 7.29). Although, I take that back, I have worked in some applications that can take hours to build the executable. In those situations, we don't want to hold the development machine hostage. The developer should be able to continue working on their next feature while the executable is building.

FIG. 7.29 *Image by xkcd.*[42]

---

[42] Image from https://xkcd.com/303/
Image licensed under Creative Commons Attribution, author says it could be included in books as long as we give credit. https://xkcd.com/license.html

We recommend that you have a machine for development, another one for building/deployment, and one more for testing.

I know what you are thinking "wait a second, are you asking me to buy three computers?"

The answer is maybe. You can opt for having three different physical machines or you can use virtual machines.

Virtual Machines

A virtual machine is software that simulates another computer. It executes the programs as the physical computer would. This is how I can be running a Windows 7 and a Windows 10 on my Mac. There are different applications you can use for this. I have used successfully Parallels, VirtualBox, and VMware. By the time this book is published, there might be other options out there, but the general concepts covered in this section will probably apply. Also, you can look at launching the virtual machines directly on your computer or having a virtual machines server in your office, or you can go completely cloud based and use Azure or Amazon instances to host your virtual machines. Just remember that cloud-based machines will not be able to work with external hardware.

For teams that work with multiple customers, having a development virtual machine per customer means they can have all the drivers and specific LabVIEW add-ons that the customer needs. It is also easier to get back to their projects when they come back after several months with new requirements.

You can still benefit from VMs if you have a single project. You can use a VM as your development machine, another VM as your build server, and finally a separate VM snapshot for your integration tests.

One of the limitations that VMs might have is their connection with hardware. If that is an issue, then you will need to have physical systems instead. We are currently evaluating a third-party hardware that allows us to connect USB hardware to the device and access it via Ethernet. It is called a USB Over IP Hub. When working with USB hardware, I have had more luck with Parallels and VMware than with VirtualBox when the virtual machine is hosted directly on my computer.

Another limitation is if you are doing benchmarking. If you need to time how long an operation lasts, you have to do it directly on a computer. VMs are not good for that.

Virtual Machine Template

It is a main VM that will be used as a seed for other clones. Normally, it cannot be powered on. It can be used as a starting point for all of your other VMs. You can choose if this template will be the bare operating system installation or if it would include also things like LabVIEW and some basic add-ons.

Virtual Machine Clone

It is an independent copy of the original machine. It saves time when you have to make multiple instances. It is independent of the template. Any changes you make to the template will not change the clone, and any changes made to the clone will not change the template.

Virtual Machine Linked Clone
Shares virtual disks with the parent. The parent allows different VMs to use the same software installation. The linked clone has to have access to the parent always. All of the public files on the parent are available in the linked clone. This makes their size a lot smaller.

Virtual Machine Snapshot
The snapshot preserves the state of a VM. The developer can return to that state at any time. It would be useful to have a snapshot of a clean installation of the operating system followed by a snapshot with the first version of the software installed. This way developers can go back to a clean OS install to test how the installer works or go back to a snapshot after a version of the software has already been installed to test upgrades.

Workstations Setup
Option 1
One option is to have a VM per customer. The VM seed has the OS that the customer uses, LabVIEW, DAQmx, VISA, and other drivers and whatever source code control client we are using. Please note that versions older than LabVIEW 2009 do require a VM per LabVIEW version. Versions 2009 and later tend to work well even if the machine has different versions installed.

When a new customer comes, you create a new VM by copying the VM seed.

Option 2
The Template VM has all the common software used in all projects, such as Notepad++ and the source code control client.

Then you create a new linked clone per client with their respective LabVIEW version. The linked clone has the operating system the client uses, their LabVIEW version, NI hardware drivers, and any additional dependencies.

Build Server
It can be a physical computer. Ideally the build server is a VM snapshot that returns to the snapshot before every build. This would ensure that indeed the installer and source code control have all the dependencies needed. You don't need to be running a continuous integration system or have automated builds to have a build server. As I mentioned earlier, having a build server is a good way to verify as part of your process that your repository always has everything it needs to build.

A more pragmatic approach is to leave the VM as it was left on the previous build and then perform the following actions:

1. Pull from source code control
2. Update from source code control
3. Apply VIPMvipc (or any other program you use to install dependencies)
4. Run unit tests
5. Run VI Analyzer with a prerelease test configuration
6. Build

If there is a failure at any of these steps, then you go back to fix them on the development computer. Note that before LabVIEW 2018 you probably did let the failures from VI Analyzer be a gate factor to build. Starting with LabVIEW 2018 you can add a bookmark for VI Analyzer to ignore certain errors;[43] in this case a failure reported by VI Analyzer should be a reason to break the build process and to go back to development.

The Delacor team has a picture of Godzilla as a backsplash on the screen of our build server with a sign that says "Build Server, Not for development, Not for testing." This is a good reminder to everyone in the team that they should not make any changes, no matter how small, directly on the build server. In case you are wondering why we have a picture of Godzilla, let's just say that some people have given Fab the nickname of LabVIEWZilla. We really don't know why.

Test Computer

It can be a physical computer. If not, a virtual machine seed that has only the operating system. It could also be a snapshot.

If you are testing that the upgrades work, then there is no need to go back to a snapshot of a clean computer. If you are testing that a clean install works, then you do go back to the clean snapshot.

LabVIEW Style Guidelines

If you don't already have a LabVIEW style guideline for your team, writing one will help you determine what are everyone's pet peeves with LabVIEW code. You will be able to discuss some idiosyncrasies that you didn't understand before. Hopefully, you can arrive at a consensus of what your team considers good LabVIEW

---

[43] Ignoring Test Failures (VI Analyzer Toolkit). zone.ni.com/reference/en-XX/help/371361R-01/lvvianalyzerhelp/ignoring_failures

programming practices, and when someone joins the team, you can show them those guidelines.

You don't have to start with a blank page; you can start with the LabVIEW Style Checklist[44] or with the LabVIEW Examples Style Guideline.[45] Get together with your team and review them; you will be surprised but will not agree with everything that NI is suggesting in those guidelines. They are guidelines, after all, and not rules.

When we work with teams, we help them write their LabVIEW style guidelines. We start with our own; however, with time, we have realized that not all LabVIEW developers read the LabVIEW style guideline for their team. We have realized that better than reinforcing these style guidelines via code reviews or during pull requests is to use VI Analyzer to enforce these guidelines. We go into more detail in the VI Analyzer Test Configuration section of the Code Review Process section of this chapter, and we show you how to create your own VI Analyzer tests in the computer-aided software engineering (CASE) tools section of this chapter.

Code Review Process

Your team can opt for using the pull request as your code review, especially if you are not colocated. Some of the concepts in this section can still apply when discussing code changes during a pull request. You can use the free labels for code to-dos, but you should still be running VI Analyzer configurations. If you have decided that you won't have formal code reviews and will use pull requests instead, then read through this section and decide what you will use during pull requests. (See Fig. 7.30.)

In order to perform a successful review, it is important that every member of the team has the following items installed:

1. VI Analyzer Toolkit.

2. Any custom VI Analyzer tests the team uses.

3. Any custom code review tools the team uses.

NOTE
If the code author has custom tests in their `C:\Users\[username]\Documents\LabVIEW Data` *folder, they will get the VI Analyzer menu option in the Tools menu in LabVIEW, even though they may not have the VI Analyzer Toolkit installed! (There is a difference between VI Analyzer and VI Analyzer Toolkit).*

---

[44] LabVIEW Style Checklist. http://zone.ni.com/reference/en-XX/help/371361R-01/lvdevconcepts/checklist/

[45] Example Programs Style Guidelines. https://forums.ni.com/t5/Using-the-NI-Community/Example-Programs-Style-Guidelines/ta-p/3698614

One way to know if the code author's computer has VI Analyzer Toolkit installed is to verify if **Tools >> VI Analyzer >> Create New Test...***exists. If that option is not in the menu, then VI Analyzer Toolkit is not installed.*

Another option is to go to NI MAX and look under **software >> <LabVIEW version >>** *see if VI Analyzer Toolkit is listed.*

We find that LabVIEW developers find the most problems while explaining their code to others.

Some of the benefits of conducting code reviews include:

- Identify bugs and potential issues earlier in the product life cycle
- Developers improve their programming skills through feedback

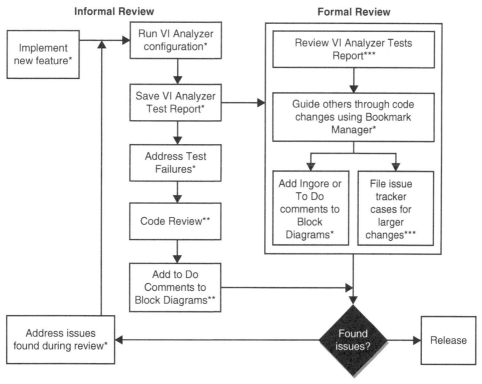

FIG. 7.30 *An example of a code review process.*

- Additional people become familiar with the code
- Ensure that the team is following the team style guidelines

Code Review Frequency

We recommend that the informal part of the code review (Run Delacor VI Analyzer tests, save test report, address high rank failures, have another developer review code) be done after implementing new features. For example, at Delacor, when we implement a new feature, we assign that JIRA case as implemented to another developer (the reviewer). If the reviewer agrees with the change, she assigns it back to developer for her to build it into the next package/release. Either both developers are present or the developer provides bookmarks for the reviewer. The developer commits and pushes both the code and the VI Analyzer Test results.

For informal code reviews, the developer can use a VI Analyzer Tests configuration (.cfg) file, which is less exhaustive than the traditional .cfg file used for formal code reviews. Schedule a formal code review for larger feature changes, architecture changes, or any other change the team decides to review via a formal code review. For this review, use the more exhaustive form of the VI Analyzer Tests configuration (.cfg) file.

Before performing a formal code review, it is important to carry out a number of "pre-review" actions. The objective is to address "low-hanging fruit," for example, issues with block diagram style, spelling, or front panel layout. Here is an example of a typical pre-review checklist; work with your team and create your own version.

Pre-Review Checklist
☐ Select VI Analyzer .cfg file for the project.
☐ Save VI Analyzer .cfg file in a folder titled *VI Analyzer* in the same directory as the LabVIEW project file.
☐ Run VI Analyzer using the selected .cfg and setting the project folder as the target.
☐ Save the VI Analyzer Report in the *VI Analyzer* folder.
☐ Commit to SCC the VI Analyzer Report.
☐ Address all of the "high-ranking" VI Analyzer test failures.
☐ Save the updated VI Analyzer Report in the *VI Analyzer* folder.
☐ Schedule code review. • One hour at the most. • Two reviewers (if more are needed, consider scheduling separate reviews).
☐ Prepare code for review. • If this is a rearchitecture or a new framework, prepare UML diagram or high-level documentation to share with reviewers before code review. • Drop bookmarks in the code. Use Bookmark Manager during code review to navigate through the code.

VI Analyzer Tests Configuration

VI Analyzer tests can check for several of your team style guidelines' checkpoints. These tests include custom tests your team created or that they downloaded from the VI Analyzer enthusiasts forum,[46] as well as the VI Analyzer toolkit built-in tests. Running all the VI Analyzer tests available will probably return many results that can be ignored (noise). This does not give a good first impression to new LabVIEW programmers in your team and will discourage them from using the tool. Instead, use a predefined VI Analyzer configuration file selected for this project. (See Fig. 7.31.)

For new projects, copy the `.viancfg`[47] file for the type of project from one of the `.viancfg` files your team has created and has installed in the `<Documents>\LabVIEW Data\CFG Templates` folder in a folder titled VI Analyzer in the same directory as the LabVIEW project file. Then make appropriate adjustments to the new project VI Analyzer configuration file.

This PC > Documents > LabVIEW Data > Delacor CFG Templates

Name	Type
Delacor Examples.cfg	CFG File
Delacor Internal.cfg	CFG File
Delacor Templates and Sample Projects.cfg	CFG File
Delacor Style Tests.viancfg	VIANCFG File

FIG. 7.31 *An example of a list of VI Analyzer configuration files.*

You can create your VI Analyzer configuration files for different stages of the code. For example, the Delacor configuration files in Figs. 7.31 and 7.32 are used as follows:

- Delacor Examples: Is based on the tests used by NI to clean up their shipping examples.[48]

- Delacor Internal: Runs a selection of VI Analyzer tests to meet the minimum style guidelines for code that will not be shared with customers.

- Delacor Templates and Sample Projects: Runs a selection of VI Analyzer tests for projects that will be used as a base for project templates. For example, Delacor uses this configuration file before releasing a new version of the DQMH toolkit.

---

[46] ni.com/vianalyzer
[47] LabVIEW 2017 or older saved VI Analyzer configuration files with the extension .cfg instead of .viancfg
[48] Example Style Guidelines—LV Add-ons forums.ni.com/t5/Developer-Center-Resources/Example-Style-Guidelines-LV-Add-ons/ta-p/3510533

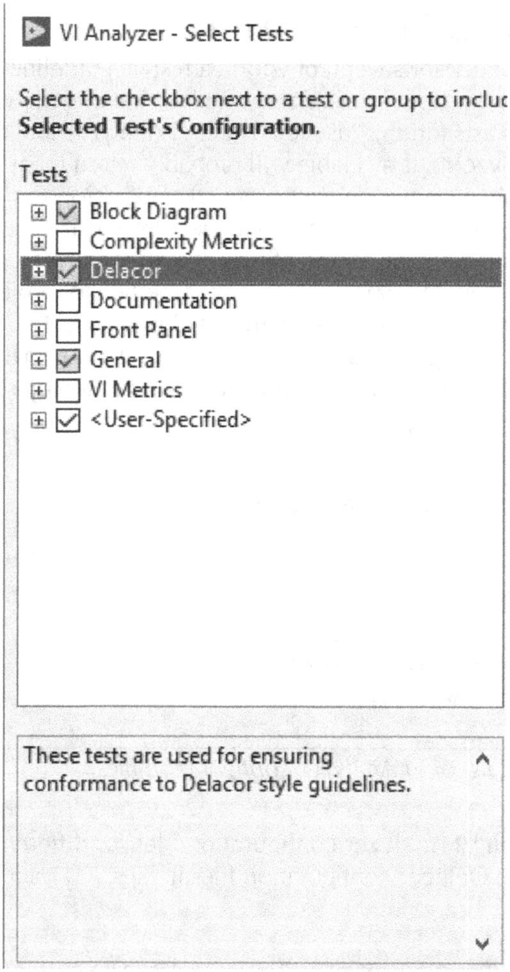

FIG. 7.32 *Example of a VI Analyzer configuration that does not select all of the VI Analyzer tests available.*

VI Analyzer Report

While preparing for a formal code review, the code author should run VI Analyzer for the project and store the VI Analyzer reports in a folder titled VI Analyzer in the same directory as the LabVIEW project. To run VI Analyzer, select **Tools >> VI Analyzer >> Analyze VIs**.

NOTE
The code author should record the report from the initial analysis and commit it to SCC repository. No need to date the report; SCC will take care of keeping the track of the version for any postupdate analysis reports.

The VI Analyzer report shows the failures that VI Analyzer found. (See Fig. 7.33.) The code author should address all of the VI Analyzer test failures before conducting a formal review. When choosing the "Sort by Test" option in the VI Analyzer Results Window, all high-ranking tests are listed at the top of the list with a "!" glyph.

It is the responsibility of the code author to run VI Analyzer tests frequently and address failures reported, not just when preparing for a formal code review.

Prepare Code and Documentation for a Code Review

For a re-architecture or a new framework, prepare a UML diagram or high-level documentation to share with reviewers before code review. For this type of code review, the code author will probably not review every VI in the framework.

For a project that uses Actor Framework, prepare a diagram of the actor tree and the communications.

For a project that uses Actor Framework, DQMH, or other actor-based architecture, prepare UML sequence diagrams to guide the developers through the sequence of events.

The code author should prepare the areas of the code they want to review. For code reviews that involve more than one VI, add bookmarks in the code to highlight the items she wants to discuss during the code review. Then use Bookmark Manager during code review to navigate through the code.

LabVIEW Compare for Code Reviews

The code author can compare previous versions of the code with the new version using LabVIEW Compare. Code reviewers might choose to use this tool to guide them when reviewing small changes. To run LabVIEW Compare, open the two versions of the VI then go to **Tools >> Compare >> Compare VIs...** or use your source code control client tool to launch LabVIEW Compare if you configured it as described in the Source Code Control section of this chapter.

LabVIEW Compare works well when changes in code are few and small. Use this tool when conducting quick peer reviews, and avoid using it when conducting one-hour formal code reviews.

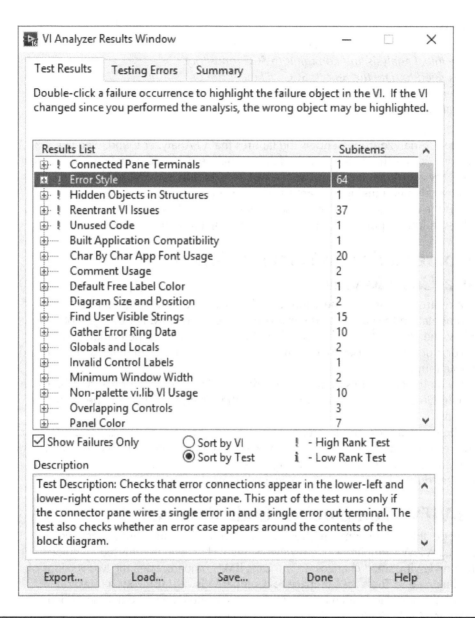

FIG. 7.33 *Failures ordered by test.*

Code Review Checklist

Use this checklist to ensure that you cover all the items your team has agreed you will cover during code reviews.

Code Review Checklist
☐ Review VI Analyzer Report
☐ Discuss any remaining VI Analyzer tests failures • If review group decides to address, add a #Code_Review_Todo bookmark with instructions on how to address it[49] • File a case in issue tracking tool for code author to address failure • If review group decides to ignore, add a #via_ignore bookmark with brief explanation why you are ignoring the VI Analyzer test[50] • If review group agrees, modify project VI Analyzer configuration file to ignore this test for this VI (or file a case in issue tracking tool to modify later)
☐ Review any diagrams or other documentation that will help code reviewers get acquainted with the code objective
☐ Use bookmarks from pre-review preparation to navigate through review
☐ Use #Code_Review_Todo tag to document in code any action items • File case to issue tracking tool for code author to address action item • Do not add #bookmarks to the front panel of VIs; add a bookmark on the block diagram containing the string ".ctl" and attach it to the appropriate control or indicator terminal
☐ Answer the following questions during code review: • Is the code maintainable? • What happens if the code returns an error? • Can the code author convert large pieces of code into subVIs? • Is the code readable?

Conducting a Code Review

In our experience, the more effective code reviews last about one hour. Take this into account when scheduling meetings for code reviews and only allocate one hour. If the code review is not complete in this time, then schedule another session.

---

[49] You can create a package of code review tools that include a VI that has on the block diagram a label with the bookmark and configure Quick Drop so you can use a key shortcut during code reviews.
[50] You can add to your code review tools package a VI that has on the block diagram a label with the bookmark and configure Quick Drop so you can use a key shortcut during code reviews.

Avoid performing a code review with an overly large group; if many people need to attend in order to review different parts of the code, then please consider holding separate reviews for each component or group of related components. Large attendance at code reviews promotes disagreement and debate rather than consensus.

The code review should begin with a review of the VI Analyzer report file. Pay close attention to all high-ranking failures, and determine how to address or at least discuss these failures during the review. If the review group agrees to ignore an issue, then you MUST add a "via_ignore"[51] bookmark-free label to the block diagram and attach it to the failure object where applicable. You can add text to the free label; however, do not remove the #via_ignore tag so the label text appears in the Bookmark Manager and if using LabVIEW 2018 or later, so the VI Analyzer toolkit can ignore the tests. If you are using LabVIEW 2017 or earlier versions, VI Analyzer will continue to report this test failure. The only way to ensure that VI Analyzer stops reporting the test failure when working with LabVIEW 2017 or earlier, is to modify the .cfg file to ignore the test that caused the test failure for this VI.

Postreview Actions

The code author and/or project manager review all the cases in the issue tracking tool that you filed during the code review and verify that they are addressed. The best code review is the one that actually happens, but even better than that code review is the one where the action items are addressed! The code author and/or project manager verifies that the code no longer has #Code_Review_Todo tags. This will leave the code ready for a future code review.

CASE Tools

CASE tools are software tools that you use to design and implement your applications. They were originally inspired by the CAD tools used to design hardware. The goal of these tools is to help you create code that is of high-quality, defect-free, and maintainable. Other tools that are considered CASE tools are the tools that you use to track your project's progress through documentation versions, tests, and tests results. I will focus on how you can create tools with LabVIEW that can help you implement code.

LabVIEW programmers use LabVIEW scripting to create CASE tools. LabVIEW scripting is an extended set of VIs that work with VI Server. LabVIEW scripting allows LabVIEW programmers to programmatically generate, edit, and inspect LabVIEW code. Please note that not all of the VI scripting functions work in executables. VI scripting was originally only used inside NI. In one of the releases of LabVIEW, someone forgot

---

[51] Ignoring Test Failures (VI Analyzer Toolkit). zone.ni.com/reference/en-XX/help/371361R-01/lvvianalyzerhelp/ignoring_failures

to password protect a `vi.lib` VI. One of the members of the LabVIEW community saw that there was a blue VI server node and started investigating more and he discovered scripting. Members of LAVA[52] started using it, and the community finally convinced NI to release LabVIEW scripting. You still need to purposely enable it in your LabVIEW options for these functions to show up on your palettes.

Why Do You Want to Use LabVIEW VI Scripting?

LabVIEW can write a VIs a lot faster than you can. The VIs will always wire VIs exactly the same way every single time. VIs wire VIs a lot faster than you, even faster than Darren Nattinger, the fastest LabVIEW programmer in the world. VIs do not get tired of writing VIs.[53]

You would want to use LabVIEW VI Scripting to create tools that would make you and your team more efficient, to enforce coding guidelines, or to automate repetitive tasks.

Common Areas That Use LabVIEW Scripting

There are several areas within LabVIEW IDE that use scripting; you might have used them without realizing that LabVIEW scripting was behind the scenes. Some of these areas are

- Quick Drop plugins
- Right-click plugins (also known as shortcut menu plugins)
- VI Analyzer tests
- Tools menu plugins (like the DQMH scripting tools)
- Custom project templates and sample project templates
- Project providers (I would stay away from those if I was you)
- Xnodes (definitely want to stay away from these; they are also known as rusty nails)

Some specific examples of tools created using LabVIEW VI Scripting are

- Convert DAQ Assistant Express VI into DAQmx code (this is the code that executes when you right-click on the DAQ Assistant express VI and select to generate code)
- Channel wires use VI Scripting to create their different implementations

---

[52] VI Scripting section in the LAVA forums https://lavag.org/forum/17-vi-scripting/
[53] De la Cueva, Fabiola. How to Train Your Monkey https://delacor.com/how-to-train-your-monkey/

- Create class accessors when you right-click on a class and select to create new VI for data access
- Create messages for Actor Framework when you right-click on an actor's method and select to create its messages
- BLT for LabVIEW from STUDIO BODs uses VI scripting to wrap your top-level VI with a splash screen
- ViBox Quick Drop from Saphir is a set of NI LabVIEW Quick Drop plugins both made by Saphir and the community; you can find it via VIPM
- Falling Shortcuts is a game to practice your Quick Drop shortcuts in preparation for the fastest LabVIEW programmer competition at NIWeek
- Watts created the LabVIEW Component builder that generates LabVIEW Components (also known as action engines or functional global variables that do more than get and set)
- DQMH productivity tools to create new DQMH modules, events, edit existing events, validate, and fix DQMH modules

Five Steps to Become a VI Scripting Ninja

LabVIEW VI Scripting is just VI server upgraded. If you already know how to use VI server, you know how to use the VI scripting properties and methods.

I have talked about VI scripting at CLA Summits[54] and at the Delacor blog.[55] It used to be that you had to go through the LAVA forums and try a lot of things before getting LabVIEW scripting to work. There were also a lot of warnings; during NI customer education courses they would even tell you that they were not supported by NI. LabVIEW scripting has been available since LabVIEW 2009, and the LabVIEW product support engineer at the time of the CLA Summit presentation assured me that if the scripting function is fully documented, then it is supported by the applications engineers at NI. NI has also added lots of documentation and tutorials to their help and shipping examples.

Step 1: Enable VI Scripting

For LabVIEW 8.6 and LabVIEW 2009, you can download LabVIEW VI Scripting.[56] For LabVIEW 2010 and later, you enable VI Scripting via **Tools >> Options >> VI Server >> VI Scripting**.

---

[54] De la Cueva, Fabiola. How to Train Your Monkey https://delacor.com/how-to-train-your-monkey/
[55] De la Cueva, Fabiola. 5 Steps to Become a VI Scripting Ninja. delacor.com/5-steps-to-become-a-vi-scripting-ninja
[56] http://sine.ni.com/nips/cds/view/p/lang/en/nid/209110

Step 2: Become Familiar with the LabVIEW VI Scripting Shipping Examples

The LabVIEW Help topic on programmatically scripting VIs in LabVIEW[57] includes a link to more information on VI server[58] as well as more details on what VI scripting is and why you should use it. The Help section on VI scripting also includes a series of exercises that guide you through how the LabVIEW programmers at NI created the shipping examples.[59] These examples are very well documented and will get you on your way on the basics of LabVIEW VI scripting.

Step 3: Join the Quick Drop Enthusiasts Group

You can use Quick Drop[60] to drop functions from the palettes faster without having to remember where the VIs are in the palettes. It also includes a feature called Quick Drop shortcuts.[61] A good ninja will not reinvent the wheel. See what tools are already out there, install the ones that you like, inspect them, and learn how they were put together. Then start creating your own tools. A good place to start looking for existing tools is the list of Community Quick Drop keyboard shortcuts[62] in the Quick Drop enthusiasts online forum.

I apologize in advance; before reading these lines you probably never cared about where the error wires were in your block diagram. One of our customers likes to push all of the block diagram error wires to the back of the diagram. We created a Quick Drop shortcut that does just that. This tool is available in the GitHub/LGP5/Teams repository.

Use Quick Drop for non–context-sensitive actions that you do dozens of times a day. The steps to create a Quick Drop shortcut follow.

1. Create a new VI from the `<LabVIEW>\resource\dialog\QuickDrop\QuickDrop Plugin Template.vit`

    - To create a new VI from a template, you can add the `.vit` file to your project and from there right-click on the `.vit` and select **New from Template**.

2. Edit the Quick Drop template. There is enough documentation on the block diagram to explain what the different areas are doing. You can download and inspect examples from the Quick Drop enthusiasts group to understand better how to modify the template.

    - The code you are writing is the code that will execute when the LabVIEW programmer presses <Ctrl-[char]> or <Ctrl-Shift-[char]> while the Quick Drop dialog box is visible <Ctrl-space>.

---

[57] http://zone.ni.com/reference/en-XX/help/371361N-01/lvconcepts/scripting_property_method/
[58] http://zone.ni.com/reference/en-XX/help/371361N-01/lvconcepts/programmatically_controlling_vis/
[59] Go to Help >> Find Examples... >> Programmatically controlling VIs >> Editing and Inspecting VIs
[60] Quick Drop enthusiasts http://ni.com/quickdrop
[61] http://www.ni.com/tutorial/7423/en/#toc4
[62] forums.ni.com/t5/Quick-Drop-Enthusiasts/List-of-Community-Quick-Drop-Keyboard-Shortcuts/gpm-p/3527206

3. Clean up any leftover template documentation that does not pertain to your tool.

4. Save a copy of the VI with a descriptive name in `<LabVIEW>\resource\dialog\QuickDrop\plugins` or in `<LabVIEW Data>\QuickDrop Plugins`.

5. Edit the VI Description to include a description of your plugin and specify a default shortcut.

Step 4: Join the LabVIEW Shortcut Plugins

LabVIEW Shortcut plugins are also known as right-click menus. The LabVIEW Shortcut Plugins group[63] has examples and help information on how to implement your own. They were introduced in LabVIEW 2015 and allow you to extend the LabVIEW IDE.

Use the right-click menu for context-sensitive actions that you do sporadically but take time to do.

The steps to create a LabVIEW Shortcut Plugin are

1. Open the generator code via `<LabVIEW>\resource\plugins\PopupMenus\Create Shortcut Menu Plug-In from Template.vi`.
   - This code generator is also available via Hidden Gems.[64]

2. Follow the instructions on the front panel and run the VI to generate your code.

3. Edit the different VIs that the generator created. You will notice that it created an LLB for your files. There is enough documentation on the block diagram to explain what the different areas are doing. You can download and inspect examples from the LabVIEW Shortcut Plugins group to understand better how to modify the templates.
   - The code you are writing is the code that will execute when the LabVIEW programmer right-clicks on an object and selects the right-click menu item that you are creating.

4. Clean up any leftover template documentation that does not pertain to your tool.

5. The generator code created the library for your plugin in `<LabVIEW Data>\PopupMenus\edit time panel and diagram\<the name of your plugin>.llb` or in `<LabVIEW Data>\PopupMenus\run time diagram\<the name of your plugin>.llb` depending on where you specified if the shortcut menu plug-in affected edit time or runtime shortcut menus.

---

[63] ni.com/lvmenus
[64] ni.com/hiddengems

Step 5: Join the VI Analyzer Enthusiasts Group

The VI Analyzer Enthusiasts group[65] has examples and help posts for custom VI Analyzer tests.

Maybe you use the old approach of having one of the old-timers in your team review your code and point out all the style errors you made. I don't know about you, but I prefer to save those finger-pointing moments for things that will teach me new code practices or help me find bugs. You probably tried VI Analyzer once and decided to never run it again because it reminded you of your spouse pointing out every little detail you forgot. I agree with you, I don't like running every single test on my code either.

We use different VI Analyzer configurations depending on the target audience for the code. We have also created a set of custom tests to have VI Analyzer do the hard job of pointing out style deviations before the code review happens. My favorite test is the one that reports when a LabVIEW developer forgot to wire the error out terminal from the File Dialog VI.

Darren Nattinger gave an excellent presentation during NIWeek 2015 called Inspecting You Code with VI Analyzer;[66] there is also a webcast available.[67]

The steps to create a custom VI Analyzer test are

1. Go to Tools >> VI Analyzer >> Create New Test…

2. Follow the wizard's instructions.

3. Edit the code to perform the actions you want to test.

4. Clean up any leftover template documentation that does not pertain to your tool.

5. Save your custom VI Analyzer tests llb in `<LabVIEW Data>\VI Analyzer Tests`.

We recommend that you keep all your VI Analyzer tests in the same repository and you package them via VIPM. We treat them like a product; they have their own issue tracking project. As part of the maintenance of the VI Analyzer tests, we create example code that we use to test each VI Analyzer test. These tests are a good way for us to train new team members. We ask them to run all of our custom VI Analyzer tests over the examples, and they can read on their own why each test marked that code as a failure.

---

[65] VI Analyzer Enthusiasts group http://ni.com/vianalyzer
[66] Nattinger, Darren. Inspecting Your LabVIEW Code with the VI Analyzer https://forums.ni.com/t5/Past-NIWeek-Sessions/Inspecting-Your-LabVIEW-Code-With-the-VI-Analyzer/ta-p/3520612
[67] Improving Code Quality through Automated Code Analysis http://www.ni.com/webcast/3797/en

When we find a bug in code we are developing, one of the questions we ask is—could this bug have been found by VI Analyzer?—If the answer is yes, we create a VI Analyzer test so this bug doesn't happen again.

Deployment, Continuous Integration, and Continuous Delivery

Continuous integration (CI) is when every code pushed into the repository is verified and built to ensure all components are working together by an automated process.

Continuous delivery (CD) is when you ship releases regularly with a push of a button.

You don't have to wait until you are ready to implement the complete continuous integration system to start making changes toward that goal. Start by creating a checklist of what your team does every time they release code. I have seen that this exercise helps the team realize the advantages of having the computer do the repetitive task automatically and does it the same way every time. It also helps identify steps that might have been forgotten. Also, doing it manually helps you fine-tune the process before you automate it.

The key members of the LabVIEW community that started moving toward bringing CI to LabVIEW are James McNally, Ching-Hwa Yu, Matthias Baudot, Allen Smith, Matt Pollock, and Benjamin Celis. More recently, Joerg Hampel has shared his Release Automation Tools for LabVIEW.[68] I hope I didn't forget anybody in this list.

Ching-Hwa Yu described the different parts of the continuous integration "machine." The two parts are the Build Engine[69] and the Build Environment.[70] You will see some variations of the VIs that he recommends, as well as how to do it using the LabVIEW CLI that ships with LabVIEW 2018 and later. If you want to go beyond what this chapter describes, visit the online Continuous Integration community.[71]

Build Engine
LabVIEW Build API

The LabVIEW Application Builder comes with a palette to automate the building process in LabVIEW.

The `Build.vi` needs the path to the LabVIEW project file and the name of the build specification.

If you plan to implement your continuous integration with a tool like Jenkins, you will need to provide an executable that accepts command-line arguments.

---

[68] Hampel Software Engineering. Release Automation Tools for LabVIEW. https://rat.hampel-soft.com.
[69] Yu, Ching-Wa. The Continuous Delivery Machine Part 1: Build Engine. https://chinghwayu.com/2016/05/cd-machine-part-1-build-engine/
[70] Yu, Ching-Wa. The Continuous Delivery Machine Part 2: Build Environment. https://chinghwayu.com/2016/06/cd-machine-part-2-build-environment/
[71] https://forums.ni.com/t5/Continuous-Integration/gp-p/5035

Chapter 7: Developing in LabVIEW for Teams **543**

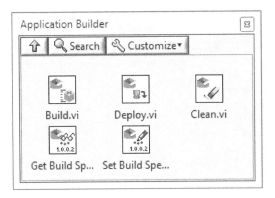

FIG. 7.34 *LabVIEW Application Builder API palette.*

That can be your own executable built with LabVIEW or just use LabVIEW itself as the executable with your build automation VI as the argument. (See Fig. 7.34.)

LabVIEW Command-Line Interface (CLI)

If you build your executable in LabVIEW and check the option to **Pass all command line arguments to application**, your executable will be able to accept arguments, and you can program your application to handle those arguments[72] (See Fig. 7.35.)

This is the mechanism that you will use to pass the arguments from your CI tool to your Build application. (See Fig. 7.36.)

If you put it all together, the application that you will use to automatically build your executables will look like Fig. 7.37.

Ching-wa Yu also modifies the build version in this step, but the bare minimum you need is to call the Build.vi to launch the build of the specified build in LabVIEW, and you can let the project keep track of the build version. Notice that if you are going to do this, you will need to add a step to your CI tool to save the project after the build is done and commit/push to your repository. If you insert the build number as part of the CI build, you don't need to commit/push the project because the only thing that changed is the version number and you provided that version. I have seen both approaches.

It used to be that we couldn't get information back from the LabVIEW CLI. Starting with LabVIEW 2018 you can run operations in LabVIEW by executing commands using the command-line interface (CLI) for LabVIEW. Before NI came out with this version that ships with LabVIEW, James McNally from Wiresmith Technology in the

---

[72] We described this approach in the DQMH chapter when we went over the DQMH CML example and how it passes the debug argument to the top-level VI.

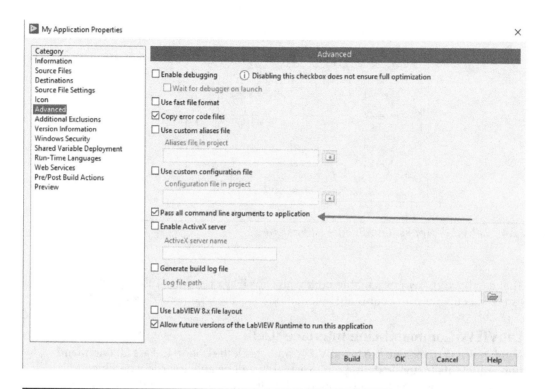

FIG. 7.35 *Enable Pass all command line arguments to application.*

FIG. 7.36 *Use the property node to handle the CLI arguments. Index 0 of the Command Line Arguments array will always include the word LabVIEW.*

UK had created his own version.[73] His version is available via GitHub,[74] and the big difference is the fact that James's version can provide continuous output while the application is running. Also you can use it in EXEs. Historically James's version couldn't work when LabVIEW was already open, but the new 2.0 version can.

---

[73] Bringin the Command Line Interface to LabVIEW by James McNally https://devs.wiresmithtech.com/blog/bringing-command-line-interface-labview/

[74] LabVIEW-CLI by Jamees McNally https://github.com/JamesMc86/LabVIEW-CLI

Chapter 7: Developing in LabVIEW for Teams **545**

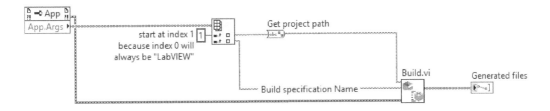

FIG. 7.37 *Programmatically build the specification in the project path given by the command-line interface.*

By the way, members of the community were very disappointed when NI did not give James his deserved recognition when they released this feature, but that is a story for another book.

You will download the CLI tool from the James McNally GitHub; you can use the VIPM package to install it. The package description says (emphasis mine):

"The LabVIEW CLI is a package that supports the client side of a two way command line interface. The package contains two parts:

- A Windows application which installs to the system path so that it can be launched from the command line

- A LabVIEW library that allows you to communicate with the application to pass data to the command line

NOTE
Since the Windows application installs to path administrator **access is required** *to install it."*

You will be prompted to authorize the installation of the Windows component. There are enough members of the community to vouch for James; you can accept this installation. (See Fig. 7.38.)

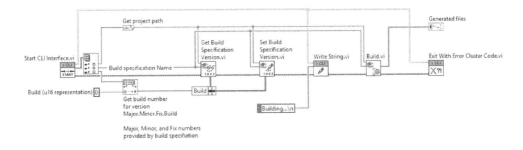

FIG. 7.38 *Integrating the James McNally LabVIEW CLI calls into the automated build VI.*

Alternatively, if you do want to use the new LabVIEW CLI that ships with LabVIEW 2018 or later,[75] all you have to do is call

```
LabVIEWCLI -OperationNameExecuteBuildSpec -ProjectPath<path to project>
-TargetName<name of target> -BuildSpecName<name of build specification>
```

For example, `LabVIEWCLI -OperationNameExecuteBuildSpec -ProjectPath "C:\temp\test.lvproj" -TargetName "My Computer" -BuildSpecName "My DLL"`

Build Environment

We mentioned earlier that in an ideal world you would be spinning a brand-new build server each time you build; this might not be feasible because it takes time and because once you have experienced the beauty of automating your builds, you will want to do it for all your projects. Then your team members will want to join in the fun.

Still, you have to ensure that before your build starts, the build server is set up for your application. This means all the dependencies are installed and ready to go.

VIPM to Manage Package Dependencies

If you are using VIPM to manage your package dependencies, then you can create a `.vipc` (VI Package configuration) file that will contain the list of all the packages your application depends on. The packages themselves can be included in the `.vipc` or they can be referred to if they are in a package repository such as LabVIEW Tools Network or your company VIPM repository. The creation of `.vipc` files requires the VIPM Pro edition. Creating packages and applying the `.vipc` file can be done with the free edition of VIPM. Having a VIPM repository for your company's packages requires that everyone in the team has the pro edition.

If you want to automate the application of a `.vipc` file, you will need to download VIPM the VIPM API package. You do need the Pro edition of VIPM to use the VIPM API. (See Figs. 7.39 and 7.40.)

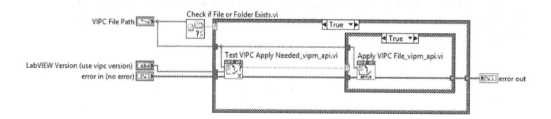

FIG. 7.39 *Create an Apply VIPC file that verifies if the .vipc file exists and if it does, checks if it needs to be applied and applies it.*

---

[75] http://zone.ni.com/reference/en-XX/help/371361R-01/lvconcepts/cli_predefined_operations/#ExecuteBuildSpec

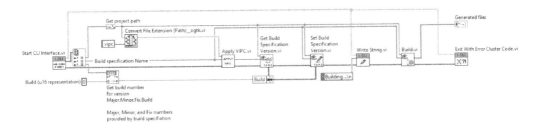

FIG. 7.40 *Automatic build including checking if a .vipc with the same project name is available at the same folder as the project and if it is there, then applying it.*

Unit Testing

You can automate running your unit tests the same way. Depending on which tool you use to run your unit tests, you can use the JKI VI Tester API, the Unit Test Framework API, or the InstaCoverage API. You can look at what the different tools do and how to use them in the unit testing chapter of this book.

Specifically for the Unit Test Framework, you can use the new LabVIEW CLI that ships with LabVIEW 2018 and later.[76] To run this operation in the CLI for LabVIEW, execute a command with the following syntax:

```
LabVIEWCLI -OperationNameRunUnitTests -ProjectPath<path to project
file> -JUnitReportPath<path to output JUnit file>
```

For example, `LabVIEWCLI -OperationNameRunUnitTests -ProjectPath "C:\temp\test.lvproj" -JUnitReportPath "C:\temp\test.xml"`

For InstaCoverage, you can check out their Jenkins demo.[77] It uses a demo project, which is a basic calculator.

VI Analyzer

You can also include VI Analyzer as part of your CI. When NI included the new CLI tools, they also included options to call VI Analyzer.[78] To run this operation in the CLI for LabVIEW, execute a command with the following syntax:

```
LabVIEWCLI -OperationNameRunVIAnalyzer -ConfigPath<path to configuration
file> -ReportPath<path to report> -ReportSaveType<file type of report>
-ConfigPassword<password of configuration file>
```

For example, `LabVIEWCLI -OperationNameRunVIAnalyzer -ConfigPath "C:\temp\test.viacfg" -ReportPath "C:\temp\output.html" -ReportSaveType "html" -ConfigPassword "abc"`

---

[76] http://zone.ni.com/reference/en-XX/help/371361R-01/lvconcepts/cli_predefined_operations/#RunUnitTests
[77] incquerylabs.com/instacoverage/demos/ci-demonstrator
[78] http://zone.ni.com/reference/en-XX/help/371361R-01/lvconcepts/cli_predefined_operations/#RunVIAnalyzer

STUDIO BODs BLT to Manage Build and Deployment

You can use BLT from STUDIO BODs[79] to cut down the deployment process to the push of a button, instead of spending hours babysitting LabVIEW while it builds and then packaging, uploading to a server, and so on. BLT also adds a splash screen to your application as well as providing the tools to track errors from your application and protect it with licensing. If you choose to include the calls for the error reporting, you can get an e-mail with the values of all the controls and indicators on the front panel of the VI as well as the error code. You can find a lot more about what caused the error and the steps before the error. For example, you could build a DQMH API tester as an executable and report the errors at that level, BLT then would include the Status indicator. The Status indicator on an API Tester includes a log of all the actions done until that point; this might be all the information you need to troubleshoot the error cause.

Using BLT Automation for Continuous Integration   BLT for LabVIEW includes an automation tool which allows you to perform basic tasks using the command line.[80] You can programmatically add a note in the Product Changes Log and build a product. When you programmatically launch the build of a product, BLT will

1. Build your project into an executable using LabVIEW
2. Compress everything in the build target directory into a single zip archive
3. If existing, it will add your splash screen and color scheme configuration to the zip file
4. Upload the zip file on the Product server via FTP
5. Generate a new product version in BLT with your last change log entries as release notes

BLT for LabVIEW leverages the low-level AppBuilder API from LabVIEW in order to automate all the build tasks. While this API is not fully supported or documented by NI, Matthias Baudot, the software architect for BLT at STUDIO BODs, gave a great presentation at NIWeek 2017 on the secrets of this API and how to use it to implement your own continuous integration workflow.[81]

Other Tips

Keep a success stories board. When architects start to share how VI Analyzer tests found an error they had on their code, the other LabVIEW programmers start to

---

[79] Build License and Track BLTforLabVIEW.com
[80] https://www.studiobods.com/help/bltforlabview/UsingBLTAutomationforContinuousI.html
[81] https://www.studiobods.com/en/niweek2017-ts721/

see the benefit of running VI Analyzer tests. If reusable code saved the day, share the story with the LabVIEW programmer who created the tool. A lot of people focus more on the bad incidents and forget to celebrate when things are working.

Conduct regular meetings with your team, where you discuss what is working, what needs improvement, and what are the highlights of the current process. Please note that you should be very specific about the wording. It is What is Working and What NEEDS IMPROVEMENT. Not the What is Working, What is not Working session. One denotes hope and a willingness to make things better; the other one is a whining session.

Prepare an onboarding new team members document and have it in a place where all team members can edit it. That document should include where to find documentation, tools, and training. Update that document every time a new member joins. Make it their first project to go through the document and verify that all the links work. Create a .vipc file that contains all the tools and packages that every team member should have in their machine. You can create a package that launches a VI upon installation that would change the LabVIEW .ini to ensure everyone has their LabVIEW system similar; for example, you can set that source code will be separated from compiled code, that you don't want to see terminals as icons, and so on.

References

Every Developer Needs a Time Machine blogpost at http://delacor.com/every-developer-needs-a-time-machine/
Source Code Control Videos at bit.ly/LVscc
Endigit Configure SourceTree to use LabVIEW Compare blogpost at https://endigit.com/2017/11/using-labviews-diff-tool-sourcetree
Atlassian Tutorial on What is Version Control at https://www.atlassian.com/git/tutorials/what-is-version-control
Git User Group at forums.ni.com at https://forums.ni.com/t5/Git-User-Group/gp-p/grp-2368?profile.language=en
Mercurial User Group at forums.ni.com at https://forums.ni.com/t5/Mercurial-User-Group/gp-p/5107?profile.language=en
Blog post on choosing Source Code Control tool for binary files. While our conclusion is different for LabVIEW purposes, the majority of the arguments do apply to us as well. Since our binary files are not as large and we do have diff/merge tools, we still think Git or Hg are better tools for LabVIEW development.
medium.com/flow-ci/github-vs-bitbucket-vs-gitlab-vs-coding-7cf2b43888a1
Piña Saez, Jarobit and De la Cueva Fabiola. DiseñatuPropio Marco de Trabajo en LabVIEW. Bilbao, España 6-7 Feb, 2018 www.ulmaembedded.com/es/noticias/curso-disenando-tu-propio-marco-de-trabajo-en-labview/no-97/
https://chinghwayu.com/2016/05/cd-machine-part-1-build-engine/
Release Automation Tools for LabVIEW at https://www.hampel-soft.com/release-automation-tools-for-labview/
Pragmatic Software Development Workshop http://www.dsh-workshops.com/

CHAPTER 8

Enterprise and IoT Messaging

Sending messages between your application's components is a powerful technique for building robust, scalable applications, whether you use a Queued Message Handler, the Delacor Queued Message Handler, the Actor Framework, or your own framework. When properly built, messaging within your application allows you to build reusable components that are easy to test, loosely coupled, and respond in predictable methods to predefined inputs. In this section, we look at inter-application messaging tools you can use to integrate LabVIEW applications with other applications using enterprise and cloud-based Internet of Things (IoT) **message brokers**.

Too often data from measurement systems is stored in proprietary formats and only specialized applications can access the data. Interoperability with other enterprise applications requires getting data out of that proprietary data silo and making it accessible to the rest of the enterprise. Whether you're building a rack-n-stack test system or a fleet of remote monitoring devices, you need to collect, analyze, and share that data in order to make data-driven decisions. Maybe you have multiple development teams working in other languages that need to interact with your LabVIEW systems. You could design a point-to-point communications system, but why reinvent the wheel? The message broker is responsible for routing messages to and from interested parties, so your application can keep doing what it is designed to do and not worry about maintaining communications with multiple endpoints. Building around a message broker allows your test and measurement systems to publish data and respond to commands, while remaining loosely coupled and asynchronous. Other applications can come online and offline to work with your data and interact with your application with no changes to your code. Messaging allows you to build applications and systems that scale.

The definitive book on messaging patterns for enterprise integration is *Enterprise Integration Patterns: Designing, Building, and Deploying Messaging Solutions* by Gregor Hohpe and Bobby Woolf, ISBN: 0-321-20068-3. The book is a deep dive into enterprise messaging patterns for back-end systems. For brevity, we focus on event-driven, **publish and subscribe** applications using a message broker. In our example applications LabVIEW publishes data, status, and error messages to topics on a message broker and listens for commands on subscribed topics. Other applications can subscribe to our outbound message and data topics to consume and act on the data and publish commands to our subscribed topics (Fig. 8.1).

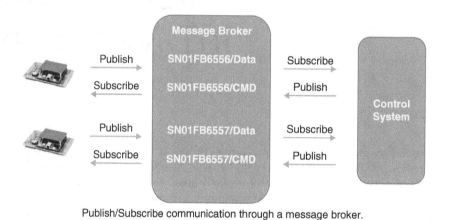

Publish/Subscribe communication through a message broker.

FIG. 8.1 *Devices publish and subscribe to messaging topics using a message broker.*

There are several key rules to keep in mind when designing an inter-application messaging system around a message broker.

Rule 1. Devices need to communicate over predefined messaging channels.

Devices and applications can't blindly dump their data into the message broker. In Fig. 8.1 two devices are publishing data and listening to commands on reserved data topics that are prefixed with the device's serial number. These are known as publish and subscribe topics. The message broker takes care of routing messages between applications based on what topics each application is subscribed to. In the example earlier two devices are publishing data to their data topic and listening for commands on their command topic. The control system subscribes to data topics and publishes commands to each device's command topic. Other applications can interact with the message broker by publishing and subscribing to topics. For example, you may have an application subscribed to every topic and logging the traffic to a database.

Rule 2. Use a flexible, universally understood message format.

To make the best use of a messaging system, every application needs to understand the content of the message. Modern message brokers can act as protocol translators, so not every application has to interact with the message broker using the same protocol; however, all applications have to understand the message payload. Some applications may use AMQP (Advanced Messaging Queueing Protocol); others may use STOMP (Simple Text Oriented Messaging Protocol) or MQTT (MQ Telemetry Transport). It really just depends on what protocols your development environment supports. That's the beauty of a message broker: many applications with different capabilities, written by different teams in different languages, can exchange messages. However, in order for the applications to communicate with each other, they all have to understand the message content.

In decades past, bandwidth was expensive and CPUs were slow, so it made sense to design an optimal binary message format. But a binary formatted message also restricts the ability of others to consume the data. Consider this quote:

> *One of the great strengths of the Unix operating system lies in its "shell," where the user has the ability to mix and match any number of program modules (also known as "utilities") to achieve the desired result. This is possible because the universal interface between all modules comprises streams of text characters, so that the output of one module can be the input of any other.*
>
> Rod Manis and Marc H. Meyer, The Unix Shell Programming Language
> (Indianapolis: Howard W. Sams Company, 1986)

Presently the most popular text-based, message interchange format is JSON (JavaScript Object Notation). In its simplest format JSON is a comma-separated array of key:value pairs separated by a colon, with the entire array enclosed in curly braces. Like this: {"key1" : "value1", "key2" : value2" }. LabVIEW doesn't have the best tools for working with JSON. LabVIEW's JSON support flattens and unflattens to and from clusters. This works OK, until the JSON packet changes and LabVIEW starts throwing errors, but it's easy enough in LabVIEW to convert to and from JSON using a 2D array of strings, with keys in the first column, and values in the second (Fig. 8.2).

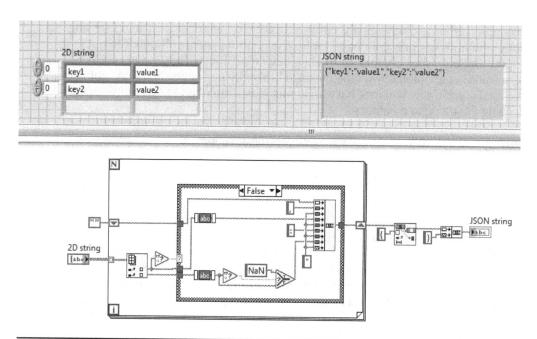

FIG. 8.2 *Encoding a 2D array of strings in JSON.*

The code snippet in Fig. 8.2 takes a 2D array of strings and formats it into a JSON-formatted string while leaving out any empty fields. It's simple, flexible, and works for basic key:value pairs. Obviously if one of the values is an array or a nested object, etc., you need to jump through some hoops to preformat the value field to match the data type. There are several third-party JSON toolkits available online. Hopefully LabVIEW will come out with better support for JSON in the future.

Rule 3. Include a description and a version number in your data packet.

Version numbers and descriptions in your data packet allow your system to age gracefully and not break at the first change. A simple description may be the make or model of a piece of equipment, or the test system the packet is coming from, and the version number is the revision of the JSON packet. Version 2.x key:value pairs may be wildly different than version 1.x as you deploy newer versions of equipment. Versioning allows downstream applications to work with multiple versions of the data payload without requiring all data generators and consumers to go through a firmware upgrade.

These guidelines are based on years of experience with mixed development teams. When you design your message format, you can make it be anything you want, but for the purposes of this chapter, all data is transferred in JSON format with all fields as UTF-8 strings. Priority is placed on readability and portability rather than compactness. Downstream applications can format the data appropriately for their needs. The cost curves for bandwidth and storage are exponentially decaying, and it's not worth putting a lot of effort into the "most efficient" payload unless you are severely bandwidth limited.

MQTT Messaging Protocol

There are a number of messaging protocols out there, and many message brokers support multiple protocols. Some message brokers will even translate between protocols for different clients. Our applications use MQTT (MQ Telemetry Transport). MQTT is a light weight, publish and subscribe protocol designed for IoT applications. We use the MQTT library from DAQ.IO (https://github.com/DAQIO/LVMQTT) written in 100 percent pure G by Peter Adelhart and kindly released as open source under the MIT license (Fig. 8.3).

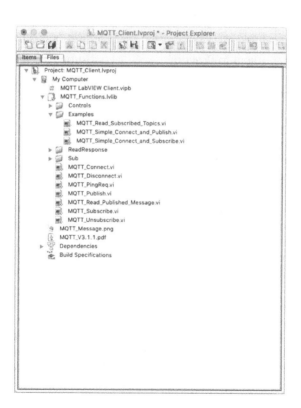

FIG. 8.3 *MQTT_Client.lvproj from https://github.com/DAQIO/LVMQTT with example VIs.*

Name	Value	Direction of Flow	Description
CONNECT	1	Client to Broker	Client request to connect
CONNACK	2	Broker to Client	Connect acknowledgment
PUBLISH	3	Client to Broker or Broker to Client	Publish message
PUBACK	4	Client to Broker or Broker to Client	Publish acknowledgment
PUBREC	5	Client to Broker or Broker to Client	Publish received (assured delivery part 1)
PUBREL	6	Client to Broker or Broker to Client	Publish release (assured delivery part 2)
PUBCOMP	7	Client to Broker or Broker to Client	Publish complete (assured delivery part 3)
SUBSCRIBE	8	Client to Broker	Client subscribe request
SUBACK	9	Broker to Client	Subscribe acknowledgment
UNSUBSCRIBE	10	Client to Broker	Unsubscribe request
UNSUBACK	11	Broker to Client	Unsubscribe acknowledgment
PINGREQ	12	Client to Broker	PING request
PINGRESP	13	Broker to Client	PING response
DISCONNECT	14	Client to Broker	Client is disconnecting

TABLE 8.1 *MQTT is a simple protocol with 14 message types.*

The MQTT protocol is an open standard, and the DAQ.IO GitHub repository includes the 3.1.1 standard the library is written against. MQTT is a binary protocol with 14 message types, a variable-length header up to 4 bytes, and depending on the message type, an optional UTF-8-encoded payload of up to 256 MB (Table 8.1). The simplicity of MQTT and its ability to run on many types of hardware has established it as a dominant IoT protocol.

Once a client has established a connection to a MQTT broker and subscribed to topics of interest, the client maintains a constant connection to the message broker. It's important to note that **MQTT topics do not have to be created in advance**. The act of publishing or subscribing to a topic creates the topic if it does not already exist. This lack of administrative overhead makes MQTT very simple to use. Just connect and pub/sub to your topics. From that point on the communication process is an event-driven, publish and subscribe message interchange over TCP/IP using the MQTT protocol. Anytime a message comes into the message broker on a subscribed topic, the broker sends the message to all subscribers in a one-to-many distribution.

Chapter 8: Enterprise and IoT Messaging **557**

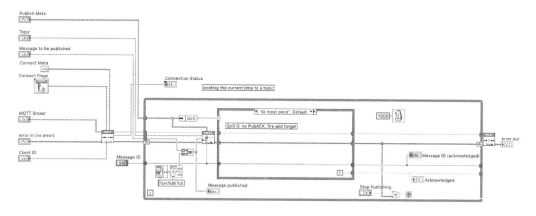

FIG. 8.4 *Simple Connect and Publish.vi.*

The DAQ.IO library comes with several examples. Simple Connect and Publish.vi (Fig. 8.4) establishes a connection to a broker and publishes the time once a second to testtopic/LVMQTT. Note that MQTT topics use a path format for the destination with a forward slash (/) as a separator. The path format lets you give some structure to your messaging architecture. You can also use wildcard characters with topic paths in order to subscribe to many topics at once. We'll see how this works later.

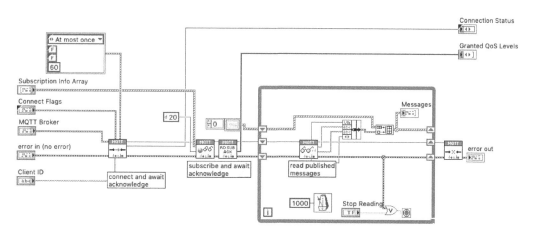

FIG. 8.5 *Simple Connect and Subscribe.vi.*

The example Simple Connect and Subscribe.vi (Fig. 8.5) establishes a connection to the broker and listens for incoming messages on testtopic/LVMQTT. It's important to note that **each connection to the broker must have a unique Client ID**. The Client ID must be between 1 and 23 UTF-8-encoded bytes in length, and can only contain following the characters: 0123456789abcdefghijklmnopqrstuvwxyzABCDEFGHIJKLMNOPQRSTUVWXYZ. In our example code we use the MAC hardware address as the Client ID on Windows and the device serial number on RT Linux.

Install a Message Broker

We're going to use Apache ActiveMQ Artemis (Fig. 8.6) as our message broker. It's open source and written in Java and can run on any platform that can run Java. Download the latest zip from https://activemq.apache.org/artemis/download.html and follow the instructions to install and launch. Be sure to **allow anonymous access** for this test case. Once Artemis is installed, navigate to http://localhost:8161/console and log in with the credentials you set up as part of the install process. If you want to install a different message broker, that's up to you, just make sure it supports MQTT 3.1.1. If you don't want to install a broker, then **test.mosquitto.org** has a broker up and running most of the time; just be sure to use a unique Client ID each time you connect.

FIG. 8.6 *Artemis console after startup.*

Now fire up LabVIEW and run the example VIs in the MQTT_Client.lvproj. First start the MQTT_Simple_Connect_and_Publish.vi (Fig. 8.7). Set the address (IP) to localhost, or to the IP of the machine where you're running Artemis, or to test.mosquitto.org. Then start writing data into the message broker. The default topic is testtopic/LVMQTT and the message is "LabVIEW says it's <time>."

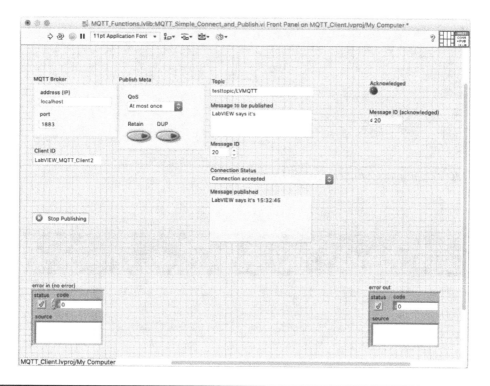

FIG. 8.7 MQTT_Simple_Connect_and_Publish.vi.

Then fire up MQTT_Simple_Connect_and_Subscribe.vi (Fig. 8.8). Set the address (IP) to localhost, or to the IP of the machine where you're running Artemis (Fig. 8.9), or to test.mosquitto.org. This VI will subscribe to the default topic of testtopic/LVMQTT and display messages in an array as they come in. That's it!

560 LabVIEW Graphical Programming

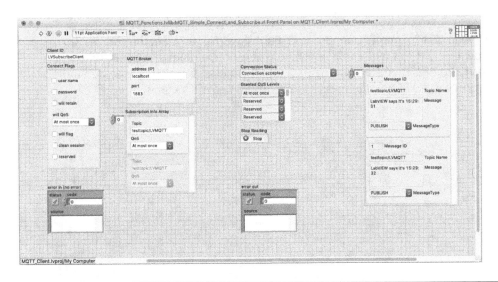

FIG. 8.8 MQTT_Simple_Connect_and_Subscribe.vi.

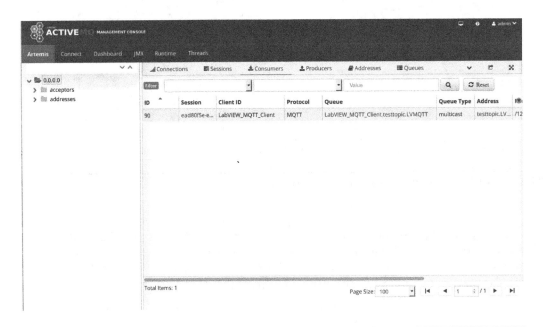

FIG. 8.9 Artemis console after connecting from LabVIEW.

MQTTDQMH Client

Now that you've seen how easy it is to publish and subscribe with a message broker, let's go over a DQMH module we built to handle communication with the message broker. DQMH provides a framework for building event-driven, testable modules and is perfect for a publish and subscribe messaging application. You can install the MQTTDQMH client as a template for use in future projects.

Getting Started

We created a new DQMH project with the create project wizard. Then we added the MQTT_Functions.lvlib to the project (Fig. 8.10). In this project we used the Singleton module. We could use the cloneable module in multiple sections of code if each clone had a unique Client ID. But for our app we used the Singleton as a single communication gateway between our application and the message broker.

Then we created four new DQMH Request events using Tools >> Delacor >> DQMH >> Event >> Create New DQMH Event (Table 8.2). The events are Connect, Publish, Subscribe, and Disconnect (Fig. 8.11). These events are the public interface to our Singleton module. We also created two Broadcast events: an "RX" event that our module uses to broadcast received messages and a "Connection OK?" event for the connection status.

Request Event	Param 1	Param 2	Description
Connect	Broker IP (string)	Client ID (string)	Establishes a MQTT connection
Publish	Topic (string)	Message (string)	Publishes a message to the broker on a specific topic
Subscribe	Topic (string)		Subscribes to a topic on the broker.
Disconnect	Boolean		Disconnects from the broker.

TABLE 8.2 *DQMH Request Events and Parameters for the MQTT Gateway*

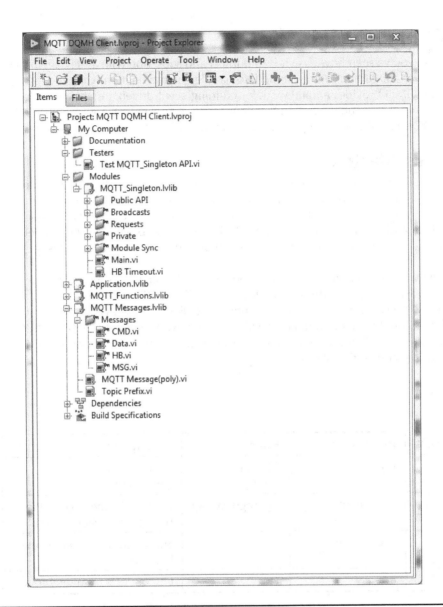

FIG. 8.10 *MQTT_DQMH_Client project.*

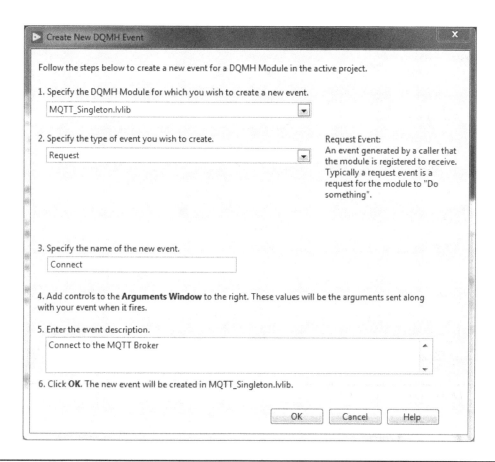

FIG. 8.11 *Create DQMH events using the DQMH Wizard.*

When events are created, the DQMH Wizard creates a new Tester UI event case for each DQMH event to make it easy to test out the code. We populated the Tester with the buttons and strings required to test out the public interface to our MQTT gateway (Fig. 8.12).

After creating the public interface, we added a helper loop to the DQMH module and created some private events to control the helper loop (Fig. 8.13). The helper loop listens to messages from the broker and broadcasts them to any listeners with our RX event. Our private events tell the helper loop to "Start Listening" once we subscribe to a topic or to "Stop Listening" if we disconnect. "Exit Listening" is called to terminate the helper loop when the module exits.

564 LabVIEW Graphical Programming

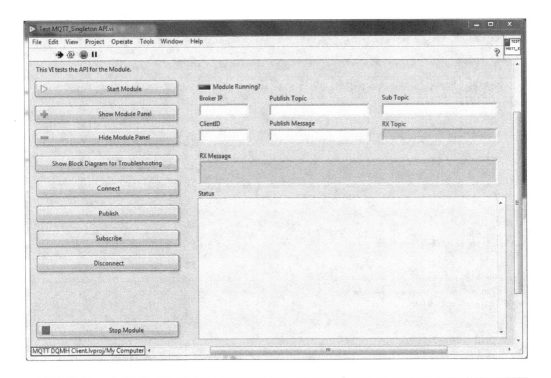

FIG. 8.12 *DQMHMQTT Tester.*

Once the Tester was ready to go, we started the Tester and DQMH module, connected the DQMH module to the broker, and subscribed to 'testtopic/LVMQTT." Then we fired up the DAQ.IO example MQTT_Simple_Connect_and_Publish.vi and started sending messages to the broker (Fig. 8.14). Immediately messages started coming in and proved that our module worked. The DQMH Tester provides any easy-to-use test harness for quickly verifying functionality. Now let's look at how we can use the DQMHMQTT Client library in an application.

Chapter 8: Enterprise and IoT Messaging

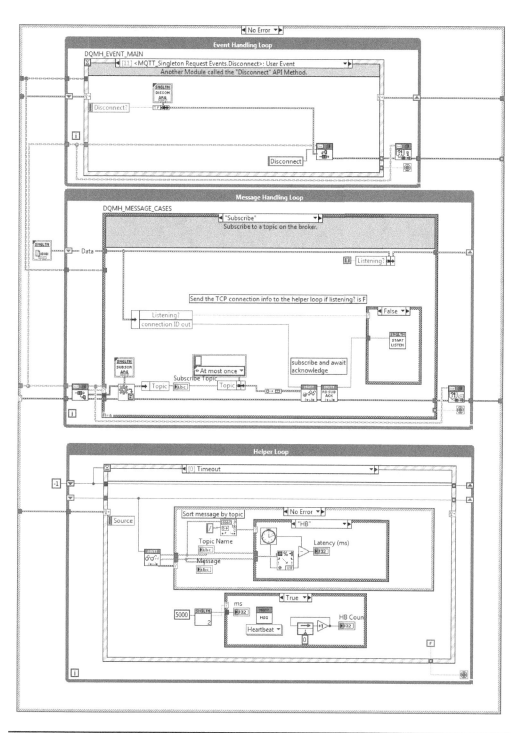

FIG. 8.13 *DQMH module Subscribe Event with helper loop listening to messages.*

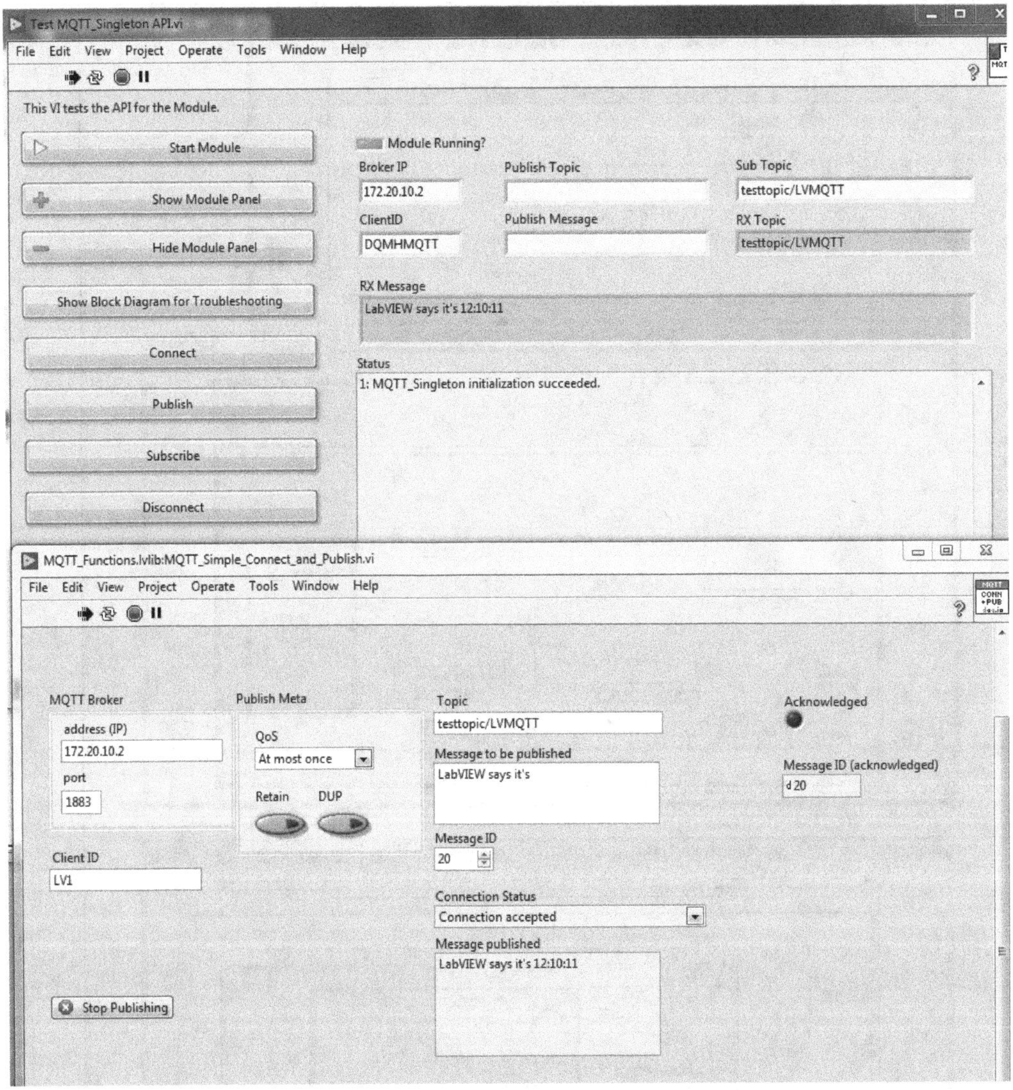

FIG. 8.14 DQMH Tester UI listening to messages from MQTT_Simple_Connect_and_Publish.vi.

MQTTDQMH Application

Our MQTTDQMH Client library has four external events used to control the DQMH module: Connect, Disconnect, Publish, and Subscribe. We also have a broadcast event (RX) to push received data out to all listeners and a "Connection OK?" event to publish the connection state. RX broadcasts the topic that the message came from and the data payload. In this implementation, any listeners need to look at the topic and decide if the message is relevant. A common design pattern is to use another module as a command parser to handle all the messages in one place. The command parser module is application specific and has state information, so it can accept a higher-level command like "START" and take care of all the tasks required to start the system. Separating these "areas of concern" keeps the MQTT Client library code generic and reusable.

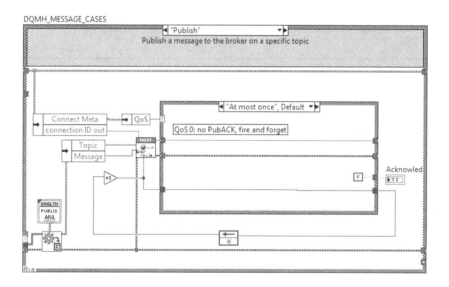

FIG. 8.15 *Publish event requires a Topic and Message.*

Our Publish event requires a Topic and a Message (Fig. 8.15). Rather than litter our application with the low-level Publish code, we created a polymorphic VI containing all our message types for use within the application (Fig. 8.16). Encapsulating the message types in subVIs makes it easy to locate where messages are coming from in

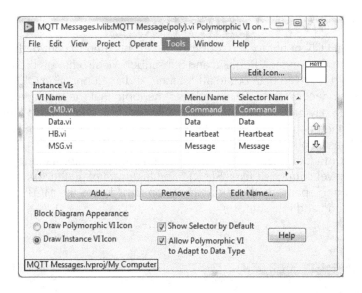

FIG. 8.16 *Polymorphic VIs as messaging inputs.*

the code, and it is easier to update the message structure later on. For our example application we defined three JSON message types: a Data message, a CMD message for commands, and MSG for system-level messages sent to the message broker for feedback and troubleshooting, similar to console.log. Included in each JSON message are the serial number and timestamp in ISO 8601 format for easy storage and retrieval in a database. We'll see how to integrate with a database later in AWS IoT. Here's what a CMD message looks like:

```
{
"SN":"000C2919ACCE",
"TS":"2019-01-15T16:21:57.785Z",
"CMD":"STOP"
}
```

Note that we have a Version and Description field in our data packet for downstream processing.

```
{
"SN":"000C2919ACCE",
"TS":"2019-01-15T16:53:17.971Z",
"Version":"0.1",
"Description":"TestBay1",
"VoltageIn":"12.1",
```

```
"CurrentIn":"1.0",
"UnitID":"0165556",
"Test1":"Passed"
}
```

We also defined a heartbeat (HB) message as plain text. The heartbeat message sends the value of the ms timer to <SN>/HB. The DQMHMQTT module automatically subscribes to the HB topic and looks at the delta ms from publish to receive to provide an indicator of the round-trip time through the messaging system. In production applications it's good practice to take appropriate actions to "fail safe" if the heartbeat message never returns indicating a failure in the larger control system. We did not format HB in JSON since this message is only used internally and not archived in a database.

The MQTT Singleton counts the number of HB messages transmitted and received and looks at the delta ms value to determine communication latency. This is displayed on the front panel of the VI along with incoming and outgoing messages (Fig. 8.17).

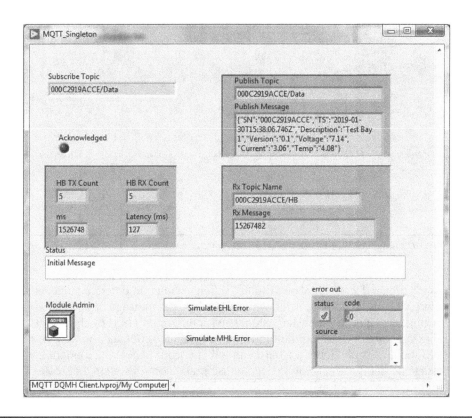

FIG. 8.17 *MQTT Singleton measures communication latency using the HB message.*

The main application is in the MQTTDQMH Client project (Fig. 8.18). Start the application, then press "Start MQTT Singleton Module" to start the communication gateway. Once the module is started, press "Connect" to connect to the broker. After the connection is established, the application will send a data packet every 2 seconds to the broker. If there are any connection errors, the MQTTDQMH module will close and reestablish the connection to the broker.

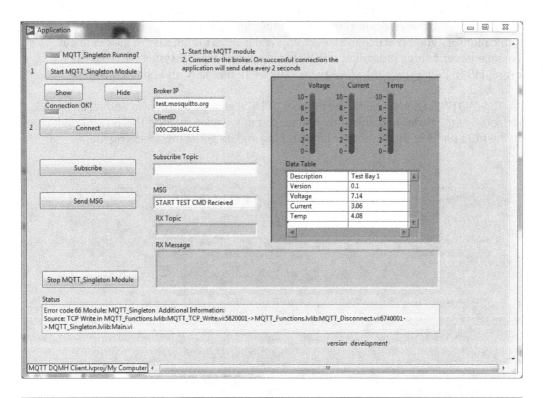

FIG. 8.18 *Main application. 1. Start the MQTT module. 2. Connect to the broker.*

A message broker (Fig. 8.19) allows multiple clients written in different languages, and by different teams, to communicate with each other. The message broker takes care of routing messages to interested listeners with minimum effort by the client. Modern message brokers can handle thousands of messages per second across hundreds to thousands of clients. Richard can tell you from personal experience that a message broker is a great way to build a distributed control system that scales. Each client only needs to know how to communicate with the broker and how to understand the message content. As the system grows, new front- and back-end interfaces can be added with little to no change to the rest of the system.

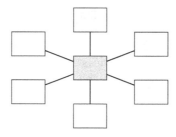

FIG. 8.19 *A message broker connects clients together in a hub and spoke, publish and subscribe messaging system.*

Messaging in the Cloud

Now that we've seen how easy it is to set up a scalable network of applications around a message broker, let's take a look at the Amazon Web Services (AWS) IoT (Internet of Things) message broker and look at some LabVIEW VIs for it. AWS IoT is a **fully managed publish and subscribe MQTT message broker** in the AWS cloud. In this section we'll show you how to integrate LabVIEW into AWS IoT and build a message-based, measurement and control system with global reach (Fig. 8.20).

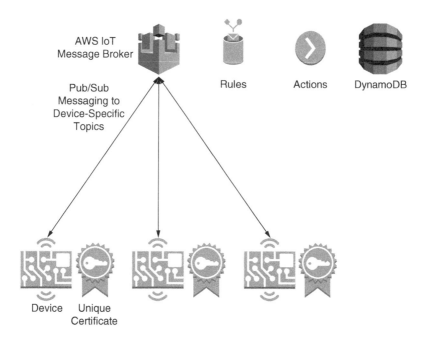

FIG. 8.20 *AWS IoT is a fully managed publish and subscribe MQTT message broker in the AWS cloud. Authorized devices publish and subscribe to topics with MQTT. Rules with actions write message traffic into DynamoDB.*

AWS IoT uses a concept called "**things**" that are JSON objects describing the desired state of remote devices. The idea is that devices will go online and offline, yet you can still interact with the "thing" and the device will update with any changes the next time it connects. Our devices are always connected and streaming data. The "thing" concept gets in the way of this, so we don't use things, but just use the IoT service as a managed message broker.

Toolkits

LabVIEW's TCP networking support has been present since the beginning, but without the SSL/TLS support needed for secure communications. NI recognizes the hole and is working to fix it. Recently LabVIEW added support for HTTPS communications by wrapping a .dll around OpenSSL. This .dll is available on all platforms, including embedded. However, the only native interface for secure communication has been through the HTTPS VIs. LVS-Tools built the Encryption Compendium on top of this .dll and created a robust toolset that we wish was native in all versions of LabVIEW. If you need a full-featured security toolset, then check out the Encryption Compendium on the LabVIEW Tools Network.

NI has an open (nonpassword protected) toolkit for interacting with AWS through a RESTful interface, but since LabVIEW does not have a security suite, it relies on some third-party LabVIEW libraries for encryption and MD5. The LabVIEW Cloud Toolkit for AWS has some good working examples and shows how to do AWS Signature 4 authentication. We recommend you check it out if you want to interface with other AWS services via a RESTful interface.

We started looking at AWS IoT service in 2016, before the Cloud Toolkit came out, as a way to monitor and control distributed LabVIEW real-time targets at multiple locations. LabVIEW's lack of security tools made it really hard to create a RESTful interface to AWS. We called on a friend, Huaxin Gong, for help. Huaxin dug around and discovered the OpenSSL .dll NI was using with the HTTPS VIs. Before too long we were able to connect via MQTT to AWS IoT and publish and subscribe to topics. We used DAQ.IO's LVMQTT library that we discussed in the first part of this chapter for the MQTT part of the code and put our own wrappers around the OpenSSL .dll. The project is on GitHub (https://github.com/Indie-Energy/AWS-IoT-RESTful) and released under the MIT license.

MIT License

Copyright (c) 2016 Indie-Energy

Permission is hereby granted, free of charge, to any person obtaining a copy of this software and associated documentation files (the "Software"), to deal in the Software without restriction, including without limitation the rights to use, copy, modify, merge, publish, distribute, sublicense, and/or sell copies of the Software, and to permit persons to whom the Software is furnished to do so, subject to the following conditions:

The above copyright notice and this permission notice shall be included in all copies or substantial portions of the Software.

THE SOFTWARE IS PROVIDED "AS IS", WITHOUT WARRANTY OF ANY KIND, EXPRESS OR IMPLIED, INCLUDING BUT NOT LIMITED TO THE WARRANTIES OF MERCHANTABILITY, FITNESS FOR A PARTICULAR PURPOSE AND NONINFRINGEMENT. IN NO EVENT SHALL THE AUTHORS OR COPYRIGHT HOLDERS BE LIABLE FOR ANY CLAIM, DAMAGES OR OTHER LIABILITY, WHETHER IN AN ACTION OF CONTRACT, TORT OR OTHERWISE, ARISING FROM, OUT OF OR IN CONNECTION WITH THE SOFTWARE OR THE USE OR OTHER DEALINGS IN THE SOFTWARE.

AWS IoT

Publishing and subscribing with the AWS IoT message broker is easy once you get it set up correctly. However, getting it set up correctly can be super-frustrating. AWS has a very strict policy to deny all actions unless there is a policy statement allowing the action. And anytime you do something wrong, AWS will close the connection with no external warnings or error messages. However, AWS is very good at logging all actions and errors in **CloudWatch** logs, so take advantage of it for troubleshooting (Fig. 8.21).

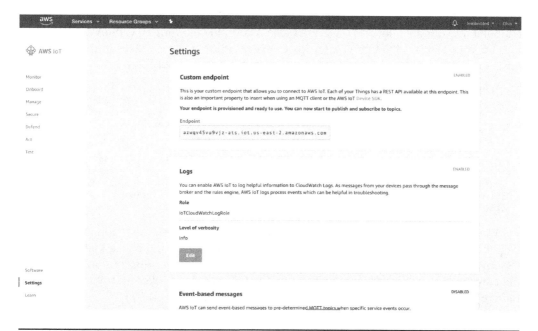

FIG. 8.21 *AWS IoT Console >> Settings This is where you enable an IoT endpoint and turn on CloudWatch logs.*

In order to use AWS IoT you have to have an AWS account, and they have a very generous free tier that should cover everything we do in this section with minimal to no charges. The general steps required are

1. Enable the service and CloudWatch logs.
2. Create certificates and keys for your devices.
3. Create a policy to allow your devices to connect.

We'll go through all of these steps in detail. AWS is constantly upgrading their services, and the web console interface may change by the time you read this; however, the concepts and steps should be the same.

AWS has some great back-end services we can integrate with IoT using **rules**. A rule is an SQL SELECT statement with a topic filter and a rule action. Rules can trigger Lambda functions, SNS (Simple Notification Service), S3 (Simple Storage Service), and DynamoDB. We use **DynamoDB** to log all the traffic coming to specific topics with a few simple rules.

Enabling the IoT Message Broker and CloudWatch

Once you have your account, log in to the console and select IoT Core from the list of services. In the bottom left of the screen select Settings. This has your custom endpoint for your account and region. It's also a good idea to turn on CloudWatch logs for your service. INFO creates a log of every action a device takes in the account. The output looks like this for a simple publish in and publish out:

```
{ "timestamp": "2019-01-11 19:51:53.988", "logLevel": "INFO", "traceId": "3a800036-
d968-a305-bdd2-bccf0e43e009", "accountId": "994370595944", "status": "Success",
"eventType": "Publish-In", "protocol": "MQTT", "topicName": "myTest", "clientId":
"iotconsole-1547236130961-0", "principalId": "994370595944", "sourceIp":
"2605:6001:f0d2:8f00:7c00:b2a3:3f5:8db", "sourcePort": 59107 }

{ "timestamp": "2019-01-11 19:51:54.002", "logLevel": "INFO", "traceId": "4c24da68-
aaf8-4785-8130-eeb81b4593f8", "accountId": "994370595944", "status": "Success",
"eventType": "Publish-Out", "protocol": "MQTT", "topicName": "myTest", "clientId":
"iotconsole-1547236130961-0", "principalId": "994370595944", "sourceIp":
"2605:6001:f0d2:8f00:7c00:b2a3:3f5:8db", "sourcePort": 59107 }
```

The Test section of the IoT console lets you publish and subscribe to topics (Fig. 8.22). It's a good idea to manually publish and subscribe and check that CloudWatch logging is working correctly. You can also use topic subscriptions in the console to watch traffic coming through the IoT message broker from your remote systems.

Create Certificates and Keys

Next we create X.509 certificates from the Secure menu (Fig. 8.23). Use the blue button in the upper-right corner to create, and select One Click create (Fig. 8.24).

Chapter 8: Enterprise and IoT Messaging **575**

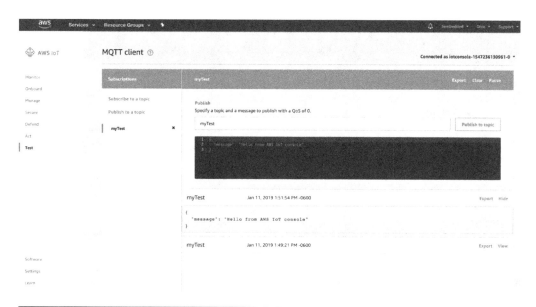

FIG. 8.22 Use AWS IoT Console >> Test to publish and subscribe to topics from the console.

FIG. 8.23 AWS IoT Console >> Secure >> Certificates. Devices need certificates in order to connect.

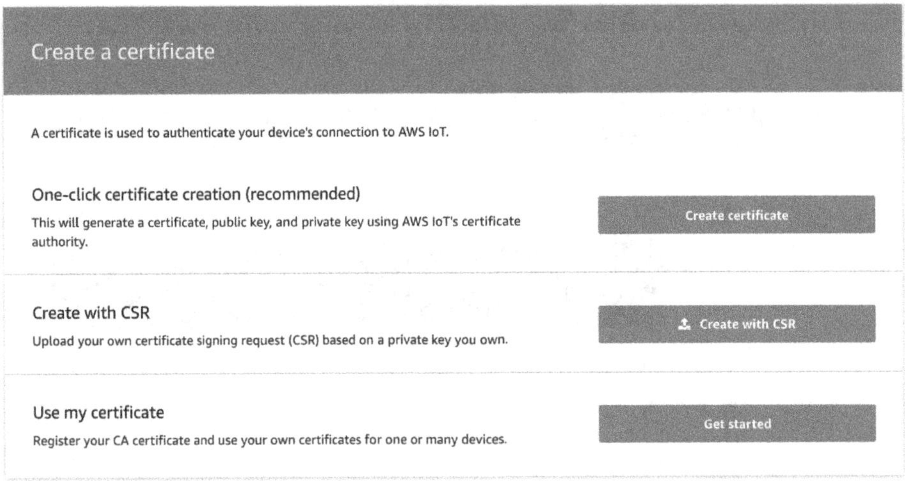

FIG. 8.24 *Create X.509 certificates with one-click creation.*

This creates the certificate, public key, and private key. Each device needs its own certificate and private key to be able to connect to IoT. Download these keys now, as this is the only time they will be available (Fig. 8.25).

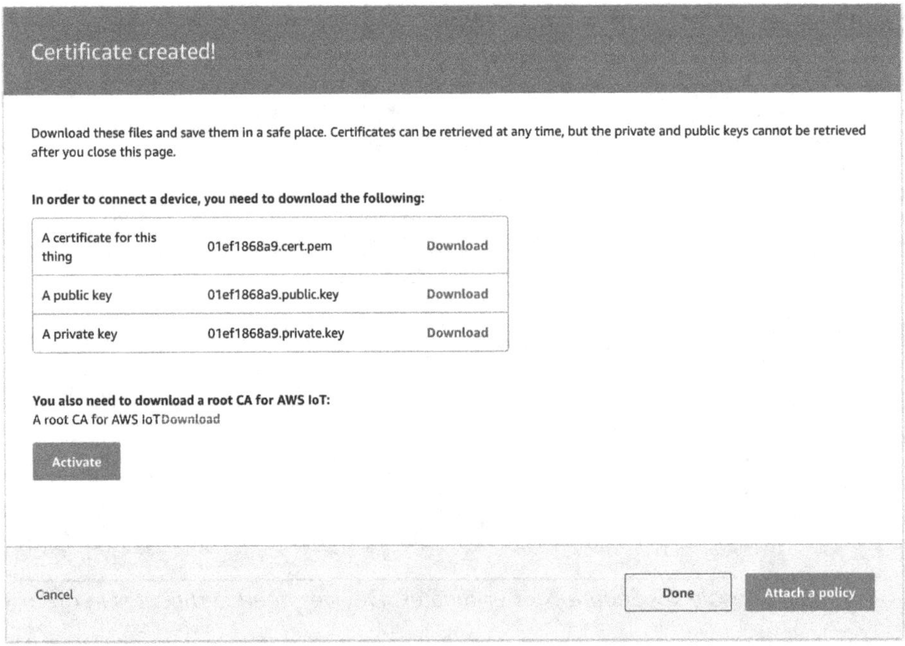

FIG. 8.25 *Download the certificate and keys at creation. They are not available later. Then activate the certificate and attach a policy.*

IoT Policies

A policy governs what your device is able to do. For this example we're going to open up IoT and give our device full access. Note the * on action and resource; this gives your device full access to the IoT resources in your account. In a production system you want to lock down and give least access permissions.

Your policy should look something like this:

```
{
  "Version": "2012-10-17",
  "Statement": [
{
"Effect": "Allow",
"Action": "iot:*",
    "Resource": "arn:aws:iot:us-east-2:your account number:*"
}
  ]
}
```

IoT policies require three statements: Effect, Action, and Resource. In the policy provided earlier, the Effect is "Allow"; if we wanted to deny access later, we could change the Effect to "Deny." Actions required to interact with the message broker are iot:Connect, iot:Publish, iot:Receive, and iot:Subscribe. If you wanted a device to only connect and publish but not subscribe, you could limit the Actions to iot:Connect and iot:Publish. In our case we want the device to publish and subscribe, so we use the wildcard "iot:*" to give full access to all Actions. We are also using the wildcard * on the Resource to allow our device to publish and subscribe to any topic. If we wanted to limit the topic range, it would look something like this "Resource": "arn:aws:iot:us-east-2:your account number:topic/Jembedded/*" to limit devices to only publishing and subscribing to topics that start with Jembedded/.

Topic paths are a mechanism you can use to structure your IoT space and separate data traffic and logging rules. For example, you might direct each customer's devices to write into topics with the customer name as the top level in the path (CustomerA/<SN>/Data), or you might segment based on location (Dallas/<SN>/Data).

This is a good time to double-check your certificate and make sure it's active with a policy that will allow a device using that certificate to connect and interact with the IoT message broker (Fig. 8.26). AWS's policy is to deny everything and not return error messages, so if you don't have everything correct, the only response will be LabVIEW complaining of an SSL connection error. However, CloudWatch logs have records of every interaction with the message broker for troubleshooting.

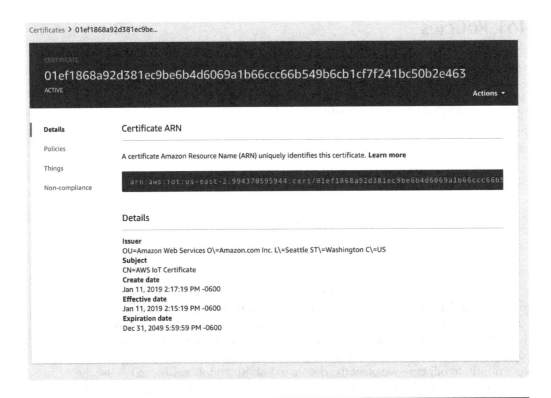

FIG. 8.26 *Verify that your certificate is active and has a valid policy.*

DynamoDB

DynamoDB is AWS's fully managed NoSQL database (Fig. 8.27). It's fast, cheap, and there are no servers to maintain. Because it's NoSQL there is no schema to set up, and items do not have to have the same formatting. At the beginning of this chapter we recommended using JSON as a message format with a version and description field so back-end systems could deal with dynamic data formats. We also added a Serial Number (SN) and a Timestamp (TS) field to our JSON packet, and this is where those fields come in handy. In DynamoDB we'll set up some tables to store our message traffic and a few simple rules to write our JSON messages into those tables using the SN field as our primary key and the TS field as our sort key.

DynamoDB

FIG. 8.27 *DynamoDB console interface.*

The core concepts for DynamoDB are tables, items, and attributes. A table is a collection of items. In our case the items are our JSON messages. Attributes are the key:value pairs in the message. In our Data message each key:value pair is an attribute with a value.

```
{
"SN":"000C2919ACCE",
"TS":"2019 01 15T16.53:17.971Z",
"Version":"0.1",
"Description":"TestBay1",
"VoltageIn":"12.1",
"CurrentIn":"1.0",
"UnitID":"0165556",
"Test1":"Passed"
}
```

DynamoDB tables use a **primary key** to uniquely identify each item in a table. The primary key is also known as the **partition key**, and DynamoDB uses the primary key and a hash function to identify the internal partition where the item is stored. The use case for using only a primary key to identify each item is valid if you have a unique Customer ID to locate data for a customer. In our case we want to store a sequence of data from our device, and multiple items will have the same primary key, so we'll use a **composite primary key** consisting of a primary key and a **sort key**. DynamoDB still uses the primary key and a hash function to identify the internal partition where the items are stored, but with a composite key the items for each primary key are stored in sorted order by the sort key. Our table uses the SN attribute as the primary key and the TS attribute as the sort key. The TS key is the timestamp in ISO 8601 format with ms precision in order to have a unique sort key. TS is also in UTC to avoid errors caused by redundant timestamps when Daylight Saving Time changes occur (Fig. 8.28).

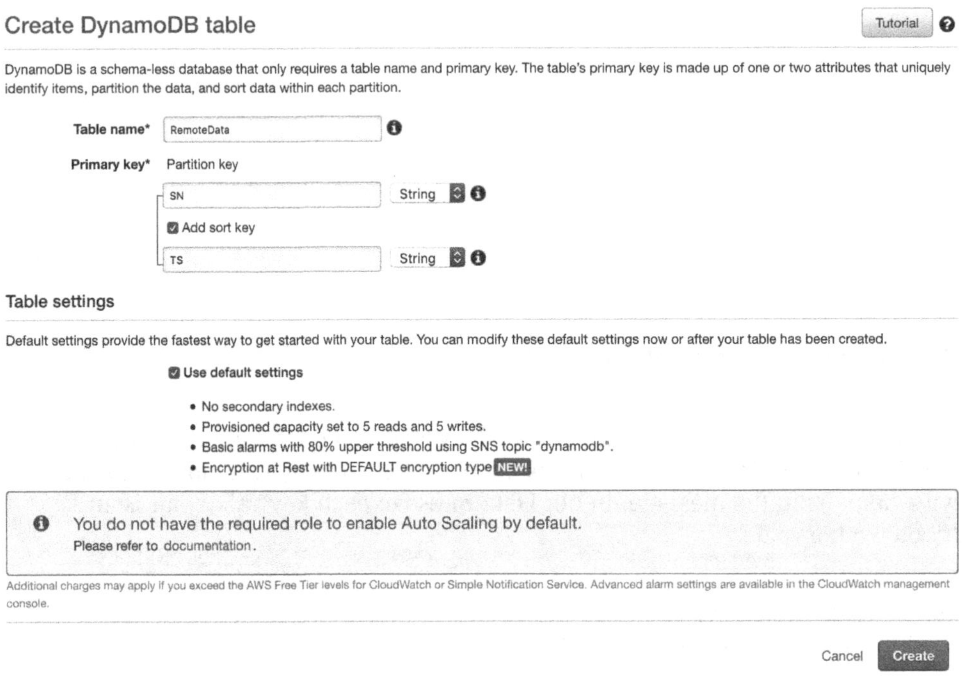

FIG. 8.28 *Create a table with a primary key and a sort key.*

Once you have created tables for each JSON message type (CMD, MSG, Data) with SN as the primary key and TS as the sort key, we can set up IoT rules to write each JSON message into DynamoDB as the message hits the IoT broker. By default tables are configured for **read and write capacity** of 5 units. This is a confusing way of saying 5 reads and 5 writes to the table per second. Scale this up or down to meet your needs.

IoT Rules

IoT rules attach actions to topics (Fig. 8.29). As messages come across the IoT message broker, AWS checks to see if a rule applies to the topic, and if so, performs the actions in the rule. The list of available actions keeps growing as AWS builds out the service. The most relevant are writing into a DynamoDB table and firing a Lambda function. Lambda is a compute-on-demand service where you can have a piece of code spun up in a container to operate on your data. Once your data triggers a Lambda function, you can process it and interact with almost any AWS service.

FIG. 8.29 *Create IoT rules from the Act section of IoT Core.*

In our case we're going to use three rules to write all Data, CMD, and MSG messages into each message's respective DynamoDB table. Log in to the AWS console and navigate to IoT Core. Make sure your rules, DynamoDB tables, and IoT endpoint are all in the same region. On the AWS IoT page select Act from the left-hand menu. Click Create to bring up the dialog for creating rules and assigning actions. Give your rule a name that makes sense for the specific payload, that is, DataRule for logging Data messages. In the rule query statement section (Fig. 8.30) enter SELECT * FROM '+/Data' to select the entire JSON payload. Note the use of the wildcard character "+", and remember that our topic format is <SN>/<TOPIC>. The wildcard will match any serial number writing to that specific topic. So if you have 10 devices or 10,000 devices writing into IoT, that one rule and one DynamoDB table will handle everything coming in to <SN>/Data. The really cool part is that AWS manages all the infrastructure and scaling for you.

The next step is to select the action for this rule. Click "Add action" to bring up the list of actions you can attach to the rule (Fig. 8.31). Select "Insert a message into a DynamoDB table," then press "Configure action" to bring up the configuration dialog (Fig. 8.32). Select the DynamoDB table you created earlier. The dialog recognizes the primary (SN) and secondary keys (TS) from when the table was created. In the "* Hash key value" field enter $SN to use the value from the SN key:value pair as the hash key. Repeat for the "Range key value" $TS. The "$" character lets the rule know that the value is text.

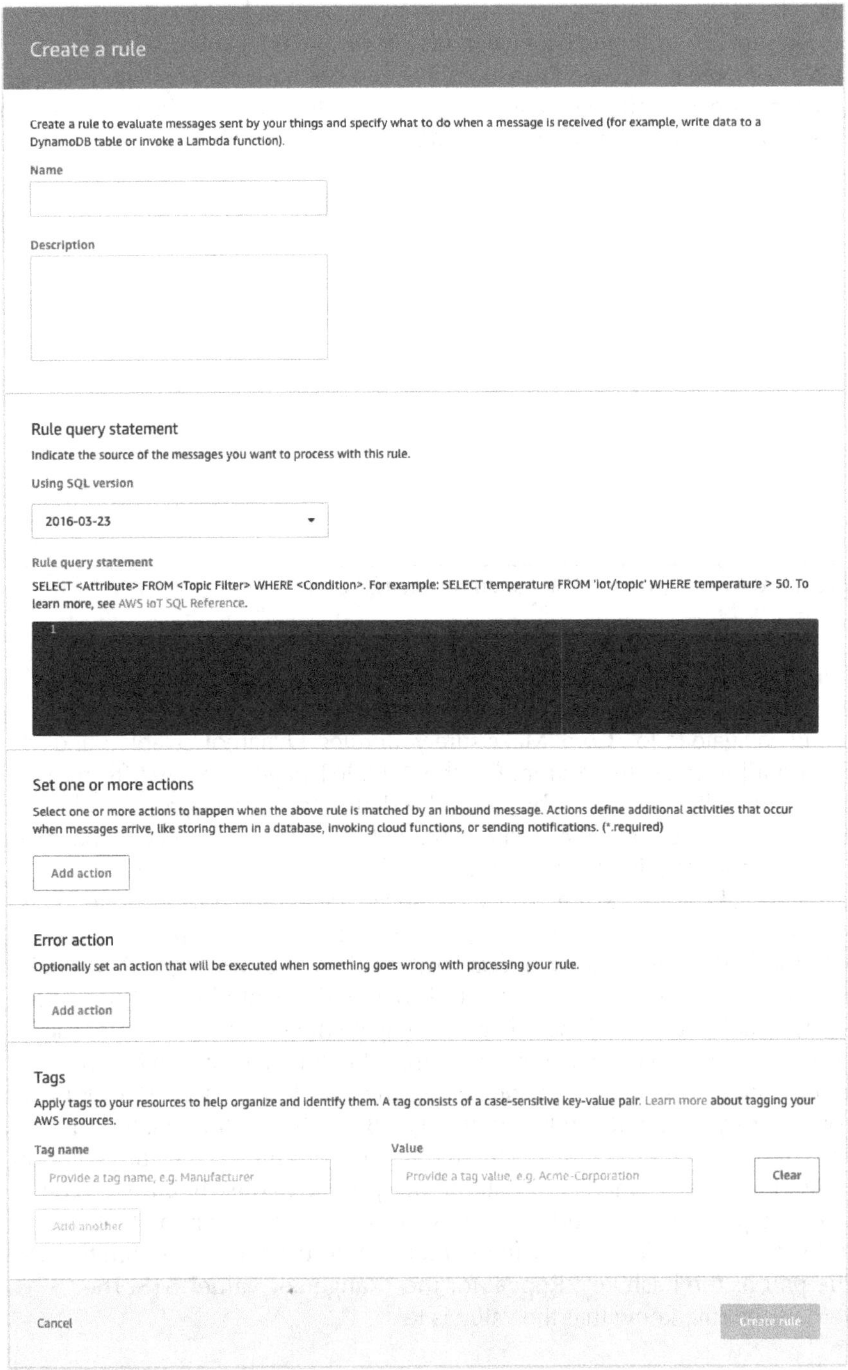

FIG. 8.30 *Create a rule to SELECT * FROM '+/Data'.*

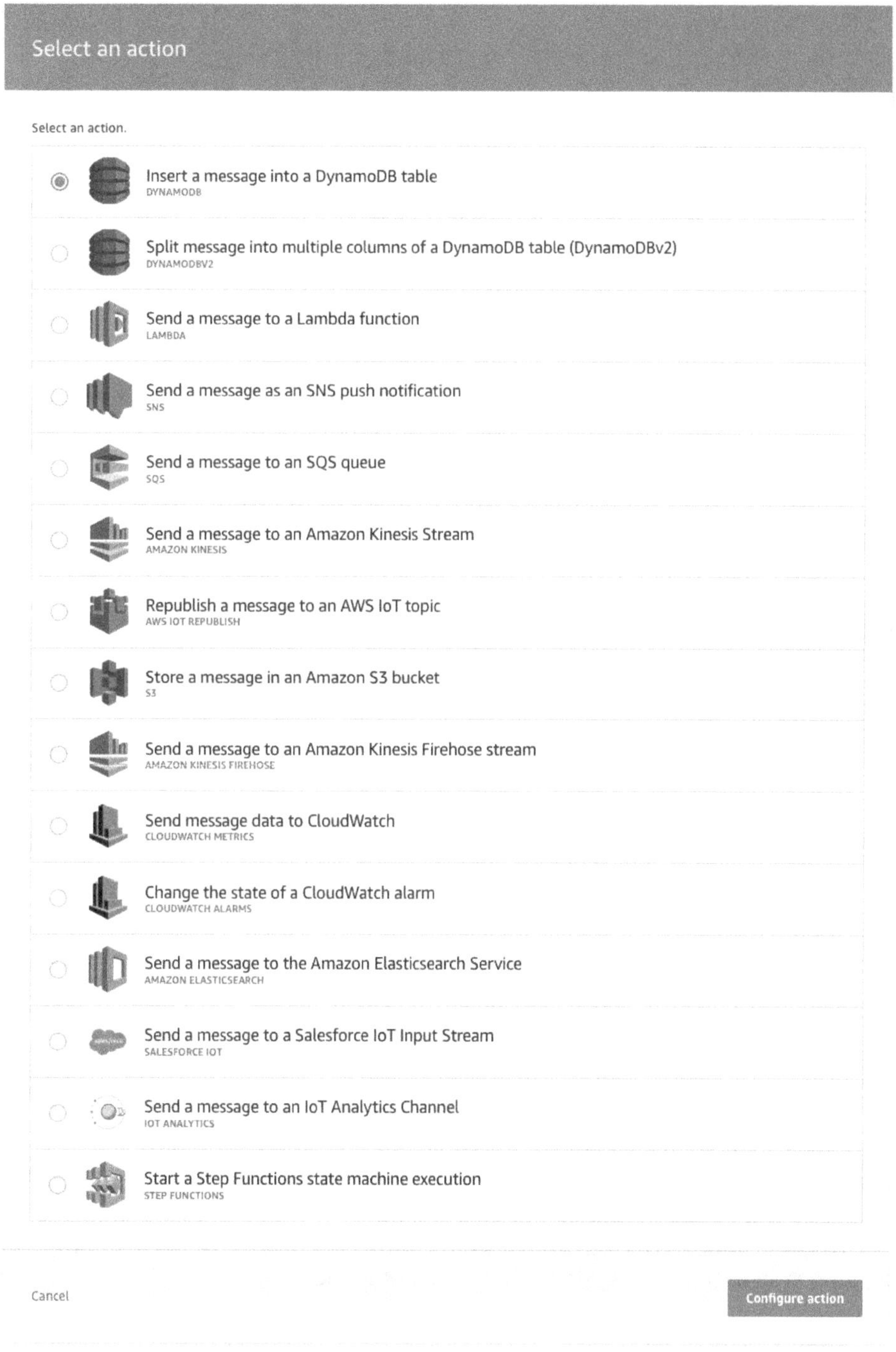

Chapter 8: Enterprise and IoT Messaging **583**

Select an action

Select an action.

- ⦿ Insert a message into a DynamoDB table
  DYNAMODB
- ○ Split message into multiple columns of a DynamoDB table (DynamoDBv2)
  DYNAMODBV2
- ○ Send a message to a Lambda function
  LAMBDA
- ○ Send a message as an SNS push notification
  SNS
- ○ Send a message to an SQS queue
  SQS
- ○ Send a message to an Amazon Kinesis Stream
  AMAZON KINESIS
- ○ Republish a message to an AWS IoT topic
  AWS IOT REPUBLISH
- ○ Store a message in an Amazon S3 bucket
  S3
- ○ Send a message to an Amazon Kinesis Firehose stream
  AMAZON KINESIS FIREHOSE
- ○ Send message data to CloudWatch
  CLOUDWATCH METRICS
- ○ Change the state of a CloudWatch alarm
  CLOUDWATCH ALARMS
- ○ Send a message to the Amazon Elasticsearch Service
  AMAZON ELASTICSEARCH
- ○ Send a message to a Salesforce IoT Input Stream
  SALESFORCE IOT
- ○ Send a message to an IoT Analytics Channel
  IOT ANALYTICS
- ○ Start a Step Functions state machine execution
  STEP FUNCTIONS

Cancel   **Configure action**

FIG. 8.31 *Give your rule an action to write data into DynamoDB.*

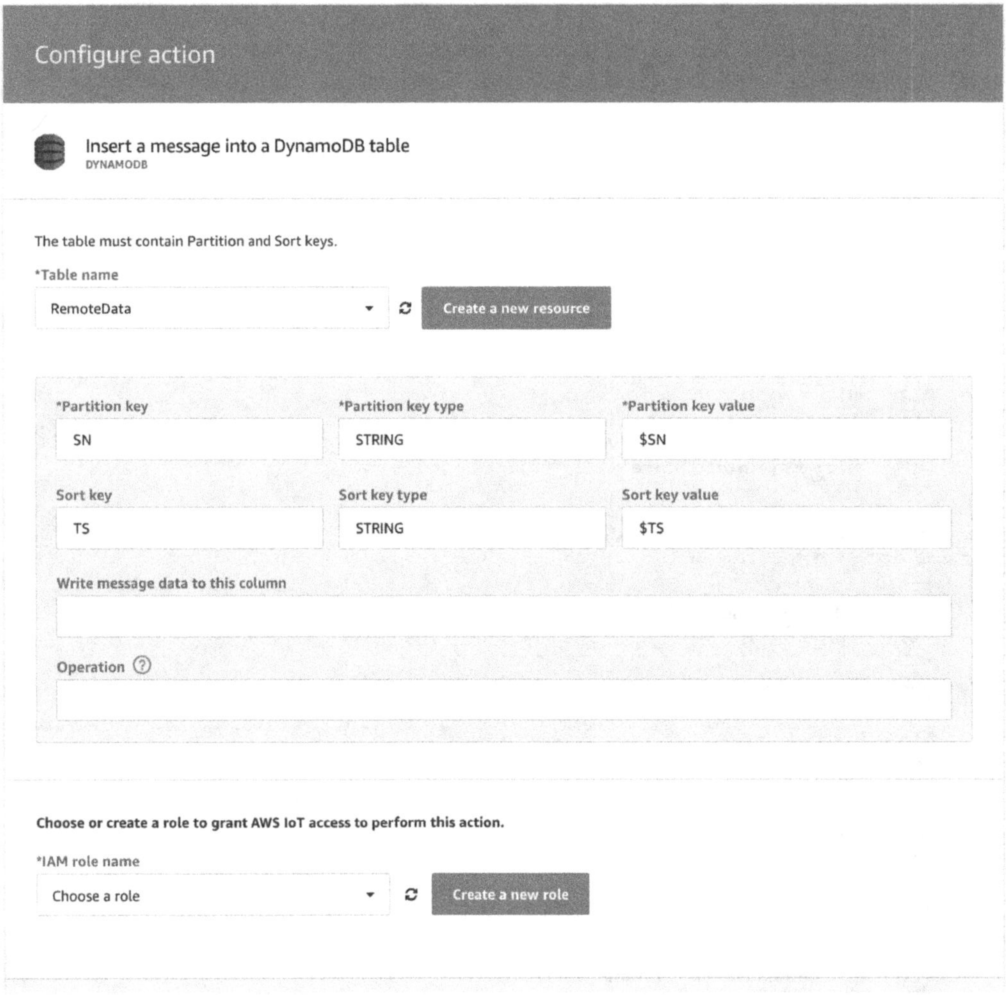

584 LabVIEW Graphical Programming

FIG. 8.32 *Define the table, partition key, and sort key for the action.*

Finally we have to give our rule a role (Fig. 8.33) to allow it to write into DynamoDB. Select "Create a new role" and give the new role a meaningful name, that is, DataRole. Update the rule to use the new role and press "Add action." That's it! Repeat for each topic and message, and you have a messaging and logging system that will scale to thousands of devices.

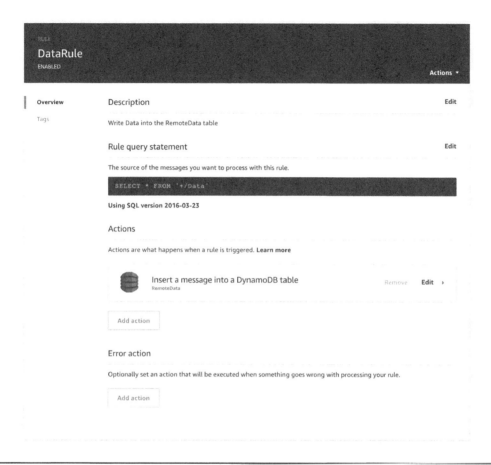

FIG. 8.33 *Verify that the rule is enabled and has a valid role.*

Now that the Data rule is configured, it's time to test it in the AWS console. Switch over to Test from the menu on the left and select "Publish to a topic" (Fig. 8.34). Specify the topic path you want to publish to. Note that you cannot use wildcards when publishing. To test out the Data rule just created, publish to the path 000C2919ACCE/Data with this JSON payload:

```
{
"SN":"000C2919ACCE",
"TS":"2019-01-31T14:03:19.894Z",
"Description":"Test Bay 1",
"Version":"0.1",
"Voltage":"5.10",
"Current":"6.12",
"Temp":"5.82"
}
```

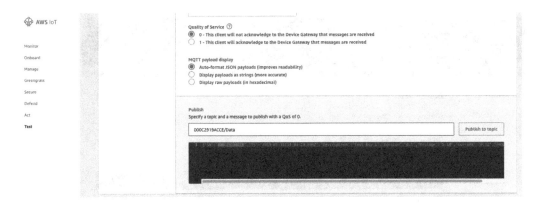

FIG. 8.34 *Publish a message to the <SN>/Data topic from the IoT console to verify the rule fires.*

Then switch over to DynamoDB and verify that the data went into the table (Fig. 8.35). If not, look into CloudWatch logs and see if there is a relevant log.

Once you verify the rules and actions are working from the console, it's time to fire up a LabVIEW application to test them out from the real world. Figure 8.36 shows the front panel of the MQTTDQMH application from Fig. 8.18 rewritten for AWS. One difference is the IP is the custom endpoint set up in Fig. 8.21. This endpoint is unique to your account and region. The second difference is the paths to the certificate and private key downloaded in Fig. 8.25. Load your parameters and fire up the Singleton module, then press "Connect." The Singleton panel should pop up and after a moment you should see heartbeat and data traffic flowing across the AWS IoT message broker. If not, troubleshoot with CloudWatch logs and see if you missed a step somewhere. AWS will not push an error message to a misbehaving client, but will just terminate the connection and write a log to CloudWatch.

The two applications discussed in this chapter were meant to give you a taste of new ways to build scalable, connected systems. Building around a message broker lets you easily interact with clients written in other languages and expand your system in creative ways. Building systems that connect via AWS IoT lets you build scalable IoT systems with global reach. We hope this gives you some inspiration to build the next big thing!

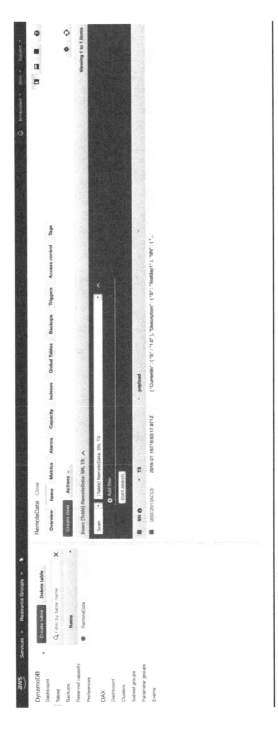

FIG. 8.35 Verify the data went into DynamoDB.

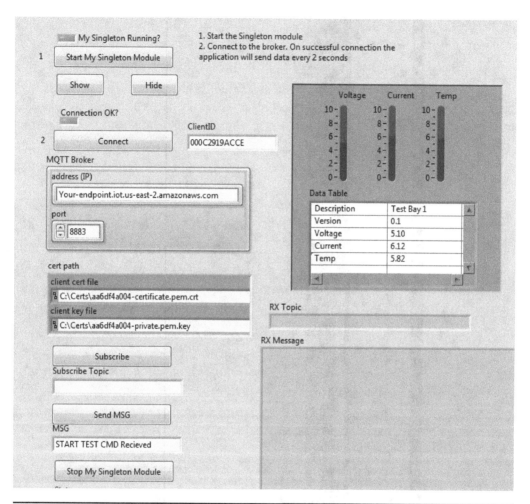

FIG. 8.36 *MQTTDQMH application rewritten for AWS IoT.*

Abbreviations

Abbreviation	Description
ADE	Application Development Environment
AE	Action Engine
API	Application Programming Interface
BD	Block Diagram (LabVIEW)
CML	Continuous Measurement and Logging
DIP	Dependency Inversion Principle (the D in SOLID principles)
DLL	Dynamic Link Library
DQMH	Delacor Queued Message Handler
DVR	Data Value Reference
EHL	Event Handling Loop
FGV	Functional Global Variable
FP	Front Panel (LabVIEW)
Hg	Mercurial (Source Code Control)
ISP	Interface Segregation Principle (the I in SOLID principles)
LCOD	LabVIEW Component-Oriented Design
LLB	LabVIEW Library File
LSP	Liskov Substitution Principle (the L in SOLID principles)
LV	LabVIEW
LVLIB	LabVIEW Project Library File
LVOOP	LabVIEW Object-Oriented Programming

Abbreviation	Description
LVPROJ	LabVIEW Project File
MHL	Message Handling Loop
NI	National Instruments
OCP	Open-Closed Principle (the O in SOLID principles)
PPL	Packed Project Library
QMH	Queued Message Handler
SEQ	Single-Element Queue
SRP	Single Responsibility Principle (the S in SOLID principles)
SVN	Subversion (Source Code Control)
TDD	Test-Driven Development
UUT	Unit Under Test
VI	Virtual Instrument
VIPC	VI Package Configuration
VIPM	VI Package Manager
WORM	Write Once Read Many (used for global variables)

Index

'\' Codes Display, 91

A

Abortable wait, 132, 133f
Absolute time, 124–126
Abstract messages, 392, 393
Abstraction, 218–219, 225, 496
Abstraction layers, 274
    hardware, 243–244, 523
    measurement, 244
AC signals:
    floating-source connections and, 167f
    measuring, 168–169
    time domain, 152
Acceptance tests, 407
Accessors, 219, 225, 226, 230, 269
    class, 538
    property nodes and, 233
Acquire Semaphore VI, 140
Acquiring data, 189
Acquisition module, 278
Acquisition Started broadcast, 330–331
Action engine (AE), 36, 222–224, 281
Active filters, 164
ActiveX objects, 108
Actor class, 358–360, 359f
Actor Core.vi, 355–356, 355f, 357f
Actor Framework, 42, 222–224, 242, 244, 282, 283, 551
    Actor class, 358–360, 359f
    advanced topics in, 392–398
    configuration files and, 388, 388f
    error handling strategies, 390–391, 391f
    feedback evaporative cooler sample project, 377–378, 377f
    fundamentals example for, 376
    helper loops, 361–362, 362f
    Initialization, 387–389
    interprocess communication and, 287, 387
    library, 356–357, 358f
    Message class, 360
    Message Maker tool, 361
    Message Queue class, 361
    module initialization in, 287–288
    OOP and, 356
    overview of, 354–366
    project structure and class hierarchy, 369–370, 370f, 371f, 372–373, 376
    project template for, 369, 369f
    queues in, 387
    scripting tools, 382, 383f
    sending messages, 364
    stopping actors, 365–366, 365f
    stopping modules, 389, 389f, 390f
    tools for, 391–392
    use cases for, 354
    user events for communication and, 374
    using, 366–387
Actor Framework Debug library, 376, 391
Actor Framework tester, 392
Actor Model, 354–355
Actor tree, 364, 365f
    adding actors and, 380
Actors, 354, 357, 357f
    adding new, 378–385
    behaviors for, 379
    designing, 379–381
    End Point, 398

Actors (*Cont.*)
  exploring, debugging, and troubleshooting, 387
  extending via inheritance, 359–360
  helper loops in, 380, 382
  launching, 363–364, 363f
  low coupling approach for, 393f
  in PPLs, 398–399
  principles of, 355
  root-level, 363, 363f, 372
  sending messages to and between, 364
  sharing reusable, 398–399
  state data for, 378
  State Pattern, 398
  stopping, 365–366, 365f
  testing, 374
  VIPM packaging of, 398
  zero coupling, 392, 394f, 395
Actuators, 150–151
Adapter pattern, 258–259
ADC. *See* Analog-to-digital converter
Add nodes, waveforms and, 107–108
Adelhart, Peter, 555
Admin object, 222
Advanced Messaging Queueing Protocol (AMQP), 553
AE. *See* Action engine
Aggregate types, 111
Agile Alliance, 487
Agile Inception Deck, 490–491
Agile Manifesto, 212, 247
Agile Samurai book, 491
Agile software development, 487
Agile Software Development (Martin), 247
Agile Warrior blog, 491
Agilent 34401, instrument driver tree for, 84, 84f
AI Single Scan, 193
Aivaliotis, Michael, 33
Aliasing, 174, 195
Aloha, 282
Amazon, 525
Amazon Web Services (AWS), 571. *See also* AWS IoT
  back-end service integrations, 574
Amplifiers, 162, 166f
  instrumentation, 161, 162f, 164
  isolation, 163
  PGIAs, 164, 182
  S/H, 178

saturation, 167
settling time and, 164, 165
signal quality and, 163
AMQP. *See* Advanced Messaging Queueing Protocol
Analog inputs:
  floating, 166–168
  ground-referenced, 166
Analog outputs, 170
Analog signals, 151, 152
  sampling, 173
Analog-to-digital converter (ADC), 3, 147, 148f, 150, 173
  code width of, 178
  error sources in, 178–180
  multiplexing and, 180–181, 180f
  PGIAs and, 182
  quantization levels of, 177
  range of, 178
  resolution of, 177, 177f
  sample clock and, 152
  waveforms and, 153f
Analysis by accident, 190
Andreas S, 513
Antialiasing, 164
Antialiasing filter, 175
Apache ActiveMQ Artemis, 558–559, 558f, 560f
Aperture time, 181
API tester, 301–302, 302f
  DQMH module debugging with, 342
  exploring, debugging, and troubleshooting with, 315, 316f, 323, 326
  as sniffer, 316, 343
  as technical asset, 492
APIs:
  AppBuilder, 548
  Build, 542–543
  DQMH public, testing, 449–450
  Interface Segregation Principle and, 255
  Queue, 221, 221f
AppBuilder API, 548
Append True/False String function, 88
Apple Macintosh, 11
  development advantages of, 14–15
  jump table, 17
  memory capacity, 15, 17
  memory manager, 20

Apple Macintosh II, 19–21
Apple Macintosh Plus, 17
Application Builder, 324
Application name, 322
Applications:
    messaging between, 552–555
    MQTTDQMH, 567–570, 586, 588f
    plug-in type, 41
    publish and subscribe, 552, 552f
    standalone, 50, 52
    team size and modularity of, 276
    timing sent to other, 125–126
Applied Research Laboratory, 10
Architecture smells, 514–515
AristosQueue, 46
Armstrong, Joe, 268
Array Size function, 97, 97f
Array Subset function, 194, 198
Array To Cluster function, 112, 112f, 116
Array To Spreadsheet String function, 92–93, 93f
Arrays, 95–98
    adding waveforms to numeric, 108
    of booleans, 111
    cluster conversion from, 112, 112f, 116
    clusters in, 106, 106f
    concatenating across loop iterations, 42
    creating, 96–97
    data types in, 96
    empty, 75, 97, 100
    empty string removal from, 75, 75f
    global, 79
    indexing of, 97, 97f
    initializing, 99–100, 99f
    in JSON, 553–554, 554f
    memory usage and performance, 101–103, 101f
    multidimensional, 97–98, 98f
    of NaN, 449
    numeric:
        conversion to clusters of Ring Indicators, 112, 112f
        waveforms added to, 108
    of objects, 215–216
    shift registers and conversion of, 94
    slicing, 98, 98f
    string conversion from, 94
Artemis. *See* Apache ActiveMQ Artemis
Assembly language, 7

Assert Type VIs, 51
AssertAPI, 406, 479, 480f, 481, 482f, 483t
Assertions, 404, 405f, 475–476
    automatic error handling and, 477, 478f
    Caraya Unit Test Framework and, 437, 477–478, 477f
Association, 212, 215, 215f
Asynchronous calls, 345
Asynchronous messaging, 376, 396f
Atlassian, 522, 523
Attribute Node, 26
Attributes, 212, 213f
    waveform, 108–109, 109f
Austin, Karen, 24f
Austin, Paul, 24f, 26f
Auto-indexing, 72, 72f
Auto-populating folders, 479, 481f
Auto-stop? boolean input, 366
Automated build processes, 524. *See also* Continuous integration
Automated test frameworks, 406
Automatic error handling, assertions and, 477, 478f
Automation:
    cost-effectiveness of, 2–3
    costs of, 3
    disadvantages of, 2
    LabVIEW and, 1–3
    operations benefiting from, 2
    of test tools, 547
Averaging, 175–177
AWS. *See* Amazon Web Services
AWS IoT, 571–572, 571f, 573, 586
    creating certificates and keys, 574, 575f, 576, 576f
    enabling message broker and CloudWatch, 574
    policies, 577
AZInterface toolkit, 260–261, 267, 268f
Azure, 525

B

Balanced signal, 161
Bandwidth, 195–196
Barber, Jack, 24f
BASIC (programming language), 8, 11, 18
Basic Averaged DCRMS Vi, 107
Batto, Deborah, 24f

Batto-Bryant, Deborah, 26f, 27
Baudot, Matthias, 542, 548
Behaviors:
    for actors, 379, 380
    object, 210
Benchmarking, 142–146
Beniwal, Ravi, 376, 392
Bias current, 166, 168
Big-endian byte ordering, 113
Binary files, 114–115, 114f
Binary message formats, 553
Bipolar converter, 183, 184t
Bitbucket, 513, 523
Black box testing, 407
Block diagram, 5–7, 12, 63, 70, 222
    moving to LVOOP, 227–228
    Quick Drop and, 37
    test coverage and, 411
Block diagramming, 14, 61
Block-mode continuous data, 192
BLT, 538, 548
Board Testing example, 212
Bogus data, 171
Bookmark Manager, 44
Bookmarks, 44
    global call documentation with, 80
    as technical insurance, 492
Boolean Array to Number function, 111
Booleans:
    arrays of, 111
    clusters of, 112f
    latching, 130
Branch merge, 509
Breakpoint Manager Window, 37
Broadcast events, 319, 320, 567
    DQMH, 302–304
        debugging, 318, 318f
        module stopping and, 349f
        obtaining with VI, 320
Broadcasts, 219
    Acquisition Started, 330–331
    RX, 567
    Status Updated, 339
Broadcasts virtual folder, 318
"Bug day" testing approach, 19
Build API, 542–543
Build Array function, 96, 97f
Build Assert VI.vi wizard, 479, 480f
Build Engine, 542–546
Build environment, 542, 546–548
Build processes:
    automated, 524
    BLT for managing, 548
    CI and, 542–546
Build servers, 524, 526–527
Build Specifications, 324
Build Timing Source Hierarchy.vi, 121
Build Waveform function, 107
Building:
    executables, 524
    PPLs, 297
    strings, 87–88, 87f
Built-in globals, 78
Bundle By Name function, 104, 105, 105f
Bundle function, 104, 105, 105f
Bus mastering, 203
Byte ordering, 113
Byte pairs, 113

C

C (programming language), 6–7, 16, 31
    compilers for, 25
    OOP and, 23
    structs in, 103
C# (programming language), 210–211
C++ (programming language), 31, 36, 210–211
Calendar origin, 124
Call Chain, 81
Call Parent Method mode, 254
Calling DQMH modules, 312–318
Capacitive coupling, 158, 159
Capacitors, 158
Caraya Unit Test Framework, 406, 410, 484t
    assertions with, 437, 477–478, 477f
    unit testing for class with, 437–438
Carriage return characters, 92
CASE. See Computer-aided software engineering
Case structures, 14f, 37, 237f
    dynamic dispatch from, 236–240
    refactoring to dynamic dispatch, 236–240
CD. See Continuous delivery
Celis, Benjamin, 542
Centralized source code control, 503, 504f, 504f–505f, 506

CERN, LabVIEW at, 58–60
Ceron, Nestor, 39
Certified LabVIEW Architect (CLA), 54
    summits of, 39–40
Chall, Steve, 24f
Chamberlain, Tom, 24f
Change-of-state detection, 76
Channel wires, 48, 49f, 287, 494, 495, 495f, 501
    VI Scripting in, 537
Channels:
    DAQ Wizard for, 27
    DAQmx hierarchy, 227f
    delay between, 181
    Event Messenger template for, 50
    messaging, 552–553
    Serial Motion Controller, 226, 227f
Charge pump-out, 181
CI. See Continuous integration
Circular buffered I/O, 203–204
Circular buffering, 192
Circular dependencies, 243
CLA. See Certified LabVIEW Architect
CLA summits, 39–40
Class accessors, scripting for, 538
Class diagrams, 211–216, 499
Class hierarchies, 219
    Actor Framework, 369–370, 370f–371f, 372–373, 376
    for low coupling approach to actors and messaging, 393f
    for zero coupling actors, 394f
Classes, 209, 210
    clusters for data of, 216–217
    extending, 359–360
    inheritance, 253
    in LabVIEW NXG, 269
    parent classes relation to, 253–255
    by reference, 240–241
    relationships between, 212, 215, 217
    unit testing, 425–426, 428–429
        creating unit tests, 430–444
        implementing code to pass unit test, 445, 446f
        maintaining tests, 447
    by value, 240
Clear Timing Source.vi, 122
CLI. See Command-line interface
Clock, 183

Cloneable modules, 300
    DQMH, 308–309, 311f, 322, 326, 347f
    DQMH helper loops in, 308–309, 311f
    Start Module VI for, 347f
    stopping, 349, 350f
Cloned virtual machines, 525–526
Close Module VI, 350f
Cloud-based instances, 525
Cloud Toolkit, 572
CloudWatch, 574, 577, 586
Clumps, 63, 64f
    execution of, 64, 65f
Cluster arrays, 106, 106f
Cluster To Array function, 111
Clusters, 96, 96f, 103–106
    array conversion to, 112, 112f, 116
    of booleans, 112f
    for class data, 216–217
    data, 356, 357f
    as data type definitions, 104
    date/time rec, 124, 125f
    local variables and, 104
    objects from, 228–234
    ring indicators in, 112, 112f
    setting size of, 116
CMD messages, 568
CML. See Continuous Measurement and Logging
CMRR. See Common-mode rejection ratio
Code, 63. See also Source code control
    reusable, 493
Code coverage, 407, 411–419
    levels of, 418–419
Code documentation, unit testing as, 403
Code merge, 509, 510f
Code reuse, 219, 225
Code reviews, 520f
    benefits of, 529–530
    checklist for, 535
    code and documentation preparation for, 533
    conducting, 535–536
    frequency of, 530
    LabVIEW Compare for, 533
    postreview actions, 536
    Pre-Review Checklist for, 530
    process for, 528–530, 529f
    VI Analyzer configuration for, 531–532
    VI Analyzer reports in, 532–533

Code width, 178
Coerce To type function, 52, 114
Coercion, 110–111
Coercion dots, 96, 96f, 111
Cohesion, 248
Command-line interface (CLI), 52, 324
    CI and, 543–546, 544f
    for UTF, 547
Command parser modules, 567
Command Pattern, 220–221, 221f, 286
Command Prompt, 324, 325f
Comments, 44
    as technical insurance, 492
Commit notes, 515–516
Common-mode noise rejection, 166
Common-mode rejection ratio (CMRR), 162
Common-mode signal, 161, 162, 162f
CompactDAQ, 173
Comparator, 184
Compile times, 524
Compilers, 25, 63
Component testing, 466
Composite primary keys, 579
Composition, 212
Computational operations, 10
Computer-aided software engineering (CASE), 528, 536–537
Computer automation, 1–3
Concatenate Strings function, 87
Concatenating arrays, 42
Conditional diagram disable structures, 467f, 468
Conditional terminal, 71
Configuration editor, 251, 254
Configuration files:
    Actor Framework using, 388, 388f
    constants and, 322
    VI Analyzer, 531, 531f
    VI Package, 546, 546f, 547f
Configuration management, 188
Connections, 156–173
Constant VIs, 322
Consumer loops, 136
Continuous data, 191–193
    block-mode, 192
    IIR filters and, 199
Continuous delivery (CD), 542
Continuous integration (CI), 475
    BLT automation for, 548
    build engine for, 542–546
    build environment for, 546–548
    defining, 542
Continuous Integration tools, 52
Continuous Measurement and Logging (CML), 276
    DQMH sample project template for, 324, 325f, 326
Continuous-time, continuous-value functions, 173
Control flow, 61
Control references, dynamic events and, 130–131, 131f
Control voltages, 170
Controls, 68
    class conversion from contents of, 229, 229f
    in clusters, 104
    grouping, 67
    laying out, 67
    local variables and, 78
    properties of, 68f
    tab, 276
Conversion, 110–114
Conversion functions, 91, 92f
Conversion speed, 178
Convert Contents of Control to Class, 229, 229f
Convolution VI, 197
Convolutions, 197
Conway, John, 36, 224
Conway's law, 488
Coordinated Universal Time (UTC), 124
Coradeschi, Tom, 24
Counters, 366, 367f, 368–369, 368f
Coupling, 515
Crash carts, 3
Create Actor Framework project template wizard, 370f
Create DQMH Module Template tool, 341f
Create DQMH sample project wizard, 319, 320f
Create New DQMH Event, 329, 330f
Create Notifier VI, 134
Create Project, 319
Create Rendezvous VI, 141
Create Semaphore VI, 139
cRIO, 58–60, 496
Critical sections, 137
    semaphores and, 140f

Current:
  bias, 166, 168
  input offset drift in, 168
Custom models, 497
Custom VI Analyzer tests, 492, 520, 520f
CVS, 503

D

DAC. *See* Digital-to-analog converters
DAQ. *See* Data acquisition
DAQ Assistant, 33–34, 201
DAQ Assistant Express VI, 537
DAQ boards, as timekeepers, 126
DAQ Channel Wizard, 27
DAQ Device calls, 333f
DAQ functions, 108
DAQ.IO, 555–557, 572
DAQmx, 33, 121, 192, 204, 226, 526
  channel hierarchy, 227f
  DAQ Assistant Express VI conversion to, 537
DAQmx Analog Output channel, 226, 227f
DAQmx Create Timing Source.vi, 121
Data, 63
  acquiring, 189
  bogus, 171
  class, clusters for, 216–217
  components of, 110
  continuous, 191–193
    block-mode, 192
    IIR filters and, 199
  private, 209–210, 219, 225–226, 253
    in actors, 380
  single-shot, 191, 192
  state, for actors, 378
  summing for waveforms, 108
  timing, in Microsoft Excel, 125
Data acquisition (DAQ), 3
  plug-in boards for, 5
  process of, 147, 148f
  signal origins, 147
  triggered, 176
Data acquisition programs, 189f
  medium-speed, 202–204
  planning, 188–189
  writing, 188–204

Data analysis, 188, 189–190
  execution time of, 194
Data cluster, 356, 357f
Data dependency, 61
  race conditions and, 78
Data duplication, global arrays and, 79
Data Manipulation palette, 113
Data packets, version numbers and descriptions in, 554–555
Data sheets, 171
Data-type conversions, 109, 110f
Data types, 84
  aggregate, 111
  in arrays, 96
  clusters as definitions for, 104
  conversion and coercion of, 110–111
  numeric types, 85–87, 85f, 96
  variant, 108
  waveforms, 106–109, 107f
Data value reference (DVR), 38, 240–241, 337
Dataflow, 61–62
  execution order and, 63
Dataflow diagrams, 12, 13
Dataflow Intermediate Representation (DFIR), 66
date/time rec cluster, 124, 125f
Date/Time To Seconds, 125
dB. *See* Decibels
DBL numeric types, 96
DC signals, 152
  floating-source connections and, 167f
DCAF. *See* Distributed Control and Automation Framework
De la Cueva, Fabiola, 39
Deadband, 150
Debugging:
  actors, 387
  with API tester, 315, 316f, 323, 326
  code calling DQMH modules, 315–316
  DQMH broadcast events, 318, 318f
  DQMH modules, 342
  DQMH request events, 317, 317f
  event calls in, 317, 317f
  global variables, 80–81
  local variables, 81
Decibels (dB), 164
Decimal String To Number function, 89
Decimate 1D Array function, 195
Decimation, 195, 200

Decorations palette, 104
Delacor Queued Message Handler
    (DQMH), 44, 46, 219, 222–224, 274,
    276, 282–284, 551
        API tester, 301–302, 302f
        helper loops, 304, 305f–307f, 308–309
        initialization, 343–345
        interprocess communication and, 343
        library template, 299
        MGI Panel Manager and, 352
        module main, 299–301, 300f
        MQTT client, 561–564
        objects in, 222
        overview of, 298
        project structure using, 319–323
        public API tests, 449–450
        state machines and, 310
        tools, 351–352
        use cases for, 298
        using, 312–342
Delacor Thermometer, 488, 488f, 490
Delete From Array function, 98, 99f
Delimiter characters, 92
DeMarco, Tom, 248
Demodulation, 160
Dependencies, 263–268
    circular, 243
    data, 61
        race conditions and, 78
    deployment process and, 524
    determining, 295
    runtime, 263–264, 265f, 266f, 267
    source code control repository, 290
Dependency Inversion Principle (DIP), 246,
    263–268, 263f
Dependency tree:
    for modules, 291
    for PPLs, 295
Deployment process:
    BLT for managing, 548
    dependencies and, 524
    improving, 524
Dequeue Message.vi, 357f
Descriptions, in data packets, 554–555
Design decisions. *See also* Test-driven design
    for parallel design, 276, 278
    project level enforcement for, 278–281
    technical debt from, 492
    as technical taxes, 492

Design patterns, 242
    Adapter, 258–259
    Factory, 218, 220f, 244
    frameworks versus, 283
Design reviews, as technical insurance, 492
Desktop Execution Trace Toolkit, 376, 391
Destroy Notifier, 135
Destroy Rendezvous VI, 141
Destroy Semaphore VI, 140
Deterministic timing, 123
Development tools, 16
DFIR. *See* Dataflow Intermediate
    Representation
Diagram disable structures, 467f
Differential connections, 161, 162, 162f,
    163f, 167f
Differential nonlinearity, 180
Digital codes, 182–183
Digital Filter Design Toolkit, 196
Digital filtering, 73, 196
Digital FIR Filter VI, 197
Digital fit, 176
Digital inputs, 169
Digital oscilloscopes, 150, 174
Digital outputs, 170, 170f
Digital signal processing (DSP), 153
Digital signals, 151
    samples, 152
Digital-to-analog converters (DACs), 182–183
Dimension size, 100
DIP. *See* Dependency Inversion Principle
Direct memory access (DMA), 5, 187, 203
Dirty dot, 40
Discrete system modeling, 73
Discrete-time, discrete-value functions, 173
Discriminator, 184
Distributed Control and Automation
    Framework (DCAF), 281
Distributed I/O, 173
Distributed source code control, 506, 507f,
    508f, 518
Dither noise, 185
DLLs. *See* Shared libraries
DMA. *See* Direct memory access
DMurrayIRL, 398
DO-178B aviation standard, 419
Documentation:
    code review preparation of, 533
    global calls, 80
    unit testing as, 403

DOS, memory limits, 15, 17
Double-buffered DMA, 187
DQMH. *See* Delacor Queued Message Handler
DQMH broadcast events, 302–304
    debugging, 318, 318f
DQMH Continuous Measurement and Logging sample project (DQMH CML), 276, 277f
    sample project template, 324, 325f, 326
DQMH Events, 219, 301, 302–304
    in functional global variables, 343
DQMH modules:
    adding new, 326–331, 334–342
    calling, 312–318
    cloneable, 308–309, 311f, 322, 326, 347f
    error handling, 351
    exploring, debugging, and troubleshooting, 342
    exploring, debugging, and troubleshooting code that calls, 315–316
    generic networking, 351
    PPL packaging of, 353–354
    project libraries, 320
    saving as templates, 340
    sharing, 352–354
    singleton, 312, 322, 326, 346f
    Start Module VI for, 346f, 347f
    Stop Module VI for, 348–349, 348f
    unit testing and, 425, 450, 452, 454–457
        with InstaCoverage, 463
        with JKI VI Tester, 458–461
        with UTF, 461–463
    validating, 340, 351
DQMH project template, 319–323, 319f
    exploring, debugging, and troubleshooting, 322–323
    project structure and, 319–322
DQMH Queue object, 222
DQMH request events, 302–303
    debugging, 317, 317f
    for MQTT gateway, 561, 561t, 563f
DQMH templates, sharing, 352
DQMH Unit Testing tools, 425, 459, 461
DQMH Wizard, 563, 563f
DQMHMQTT Tester, 564, 564f

Driessen, Vincent, 518
DSP. *See* Digital signal processing
DVRs. *See* Data value reference
Dye, Rob, 24f, 116
Dynamic dispatch, 218, 221f, 225
    in Actor Core.vi, 357f
    case structures refactored to, 236–240
Dynamic Events, 130–131, 131f
DynamoDB, 574, 578–580, 579f, 580f
    inserting messages into, 584f
    IoT rules and, 581, 583f, 584, 587f

E

Edit Format String command, 92
EHL. *See* Event Handling Loop
Elapsed time, 76, 76f
Electromagnetic fields, 158–159
Electromagnetic noise sources, 159
Electrostatic shield, 159
Embedded systems, LabVIEW RT running on, 5
Empty arrays, 75, 97, 100
Empty fields, 554
Empty strings, removing from incoming arrays, 75, 75f
Encapsulation, 216–217, 224, 225, 230
Encryption Compendium, 572
End Point Actors, 398
Endevo, 37, 241
Endigit, 513
Engeling, Thad, 26f
Engineering block diagrams, 12
Enterprise Integration Patterns (Hohpe and Woolf), 552
Enthought, 52
Enumerated types (enums), 115–116, 115f
    class hierarchies from, 234–235, 235f, 236f
Epoch time values, 124
Equivalent-time sampling, 174
Error, 151
    ADC sources of, 178–180
    gain, 179
    offset, 179
    settling time and, 164
Error 43, 390–391, 391f

Error handling, 405f
    assertions and, 477, 478f
    DQMH modules, 351
Error handling strategies, 288–289
    Actor Framework and, 390–391, 391f
Error registers, 52, 53f
Escape sequences, 91, 91f
Ethernet, USB hardware connection over, 525
Event calls:
    in debugging, 317, 317f
    VIs wrapping, 315
Event Handling Loop (EHL), 299, 300, 317
    different rates in multiple, 310f
    Stop Module request handling by, 348f
Event Messenger channel template, 50
Event Refnum, 345
Event Structures, 37, 42, 129
    scripting functions in, 44
Event timer, 43–44
Events, 43–44, 128–131
    broadcast, 319, 320, 567
        DQMH, 302–304, 318, 318f
        module stopping and, 349f
        obtaining with VI, 320
    DQMH, 219, 301, 302–304
        in functional global
            variables, 343
    dynamic, 130–131, 131f
    Filter, 129, 130f
    fixed cases in helper loops,
        337–338, 338f
    hardware-timed DAQ, 121
    interprocess communication with, 343
    Notify, 129, 129f
    Publish, 567, 567f
    Put Helper Loop to Sleep, 308, 309, 338
    registering for, 344, 345
    request, 299, 319, 320, 328
    Request and Wait for Reply DQMH,
        303, 326
    Stop Acquiring, 337, 337f
    Stop Module, 308
    for synchronization, 128–134
    user:
        Actor Framework
            communication with, 374
        generating, 315f
        interprocess communication
            with, 287
    user-interface, 128, 128f

Excel, test vector editing with, 424
Excitation, 165
Executables, 324
    building, 524
Execute Command method, 221
Execution:
    of clumps, 64, 65f
    data analysis and time of, 194
    desktop, for FPGAs, 469, 469f,
        472–473, 473f, 474f
    multithreaded, 29
    of nodes, data dependency and, 61
    order of, 63
        global variables and, 79
    parallel:
        file access and, 138
        shared resources and, 136
    subVIs and, 65, 66f
    timing and, 123
Execution engine, 63, 64
EXEs. *See* Standalone applications
Express VIs, 33
External triggers, 184
Extract Single Tone Information VI, 126

F

Factory pattern, 218, 220f, 244
Failures, testing, 449
Falling Shortcuts, 538
Faraday cage, 159
Fast Fourier Transform (FFT), 153, 198
Feedback control loops, 150
Feedback Nodes, 76
    in benchmarking, 143f
FFT. *See* Fast Fourier Transform
Fiducial pulses, 190
Field-programmable gate arrays (FPGAs),
    5, 34
    unit testing and, 466–473
        with FPGA desktop execution
            node, 472–473, 473f, 474f
        with FPGA Simulation Mode,
            471, 471f
        with UTF, 470, 470f
FieldDAQ, 173
FIFO. *See* First-in, first-out
File access, parallel execution and, 138
File Dialog Express VI, 33

Files:
    binary, 114–115, 114f
    configuration:
        Actor Framework using, 388, 388f
        constants and, 322
        VI Analyzer, 531, 531f
        VI Package, 546, 546f, 547f
    ignore, 517–518
    metadata XML, 342
    naming, 278
Filter events, 129, 130f
Filter functions, 196
Filtering, 175–177
    digital, 73, 196
    input, 194
Filters, 162, 166f
    active, 164
    antialiasing, 175
    Digital FIR, 197
    FIR, 73, 73f, 197, 197f
    IIR, 198–199
    low-pass, 165, 175
    median, 199
    Order of filter transfer functions, 176
    reconstruction, 182
    transfer function of, 176
Fine-resolution timers, 117
Finite impulse response filter (FIR filter), 73, 73f, 197, 197f
Finite-sequence length artifacts, 198
FIR filter. *See* Finite impulse response filter
First-in, first-out (FIFO), 203
Fixed event cases, in helper loops, 337–338, 338f
Flatten To String function, 114–115, 114f
Fletcher, Meg, 24f
Floating analog inputs, 166–168, 167f
For Loops, 13f, 38, 71, 72, 72f
    array conversion to string with, 94
    array creation with, 96, 96f
Format And Precision pop-up menu, 114
Format Date/Time String, 124
Format Into String function, 88, 90, 92, 115
Format Value function, 88
Fortran (programming language), 7
Forward coefficients, 198
Fowler, Martin, 212

FPGA desktop execution nodes, 469, 469f, 472–473, 473f, 474f
FPGA Simulation Mode, 468, 468f, 471, 471f
FPGAs. *See* Field-programmable gate arrays
Fract/Exp String To Number function, 90
Frame of reference, 275
Frameworks, 273. *See also* Actor Framework; Caraya Unit Test Framework; Delacor Queued Message Handler; Unit Test Framework
    advantages of, 282–283
    automated test, 406
    available, 281–282
    complexity of, 285
    defining, 275
    design patterns versus, 283
    disadvantages of, 283
    Distributed Control and Automation Framework, 281
    error handling strategies, 288–289
    evaluation criteria for, 285–286
    initialization of, 343–345
    interprocess communication and, 287
    key components of, 286–289
    learning curves for, 285–286
    making, 284
    module initialization for, 287–288
    PPLs and, 297–298
    programmer contract with, 283–284
    sharing modules and, 289–290
    stopping processes gracefully, 288
    as technical assets, 492
    technical support for, 286
    tools available within, 286
Frequency domain, 153
Frequency foldback, 174
Front panel, 63, 67
fs. *See* Sampling frequency
Functional global, 71
Functional global variables, 281
    DQMH events in, 343
Functional Global Variables, 36

G

G (programming language), 6, 54, 494
    dataflow and, 61
    variables in, 77–81

G compiler, 63
G Object-Oriented Programming (GOOP), 37, 241–243
G Systems, 20
Gabor spectrogram, 153
Gain errors, 179
Gang of Four book, 258
GDevCon, 53
GDS. *See* GOOP Development Suite
GDS Templates, 242
GDS UML Diagram, 501
GDS UML Modeler, 243
Gee, Trisha, 430
General-Purpose Interface Bus (GPIB), 8, 16, 249
Generate Occurrence function, 131
Generate User Event.vi, 315f
Get Components VI, 215
Get Date/Time In Seconds, 117, 124
Get Date/Time String, 117, 124
Get Waveform Attribute function, 109
Get Waveform Components function, 107
Getting Started Window, creating projects from, 319
Git, 506, 523
    commit message guidelines for, 516
Git-flow, 518–521, 519f, 523
Git submodules, 290
GitHub, 523
GitHub-flow, 518, 521–522, 521f, 523
Gitlab, 517, 523
"Given-When-Then" (McNally), 430
Global arrays, 79
Global Positioning System (GPS), 126
Global variables, 78–79, 138–139, 139f
    access protection for, 140
    debugging, 80–81
    functional, 281, 343
    within project libraries, 280–281
    WORM, 79
Global VIs, 322
Goeres, Justin, 43, 219, 308
Gong, Huaxin, 572
Goodhart's law, 419
GOOP. *See* G Object-Oriented Programming
GOOP Development Suite (GDS), 37, 208, 212, 241
    NXG and, 269
Goulet, Andrea, 491

GPIB. *See* General-Purpose Interface Bus
GPS. *See* Global Positioning System
Graphical user interfaces (GUIs), 11, 67, 276
    unit testing and, 475
Ground loops, 159, 163f
Ground-referenced analog inputs, 166
Grounding, 157–158, 158f
Gruggett, Lynda, 132
GUIs. *See* Graphical user interfaces

H

HAL. *See* Hardware abstraction layers
Hampel, Joerg, 351
Handle blocks, 114
Hard-coded constants, 322
Hardware abstraction layers (HAL), 243–244, 523
Hardware-timed, circular buffered I/O, 203
Hardware-timed DAQ events, 121
Harrell, Lynda, 20
Harvey, Audrey, 20, 27
Hashtags, global call documentation with, 80
Heartbeat messages, 374, 569
Helper loops, 309f
    in Actor Framework, 361–362, 362f
    in actors, 380, 382
    in cloneable DQMH modules, 308–309, 311f
    DQMH, 304, 305f–307f, 308
    fixed event cases in, 337–338, 338f
    parallel, 308
    Wake Up Helper Loop request and, 334–335, 336f
Henson, Nancy, 39
Hex Display, 91
Hg (Mercurial), 506, 513, 517
Hg-flow, 518, 520, 520f
Hidden responsibilities, 254
Hide Front Panel Control command, 100
High-resolution and high-accuracy timing, 126–127
High-Resolution Relative Seconds.vi, 143, 143f
Histogram function, 195
Hohpe, Gregor, 552
Holmstrom, Mikael, 241, 243
Horn, Peter, 406, 479
Hudson, Duncan, 26f

HW ID string, 329
HyperDrives, 15
Hysteresis, 150

I

I/O operations. *See* Input/output operations
I/O subsystems, choosing, 171–173
I16 integers, 113
I16 numeric types, 96
ICs. *See* Integrated circuits
Idea exchange, 40, 41, 44
IEC 61508, 419
IEEE-488. *See* General-Purpose Interface Bus
IEEE-488 control ports, 20
IEEE-754, 86
IEEE floating-point numbers, 86
IFFT. *See* Inverse Fast Fourier Transform
Ignore files, 517–518
Ignore previous, 133
IIR filters. *See* Infinite Impulse Response filters
Impulse response, 176
In-Place Element, 38
Inclinometers, 475–476
IncqueryLabs, 418, 419, 475
Incremental circulation, 193
Index Array function, 97
Indexing:
    of arrays, 97, 97f
    auto-indexing, 72, 72f
Indicators:
    in clusters, 104
    Ring, 90, 112, 112f
    status, 112, 315, 316
Inductance, 159
Inductive coupling, 159
inf, 86
Infinite Impulse Response filters (IIR filters), 198–199
Infinite loops, 71
Info-LabVIEW mailing list, 24
Inheritance, 212, 213, 217
    actor extension via, 359–360
    classes, 253
    Interface Substitution Principle and, 257
    message extension via, 360
    multiple, 257
    UML for diagramming, 213f–215f

Initialization:
    Actor Framework, 387–389
    of actors, 380–381
    arrays, 99–100, 99f
    DQMH and, 343–345
    frameworks, 343–345
    modules, 287–288, 343–345
    nested actors, 389, 389f
    root-level actor, 387–388, 388f
Initialize Array function, 100
Inlining, 144, 145f, 281
Input filtering, 194
Input offset current drift, 168
Input/output operations (I/O operations), 10
Input signals, connecting, 166
Inputs, 147
    analog:
        floating, 166–168
        ground-referenced, 166
    digital, 169
    TTL, 169, 169f
    unipolar, 183
Insert Into Array function, 98
InstaCoverage, 406, 410, 483t, 484t
    automating, 547
    code coverage with, 414, 416f, 417f, 418
    DQMH module testing with, 463
    Jenkins and, 547
    LabVIEW RT and, 464, 465f
    project coverage and, 419
    unit test reporting and, 475
    unit testing for class with, 438, 440f, 441, 443, 444f
An Instrument That Isn't Really (Santori), 8
Instrumentation amplifier, 161, 162f
    programmable-gain, 164
Instruments. *See also* Virtual instrument
    data sheets for, 171
    driver trees for, 84, 84f
    serial, 249
    standalone, 4
    virtual versus real, 4–6
Integral nonlinearity, 180
Integrated circuits (ICs), 10
Integration tests, 425, 466
Inter-application messaging systems, key rules for designing, 552–555
Interchannel delay, 181

Interface boards, 20
Interface Segregation Principle (ISP), 246, 255–262, 256f
Interfaces, 258, 267–268. See also Command-line interface; Graphical user interfaces; User interfaces
    AZInterface toolkit for, 260–261, 267, 268f
    hardware, 8, 16, 19, 20, 249
    man-machine, 67
    RESTful, 572
Interference:
    radio-frequency, 159–160
    sources of, 159
Internal triggering, 184
Internet of Things (IoT), 551. See also AWS IoT
    policies, 577
    rules, 580–581, 581f–583f, 584–586, 585f
Interprocess communication:
    Actor Framework and, 287, 387
    DQMH and, 343
    with events, 343
    messages and, 387
    methods of, 287
    with queues, 287, 387
Intervals, 118–119
Intricate conversions, 111–114
Inverse Fast Fourier Transform (IFFT), 153
IoT. See Internet of Things
Isolated sections, 404
Isolation amplifier, 163
ISP. See Interface Segregation Principle
Issue trackers, 516
iTacho serial instrument, 249, 426, 427t, 429
Iteration timers, 143

J

Java (programming language), 36
Javascript (programming language), objects in, 103
Jenkins, 475, 542
    InstaCoverage and, 547
JKI VI Tester, 406, 410, 483t, 484t
    automating, 547
    DQMH module testing with, 458–461
    LabVIEW RT and, 464
    unit test reporting and, 475
    unit testing for class with, 434–437
Join Number function, 114
Joint time-frequency domain (JTF), 153, 154f
JSON, 553–554, 554f, 568, 578
JTF. See Joint time-frequency domain
JUnit, 475

K

Kay, Meg, 26f
Kirchner, Norm, 43, 281, 308
Knights of NI, 46–47
Knutson, Dennis, 47, 47f
Kodosky, Jeff, 8, 9–18, 9f, 20, 23, 24f, 25–26, 28–30, 47, 50, 54, 56, 494, 496
Kring, James, 36

L

LabVIEW:
    advantages of, 6–7
    Automation and, 1–3
    Build API, 542–543
    building prototype of, 14–15, 16f
    at CERN, 58–60
    development of, 15–19, 18f, 22f, 28f
    environment, 67–70
    future versions of, 54
    under the hood, 62–66
    interface boards and, 20
    multithreaded execution in, 29
    OOP history and, 36–37
    origins of, 7–8
    porting to Windows and Sun, 25
    programming technique for, 12–14
    release timeline, 56–58
    shipping first version of, 19
    TCP networking support, 572
LabVIEW 1.2, 20–21, 21f
LabVIEW 2, 23, 24f
LabVIEW 2.5, 25, 26f
LabVIEW 3, 26–27, 26f
LabVIEW 4, 27, 29, 493
LabVIEW 5, 28–30
LabVIEW 6i, 31–32
LabVIEW 8, 34–35
LabVIEW 8.2, 36–37, 241

LabVIEW 8.6, 37–38
LabVIEW 10, 27, 29
LabVIEW 2009, 38–39, 66
LabVIEW 2010, 40–41
LabVIEW 2011, 41–42
LabVIEW 2012, 42–43
LabVIEW 2013, 43–44
LabVIEW 2014, 45
LabVIEW 2016, 48
LabVIEW 2017, 49–50
LabVIEW 2018, 51–52
LabVIEW Advanced Virtual Architects (LAVA), 33, 538
LabVIEW Application Builder, 27, 39
LabVIEW Base Package, 42
LabVIEW Champions, 34, 54
LabVIEW Compare (LVCompare), 39, 509, 511–513
    for code reviews, 533
LabVIEW crash carts, 3
LabVIEW Examples Style Guideline, 528
LabVIEW for Everyone (Travis and Kring), 36
LabVIEW FPGA, 30–31, 34, 496
LabVIEW Merge (LVMerge), 511, 513–514
LabVIEW Messenger Library, 281
LabVIEW NXG, 49
    classes in, 269
    objects in, 217
    unit testing and, 482
LabVIEW NXG 1.0-3.0, 54–56
LabVIEW Object, 217–219, 217f, 218f
LabVIEW object-oriented programming (LVOOP), 207, 208
    caveats, 268–269
    GOOP differences from, 241–242
    moving projects to, 227–228
    other language differences from, 210–211
    reasons for using, 224–226
    refactoring to, 228–240
    by reference, 240–241
    where and when to use, 219–224
LabVIEW RT, 5, 30
    unit testing and, 464, 465f
LabVIEW Run-Time Engine, 49, 50
LabVIEW scripting, 536
    common areas using, 537–538
    enabling, 538
    learning, 538–542
    Quick Drop shortcuts, 539–540
    reasons for using, 537
    shipping examples for, 539
    Shortcut Plugins, 540
    VI Analyzer Enthusiasts group and, 541–542
LabVIEW Shortcut Plugins, 540
LabVIEW Signal Processing Toolkit, 153
LabVIEW State Diagram Toolkit, 501
LabVIEW Style Checklist, 528
LabVIEW Style Guidelines, 527–528
LabVIEW 7, 33–34
LabVIEW Task Manager, 376, 392
LabVIEW Tools Network, 37, 46, 52, 212
LabVIEW Wiki, 33
Ladolcetta, David, 46
Lambda functions, 574
    IoT rules and, 581
LapDog, 282
Large Hadron Collider, 59–60
Last Ack message, 357
Latching booleans, 130
Launch Actor.vi, 358
Launch Nested Actor.vi, 389, 389f
Launch Root Actor.vi, 359, 366, 367f, 373f, 388–389, 388f
LAVA. *See* LabVIEW Advanced Virtual Architects
LAVA code repository, 37
LAVA forums, 33
Leak resistors, 166, 168
Least-significant bit (LSB), 185
Lemmens, Stefan, 366, 377
LGP5 Acquisition library, 329f
LGP5 Acquisition module, 337
Libraries. *See also* Packed project libraries
    Actor Framework, 356–357, 358f
    Actor Framework Debug, 376
    DQMH template for, 299
    global variables in, 280–281
    LabVIEW Messenger, 281
    LVMQTT, 572
    Point-by-Point, 193, 193f
    project, 36, 251–252, 279–280, 320
    shared, 31, 50
Line feed characters, 92
Linear test vectors, 420
Linearity, 179
Linked clone virtual machines, 526

Liskov, Barbara, 252
Liskov Substitution Principle (LSP), 246, 252–255, 252f, 360
LLVM. *See* Low-Level Virtual Machine
Local variables, 26, 77–78
    clusters and, 104
    controls and, 78
    debugging, 81
Localized Names, 233, 233f
Loftus-Mercer, Stephen, 42, 48, 50, 209, 216, 224, 225, 249, 253, 255, 283, 354, 366
Log Event message, 374, 375f
Logging module, 278
Looping:
    For Loops, 72, 72f
    While Loops, 71, 72f
Loops, 12–13. *See also* Helper loops
    concatenating arrays with, 42
    consumer, 136
    Event Handling, 299, 300, 310f, 317, 348f
    feedback control, 150
    For Loop, 13f, 38, 71, 72, 72f, 94, 96, 96f
    ground, 159, 163f
    infinite, 71
    Message Handling, 299–301, 348f, 355
    producer, 136
    single-cycle timed, 31
    stopping, 71
    timed, 117, 120f, 121, 121f
        with frames, 120, 122
        priority and execution, 123
        while, 120
    While Loop, 13f, 71, 72f
Low coupling approach for actors and messaging, 393f
Low cutoff frequency, 197
Low-Level Virtual Machine (LLVM), 66
Low-pass filters, 165, 175
LSB. *See* Least-significant bit
LSP. *See* Liskov Substitution Principle
Luick, Dean, 26f
LVCompare. *See* LabVIEW Compare
LVMerge. *See* LabVIEW Merge
LVMQTT library, 572
LVOOP. *See* LabVIEW object-oriented programming
LVS-Tools, 572

M

MacCrisken, Jack, 8, 9, 9f, 10, 16, 25
MacPaint, 11, 14
Magnetic fields, 163f
Magnetic shielding, 159
Maharjan, Kabul, 351
Make Current Value Default command, 100
MAL. *See* Measurement abstraction layers
Malleable VIs, 50, 51, 85
Man-machine interfaces (MMIs), 67
Manis, Rod, 553
Martin, Robert C., "Uncle Bob," 247, 248
    on Dependency Inversion Principle, 263
    on Interface Segregation Principle, 256–257
Match First String function, 90
Match Pattern function, 89
Matrix, 93–94, 94f
McBee, Jon, 209
McKaskle, Greg, 24, 26f
McNally, James, 52, 430, 431, 466–468, 513, 542–545
Measurement abstraction layers (MAL), 244
Median Filter VI, 199
Median filters, 199
Medical systems, 165
Medium-speed acquisition and processing, 202–204
Member VIs, 209
Memory:
    array usage of, 101–103, 101f
    direct access to, 5, 187, 203
    state, 76, 80
Memory leaks, queues and, 343
Memory management, 20
Memory manager, 103
Memory usage, arrays and, 101–103, 101f
Mengisen, Jean-Claude, 284, 351
Mercurial (Hg), 506, 513, 517, 523
Mercurial subrepositories, 290
Merge requests, 516–517
Merging pull requests, 517
Message brokers, 551, 552f, 570
    AWS IoT, 571–572, 571f
    cloud, 571–572, 571f
    installing, 558–559, 558f, 560f

inter-application messaging system
    design and, 552–555
  MQTT, 556
Message class, 221, 221f, 360
Message handlers, 356, 358
Message Handling Loop (MHL), 299–301
  exit message handling by, 348f
  in NI QMH, 355
  Start Acquiring case, 336, 336f
  Stop Acquiring case, 334f
Message Maker tool, 361
Message Queue class, 301, 361
Message transport, 354
Message Transport class, 221, 221f
Messages, 354–357, 551
  abstract, 392, 393
  binary formats, 553
  class hierarchy for, 371f
  CMD, 568
  creating, 382
  extending via inheritance, 360
  formats for, 553
  heartbeat, 374, 569
  interprocess communication
      and, 387
  Last Ack, 357
  Log Event, 374, 375f
  low coupling approach for, 393f
  sending, 364
  text formats, 553
  version numbers and descriptions in,
      554–555
  zero coupling actors and, 394f
Messaging channels, 552–553
Metadata XML files, 342
Methods, 210, 248
Meyer, Bertrand, 250
Meyer, Marc, 553
MGI Monitored Actor Toolkit, 376, 391
MGI Panel Manager, 352, 392
MHL. *See* Message Handling Loop
Microsoft Excel, timing data in, 125
Microvolt signals, 164
Mihura, Bruce, 26f
MMIs. *See* Man-machine interfaces
Modeling, 494–497. *See also* Unified
    Modeling Language
  discrete system, 73
  tools for, 501

Models, 494
  Actor, 354–355
  custom, 497
  reasons and situations for using, 497
  types of, 497–499
Modules. *See also* DQMH modules
  acquisition, 278
  cloneable, 300
      DQMH, 322, 326
      DQMH helper loops in,
        308–309, 311f
      Start Module VI for, 347f
      stopping, 349, 350f
  command parser, 567
  communication between, 495–496,
      495f, 496f
  dependency tree for, 291
  initialization of, 287–288, 343–345
  logging, 278
  MQTT Singleton, 569, 569f
  naming, 326, 328, 328f
  packaging in PPLs, 292–294
  packaging with VIPM, 290–291
  sharing, 289–290
  singleton, 300
      Start Module VI for, 346f
  stopping, 389, 389f, 390f
      broadcast events and, 349f
Moore Good Ideas, 376, 391, 392
MOSFETs, 170
Motorola 68000 series microprocessors, 15
Moving averager, 176, 199–201
Moving Avg Array function, 199, 200f
MQ Telemetry Transport (MQTT), 553,
    555–558
  example VIs for, 557
  message types, 556t
  topics, 556, 557
MQTT Simple Connect and Publish.vi, 559f
MQTT Simple Connect and Subscribe.vi,
    559, 560f
MQTT Singleton module, 569, 569f
MQTTDQMH application, 567–570,
    586, 588f
MQTTDQMH client, 561–564
ms time-out input, 132
ms Timer, 120, 143, 143f, 144f
Multidimensional arrays, 97–98, 98f
Multiframe Timed structures, 122, 122f
Multiple buffering, 192

Multiple inheritance, 257
Multiplexing, 165, 166f
    ADCs and, 180–181, 180f
    settling time and, 181
Multitasking, 61, 123
    cooperative, 64
Multithreaded execution, 29
Multithreading, 64, 65f
    semaphores and, 140
Mutexes, 139

N

Named access, 104
NaN (not-a-number), 86, 87
    arrays of, 449
    testing and, 449
National Electrical Code, 157
National Instruments (NI), 5, 8, 10, 19, 27
    FPGA use by, 31
Nattinger, Darren, 42, 44–46, 289, 344, 541
Nested actors, 363, 363f, 364
    initialization of, 389, 389f
    queues in, 387
Network connections, 114, 572
    I/O subsystems and, 172–173
NI. *See* National Instruments
NI-9467 C Series Synchronization Module for CompactDAQ and CRIO, 126
NI-DAQ, 30
NI forums, 33
    Knights in, 46–47
NI QMH Project Template, 344
NIWeek 2015, 45
NIWeek 2017, 54
Nodes:
    add, 107–108
    Attribute, 26
    data dependency and execution of, 61
    Feedback, 76, 143f
    FPGA desktop execution, 469, 469f, 472–473, 473f, 474f
    property, 69–70, 69f
        accessors and, 233
        CI arguments and, 544f
        subVIs and, 70
        VI Server, 374

Start Asynchronous Call, 345
Xnodes, 537
Noise:
    capacitive coupling and, 158
    common-mode rejection of, 166
    digital conversion performance and, 185–187
    dither, 185
    electromagnetic sources of, 159
    magnetic fields and, 163f
    sources of, 157
    spectral floor for, 175
Noise power, 175
Nonlinear phase distortion, 199
Nonlinearity, 180
Normal-mode signal, 161
NoSQL databases, 578
Notepad++, 526
Notifiers, 135f
    creating, 134
    destroying, 135
    interprocess communication with, 287
    for synchronization, 134–136
Notify events, 129, 129f
NuBus interface boards, 20
Null terminators, 116
Numeric arrays:
    conversion to clusters of Ring Indicators, 112, 112f
    waveforms added to, 108
Numeric types, 85–87, 85f
    in arrays, 96
NXG, 37, 208
Nyquist rate, 173–175, 194

O

Object behaviors, 210
Object Management Group (OMG), 211, 498
Object-oriented design, SOLID principles for, 245–268
Object-oriented programming (OOP), 17, 23
    Actor Framework and, 356
    caveats, 268–269
    concepts for, 208–217
    LabVIEW history, 36–37
    principles of, 216–219

The Object-Oriented Thought Process (Weisfeld), 208
Objects, 209, 210
    ActiveX, 108
    Admin, 222
    arrays of, 215–216
    clusters converted to, 228–234
    in DQMH, 222
    DQMH Queue, 222
    in Javascript, 103
    LabVIEW, 217–219, 217f, 218f
    in LabVIEW NXG, 217
Obtain Broadcast Events for Registration VI, 320
Occurrences, 131–134, 132f
OCXO. *See* Oven-controlled crystal oscillator
Offset errors, 179
Omega Engineering, 168
OMG. *See* Object Management Group
One-time experiments, automating, 3
OO tools, 35
OOP. *See* Object-oriented programming
Open-Closed Principle, 246, 250–252, 250f, 360
Open standards, 211n7
OpenGDS, 241
    UML Modeler, 500f
Options window, 250–251, 251f, 254
Order of filter transfer functions, 176
Outliers, 199
Outputs, 147
    analog, 170
    digital, 170, 170f
    NaN testing of, 449
    signal conditioning for, 169–171
Oven-controlled crystal oscillator (OCXO), 127
Overflow, 86
Overloads, safety issues and, 164–165
Oversampling, 176, 186, 196

P

Packaging:
    actors, with VIPM, 398
    modules:
        in PPLs, 292–294
        with VIPM, 290–291

Packed project libraries (PPLs), 41, 50, 251–252
    actors in, 398–399
    building, 297
    dependency tree for, 295
    determining directory for, 296
    DQMH modules in, 353–354
    framework considerations for, 297–298
    module packaging in, 292–294
Page-Jones, Meilir, 248
Parallel design, design decisions for, 276, 278
Parallel execution:
    file access and, 138
    shared resources and, 137
Parallel helper loops, 308
Parallel processing, 61
Parallel tasks, 275–276
Parallels, 525
Parasitic detection, 160
Parker, Jeff, 20, 23
Parsing strings, 88–91, 89f, 90f
Partition keys, 579
Patterns:
    Command, 220–221, 221f, 286
    Factory, 218, 220f, 244
    matching, 89
PCI-6810/PXI-6810 Serial Data Analyzer board, 31
Peak Detector function, 195
Pearson Electronics, 169
Perforce, 503
Performance:
    arrays and, 101–103, 101f
    benchmarking, 142–146
    of digital conversion, noise and, 185–187
PERL (programming language), 29
PFI. *See* Programmable function input
PGIA. *See* Programmable-gain instrumentation amplifier
PharLap TNT Embedded, 30
Phase distortion:
    nonlinear, 199
Phase response, 176
Pick Line function, 87
PID Toolkit control blocks, 76
Plug-in data acquisition boards, 5

Plug-in expansion boards, 19
Plug-in type applications, 41
Point-by-Point library, 193, 193f
Policies:
 IoT, 577
 unit testing, 492
Polling, 127–128, 128f
Pollock, Matt, 376, 542
Polymorphic shift registers, 73
Polymorphic VIs, 50
 as messaging inputs, 567, 568f
Polymorphism, 85, 109, 110f, 218, 225
 waveforms and, 107
Potentiometers, 165
Powell, Brian, 7, 24f, 26f, 39, 491
Powell, James, 281
Power spectrum, 174, 175, 175f
Power supplies, 243–244
PPLs. *See* Packed project libraries
Pre-Review Checklist, 530
Precision, 86
 Format And Precision pop-up menu, 114
 loss of, 111
Precision Acoustics, 520
Primary keys, 579
Priority Dequeue.vi, 357f
Priority queues, 361
Private data, 209–210, 219, 225–226, 253
 in actors, 380
Private methods, testing and, 425
Private VIs, 280
 testing and, 424–425
Process auditing, 488, 489f
Process smells, 514–515
Processes. *See also* Build processes; Deployment process; Interprocess communication
 defining, 274
 stopping gracefully, 288
Producer loops, 136
Producer–Consumer template, 220, 220f, 221f, 389
Programmable function input (PFI), 127
Programmable-gain instrumentation amplifier (PGIA), 164
 ADCs and, 182
Project Explorer, 46
 right-click DQMH menu for, 351–352

Project libraries, 36, 251–252, 279–281
 DQMH modules, 320
Project providers, 537
Project Risk Registers, 490
Project structure:
 Actor Framework, 369–370, 371f, 372–373, 376
 DQMH and, 319–323
 UTF and, 408–409, 408f
Project templates, 43
 for Actor Framework, 369, 369f
 DQMH, 319–323, 319f
  exploring, debugging, and troubleshooting, 322–323
  project structure and, 319–322
 as technical assets, 492
Projects:
 creating, 319
 design decision enforcement in, 278–281
 problem solved by, 490–491
Property nodes, 69–70, 69f
 accessors and, 233
 CI arguments and, 544f
 subVIs and, 70
 VI Server, 374
Property pages, 68f
Public API virtual folders, 320
Public VIs, 280
 testing and, 424–425
Publish and subscribe applications, 552, 552f
Publish events, 567, 567f
Pull requests, 516–517, 522
Pull-up resistors, 169, 169f
Pulse ring indicator, 90
Pulse trains, 151
 digital outputs and, 170
Put Helper Loop to Sleep event, 308, 309, 338
PXI, 58–60
PXI-6608 counter/timer module, 127
Python (programming language), 52
Python Integration Toolkit, 52

Q

QMH. *See* Queued Message Handler
Quantization levels, 177
Queue API, 221, 221f

Queued Message Handler (QMH), 137,
  138f, 222–224, 283, 324, 551
    MHL for, 355
Queues, 129, 136–137, 137f, 225. *See also*
  Delacor Queued Message Handler
    in Actor Framework, 356, 387
    AristosQueue, 46
    in DQMH modules, 343
    interprocess communication with,
      287, 387
    memory leaks and, 343
    Message Queue class, 301, 361
    in nested actors, 387
    priority, 361
    Scheduler, 119
    Send-to-Self Enqueuer, 357f
    single element, 241
    for synchronization, 136–137
Quick Drop, 37–39, 41, 45
    creating shortcuts for, 539–540
    plugins, 537, 538
    ViBox, 538
QuickDraw, 14–15

R

Race conditions, 80, 139
  global variables and, 78
Radio-frequency interference (RFI),
  159–160
Range, 86, 178
Rank, 199
Rayner, Ben, 498
RCS. *See* Revision control system
Real time:
  defining, 190
  unit testing and, 464
Real-time analysis and display, 190–195
Real-time operating system (RTOS), 30
  Timed structures on, 123
  timer resolution in, 117
Real-time system integration (RTSI), 127
Receive Message.vi, 357f
Reconstruction filter, 182
Recursion, 39
Reentrant subVIs, 138
Refactoring:
  to dynamic dispatch, 236–240
  to LVOOP, 228–240

Reference junction, 168
Reference potential, 157
Referenced single-ended input (RSE),
  166, 167f
Register for Events VI, 344, 345
Regression testing, 408
  as technical insurance, 492
Regular expressions, 89
Relative timestamps, 143
Relays, 170
Release Semaphore VI, 140
Remote I/O, 172
Rendezvous VIs, 141, 142f
Replace Array Element function, 101, 102f
Replace Array Subset function, 74f, 98, 99f
Representation pop-up menu, 111
Request:
  Show Panel, 317
  Wake Up Helper Loop, 334–335,
    335f, 336f, 338
Request and Wait for Reply DQMH event,
  303, 326
Request events, 299, 319, 320
  Start Acquiring, 328
Requests, 219
Requests virtual folder, 317, 320
Resampling, 200
Resistance temperature detectors
  (RTDs), 165
Resolution:
  of ADCs, 177, 177f
  of timing, 117–118, 126–127
RESTful interfaces, 572
Return, 157
Return on investment (ROI), 488, 490
Reusable code, 493
Reverse Array function, 114
Reverse coefficients, 198
Reverse String function, 114
Revision control system (RCS), 501
RF signals, 160
RFI. *See* Radio-frequency interference
Rhia, Stepan, 134
Richardson, Greg, 26f
Right-click plugins, 537
Ring indicator:
  pulse, 90
Ring Indicators, 112, 112f
Rogers, Steve, 20, 24f, 25, 26f, 63
ROI. *See* Return on investment

Root-level actor, 363, 363f, 372
    initialization of, 387–388, 388f
RS-422 serial ports, 19
RSE. *See* Referenced single-ended input
RTDs. *See* Resistance temperature detectors
RTOS. *See* Real-time operating system
RTSI. *See* Real-time system integration
Rules, 574
Runtime dependency, 263–264, 265f, 266f, 267
RX broadcasts, 567

S

S/H amplifier. *See* Sample-and-hold amplifier
S3, 574
S5 Solutions, 282
Safety critical systems, code coverage and, 419
Safety ground, 157
Sagatelyan, Dmitry, 209, 247
Sample-and-hold amplifier (S/H amplifier), 178
Sample clock, 152, 153f
Samples, 152, 153f
Sampling, 195
    analog signals, 173
    digital signals, 152
    equivalent-time, 174
    oversampling, 176, 186, 196
    resampling, 200
    undersampling, 174
Sampling frequency (fs), 197
Sampling rates, 174, 174f
Sampling theorem, 173–175
Santori, Michael, 8
Saphir, 538
sbRIO-9607, 145
Scalar numerics, 85f
Scalar types, 84
Scaled Window VI, 198
Scan From String function, 89, 90, 90f, 92, 116
Scheduler queues, 119
Schmidt, Steen, 43, 308
SCM. *See* Source code management
Scripting. *See also* LabVIEW scripting
    for class accessors, 538
    functions in Event Structure, 44
    VIs, 33, 536

SCRs. *See* Silicon-controlled rectifiers
SCSI (Small Computer Systems Interface) ports, 19
SCTL. *See* Single-cycle timed loop
Search 1D Array function, 195
Seconds To Date/Time, 117–118, 125
Seebeck effect, 160
Select Item, 104
Self-inductance, 158
Semaphores, 137–140
Send Message VIs, 366
Send Notification VI, 135
Send-to-Self Enqueuer queue, 357f
Sensors, 147, 148–150, 148f, 149t
Sequence diagrams, 379, 379f, 499, 500f
Sequence Structure, 13, 14f
Sequence test vectors, 420
Serial commands, 249
Serial Instruments, 249, 426, 427t, 429
Serial links, 114
Serial Motion Controller Channel, 226, 227f
Set Cluster Size pop-up setting, 116
Set Occurrence function, 131
Set Waveform Attribute, 108
Settling time, 164, 165
    DACs and, 182
    multiplexing and, 181
Setup, 447, 461
SGL numeric types, 96
Shah, Darshan, 26f, 30
Shannon sampling theorem, 173
Shared libraries (DLLs), 31, 50
Shared resources, 137
Sharing:
    DQMH modules, 352–354
    DQMH templates, 352
    modules, 289–290
    reusable actors, 398–399
    templates, 342
Shielding, 157
    electric field, 158, 159
    magnetic, 159
Shift registers, 13f, 36, 38, 73–74, 74f
    array conversion with, 94
    empty string removal with, 75, 75f
    Feedback Nodes, 76

moving average with, 73f
uninitialized, 75–76
Shortcut Plugins, 540
Show Block Diagram for Troubleshooting option, 315
Show Buffer Allocations window, 103
Show Panel request, 317
Sieve of Eratosthenes, 18
Sign bit, 183
Signal bandwidth, 195–196
Signal common, 157, 158f
Signal conditioning, 147, 148f, 156–157, 162, 164, 165f
    outputs and, 169–171
Signal-to-noise ratio (SNR), 164
Signals:
    AC:
        floating-source connections and, 167f
        measuring, 168–169
        time domain, 152
    amplifiers and quality of, 163
    analog, 151, 152
        sampling, 173
    balanced, 161
    categories of, 151–155, 155f, 156f
    common-mode, 161, 162, 162f
    conditioning for outputs, 169–171
    DAQ and origins of, 147
    digital, 151
        samples, 152
    input, connecting, 166
    microvolt, 164
    normal-mode, 161
    origins of, 147
    RF, 160
Signed integers, 86
Significant figures, 86
Silicon-controlled rectifiers (SCRs), 170
Simple Connect and Publish.vi, 557, 557f
Simple Connect and Subscribe.vi, 557, 557f
Simple Text Oriented Messaging Protocol (STOMP), 553
Simulated I/O for Digital Pattern Generator.vi, 471
Sine Waveform VI, 107
Single-cycle timed loop (SCTL), 31
Single element queues, 241
Single-ended connections, 161, 161f, 163f
Single Responsibility Principle (SRP), 246, 247–249, 248f
Single-shot data, 191, 192
Singleton modules, 300
    DQMH, 312, 322, 326
    Start Module VI for, 346f
Skew, 180–181, 180f
Skin effect, 158
Slew rate, 182
Slicing arrays, 98, 98f
Small Computer Systems Interface. See SCSI ports
Smart sensors, 150
Smith, Allen, 46, 209, 283, 354, 361, 374, 376, 542
Snapshots, of virtual machines, 526
Sniffers, 316
SNR. See Signal-to-noise ratio
SNS, 574
Snyder, Dave, 282
Software-diagramming techniques, 12
Software engineering, 491
Software Engineering Approach to LabVIEW (Watts and Conway), 36, 224
Software engineering practices, 488
Software-generated triggers, 185
SOLID principles, 245–268, 360
Solid-state relays, 170
Sort keys, 579
Source code control, 488, 501
    actor sharing and, 398
    branch merge versus code merge, 509
    build servers pulling from, 527
    centralized, 503, 504f, 505f, 506
    commit notes, 515–516
    defining, 502
    distributed, 506, 507f, 508f, 518
    DQMH module sharing and, 353
    good practices for, 514–518
    ignore files, 517–518
    LabVIEW Compare with, 511–513
    LabVIEW Merge with, 513–514
    pull and merge requests, 516–517
    repository dependencies, 290
    as technical insurance, 492
    tool and workflow selection, 522–523
    tools, 39
Source code dependency, 263–264, 265f, 266f, 267

Source code management (SCM), 501
Source code workflows, 518–522
SourceTree, 513, 521
SpaceX, 54
Spectral leakage, 198
Spectral noise floor, 175
Spectrogram, 153
Split Array function, 194
Split Number function, 114
Spreadsheet String To Array function, 93, 93f
Spreadsheets, 12
       strings from, 92–94, 93f–95f
SRP. *See* Single Responsibility Principle
SSL/TLS support, 572
Standalone applications (EXEs), 50, 52
Standalone instruments, 4
Standalone VIs, unit testing with, 430–431, 456–457, 483t
Standard Deviation and Variance function, 192, 192f
Start Acquiring MHL case, 336, 336f
Start Acquiring request event, 328, 330, 334
Start Asynchronous Call node, 345
Start Module VI, 320, 344, 344f, 346f, 347f
State, 253
State data, for actors, 378
State Diagram Editor, 498, 499f
State diagrams, 498
State machines, 76, 81–83, 82f, 498f
       DQMH and, 310
State memory, 76
       in subVIs, 80
State Pattern Actor, 398
State transformation, 224
Statistical functions, 192, 192f
Status indicators, 112, 315, 316
Status Updated broadcast, 339
STC. *See* System timing controller
STFT Spectrogram PtByPt VI, 154f
STOMP. *See* Simple Text Oriented Messaging Protocol
Stop Acquiring event, 337, 337f
Stop Acquiring MHL case, 334f
Stop conditions, 71, 72f
Stop Module event, 308
Stop Module VI, 348–349
Stop Msg, 357, 389, 389f, 390f
Strain gauges, 164, 165, 165f
Strict typedef, 103

String Subset function, 90
Strings:
       array conversion to, 94
       building, 87–88, 87f
       concatenating, 87
       formatting into, 88, 90, 92, 115
       matching, 90
       parsing, 88–91, 89f, 90f
       reversing, 114
       spreadsheet interaction with, 92–94, 93f–95f
       unprintable characters in, 91–92, 92f
Structured types, 84
STUDIO BODs, 538, 548
Style guidelines, 492, 527–528
Subpanels, 276
Subroutines, 10
subVIs, 4, 19, 41, 42, 83–84
       call overhead for, 144
       execution systems and, 65, 66f
       inlining, 144, 145f
       Property nodes and, 70
       reentrant, 138
       state memory in, 80
Sun SPARCstation, 25
SVN, 503, 523
SVN Externals, 290
Swap Bytes function, 113
Swap Words function, 113
Symbio, 241
Synchronization, 127–141
       Events for, 128–134
       notifiers for, 134–136
       occurrences for, 131–134, 132f
       polling for, 127–128
       queues for, 136–137
Synchronize for Module Events VI, 344
Synchronize Module Events VI, 320, 344f
Synchronous communication, 396f
System clock/calendar, 117
System timer, 117
System timing controller (STC), 185

T

Tab controls, 276
Tab-delimited text, 92
Taggart, Sam, 361, 374, 379, 392
Target to Host FIFO, 467

Index **615**

TCP networking support, 572
TDD. *See* Test-driven design; Test-driven development
Team size, application modularity and, 276
Teardown, 447, 461
Technical assets, 492
Technical bankruptcy, 493
Technical debt, 491, 492
Technical depreciation, 493
Technical insurance, 492
Technical investments, 492
Technical tax, 492
Technical wealth, 491–493
Technova, 209
Temperature Handbook (Omega Engineering), 168
Temperature set point, 312
Templates:
    adding DQMH module from, 339–342
    DQMH, 352
    DQMH CML sample project, 324, 325*f*, 326
    DQMH library, 299
    Event Messenger channel, 50
    GDS, 242
    LabVIEW scripting in, 537
    Producer–Consumer, 220, 220*f*, 221*f*, 389
    project, 43
        for Actor Framework, 369, 369*f*
        DQMH, 319–323, 319*f*, 322–323
        as technical assets, 492
    saving DQMH modules as, 340
    sharing, 342
    for virtual machines, 525
Test cases, designed to fail, 449
Test computers, 527
Test coverage, 411–419
    levels of, 418–419
Test-driven design (TDD), 278
Test-driven development (TDD), 429
    for classes:
        creating unit test, 430–444
        implementing code to pass unit test, 445, 446*f*
        maintaining unit tests, 447
    defining, 407

Test harness, 406
Test vectors, 419–424
    defining, 420, 421*f*
    Excel editing of, 424
    FPGA VI testing and, 468, 469*f*
    linear, 420
    sequence, 420
    using, 422–424
    UTF and, 419, 420
Tester, 301–302, 302*f*, 319
Testing:
    actors, 374
    NaN output and, 449
Testing pyramid, 401, 401*f*
Text compare tools, 515
Text message formats, 553
Thermal Chamber Controller example, 312–318, 313*f*, 314*f*, 316*f*
Thermistors, 165
Thermocouples, 164, 168
    I/O subsystem design and, 171–172
Thermojunction voltages, 160
Threads, 64, 65*f*
    multithreaded execution, 29
    semaphores and, 140
Throughput, 187–188, 195
Tian, Wei, 24*f*
Tick Count (ms), 117, 143
Time:
    absolute, 124–126
    aperture, 181
    elapsed, 76, 76*f*
    epoch values for, 124
    real:
        defining, 190
        unit testing and, 464
    settling, 164, 165
        DACs and, 182
        multiplexing and, 181
Time base, 183
Time-Delayed Send Message.vi, 374, 375*f*
Time window, 198
Timed buffered I/O, 203
Timed Loop With Frames, 120, 122
Timed Loops, 117, 120, 120*f*, 121, 121*f*
    priority and execution, 123
Timed Sequence Structure, 117
Timed Sequences, 120

Timed structures, 120–122
    execution and priority, 123
    multiframe, 122, 122f
Timeout case, 337, 337f
Timer test, 118, 118f
Timestamps, 124
    in benchmarking, 143
    relative, 143
Timing, 116–127
    data acquisition and techniques for, 201
    deterministic, 123
    execution and priority, 123
    guidelines for, 123–124
    high-resolution and high-accuracy, 126–127
    intervals, 118–119
    resolution of, 117–118
    sending to other applications, 125–126
    triggering and, 183–185
    using built-in functions for, 118
Timing skew, 180–181, 180f
Timing sources, 121–122
Tip strips, 27
To Upper Case function, 89
Tools menu plugins, 537
Top Level Baseline, 281
Top Level Baseline Prime, 281
Tortoise SVN Diff Viewer, 512f
TortoiseGit, 513, 514
TortoiseHg, 513, 523
TortoiseSVN, 523
Transducers, 148–150, 148f, 165f
Transfer function, 176
Transpose 2D Array function, 93
Travis, Jeffrey, 36
Trepanier, Derek, 352, 376, 391, 392
Triacs, 170
Trigger circuits, 152
Trigger pulses, 152, 153f
Triggered data acquisition, 176
Triggering, 183–185
Truchard, Jim, 8, 9, 9f, 10, 13, 17, 25, 56
TTL inputs, 169, 169f
Two-dimensional array (matrix), 93–94, 94f, 98f
    in JSON, 553–554, 554f
Type Cast function, 113
Type casting, 111–114
Type descriptor, 110
    type casting and, 113
Type Specialization structure, 51
Type specifier, 93
Typedefs, 103, 104
    converting to classes, 229

U

UML. *See* Unified Modeling Language
UML Class Editor tool, 208, 269
UML Distilled (Fowler), 212
UML Modeler, 208, 500f
Unbundle By Name function, 104, 105f
Unbundle function, 104, 105f
Underflow, 86
Undersampling, 174
Undo, 29
Unflatten From String function, 114, 114f
Unified Modeling Language (UML), 211–216, 243
    actor design with, 379, 379f
    association diagrams with, 215f
    class diagrams, 499
    inheritance diagrams with, 213f–215f
    sequence diagrams, 499, 500f
    state diagrams and, 498
    using class diagrams from, 226–227
Uninitialized shift registers, 75–76
Unipolar converters, 183t
Unipolar inputs, 183
Unit of work, 402
Unit Test Framework (UTF), 402, 406, 483t, 484t
    automating, 547
    CLI for, 547
    code coverage with, 412–414, 415f
    creating unit test with, 431–434
    DQMH module testing with, 461–463
    FPGA VIs and, 467f, 470, 470f
    LabVIEW RT and, 464
    project coverage metric, 419
    project structure and, 408–409, 408f
    test vectors and, 419, 420
    unit test reporting and, 475
Unit testing, 401
    automating, 547

black box and white box, 407
CI and, 547
classes, 425–426, 428–429
  with Caraya Unit Test Framework, 437–438
  creating unit tests, 430–444
  with InstaCoverage, 438, 440f, 441, 443, 444f
  with JKI VI Tester, 434–437
  standalone VIs, 430–431
  with UTF, 431–434
as code documentation, 403
defining, 402–403
DQMH modules and, 450, 452, 454–457
FPGA VIs and, 466–473
getting started with, 408–410
GUIs and, 475
LabVIEW NXG and, 482
LabVIEW RT and, 464, 465f
policies for, as technical assets, 492
reporting, 475
test vectors and, 419–424
tool comparison for, 483t, 484t
VI writing for, 405–406
Unit tests, 405f, 419
  build servers running, 527
  describing, 430
  for DQMH public APIs, 449–450
  in Git-flow process, 520
  guidelines for, 431
  properties of good, 403–404
  setup and teardown, 447
  as technical insurance, 492
Unit under test (UUT), 312–313
UNIX, 11
The Unix Shell Programming Language (Manis and Meyer, M.), 553
Unprintable characters, 91–92, 92f
Unsigned integers, 86
USB data acquisition, 58, 59
USB Over IP Hubs, 525
User events:
  Actor Framework communication with, 374
  generating, 315f
  interprocess communication with, 287
User interfaces:
  events, 128, 128f
  state machines and, 83
  VI components for, 11
UTC. *See* Coordinated Universal Time
UTF. *See* Unit Test Framework
UUT. *See* Unit under test

V

Validate DQMH utility, 340
Vargo, Tim, 376, 392
Variables:
  functional global, 281, 343
  global, 78–81, 138–139, 139f, 140, 280–281
  local, 26, 77–78, 81, 104
Variant data types, 108
VCS. *See* Version control systems
Velick, Henry, 24f
Vento, Tony, 47, 47f
Version control systems (VCS), 501
Version numbers, 554–555
VHDL, 31
VI. *See* Virtual instrument
VI Analyzer:
  build servers and, 527
  CI and, 547
  in code review, 528–530
  configuration of, 531, 531f, 532f
  custom tests in, 492, 520, 520f
  LabVIEW scripting in tests, 537
  packaging tests for, 541
  reports from, 532–533
VI Analyzer Enthusiasts group, 541–542
VI Package configuration (vipc), 546, 546f, 547f
VI Scripting, 33
  in channel wires, 537
VI Server property nodes, 374
VI Technologies, 208, 269
ViBox Quick Drop, 538
Viewpoint Systems SVN Toolkit, 511
vipc. *See* VI Package configuration
VIPM, 290–291
  actor packaging with, 398
  DQMH module packaging with, 353
  package dependency management with, 546
  VI Analyzer test packaging with, 541
VIPM packages, 342

Virtual folders:
    Broadcasts, 318
    Public API, 320
    Requests, 317, 320
Virtual instrument (VI), 3, 4f
    compilation of, 63–66, 64f
    hierarchy of, 10
    load times for, 294f
    parts of, 62–63, 62f
    real instrumentation versus, 4–6
    scripting, 536
    unit testing and writing, 405–406
    user interface components, 11
    user research and, 67
Virtual machines, 525–526
VirtualBox, 525
VISA, 249, 526
VITAC, 431
VMware, 525
Voltage, 157
    thermojunction, 160
Voltage—Continuous Input.vi, 204, 204f

W

Wachno, Oli, 361, 393
Wait (ms), 117, 119
Wait At Rendezvous VI, 141
Wait on Notification VI, 134, 136
Wait on Occurrence function, 131, 132f
    abortable wait with, 132, 133f
    ignore previous input, 133
    last occurrence triggering, 133–134
    ms time-out input on, 132
Wait Until Next ms Multiple, 117–119, 119f, 201
Wake Up Helper Loop request, 334–335, 335f, 336f, 338
Watcom C compiler, 25
Watts, Steve, 36, 53, 224, 289, 490, 538
Waveform Generation VI, 107
Waveform Measurement utilities, 107
Waveforms, 106–109, 107f
    ADCs and, 153f
    attributes, 108–109, 109f
    numeric arrays added to, 108
    summing data for, 108
    time domain signals as, 152
Web services, 43
Websequence diagrams, 501
Weisfeld, Matt, 208
While Loops, 13f, 71, 72f
    timed, 120
White box testing, 407
White Rabbit, 60
Windows:
    Timed structures and, 123
    timer resolution in, 117
Windows 3.0, 25
Wiresmith Technology, 52, 543
Wiring, noise sources and, 157
Woolf, Bobby, 552
Woram, Kevin, 26f
Work stations, 523–525
    setups for, 526
Workflows:
    selecting, 522–523
    source control and, 518–522
WORM. *See* Write Once Read Many
WORM globals, 79
Wrapper Pattern. *See* Adapter Pattern
Write Once Read Many (WORM), 280–281

X

X Windows, 25
X.509 certificates, 574, 576f
XControls, 475
Xnodes, 537

Y

Yu, Ching Hwa, 467, 468, 471, 475, 542, 543

Z

Zagorodni, Andrei, 260
Zero coupling actors, 392, 394f, 395
Zuehlke, 351
Zuehlke Project Explorer, 284